ENCYCLOPEDIA OF CORROSION TECHNOLOGY

ENCYCLOPEDIA OF CORROSION TECHNOLOGY

PHILIP A. SCHWEITZER
Consultant
Fallston, Maryland

MARCEL DEKKER
MARCEL DEKKER, INC. NEW YORK • BASEL • HONG KONG

Library of Congress Cataloging-in-Publication Data

Schweitzer, Philip A.
Encyclopedia of corrosion technology/Philip A. Schweitzer.
p. cm.
Includes biographical references and index.
ISBN 0-8247-0137-2 (alk. paper)
1. Corrosion and anti-corrosives—Encyclopedias. I. Title.
TA418.74.S383 1998
620.1'1223—dc21

98-5181
CIP

The publisher offers discounts on this book when ordered in bulk quantities. For more information, write to Special Sales/Professional Marketing at the address below.

This book is printed on acid-free paper.

MARCEL DEKKER, INC.
270 Madison Avenue, New York, New York 10016
http://www.dekker.com
Current printing (last digit):
10 9 8 7 6 5 4 3 2 1

PRINTED IN THE UNITED STATES OF AMERICA

Preface

Corrosion is expensive and can be hazardous. It is costly to replace and/or repair equipment, structures, and other miscellaneous items that have been damaged as a result of corrosion. It can be hazardous when corrosion has weakened a portion of a vessel, bridge, or other structure causing it to fail, resulting in injury to persons and/or fires or explosions.

Materials are capable of corroding as the result of prolonged exposure to the atmosphere as well as contact with aggressive media. It is the purpose of this book to explain the many terms associated with corrosion, including the various types and forms of corrosion, and metallurgical and other terms as they relate to the corrosion process.

All the most commonly used materials of construction have been included because the various forms and types of corrosion affect different materials in different ways.

Methods whereby corrosion can be controlled or prevented are explained. Information regarding areas of application, conditions of protection, and conditions under which they are useful have been included.

Ample references are supplied to permit more detailed study of many of the topics.

This encyclopedia will provide insight into the causes and problems of corrosion and offer some assistance in solving these problems.

Philip A. Schweitzer

Contents

ENCYCLOPEDIA OF CORROSION TECHNOLOGY

A

ABRASION CORROSION

See "Erosion Corrosion."

ABSORPTION

See also "Sheet Linings."

Unlike metals, polymers will absorb varying quantities of the corrodents they come into contact with, especially organic liquids. This can result in swelling, cracking, and penetration to the substrate. Swelling can cause softening of the polymer. If the polymer has a high absorption rate, permeation will probably occur. An approximation of the expected permeation and/or absorption of a polymer can be based on the absorption of water. Some typical rates are shown in Table A-1. Table A-2 shows the absorption of selected liquids in FEP and Table A-3 shows the absorption of selected liquids in PFA.

ACID

Any chemical compound containing hydrogen capable of being replaced by positive elements or radicals to form salts. In terms of the dissociation theory, it is a compound which on dissociation in solution yields excess hydrogen ions.

ACID BRICK

Acid brick is brick made from selected clays having a higher silica content than ordinary firebrick and containing little acid soluble components. It is used to line vessels to impart corrosion resistance against hot acid or erosion–corrosion attack. It is fired at higher temperatures and for longer periods of time than the same clay when used to make "common" brick. Acid resistant brick is covered by ASTM Specification C-279. The two most commonly used bricks are red shale and fireclay. These are used for all applications except those where the exposure is to strong alkali or hydrofluoric acid. Of the two, the most frequently used is red shale.

Red Shale Brick

Red shale brick is usually described as meeting Type L in ASTM C-279. These bricks provide a lower absorption rate than fireclay and are usually somewhat more brittle. They are applied in those areas where lowest absorption masonry is desired.

Fireclay

Fireclay brick is usually described as meeting type H in ASTM C-279. They contain a higher proportion of alumina and lower percentages of silica and iron than do shale brick. Fireclay bricks have a higher absorption rate than shale bricks although some manufacturers will provide a denser brick that will meet type L for absorption. These brick are usually selected for outdoor exposures where rapid thermal changes occur, since they are less brittle than the shale brick. Since they also have a low iron content they are used in process equipment where this characteristic is important in maintaining product purity.

Carbon Brick

Carbon Brick is used in areas exposed to strong alkali (pH 12.5+) and hydrofluoric acid, or fluoride salts in an acid medium. These bricks are more shock resistant than either red shale or fireclay brick, permitting them to be used in areas where rapid pressure changes take place, a condition that can cause shale or fireclay to spall.

Silica Brick

All silica brick is used in very high acid concentrations, particularly in phosphoric acid.

See Ref. 1.

Table A.1 Water Absorption Rates of Polymers

Polymer	Water absorption 24 h at 73°F (23°C) (%)
PVC	0.05
CPVC	0.03
PP (Homo)	0.02
PP (Co)	0.03
PE (EHMW)	< 0.01
E-CTFE	< 0.1
PVDF	< 0.04
Saran	nil
PFA	< 0.03
ETFE	0.029
PTFE	< 0.01
FEP	< 0.01

Table A.2 Absorption of Selected Liquids[a] by FEP

Chemical	Temperature (°F/°C)	Range of weight gains (%)
Aniline	365/185	03.–0.4
Acetophenone	394/201	0.6–0.8
Benzaldehyde	354/179	0.4–0.5
Benzyl alcohol	400/204	0.3–0.4
n-Butylamine	172/78	0.3–0.4
Carbon tetrachloride	172/78	2.3–2.4
Dimethyl sulfoxide	372/190	0.1–0.2
Nitrobenzene	410/210	0.7–0.9
Perchlorethylene	250/121	2.0–2.3
Sulfuryl chloride	154/68	1.7–2.7
Toluene	230/110	0.7–0.8
Tributyl phosphate	392/200[b]	1.8–2.0

[a]168-hour exposure at their boiling points.
[b]Not boiling.

Table A.3 Absorption of Representative Liquids in PFA

Liquid[a]	Temperature (°F/°C)	Range of weight gains (%)
Aniline	365/185	0.3–0.4
Acetophenone	394/201	0.6–0.8
Benzaldehyde	354/179	0.4–0.5
Benzyl alcohol	400/204	0.3–0.4
n-Butylamine	172/78	0.3–0.4
Carbon tetrachloride	172/78	2.3–2.4
Dimethyl sulfoxide	372/190	0.1–0.2
Freon 113	117/47	1.2
Isooctane	210/99	0.7–0.8
Nitrobenzene	410/210	0.7–0.9
Perchlorethylene	250/121	2.0–2.3
Sulfuryl chloride	154/68	1.7–2.7
Toluene	230/110	0.7–0.8
Tributyl phosphate	392/200[b]	1.8–2.0
Bromine, anhydrous	–5/–22	0.5
Chlorine, anhydrous	248/120	0.5–0.6
Chlorosulfonic acid	302/150	0.7–0.8
Chromic acid 50%	248/120	0.00–0.01
Ferric chloride	212/100	0.00–0.01
Hydrochloric acid 37%	248/120	0.00–0.03
Phosphoric acid, concentrated	212/100	0.00–0.01
Zinc chloride	212/100	0.00–0.03

[a]Liquids were exposed for 168 hours at the boiling point of the solvents. The acidic reagents were exposed for 168 hours.
[b]Not boiling.

ACID MINE WATERS

These are waters that are present in some underground coal mines. They are extremely corrosive because of their free acidity and the presence of high concentrations of ferric and sulfate ions. Their corrosiveness is a result of the aerial and microbial oxidation of pyrite sulfur present in the coal seams or related strata.

ACID RAIN

When rain has a pH less than 5.6 it is classified as acid rain. It is the result of atmospheric moisture coming into contact with sulfur dioxide gases from industrial emissions and/or with nitrogen oxide gases from car exhausts.

ACRYLATE-BUTADIENE RUBBER (ABR) and ACRYLIC ESTER–ACRYLIC HALIDE (ACM) RUBBERS

Acrylate-butadiene and acrylic ester–acrylic halide rubbers are very similar to ethylene-acrylic rubbers.

Physical and Mechanical Properties

The ABRs and ACM rubbers exhibit good resilience and tear resistance but poor impact resistance. Abrasion resistance and compression set are good. The maximum temperature rating is 340°F (170°C), the same as for the EA rubbers.

Table A-4 lists the physical and mechanical properties of the ACM rubbers.

Table A.4 Physical and Mechanical Properties of Acrylic Ester-Acrylic Halide (ACM) Rubbers[a]

Specific gravity	1.1
Hardness range, Shore A	45–90
Tensile strength, psi	2175
Elongation, % at break	400
Compression set, %	Fair
Tear resistance	Good
Maximum temperature, continuous use	340°F (170°C)
Electrical properties	Poor to fair
Abrasion resistance	Fair to good
Permeability to gases	Low
Resistance to sunlight	Good
Resistance to heat	Excellent

[a]These are representative values since they may be altered by compounding.

Resistance to Sun, Weather, and Ozone

Acrylate-butadiene and acrylic ester–acrylic halide rubbers exhibit good resistance to sun, weather, and ozone.

Chemical Resistance

The acrylate-butadiene and ACM rubbers have excellent resistance to aliphatic hydrocarbons (gasoline, kerosene) and offer good resistance to water, acids, synthetic lubricants, and silicate hydraulic fluids. They are unsatisfactory for use in contact with alkali, aromatic hydrocarbons (benzene, toluene), halogenated hydrocarbons, alcohol, and phosphate hydraulic fluids.

Applications

These rubbers are used where resistance to atmospheric conditions and heat is required.

See Refs. 2 and 3.

ACRYLIC ESTER-ACRYLIC HALIDE RUBBERS

See "Acrylate-Butadiene Rubber."

ACRYLONITRILE-BUTADIENE-STYRENE (ABS)

ABS is a vast family of compounds whose properties can be varied extensively, depending on the ratio of acrylonitrile to other components. Higher strength, better toughness, greater dimensional stability, and other properties can be obtained at the expense of other characteristics. Typical physical and mechanical properties that can be obtained are shown in Table A-5. The ABS material has limited heat tolerance, with a maximum operating temperature of approximately 195°F (90°C), with relatively low strength and limited chemical resistance. But its low price and ease of fabrication and joining makes it attractive for use as distribution piping for gas, water, and waste. It also finds application for vent lines, automotive parts, and other consumer items.

ABS plastic will be attacked by oxidizing agents, strong acids, and will stress crack in the presence of certain organic compounds. The compatibility of ABS plastic with selected corrodents is shown in Table A-6. Because manufacturers can vary the properties so greatly the corrosion resistance of the specific material to be used should be verified with the manufacturer.

ADSORPTION

Adsorption is a surface phenomenon exhibited by solids which consists of the adhesion, in an extremely thin layer of the molecules of gases, of liquids or of dissolved substances with which they are in contact. There are two types depending on the nature of forces

Table A.5 Physical and Mechanical Properties of ABS

Specific gravity	1.03
Water absorption 24 h at 73°F (23°C), %	0.2–0.4
Tensile strength at 73°F (23°C), psi	5,350
Modulus of elasticity in tension at 73°F (23°C) × 10^5	2.4
Flexural strength, psi	9400
Izod impact strength, notched at 73°F (23°C)	8.5
Coefficient of thermal expansion	
in./in.–°F × 10^{-5}	5.6
in./10°F/100 ft.	0.056
Thermal conductivity, Btu/h/sq ft/°F/in.	1.7
Heat distortion temperature at 66 psi, °F/°C	204/94
Resistance to heat at continuous drainage, °F/°C	140/60
Limiting oxygen index, %	19
Flame spread	Not applicable
Underwriters lab rating (Sub 94)	94 HB

involved. In chemisorption, a single layer of molecules, atoms, or ions is attached to the surface by chemical bonds and is essentially irreversible. In physical adsorption, attachment is by the weaker Van der Waal's forces whose energy levels approximate those of condensation.

Compare with "Absorption."

ALIPHATIC HYDROCARBONS

Aliphatic hydrocarbons are straight chain organic compounds that are either alkanes, alkenes, alkynes, or their derivatives. All moncyclic organic compounds are aliphatic and cyclic aliphatic compounds are alecyclic.

ALKALINE

Alkaline describes a solution with an excess of hydroxyl ions having a pH greater than 7.

ALLIGATORING

Alligatoring is a rupture of an organic coating film, usually caused by application of a hard brittle film over a more flexible film having an appearance similar to an alligator hide. It is a form of checking in which the surface hardens and shrinks at a much faster rate than the body of the coating.

Table A.6 Compatibility of ABS with Selected Corrodents[a]

Chemical	Maximum temp. °F	Maximum temp. °C	Chemical	Maximum temp. °F	Maximum temp. °C
Acetaldehyde	x	x	Benzyl chloride	x	x
Acetic acid 10%	100	38	Borax	140	60
Acetic acid 50%	130	54	Boric acid	140	60
Acetic acid 80%	x	x	Bromine liquid	x	x
Acetic acid, glacial	x	x	Butadiene	x	x
Acetic anhydride	x	x	Butyl acetate	x	x
Acetone	x	x	Butyl alcohol	x	x
Acetyl chloride	x	x	Butyric acid	x	x
Adipic acid	140	60	Calcium bisulfite	140	60
Allyl alcohol	x	x	Calcium carbonate	100	38
Allyl chloride	x	x	Calcium chlorate	140	60
Alum	140	60	Calcium chloride	140	60
Aluminum chloride, aqueous	140	60	Calcium hydroxide, sat.	140	60
Aluminum fluoride	140	60	Calcium hypochlorite	140	60
Aluminum hydroxide	140	60	Calcium nitrate	140	60
Aluminum oxychloride	140	60	Calcium oxide	140	60
Aluminum sulfate	140	60	Calcium sulfate 25%	140	60
Ammonia gas dry	140	60	Carbon bisulfide	x	x
Ammonium bifluoride	140	60	Carbon dioxide, dry	90	32
Ammonium carbonate	140	60	Carbon dioxide, wet	140	60
Ammonium chloride, sat.	140	60	Carbon disulfide	x	x
Ammonium fluoride 10%	x	x	Carbon monoxide	140	60
Ammonium fluoride 25%	x	x	Carbon tetrachloride	x	x
Ammonium hydroxide 25%	90	32	Carbonic acid	140	60
Ammonium hydroxide, sat.	80	27	Cellosolve	x	x
Ammonium nitrate	140	60	Chloracetic acid	x	x
Ammonium persulfate	140	60	Chlorine gas, dry	140	60
Ammonium phosphate	140	60	Chlorine gas, wet	140	60
Ammonium sulfate 10–40%	140	60	Chlorine, liquid	x	x
Ammonium sulfide	140	60	Chlorobenzene	x	x
Amyl acetate	x	x	Chloroform	x	x
Amyl alcohol	80	27	Chlorosulfonic acid	x	x
Amul chloride	x	x	Chromic acid 10%	90	32
Aniline	x	x	Chromic acid 50%	x	x
Antimony trichloride	140	60	Citric acid 15%	140	60
Aqua regia 3:1	x	x	Citric acid, 25%	140	60
Barium carbonate	140	60	Copper chloride	140	60
Barium chloride	140	60	Copper cyanide	140	60
Barium hydroxide	140	60	Copper sulfate	140	60
Barium sulfate	140	60	Cresol	x	x
Barium sulfide	140	60	Cyclohexane	80	27
Benzaldehyde	x	x	Cyclohexanol	80	27
Benzene	x	x	Dichloroacetic acid	x	x
Benzene sulfonic acid 10%	80	27	Dichloroethane (ethylene dichloride)	x	x

Table A.6 Continued

Chemical	Maximum temp. °F	Maximum temp. °C	Chemical	Maximum temp. °F	Maximum temp. °C
Benzoic acid	140	60	Ethylene glycol	140	60
Benzyl alcohol	x	x	Ferric chloride	140	60
Ferric nitrate, 10–50%	140	60	Phosphoric acid 50–80%	130	54
Ferrous chloride	140	60	Picric acid	x	x
Fluorine gas, dry	90	32	Potassium bromide 30%	140	60
Hydrobromic acid 20%	140	60	Sodium carbonate	140	60
Hydrochloric acid 20%	90	32	Sodium chloride	140	60
Hydrochloric acid 38%	140	60	Sodium hydroxide 10%	140	60
Hydrofluoric acid 30%	x	x	Sodium hydroxide 50%	140	60
Hydrofluoric acid 70%	x	x	Sodium hydroxide, concentrated	140	60
Hydrofluoric acid 100%	x	x	Sodium hypochlorite 20%	140	60
Hypochlorous acid	140	60	Sodium hypochlorite, concentrated	140	60
Ketones, general	x	x	Sodium sulfide to 50%	140	60
Lactic acid 25%	140	60	Stannic chloride	140	60
Magnesium chloride	140	60	Stannous chloride	100	38
Malic acid	140	60	Sulfuric acid 10%	140	60
Methyl chloride	x	x	Sulfuric acid 50%	130	54
Methyl ethyl ketone	x	x	Sulfuric acid 70%	x	x
Methyl isobutyl ketone	x	x	Sulfuric acid 90%	x	x
Muriatic acid	140	60	Sulfuric acid 98%	x	x
Nitric acid 5%	140	60	Sulfuric acid 100%	x	x
Nitric acid 20%	130	54	Sulfuric acid, fuming	x	x
Nitric acid 70%	x	x	Sulfurous acid	140	60
Nitric acid, anhydrous	x	x	Thionyl chloride	x	x
Oleum	x	x	Toluene	x	x
Perchloric acid 10%	x	x	White liquor	140	60
Perchloric acid 70%	x	x	Zinc chloride	140	60
Phenol	x	x			

[a]The chemicals listed are in the pure state or in a saturated solution unless otherwise indicated. Compatibility is shown to the maximum allowable temperature for which data are available. Incompatibility is shown by an x. A blank space indicates that data are unavailable.
Source: PA Schweitzer. *Corrosion Resistance Tables*. 4th ed. Vols. 1–3. New York: Marcel Dekker, 1995.

ALLOY

An alloy is a mixture of two or more metallic elements to produce a single phase. Alloys can be either heterogenous where the alloy is a mixture of two or more separate phases, or they may be homogenous, where the components are completely soluble in one another.

The term alloy is also used to describe resin, polymer, and plastic mixtures formed from two or more immiscible polymers united by another component and having improved performance properties.

Table A.7 Mechanical and Physical Properties of Alloy B-2

Modulus of elasticity × 10^6, psi	31.4
Tensile strength × 10^3, psi	110
Yield strength 0.2% offset × 10^3, psi	60
Elongation in 2 in., %	60
Hardness, Brinell	210
Density, lb/in.^{3}	0.333
Specific gravity	9.22
Specific heat, at 212°F, Btu/lb °F	0.093
Thermal conductivity, Btu/ft^2/in. h °F	
at 32°F	77
at 212°F	85
at 392°F	93
at 572°F	102
at 752°F	111
at 932°F	120
at 1112°F	130
Coefficient of thermal expansion, in./in. °F × 10^{-6}	
at 68–200°F	5.7
at 68–600°F	6.2
at 68–1000°F	6.5

ALLOY B-2

Within the nickel–molybdenum series, one major alloy—alloy B-2—is a low-carbon and low-silicon (0.02%, 0.08% maximum) version of alloy B. The alloy is uniquely different from other corrosion-resistant alloys because it does not contain chromium. Molybdenum is the primary alloying ingredient and provides a significant corrosion resistance to reducing environments. Table A-7 gives the mechanical and physical properties of alloy B-2.

Since alloy B-2 was developed to handle hydrochloric acid, it is recommended for the handling of all concentrations in the temperature range of 158–212°F (70–100°C) and for the handling of wet hydrogen chloride gas.

Alloy B-2 has excellent resistance to pure sulfuric acid at all concentrations and temperatures below 60% acid and good resistance to 212°F (100°C) above 60% acid. It is also resistant to a number of nonoxidizing environments, such as hydrofluoric and phosphoric acids, and numerous organic acids, such as acetic, formic, and crysylic acids.

Alloy B-2 is also resistant to many nonoxidizing chloride-bearing salts such as aluminum chloride, magnesium chloride, and antimony chloride.

Because alloy B-2 is nickel rich (approximately 70%), it is resistant to chloride-induced stress corrosion cracking. By virtue of the high molybdenum content it is highly resistant to pitting attack in acid chloride environments.

Alloy B-2 is not recommended for elevated temperature service. Service is limited to 650°F (343°C) operation. The major factor that limits the use of alloy B-2 is the poor corrosion resistance in oxidizing environments. It has virtually no corrosion resistance to oxidizing acids such as nitric and chromic or to oxidizing salts such as ferric chloride or cupric chloride. The presence of oxidizing salts in reducing acids must be considered. Oxidizing salts such as ferric chloride, ferric sulfate, or cupric chloride, even when present in the part per million range, can accelerate the attack in hydrochloric or sulfuric acids. Even dissolved oxygen has sufficient oxidizing power to affect the corrosion rates in hydrochloric acid. Table A-8 provides corrosion data of alloy B-2 in the presence of selected corrodents.

See Refs. 3 and 4.

ALLOY C-276

Alloy C-276 is a nickel–molybdenum–chromium–tungsten alloy with outstanding corrosion resistance that is maintained even in welded condition. The high nickel and molybdenum contents provide good corrosion resistance in reducing environments, while the chromium imparts resistance to oxidizing media. The molybdenum also aids resistance to localized corrosion, such as pitting. The low carbon content minimizes carbide precipitation during welding to maintain resistance to corrosion (intergranular attack) in heat-affected zones of welded joints. Alloy C-276 also has high strength at room and elevated temperatures. Refer to Table A-9 for the mechanical and physical properties of alloy C-276.

Alloy C-276 is extremely versatile because it possesses good resistance in both oxidizing and reducing media, and this includes conditions with halogen ion contamination. The pitting and crevice corrosion resistance of alloy C-276 makes it an excellent choice when dealing with acid chloride salts.

Alloy C-276 has exceptional corrosion resistance to many chemical process materials, including highly oxidizing neutral and acid chlorides, solvents, chlorine, formic and acetic acid, and acetic anhydride. It also resists highly corrosive agents such as wet chlorine gas, hypochlorites, and chlorine solutions.

Alloy C-276 has excellent resistance to phosphoric acid. At all temperatures below the boiling point of phosphoric acid, when concentrations are less than 65% by weight, tests have shown corrosion rates of less than 5 mpy. At acid concentrations above 65% by weight and up to 85%, alloy C-276 displays similar corrosive behavior, except at temperatures between about 240°F (116°C) and the boiling point, where corrosion rates may be erratic and may reach 25 mpy. Refer to Table A-8 for the compatibility of alloy C-276 with selected corrodents.

See Refs. 3 and 4.

ALLOY DESIGNATION

See "Unified Numbering System."

ALLOY G AND G-3

Alloy G and its modification, alloy G-3, are nickel–chromium–iron–molybdenum–copper alloys exhibiting excellent resistance to a variety of corrosive conditions. The alloys are highly resistant to pitting and stress corrosion cracking in both acid and alkaline environments, including hot sulfuric and phosphoric acids, hydrofluoric and contaminated nitric acids, mixed acids, and sulfate compounds.

Table A.8 Compatibility of Alloy B-2 and Alloy C-276 with Selected Corrodents[a]

	Maximum temperature (°F/°C)	
Chemical	Alloy B-2	Alloy C-276
Acetaldehyde	80/27	140/60
Acetamide		60/16
Acetic acid 10%	300/149	300/149
Acetic acid 50%	300/149	300/149
Acetic acid 80%	300/149	300/149
Acetic acid, glacial	560/293	560/293
Acetic anhydride	280/138	280/138
Acetone	200/93	200/93
Acetyl chloride	80/27	
Acrylic acid	210/99	
Acrylonitrile	210/99	210/99
Adipic acid		210/99
Allyl alcohol		570/299
Allyl chloride	200/93	
Alum	150/66	150/66
Aluminum acetate	60/16	60/16
Aluminum chloride, aqueous	300/149	210/99
Aluminum chloride, dry	210/99	210/99
Aluminum fluoride	80/27	80/27
Aluminum sulfate	210/99	210/99
Ammonia gas	200/93	200/93
Ammonium bifluoride		380/193
Ammonium carbonate	300/149	300/149
Ammonium chloride 10%	210/99	210/99
Ammonium chloride 50%	210/99	210/99
Ammonium chloride, sat.	570/299	570/299
Ammonium fluoride 10%	210/99	210/99
Ammonium fluoride 25%		210/99
Ammonium hydroxide 25%	210/99	570/299
Ammonium hydroxide, sat.	210/99	570/299
Ammonium persulfate	x	
Ammonium sulfate 10–40%	80/27	200/93
Ammonium sulfite		100/38
Amyl acetate	340/171	340/171
Amyl alcohol		180/82
Amyl chloride	210/99	90/32
Aniline	570/299	570/299
Antimony trichloride	210/99	210/99
Aqua regia 3:1	x	x
Barium carbonate	570/299	570/299
Barium chloride	570/299	210/99
Barium hydroxide	270/132	270/132
Barium sulfate	80/27	
Benzaldehyde	210/99	210/99
Benzene	210/99	210/99

Table A.8 Continued

Chemicals	Maximum temperature (°F/°C) Alloy B-2	Alloy C-276
Benzene sulfonic acid 10%	210/99	210/99
Benzoic acid	210/99	
Benzyl alcohol	210/99	210/99
Benzyl chloride	210/99	
Borax	120/49	120/49
Boric acid	570/299	570/299
Bromine gas, dry	60/16	60/16
Bromine gas, moist		60/16
Bromine liquid		180/82
Butadiene	300/149	300/149
Butyl acetate	200/93	200/93
Butyl alcohol	210/99	200/93
n-Butylamine	210/99	210/99
Butyric acid	280/138	280/138
Calcium bisulfite		80/27
Calcium carbonate	210/99	210/99
Calcium chlorate		210/99
Calcium chloride	350/177	350/177
Calcium hydroxide 10%	210/99	170/77
Calcium hydroxide, sat.	210/99	
Calcium hypochlorite	x	
Calcium nitrate	210/99	210/99
Calcium oxide		90/32
Calcium sulfate 10%	320/160	320/160
Caprylic acid	300/149	300/149
Carbon bisulfide	180/82	210/99
Carbon dioxide, dry	570/299	570/299
Carbon dioxide, wet	570/299	200/93
Carbon disulfide	180/82	300/149
Carbon monoxide	570/299	570/299
Carbon tetrachloride	300/149	300/149
Carbonic acid	80/27	80/27
Cellosolve	210/99	210/99
Chloracetic acid, 50% water		210/99
Chloracetic acid	370/188	300/149
Chlorine gas, dry	200/93	570/299
Chlorine gas, wet	x	220/104
Chlorine, liquid		110/43
Chlorobenzene	350/177	350/177
Chloroform	210/99	210/99
Chlorosulfonic acid	230/110	230/110
Chromic acid 10%	130/54	210/99
Chromic acid 50%	x	210/99
Chromyl chloride	210/99	210/99
Citric acid 15%	210/99	210/99
Citric acid, concentrated	210/99	210/99

Table A.8 Continued

Chemicals	Maximum temperature (°F/°C) Alloy B-2	Alloy C-276
Copper acetate	100/38	100/38
Copper carbonate	90/32	90/32
Copper chloride	200/93	200/93
Copper cyanide	150/66	150/66
Copper sulfate	210/99	210/99
Cresol	210/99	210/99
Cupric chloride 5%	60/16	210/99
Cupric chloride 50%	210/99	210/99
Cyclohexane	210/99	210/99
Cyclohexanol	80/27	80/27
Dichloroethane	230/110	230/110
Ethylene glycol	570/299	570/299
Ferric chloride	90/32	90/32
Ferric chloride 50% in water	x	
Ferric nitrate 10–50%	x	
Ferrous chloride	280/138	280/138
Fluorine gas, dry	80/27	150/66
Fluorine gas, moist		570/299
Hydrobromic acid, dilute	210/99	
Hydrobromic acid 20%	210/99	90/32
Hydrobromic acid 50%	260/127	90/32
Hydrochloric acid 20%	140/60	150/66
Hydrochloric acid 38%	140/60	90/32
Hydrofluoric acid 30%	140/60	210/99
Hydrofluoric acid 70%	110/43	200/93
Hydrofluoric acid 100%	80/27	210/99
Hypochlorous acid	90/32	80/27
Iodine solution 10%		180/82
Ketones, general	180/82	100/38
Lactic acid 25%	250/121	210/99
Lactic acid, concentrated	250/121	210/99
Magnesium chloride	300/149	300/149
Malic acid	210/99	210/99
Manganese chloride 40%	210/99	210/99
Methyl chloride	210/99	90/32
Methyl ethyl ketone	210/99	210/99
Methyl isobutyl ketone	200/93	200/93
Muriatic acid	90/32	90/32
Nitric acid 5%	x	210/99
Nitric acid 20%	x	160/71
Nitric acid 70%	x	200/93
Nitric acid, anhydrous	x	80/27
Nitrous acid, concentrated	x	x
Oleum to 25%	110/43	140/60
Perchloric acid 70%		220/104
Phenol	570/299	570/299

Table A.8 Continued

Chemicals	Maximum temperature (°F/°C) Alloy B-2	Alloy C-276
Phosphoric acid 50–80%	210/99	210/99
Picric acid	220/104	300/149
Potassium bromide 30%	90/32	90/32
Salicylic acid	80/27	250/121
Silver bromide 10%	90/32	90/32
Sodium carbonate	570/299	210/99
Sodium chloride to 30%	210/99	210/99
Sodium hydroxide 10%[b]	240/116	230/110
Sodium hydroxide 50%	250/121	210/99
Sodium hydroxide, concentrated	200/93	120/49
Sodium hypochlorite 20%	x	x
Sodium hypochlorite, concentrated	x	x
Sodium sulfide to 50%	210/99	210/99
Stannic chloride to 50%	210/99	210/99
Stannous chloride[c]	570/299	210/99
Sulfuric acid 10%	210/99	200/93
Sulfuric acid 50%	230/110	230/110
Sulfuric acid 70%	290/143	290/143
Sulfuric acid 90%	190/88	190/88
Sulfuric acid 98%	280/138	210/99
Sulfuric acid 100%	290/143	190/88
Sulfuric acid, fuming	210/99	90/32
Sulfurous acid	210/99	370/188
Toluene	210/99	210/99
Trichloroacetic acid	210/99	210/99
White liquor	100/38	100/38
Zinc chloride	60/16	250/121

[a]The chemicals listed are in the pure state or in a saturated solution unless otherwise indicated. Compatibility is shown to the maximum allowable temperature for which data are available. Incompatibility is shown by an x. A blank space indicates that data are unavailable. When compatible, corrosion rate is < 20 mpy.
[b]Alloy B-2 is subject to stress cracking.
[c]Alloy B-2 is subject to pitting.
Source: PA Schweitzer. *Corrosion Resistance Tables,* 4th ed. Vols. 1–3. New York: Marcel Dekker, Inc. 1995.

Alloy G-3 was developed with a lower carbon content to prevent precipitation at the welds. The corrosion resistance of alloy G and alloy G-3 are approximately the same. However, the thermal stability of alloy G-3 is greater.

The mechanical and physical properties of alloy G and alloy G-3 are given in Table A-10.

Alloy G has found many applications in pollution control equipment. Excellent performance has been demonstrated in municipal garbage incinerator systems, including

Table A.9 Mechanical and Physical Properties of Alloy C-276

Modulus of elasticity $\times 10^6$, psi	29.8
Tensile strength $\times 10^3$, psi	100
Yield strength 0.2% offset $\times 10^3$, psi	41
Elongation in 2 in., %	40
Hardness, Brinell	190
Density, lb/in.^{3}	0.321
Specific gravity	8.89
Specific heat, Btu/lb °F	0.102
Thermal conductivity, Btu/ft^2/h °F/in.	
at –270°F	50
at 0°F	65
at 100°F	71
at 200°F	77
at 400°F	90
at 600°F	104
at 800°F	117
at 1000°F	132
at 1200°F	145
Coefficient of thermal expansion, in./in. °F $\times 10^{-6}$	
at 75–200°F	6.2
at 75–400°F	6.7
at 75–600°F	7.1
at 75–800°F	7.3
at 75–1000°F	7.4
at 75–1200°F	7.8
at 75–1400°F	8.3
at 75–1600°F	8.8

fans, ducts, and scrubbing equipment. The SO_2 scrubber systems for power plants that use water or alkaline quench have been incorporating alloy G as a major material of construction because of its resistance to sulfuric acid and conditions during which the chloride ion can concentrate.

See Refs. 3 and 4.

ALUMINUM

Aluminum represents approximately 8% of the metallic elements found in the earth's crust, making it the most prevalent metal. The principal ore is bauxite.

The high resistance of aluminum to corrosion in many environments is the result of the very rapid formation of a thin, compact, and adherent oxide film. This film, which is formed in air, prevents the further corrosion of the underlying metal and is stable in the pH range of approximately 4 to 9. At lower and higher pH values the film will dissolve with some notable exceptions. It is stable in concentrated nitric and which has a pH of 1 and concentrated ammonium hydroxide which has a pH of 13.

The term aluminum is used for both pure aluminum and aluminum alloys. However,

Table A.10 Mechanical and Physical Properties of Alloy G and Alloy G-3

Property	G	G-3
Modulus of elasticity × 10^6, psi	27.8	27.8
Tensile strength × 10^3, psi	90	90
Yield strength 0.2% offset × 10^3, psi	35	35
Elongation in 2 in., %	35	45
Hardness, Brinell	169	885(Rb)
Density, lb/in.^{3}	0.30	0.30
Specific gravity	8.31	8.31
Specific heat, J/kg K	456	464
Thermal conductivity, W/mK	10.1	10.0
Coefficient of thermal expansion, in./in. °F × 10^{-6}		
at 70–200°F	7.5	7.5
at 70–400°F	7.7	7.7
at 70–600°F	7.9	7.9
at 70–800°F	8.3	8.3
at 70–1000°F	8.7	8.7
at 70–1200°F	9.1	9.1

practically all commercial applications make use of aluminum alloys. An aluminum content of 99% or greater is considered unalloyed aluminum.

The Aluminum Association has developed an Alloy Designation System for wrought aluminum products. Wrought alloys of aluminum are of two types: non–heat treatable of the 1XXX, 3XXX, 4XXX, and 5XXX series. The heat treatable alloys are of the series 2XXX, 6XXX and 7XXX. The general system is shown in the table. The primary alloying element is shown first, followed by other usual elements.

Series Designation	Alloying Materials
1XXX	99.9% minimum aluminum
2XXX	Al-Cu, Al-Cu-Mg, Al-Cu-Mg-Li, and Al-Cu-Mg-Si
3XXX	Al-Mn and Al-Mn-Mg
4XXX	Al-Si
5XXX	Al-Mg and Al-Mg-Mn
6XXX	Al-Mg-Si, Al-Mg-Si-Mn and Al-Mg-Si, Cu
7XXX	Al-Zn, Al-Zn-Mg, Al-Zn-Mg-Mn, and Al-Zn-Mg-Cu

The mechanical and physical properties are shown in Table A-11.

All of the non–heat treatable alloys have a high degree of corrosion resistance. These alloys, which do not contain copper as a major alloying ingredient, have a high resistance to corrosion by many chemicals. They are compatible with dry salts of most inorganic chemicals and, within their passive range of pH 4–9, is aqueous solutions, with most halide salts, under conditions at which the alloys are polarized to their pitting potentials. In most other solutions where conditions are less likely to occur that will polarize the alloys to these potentials, pitting is not a problem.

Table A.11 Mechanical and Physical Properties of Aluminum Alloys

Property	Aluminum alloy 3003-3	5052-0	6061-T6	6063-T6
Modulus of elasticity × 10^6, psi	10	10.2	10	10
Tensile strength × 10^3, psi	17	41	45	35
Yield strength 0.2% offset × 10^3, psi	8	36	40	31
Elongation in 2 in., %	40	25	12	12
Density, lb/in.^{3}	.099	.097		.098
Specific gravity	2.73	2.68		2.70
Specific heat, Btu/h °F	0.23	0.23		
Thermal conductivity, Btu/h/ft^2/°F/in.	1070	960	900	1090
Coefficient of thermal expansion, in./°F/in. × 10^{-6}				
at −58–68°F	12		12.1	12.1
at 68–212°F	12.9	13.2	13.0	13.0
at 68–392°F	13.5		13.5	13.6
at 68–572°F	13.9		14.1	14.2

Aluminum alloys are not compatible with most inorganic acids, bases, and salts with pH outside the passive range of the alloys (pH 4–9).

Aluminum alloys are resistant to a wide variety of organic compounds, including most aldehydes, esters, ethers, hydrocarbons, ketones, mercaptans, other sulfur-containing compounds, and nitro compounds. They are also resistant to most organic acids, alcohols, and phenols, except when these compounds are nearly dry and near their boiling points. Carbon tetrachloride also exhibits this behavior.

Aluminum alloys are most resistant to organic compounds halogenated with chlorine, bromine, and iodine. They are also resistant to highly polymerized compounds.

It should be noted that the compatibility of aluminum alloys with mixtures of organic compounds cannot always be predicted from their compatibility with each of the compounds. For example, some aluminum alloys are corroded severely in mixtures of carbon tetrachloride and methyl alcohol, even though they are resistant to each compound alone. Caution should also be exercised in using data for pure organic compounds to predict performances of the alloys with commercial grades that may contain contaminants. Ions of halides and reducible metals, commonly chloride and copper, frequently have been found to be the cause of excessive corrosion of aluminum alloys in commercial grades of organic chemicals that would not have been predicted from their resistance to pure compounds.

Regardless of environment, pure aluminum has the greatest corrosion resistance, followed by the non–heat-treatable alloys and finally the heat treatable alloys. The two most frequently used alloys are 3003 and 3004. The 3XXX series of alloys are not susceptible to the more drastic forms of localized corrosion. The principal type of corrosion encountered is pitting corrosion. With a low copper content of $< 0.05\%$ the 3003 and 3004 alloys are almost as resistant to pitting as pure aluminum.

The major application of aluminum is for products to be used in outdoor service. In these atmospheres, whether they be seacoast, urban, industrial, or rural, the corrosion of aluminum is self-limiting. Except for the alloys and tempers that are subject to exfoliation, the corrosion rate becomes very slow within five years.

Refer to Table A-12 for the compatibility of aluminum with selected corrodents.

Reference 3 provides a wider comprehensive listing of the compatibility of aluminum with a variety of corrodents.

See also Refs. 5–7.

ALUMINUM BRONZE

See "Copper-Aluminum Alloys."

AMBIENT TEMPERATURE

Ambient temperature is the temperature of the surrounding medium coming into contact with a material or apparatus. It is not necessarily room temperature.

ANAEROBIC CORROSION

Anaerobic corrosion is usually caused by the sulfide metabolic reaction products (biogenic sulfides) of sulfate reducing bacteria. It occurs where there is an abundance of sulfate and the reaction of the metal substrate is between pH 5.5 and 8.5.

Corroded steel is characterized by a coating of strongly reduced black sulfide-containing corrosion products. Cast iron migrates from the metal, leaving a soft residue of largely carbon.

ANNEALING

Annealing is a heating and cooling operation of a metal or alloy that usually implies relatively slow cooling. The purpose of such a heat treatment may be (a) to induce softness, (b) to remove stress, (c) to alter ductility, toughness, electrical, magnetic, or other physical properties, (d) to refine the crystalline structure, (e) to remove gases, or (f) to produce a definite microstructure. The temperature of the operation and the rate of cooling depend upon the material being heat treated and the purpose of the treatment. Certain specific heat treatments coming under the comprehensive term annealing are as follows:

Process annealing

Heating iron-based alloys to a temperature below or close to the lower limit of the critical temperature, generally 1000 to 1300°F (540 to 750°C).

Normalizing

Heating iron-based alloys to approximately 100°F (50°C) above the critical temperature range, followed by cooling to below that range in still air at ordinary temperatures.

Table A.12 Compatibility of Aluminum Alloys with Selected Corrodents[a]

Chemical	Maximum temp. °F	Maximum temp. °C	Chemical	Maximum temp. °F	Maximum temp. °C
Acetaldehyde	360	182	Barium sulfide	x	x
Acetamide	340	171	Benzaldehyde	120	49
Acetic acid 10%	110	43	Benzene	210	99
Acetic acid 50%	130	54	Benzene sulfonic acid 10%	x	x
Acetic acid 80%	90	32	Benzoic acid 10%	400	204
Acetic acid, glacial	210	99	Benzyl alcohol	110	43
Acetic anhydride	350	177	Benzyl chloride	x	x
Acetone	500	260	Borax	x	x
Acetyl chloride	x	x	Boric acid	100	38
Acrylonitrile	210	99	Bromine gas, dry	60	16
Adipic acid	210	99	Bromine gas, moist	x	x
Allyl alcohol	150	66	Bromine liquid	210	99
Allyl chloride	x	x	Butadiene	110	43
Alum	110	43	Butyl acetate	110	43
Aluminum acetate	60	16	Butyl alcohol	210	99
Aluminum chloride, aqueous	x	x	*n*-Butylamine	90	32
Aluminum chloride, dry	60	16	Butyl phthalate	x	x
Aluminum fluoride	120	49	Butyric acid	180	82
Aluminum hydroxide	80	27	Calcium bisulfite	x	x
Aluminum nitrate	110	43	Calcium carbonate	x	x
Aluminum sulfate	x	x	Calcium chlorate	140	60
Ammonia gas	x	x	Calcium chloride 20%	100	38
Ammonium carbonate	350	177	Calcium hydroxide 10%	x	x
Ammonium chloride 10%	x	x	Calcium hydroxide, sat.	x	x
Ammonium chloride 50%	x	x	Calcium hypochlorite	x	x
Ammonium chloride, sat.	x	x	Calcium nitrate	170	77
Ammonium fluoride 10%	x	x	Calcium oxide	90	32
Ammonium fluoride 25%	x	x	Calcium sulfate[b]	210	99
Ammonium hydroxide 25%	350	177	Caprylic acid	300	149
Ammonium hydroxide, sat.	350	177	Carbon bisulfide	210	99
Ammonium nitrate	350	177	Carbon dioxide, dry	570	299
Ammonium persulfate	350	177	Carbon dioxide, wet	170	77
Ammonium phosphate	x	x	Carbon disulfide	210	99
Ammonium sulfate 10–40%	x	x	Carbon monoxide	570	299
Ammonium sulfide	170	77	Carbon tetrachloride	x	x
Ammonium sulfite	x	x	Carbonic acid	80	27
Amyl acetate	350	177	Cellosolve	210	99
Amyl alcohol	170	77	Chloracetic acid, 50% water	x	x
Amyl chloride	90	32	Chloracetic acid	x	x
Aniline[b]	350	177	Chlorine gas, dry	210	99
Antimony trichloride	x	x	Chlorine gas, wet	x	x
Aqua regia 3:1	x	x	Chlorobenzene	150	66
Barium carbonate	x	x	Chloroform, dry	170	77
Barium chloride 30%	180	82	Chlorosulfonic acid, dry	170	77
Barium hydroxide	x	x	Chromic acid 10%	200	93
Barium sulfate	210	99	Chromic acid 50%	100	38

Table A.12 Continued

Chemical	Maximum temp. °F	°C	Chemical	Maximum temp. °F	°C
Chromyl chloride	210	99	Muriatic acid	x	x
Citric acid 15%	210	99	Nitric acid 5%	x	x
Citric acid, concentrated	70	21	Nitric acid 20%	x	x
Copper acetate	x	x	Nitric acid 70%	x	x
Copper carbonate	x	x	Nitric acid, anhydrous	90	32
Copper chloride	x	x	Nitrous acid, concentrated	x	x
Copper cyanide	x	x	Oleum	100	38
Copper sulfate	x	x	Perchloric acid 10%	x	x
Cresol	150	66	Perchloric acid 70%	x	x
Cupric chloride 5%	x	x	Phenol	210	99
Cyclohexane	180	81	Phosphoric acid 50–80%	x	x
Cyclohexanol	x	x	Picric acid	210	99
Dichloroethane (ethylene dichloride)	110	43	Potassium bromide 30%[b]	80	27
Ethylene glycol	100	38	Salicylic acid	130	54
Ferric chloride	x	x	Silver bromide 10%	x	x
Ferric chloride 50% in water	x	x	Sodium carbonate	x	x
Ferric nitrate 10–50%	x	x	Sodium chloride	x	x
Ferrous chloride	x	x	Sodium hydroxide 10%	x	x
Fluorine gas, dry	470	243	Sodium hydroxide 50%	x	x
Fluorine gas, moist	x	x	Sodium hydroxide, concentrated	x	x
Hydrobromic acid, dilute	x	x	Sodium hypochlorite 20%	80	27
Hydrobromic acid 20%	x	x	Sodium hypochlorite, concentrated	x	x
Hydrobromic acid 50%	x	x	Sodium sulfide to 50%	x	x
Hydrochloric acid 20%	x	x	Stannic chloride	x	x
Hydrochloric acid 38%	x	x	Stannous chloride, dry	x	x
Hydrocyanic acid 10%	100	38	Sulfuric acid 10%	x	x
Hydrofluoric acid 30%	x	x	Sulfuric acid 50%	x	x
Hydrofluoric acid 70%	x	x	Sulfuric acid 70%	x	x
Hydrofluoric acid 100%	x	x	Sulfuric acid 90%	x	x
Hypochlorous acid	x	x	Sulfuric acid 98%	x	x
Iodine solution 10%	x	x	Sulfuric acid 100%	x	x
Ketones, general	100	38	Sulfuric acid, fuming	90	32
Lactic acid 25%	80	27	Sulfurous acid	370	188
Lactic acid, concentrated[c]	100	38	Thionyl chloride	x	x
Magnesium chloride	x	x	Toluene	210	99
Malic acid	210	99	Trichloroacetic acid	x	x
Methyl chloride	x	x	White liquor	100	38
Methyl ethyl ketone	150	66	Zinc chloride	x	x
Methyl isobutyl ketone	150	66			

[a]The chemicals listed are in the pure state or in a saturated solution unless otherwise indicated. Compatibility is shown to the maximum allowable temperature for which data are available. Incompatibility is shown by an x. A blank space indicates that data are unavailable. When compatible, corrosion rate is < 20 mpy.
[b]Material subject to pitting.
[c]Material subject to intergranular corrosion.
Source: PA Schweitzer. *Corrosion Resistance Tables*. 4th ed. Vols. 1–3. New York: Marcel Dekker, 1995.

Patenting

Heating iron-based alloys above the critical temperature range followed by cooling below that range in air, molten lead, or a mixture of nitrates and nitrites maintained at a temperature usually between 800 to 1050°F (425 to 555°C), depending on the carbon content of the steel and the properties required in the finished product. This treatment is applied in the wire industry to medium or high carbon steel as a treatment to precede further wire drawing.

Spheradizing

Any process of heating and cooling steel that produces a rounded or globular form of carbide. The following spheradizing methods are used: (a) prolonged heating at a temperature just below the lower critical temperature, usually followed by relatively slow cooling, (b) for small objects of high carbon steels the spheradizing result is achieved more rapidly by prolonged heating to temperatures alternately within and slightly below the critical temperature range, (c) tool steel is generally spheradized by heating to a temperature of 1380 to 1480°F (750 to 805°C) for carbon steels and higher for many alloy tool steels, holding at heat from 1 to 4 h and cooling slowly in the furnace.

Tempering (Drawing)

Reheating carbon steel to a temperature below the lower critical temperature followed by any desired rate of cooling. Although the terms tempering and drawing are practically synonymous as used in commercial practice, the term tempering is preferred.

The term annealing is also applied to the heat treatment of polymer alloys to effect similar benefits.

See Ref. 8

ANODE

An anode is an electrode of an electrolytic cell where oxidation is the principal reaction. It is also the electrode where corrosion usually occurs and from where metal ions enter into solution.

A sacrificial anode is a chemically active metal that when electrically connected will provide energy needed to cathodically protect a less anodic metal. Zinc, aluminum, and magnesium are commonly used as sacrificial anodes. See "Cathodic Protection."

ANODIC PROTECTION

Anodic protection is a technique used to reduce the corrosion rate of a metal by polorizing it into its passive region. For its corrosion resistance the anodic metal is dependent upon an insoluble film that can be reinforced and maintained by the anodic effect of an impressed anodic polorization.

A typical example of anodic protection is found in steel storage tanks used to store sulfuric acid. Anodic protection systems require careful supervision, because if the proper potentials are not maintained continuously, corrosion by electrolysis may take place.

See Refs. 9 and 10.

ANODIC UNDERMINING

Anodic undermining is a form of corrosion that takes place underneath an organic coating. A typical example is the dissolution of the tin coating between the organic coating and the steel substrate in a food container. A cathodic reaction develops that may involve a component in the foodstuff, or a defect in the tin coating that may expose iron, which then serves as a cathode. Under these circumstances the tin is selectively dissolved and the coating separates from the metal and loses its protective character.

ANODIZING

Anodizing is one commercial method whereby a conversion coating is formed by electrolytic methods. By means of anodic oxidation a thin, dense, and durable oxide film is formed on a metal surface. The predominant application is for the protection of aluminum.

Two types of oxide films have been produced on aluminum by anodic oxidation. They are porous and nonporous films. The porous oxide films are widely used for corrosion protection. They consist of a duplex layered structure having an outer porous layer and an inner nonporous layer (barrier type).

The porous structure has a strong adsorbing ability permitting the surface of an anodic film to be dyed, but it may be contaminated. Since this property of a porous layer causes the formation of corrosion cells, a sealing process to seal the pore is a very important post treatment.

Sealing is accomplished using hot water or steam. This process seals the pores of the aluminum by formation of boehmite (Al_2O_3-H_2O) or bayerite (Al_2O_3-$3H_2O$). In practice sealing is conducted after dyeing, since the sealed film will not absorb dye.

An anodized coating has desirable protective, decorative, or functional properties. Titanium, stainless steel, and zirconium are also subject to anodizing.

See Ref. 9.

ARAMID FIBERS

Aramid fibers are high strength fibers used as reinforcing in FRP structures. See "Thermostat Reinforcing Materials."

AROMATIC COMPOUNDS

Aromatic compounds are organic compounds that contain a benzene ring in its molecule.

ATMOSPHERIC CORRODENTS

See "Atmospheric Corrosion."

ATMOSPHERIC CORROSION

Atmospheric corrosion, though not a separate form of corrosion, has received considerable attention because of the staggering associated costs that result. With the large number of outdoor structures such as buildings, fences, bridges, towers, automobiles, ships, and innumerable other applications exposed to the atmospheric environment there is no wonder that so much attention has been given to the subject.

Atmospheric corrosion is a complicated electrochemical process taking place in corrosion cells consisting of base metal, metallic corrosion products, surface electrolyte, and the atmosphere. Many variables influence the corrosion characteristics of an atmosphere. Relative humidity, temperature, sulfur dioxide content, hydrogen sulfide content, chloride content, amount of rainfall, dust, and even position of the exposed metal exhibit marked influence on corrosion behavior. Geographical location is also another factor.

Because this is an electrochemical process, an electrolyte must be present on the surface of the metal for corrosion to occur. In the absence of moisture, which is the common electrolyte associated with atmospheric corrosion, metals corrode at a negligible rate. For example, carbon steel parts left in a desert remain bright and tarnish free over long periods. Also in climates where the air temperature is below the freezing point of water or of aqueous condensation on the metal surface, rusting is negligible because ice is a poor conductor and does not function effectively as an electrolyte.

Atmospheric corrosion depends not only on the moisture content present but also on the dust content and the presence of other impurities in the air, all of which have an effect on the condensation of moisture on the metal surface and the resulting corrosiveness. Air temperature can also be a factor.

Atmospheric Types

Since corrosion rates are affected by local conditions, atmospheres are generally divided into rural, industrial, and marine.

Additional subdivisions such as urban, arctic, and tropical (wet or dry) can also be included. But the three major categories are of main concern.

For all practical purposes, the more rural the area, with little or no heavy manufacturing operations, or with very dry climatic conditions, the less will be the problems of atmospheric corrosion.

In an industrial atmosphere, all types of contamination by sulfur in the form of sulfur dioxide or hydrogen sulfide are usually the most important. The burning of fossil fuels generates large amounts of sulfur dioxide, which is converted to sulfuric and sulfurous acids in the presence of moisture. Theoretically the combustion of these fossil fuels and hazardous waste products should produce only carbon dioxide, water vapor, and inert gas as combustion products. This is seldom the case, however. Depending upon the impurities contained in the fossil fuel, the chemical composition of the hazardous waste materials incinerated, and the combustion conditions encountered, a multitude of other compounds may be formed.

In addition to the most common contaminants, previously mentioned pollutants such as hydrogen chloride, chlorine, hydrogen fluoride, and hydrogen bromide are produced as combustion products from the burning of chemical wastes. When organophosphorous compounds are incinerated, corrosive phosphorous compounds are produced. Chlorides are also a product of municipal incinerators.

Road traffic and energy production lead to the formation of NO_X which may be oxidized to HNO_3. This reaction has a very low rate, therefore in the vicinity of the emission source the contents of HNO_3 and nitrates are very low. The antipollution regulations that have been enacted do not prevent the escape into the atmosphere of quantities of these materials sufficient to prevent corrosion problems. The corrosivity of an industrial atmosphere diminishes with increasing distance from the city.

Marine environments are subject to chloride attack resulting from the deposition of fine droplets of crystals formed by evaporation of spray that has been carried by the wind from the sea. The quantity of chloride deposition from marine environments is directly proportional to the distance from the shore. The closer to the shore, the greater the deposition and corrosive effect. The atmospheric test station at Kure Beach, NC shows that steels exposed 80 feet from the ocean corrode ten to fifteen times faster than steels exposed 800 feet from the ocean.

In addition to these general air contaminants, there may also be specific pollutants found in a localized area. These may be emitted from a manufacturing operation on a continuous or spasmodic basis and can result in a much more serious corrosion problem than that caused by the presence of general atmospheric pollutants.

Because of these varying conditions, a material that is resistant to atmospheric corrosion in one area may not be satisfactory in another. For example, galvanized iron is perfectly suitable for application in rural atmospheres, but it is not suitable when exposed to industrial atmospheres.

Factors Affecting Atmospheric Corrosion

As previously described, atmospheric corrosion is an electrochemical process and as such depends upon the presence of an electrolyte. The usual electrolyte associated with atmospheric corrosion is water resulting from rain, dew, fog, melting snow, and/or high humidity. Since an electrolyte is not always present, atmospheric corrosion is considered a discontinous process. Corrosion only takes place during the "time of wetness."

Time of Wetness

This term refers to the length of time during which the metal surface is covered by a film of water that renders significant atmospheric corrosion possible. The "time of wetness" is dependent upon local climatic conditions such as the frequency of rain, fog, and dew, the temperature of the metal surface, the temperature of the air, the relative humidity of the atmosphere, the wind speed, and the hours of sunshine.

The "time of wetness" can be determined by either meteorological measurements of temperature and relative humidity or by electrochemical cells. The "time of wetness" determined by meteorological measurements may not necessarily be the same as the actual "time of wetness," because wetness is influenced by the type of metal, pollution of the atmosphere, presence of corrosion products, and degree of coverage against rain. However, the results from these measurements usually show a good correlation with corrosion data from field tests under ordinary outdoor conditions.

Adsorption Layers

The adsorption of water on the metal surface may be the result of the relative humidity of the atmosphere, of the chemical and physical properties of the corrosion products, of the properties of the materials deposited from the air, or of a combination of all three. Industrial atmospheres contain suspended particles of carbon, carbon compounds, metal oxides, sulfuric acid, sodium chloride, and ammonium sulfate. When these substances combine with moisture or when because of their hygroscopic nature they form an electrolyte on the surface, corrosion is initiated.

When hygroscopic salts which are deposited or formed by corrosion absorb moisture from the atmosphere, the metal surface may become wetted. Such absorption occurs above a certain relative humidity, called the critical relative humidity, which corresponds to the vapor pressure above a saturated solution of the salt present. The amount of water on the surface has a direct effect on the corrosion rate. The more water present, the greater the corrosion rate.

Phase Layers

Phase layers are the result of the formation of dew by condensation on a cold metallic surface, or of precipitation in the form of rain or fog, and wet or melting snow. The rate of corrosion will be dependent upon the concentration and nature of the corrodents in the electrolyte, which will vary depending upon the deposition rates, frequency of wetting, drying conditions, and degree of rain protection provided.

If the surface is wetted after a long dry spell during which there has been a large accumulation of surface contamination, the corrosion rate will be greater than that for a smaller amount accumulated during a shorter dry period. Corrosion will also be affected by the quantity of electrolyte present.

Dew is an important source of atmospheric corrosion—more so than rain—and particularly under sheltered conditions. Dew forms when the temperature of the metal surface falls below the dew point of the atmosphere. This can occur outdoors during the night when the surface temperature of the metal is lowered as a result of radiant heat transfer between the metal and the sky. It is also quite common for dew to form during the early morning hours when the air temperature rises faster than the metal temperature. Dew may also form when metal products are brought into warm storage after cold shipment.

Under sheltered conditions dew is an important cause of corrosion. The high corrosivity of dew is a result of several factors. Relatively speaking, the concentration of contaminants in dew is higher than in rainwater, which leads to lower pH values. Heavily industrialized areas have reported pH values of dew in the range of 3 and lower.

The washing effect, which occurs with rain, is usually slight or negligible. With little or no runoff, the pollutants remain in the electrolyte and continue their corrosive action. As the dew dries, these contaminants remain on the surface to repeat their corrosive activity with subsequent dew formations.

Depending upon the conditions, rain can either increase or decrease the effects of atmospheric corrosion. Corrosive action is caused by rain when a phase layer of moisture is formed on the metal surface. This activity is increased when the rain washes corrosive promoters such as H^+ and SO_4^{2-} from the air (acid rain). Rain has the ability to decrease corrosive action on the surface of the metal as a result of washing away the pollutants that have been deposited during the preceding dry spell.

Whether the rain will increase or decrease the corrosive action is dependent upon the ratio of deposition between the dry and wet contaminants. When the dry period deposition of pollutants is greater than the wet period deposition of sulfur compounds, the washing effect of the rain will dominate and the corrosive action will be decreased.

In areas where the air is not as heavily polluted, the corrosive action of the rain will assume a much greater importance because it will increase the corrosion rate.

High concentrations of sulfate and nitrate, and high acidity, will be found in areas having an appreciable amount of air pollution. The pH of fog water has been found to be in the range of 2.2 to 4.0 in highly contaminated areas. This leads to increased corrosivity.

Dust

On a weight basis, in many locations, dust is the primary air contaminant. When in contact with metallic surfaces and combined with moisture, this dust can promote corrosion by forming galvanic or differential aeration cells that, because of their hygroscopic nature, form an electrolyte on the surface. This is particularly true if the dust consists of water-soluble particles, or particles on which sulfuric acid is absorbed. Dust-free air therefore is less likely to cause corrosion.

Temperature

During long-term exposure in a temperate climatic zone, temperature appears to have little or no effect on the corrosion rate. The overall effect of temperature on the corrosion rate is complex. As the temperature increases, the rate of corrosive attack is increased as the result of an increase in the rate of electrochemical and chemical reactions as well as the diffusion rate. Consequently, under constant humidity conditions, a temperature increase will promote corrosion.

By the same token, an increase in temperature can cause a decrease in the corrosion rate by causing a more rapid evaporation of the surface moisture film created by rain or dew. This reduces the time of wetness, which in turn reduces the corrosion rate. In addition, as the temperature increases, the solubility of oxygen and other corrosive gases in the electrolyte film is decreased.

When the air temperature falls below 32°F (0°C) the electrolyte film may freeze. When freezing occurs, there is a pronounced decrease in the corrosion rate, which is illustrated by the low corrosion in subarctic or arctic regions.

In general, temperature is a factor influencing corrosion rates, but it is of importance only under extreme conditions.

Specific Atmospheric Corrodents

The electrolyte film on the surface will contain various materials deposited from the atmosphere or originating from the corroding metal. The composition of the electrolyte is often the factor that determines the rate of corrosion.

The primary contaminants in the air that lead to atmospheric corrosion are SO_x, NO_x, chlorides, and oxygen.

SO_x

Sulfur dioxide, which results from the burning of fossil fuels (such as coal and oil) and the combustion products from the incineration of organic and hazardous wastes, is the most important corrosive contaminant found in industrial atmospheres. Most of the sulfur derived from the burning of fossil fuels is emitted in the form of gaseous SO_2. Once in the atmosphere their physical and chemical state undergoes change. The sulfur dioxide is oxidized on moist particles or in droplets of water to sulfuric acid:

$$SO_2 + H_2O + \tfrac{1}{2}O_2 \rightarrow H_2SO_4$$

The sulfuric acid can be partially neutralized, particularly with ammonia resulting from the biological decomposition of organic matter. This neutralization forms particles containing $(NH_4)_2SO_4$ and forms of acid ammonium sulfate such as NH_4HSO_4 and $(NH_4)_3H(SO_4)_2$.

Atmospheric corrosion results from the deposition of these various materials on metallic surfaces. Deposition of these sulfur compounds is accomplished by:

1. Dry deposition
   a. Absorption of sulfur dioxide gas on metal surfaces
   b. Impaction of sulfate particles
2. Wet deposition
   a. Removal of gas from the atmosphere by precipitation in the form of rain or fog.

The primary cause of atmospheric corrosion is the dry deposition of sulfur dioxide on metallic surfaces. This type of corrosion is usually confined to areas having a high population, many structures, and severe polution. Therefore the atmospheric corrosion caused by sulfur pollutants is usually restricted to the source.

NO_x

These emissions originate from combustion processes other than those emitting SO_x. Road traffic and energy production are the primary sources. Most of the nitrogen oxides are emitted as NO in combustion processes. In the atmosphere, oxidation to NO_2 takes place successfully according to:

$$2\ NO + O_2 \rightarrow 2\ NO_2$$

As the pollutant moves further from the source it is further oxidized by the influence of ozone:

$$NO + O_3 \rightarrow NO_2 + O_2$$

Near the emission source nitrogen dioxide is considered to be the primary pollutant. The NO_2/NO ratio in the atmosphere varies with time and distance from the source. Allowed enough time, the NO_x may be further oxidized according to the reaction:

$$2NO + H_2O + \frac{3}{2}O_2 \rightarrow 2\ HNO_3$$

Since this reaction has a very slow rate, the amounts of HNO_3 and nitrates in the vicinity of the source are very low.

Chlorides

In marine environments chloride deposition is in the form of droplets or crystals formed by evaporation of spray that has been carried by the wind from the sea. As distance from the shore increases, this deposition decreases as the droplets and crystals are filtered off when the wind passes through vegetation, or when the particles are settled by gravity.

Gaseous HCl is a combustion product derived from the burning of coal and municipal incinerators. This gaseous HCl is very soluble in water and forms hydrochloric acid, which is extremely corrosive.

Oxygen

Oxygen is a natural constituent of air and is readily absorbed from the air into the water film on the metal surface, which may be considered saturated, thus promoting any oxidation reactions.

Hydrogen Sulfide

Trace amounts of hydrogen sulfide are present in some contaminated atmospheres. This can cause the tarnishing of silver and copper by the formation of tarnish films.

Effects on Metals Used for Outdoor Applications

Carbon steel is the most widely used metal for outdoor applications although large quantities of zinc, aluminum, copper, and nickel-bearing alloys are also used. Metals customarily used for outdoor installations will be discussed.

Carbon Steel

Except in a dry, clean atmosphere, carbon steel does not have the ability to form a protective coating, as some other metals do. In such an atmosphere a thick oxide film will form that prevents further oxidation. Solid particles on the surface are responsible for the start of corrosion. The settled airborne dust promotes corrosion by absorbing SO_2 and water vapor from the air. Even greater corrosive effects result when particles of hygroscopic salts, such as sulfates or chlorides, settle on the surface and form a corrosive electrolyte.

To protect the surface of unalloyed carbon steel, an additional surface protection must be applied. This protection usually takes the form of an antirust paint or other type of paint formulated for resistance against a specific type of contaminant known to be present in the area. On occasion, plastic or metallic coatings are used.

Weathering Steels

These are steels to which small amounts of copper, chromium, nickel, phosphorus, silicon, manganese, or various combinations thereof have been added. This results in a low alloy carbon steel that has improved corrosion resistance in rural areas, or in areas exhibiting relatively low pollution levels. Factors that affect the corrosion resistance of these steels are climatic conditions, pollution levels, degree of sheltering from the atmosphere, and specific composition of the steel. Exposure to most atmospheres results in a corrosion rate that becomes stabilized in 3 to 5 years. Over this period of time a protective film, or patina, dark brown to violet in color, forms. This patina is a tightly adhering rust formation on the surface of the steel that cannot be wiped off. Since the formation of this film is dependent upon the pollution in the air, in rural areas where there may be little or no pollution a longer period of time may be required to form this film. In areas that have a high pollution level of SO_2 loose rust particles are formed with a much higher corrosion rate. This film of loose particles offers little or no protection against continued corrosion.

When chlorides are present, such as in a marine environment, the protective film will not be formed. Under these conditions corrosion rates of the weathering steels are equivalent to those of unalloyed carbon steel.

In order to form the patina, periodic flushing followed by a dry period is required. If the steel is installed in such a manner as to be sheltered from the rain, the dark patina will not be formed. Instead, a rust lighter in color forms, which provides the same resistance. The corrosion rate of the weathering steels will be the same as the corrosion rate of unalloyed steel when it is continuously exposed to wetness, such as in water or soil.

Since the patina formed has a pleasant aesthetic appearance, the weathering steels can be used without the application of any protective coating of antirust paint, zinc, or aluminum. This is particularly true in urban and rural areas.

In order to receive the maximum benefit from weathering steels, consideration must be given to the design. The design should eliminate all possible areas where water, dirt, and corrosion products can accumulate. When pockets are present, the time of wetness increases, which leads to the development of corrosive conditions. The design should make maximum use of exposure to the weather. Sheltering from rain should be avoided.

While the protective film is forming, rusting will proceed at a relatively high rate, during which time rusty water is produced. This rusty water may stain masonry, pavements and the like. Consequently, steps should be taken to prevent detrimental staining effects, such as coloring the masonry brown, so that any staining will not be obvious.

Zinc

Galvanized steel (zinc-coated steel) is used primarily in rural or urban atmospheres for protection from atmospheric corrosion. Galvanizing will also resist corrosion in marine atmospheres, providing saltwater spray does not come into direct contact. In areas where SO_2 is present in any appreciable quantity, galvanized surfaces will be attacked.

Aluminum

Except for those aluminum alloys that contain copper as a major alloying ingredient, aluminum alloys have a high resistance to weathering in most atmospheres. When exposed to air, the surface of the aluminum becomes covered with an amorphous oxide film that provides protection against atmospheric corrosion, particularly SO_2.

The shiny metal appearance of aluminum gradually disappears and becomes rough when exposed to SO_2. A gray patina of corrosion products forms on the surface. If aesthetics are a consideration the original surface luster can be retained by anodizing. This anodic oxidation strengthens the oxide coating and improves its protective properties.

It is important that the design utilizing aluminum should eliminate rain sheltered pockets on which dust and other pollutants may collect. The formation of the protective film will be disturbed and corrosion accelerated by the presence of these pollutants.

Copper

When exposed to the atmosphere over long periods of time, copper will form a coloration on the surface known as patina, which in reality is a corrosion product that acts as a protective film against further corrosion. The length of time required to form the patina depends upon the atmospheres, because the color is due to the formation of copper hydroxide compounds. Initially the patina has a dark color; gradually it turns green. In urban or industrial atmospheres the compound is a mixture of copper/hydroxide/chloride. It takes approximately 7 years for these compounds to form. When exposed to clean or rural atmospheres tens or hundreds of years may be required to form the patina.

The corrosion resistance of copper is the result of the formation of this patina or protective film. Copper roofs are still in existance on many castles and monumental buildings that are centuries old.

Nickel 200

When exposed to the atmosphere a thin corrosion film (usually a sulfate) forms dulling the surface. The rate of corrosion is extremely slow but increases as the SO_2 content of the atmosphere increases. When exposed to marine or rural atmospheres the corrosion rate is very low.

Monel Alloy 400

The corrosion of monel is negligible in all types of atmospheres. When exposed to rain a thin gray-green patina forms. In sulfurous atmospheres, a smooth brown adherent film forms.

Inconel Alloy 600

In rural atmospheres inconel alloy 600 will remain bright for many years. When exposed to sulfur-bearing atmospheres a slight tarnish is apt to develop. It is desirable to expose this alloy to atmospheres where the beneficial effects of rain in washing the surface and sun and wind in drying can be utilized. It is not recommended to design on the basis of sheltered exposure.

See Refs. 6, 11–15.

AUSTENITE

Austenite is a form of carbon steel with a face centered cubic crystal structure. This form of carbon steel cannot exist below 1333°F (710°C). During heat treatment the holding temperature and time is specified so that the alloy becomes fully austenitic. For common carbon steels the holding temperature is typically specified at 1650°F (900°C). This will make the alloy fully austenitic.

Since austenite has a higher solubility for carbon than the lower temperature forms of carbon steel, heating the steel to an austenizing temperature causes any carbides that may have formed at the lower temperature to dissolve. Alloys that can form austenite at high temperatures, but transform to other crystal forms at lower temperatures, are capable of being hardened by heat treatment. Martensitic steels are an example.

The austenitic microstructure can be made to be stable at low temperatures by alloying with nickel or manganese. See "Austenitic Stainless Steels."

AUSTENITIC DUCTILE CAST IRONS

These cast irons are similar to the austenitic gray cast irons except that they have been treated with magnesium to produce a nodular graphite structure, thereby producing a ductile material. Several different grades are produced with grade D-2 being the most commonly used grade.

These alloys find use in mildly oxidizing acids, alkalies, salts, seawater, water, foods, plastics, and synthetic fiber manufacturing.

See Refs. 16 and 17.

AUSTENITIC GRAY CAST IRONS

The austenitic gray cast irons are gray irons that have been alloyed with nickel and sometimes copper to produce an austenitic matrix similar to that of the 300 series stainless steels. These cast irons offer better corrosion and wear resistance, toughness, and high temperature properties than the standard gray cast irons.

These alloys find use in mildly oxidizing acids, alkalies, salts, seawater, foods, plastics, and synthetic fiber manufacturing.

See Refs. 16 and 17.

AUSTENITIC STAINLESS STEELS

This series of stainless steels is the most widely used for corrosion resistance of any of the stainless steels and therefore is the most important series to the process industry. The addition of substantial quantities of nickel to high chromium alloys stabilizes the austenite at room temperatures. This group of alloys contain 16–26% and 6–22% nickel. The carbon content is kept low (0.08%) to minimize carbide precipitation. The types of stainless comprising the austenitic group are as follows:

201	304N	316SF	20Cb3
202	305N	316L	904L
205	308	316N	20Mo4
301	309	317L	20Mo6
302	309S	317LN	A1-6XN
302B	310N	321	25-6Mo
303	310S	330	ALLOY 31
304	314S	347	ALLOY 654
304L	316	348	

The corrosion resistance of the austenitic stainless steels is the result of the formation of an oxide film on the surface of the metal. Consequently they perform best under oxidizing conditions, since reducing conditions and chloride ions destroy the film, causing rapid attack. Chloride ions, combined with high tensile stresses, cause stress-corrosion cracking.

Types 304 and 304L

These grades of stainless, although possessing a wide range of corrosion resistance, have the least overall resistance of any of the austenitic grades we are considering. The physical and mechanical properties are shown in Table A-13. Type 304 is subject to intergranular corrosion as a result of carbide precipation. Welding can cause this phenomenon, but competent welders, using good welding techniques, can control the problem. Depending upon the particular corrodent being handled, the effect of the carbide precipitation may or may not present a problem. If the corrodent being handled will attack through intergranular corrosion, another alloy should be used.

If the carbon content in the alloy is not allowed to exceed 0.03%, carbide precipitation can be controlled. Type 304L is such an alloy. This alloy can be used for welded sections without the danger of carbide precipitation.

Type 304N has nitrogen added to the alloy which improves its resistance to pitting and crevice corrosion.

Table A.13 Mechanical and Physical Properties of Types 304 and 304L Stainless Steel

Property	Type of alloy 304	304L
Modulus of elasticity × 10^6, psi	28.0	28.0
Tensile strength × 10^3, psi	85	80
Yield strength 0.2% offset × 10^3, psi	35	30
Elongation in 2 in., %	55	55
Hardness, Rockwell	B-80	B-80
Density, lb/in.^{3}	0.29	0.29
Specific gravity	8.02	8.02
Specific heat (32–212°F), Btu/lb°F	0.12	0.12
Thermal conductivity, Btu/h ft^2 °F at 212°F	9.4	9.4
Thermal expansion coefficient (32–212°F) × 10^{-6} in./in.°F	9.6	9.6
Izod impact, ft-lb	110	110

Types 304 and 304L stainless steel exhibit good overall corrosion resistance. They are used extensively in handling nitric acid and most organic acids. Refer to Table A-14 for the compatibility of these stainless steels with selected corrodents.

Types 316 and 316L

These chromium nickel grades of stainless steel have molybdenum added in the range of 2–3%. The molybdenum substantially increases resistance to pitting and crevice corrosion in systems containing chlorides and improves overall resistance to most types of corrosion in chemically reducing and neutral solutions. They may be susceptible to chloride stress corrosion cracking. The mechanical and physical properties are shown in Table A-15.

In general these alloys are more corrosion resistant than the type 304 stainless steels. With the exception of oxidizing acids such as nitric acid, the type 316 alloys will provide satisfactory resistance to corrodents handled by type 304 with the added ability to handle some corrodents that the type 304 alloys cannot handle. Refer to Table A-16 for the compatibility of type 316 and 316L stainless steel with selected corrodents.

Type 316L stainless steel is the low carbon version of type 316 and offers the additional feature of preventing excessive intergranular precipitation of chromium carbides during welding and stress relieving.

Nitrogen is another element that improves resistance to pitting and crevice corrosion, and it is considerably less expensive than molybdenum. AOD and other processing methods makes possible the addition of substantial amounts of nitrogen to steels. As a result, nitrogen-bearing types 316LN and 316L(Hi)N are now available.

Types 317 and 317L

Containing a greater amount of molybdenum, chromium, and nickel, type 317 stainless steels offer higher resistance to pitting and crevice corrosion than type 316 in various process environments encountered in the chemical process industry. However they may still be susceptible to chloride stress corrosion cracking.

Type 317L, like 316L, is a low carbon variation of the basic alloy, which offers the

Table A.14 Compatibility of Types 304, 304L, and 347 Stainless Steel with Selected Corrodents[a]

Chemical	Maximum temp. °F	Maximum temp. °C	Chemical	Maximum temp. °F	Maximum temp. °C
Acetaldehyde	200	93	Benzaldehyde	210	99
Acetamide	100	38	Benzene	230	110
Acetic acid 10%	200	93	Benzene sulfonic acid 10%	210	99
Acetic acid 50%	170	77	Benzoic acid	400	204
Acetic acid 80%	170	77	Benzyl alcohol	90	32
Acetic acid, glacial	210	99	Benzyl chloride	210	99
Acetic anhydride	220	104	Borax	150	66
Acetone	190	88	Boric acid[b]	400	204
Acetyl chloride	100	38	Bromine gas, dry	x	x
Acrylic acid	130	54	Bromine gas, moist	x	x
Acrylonitrile	210	99	Bromine liquid	x	x
Adipic acid	210	99	Butadiene	180	82
Allyl alcohol	220	104	Butyl acetate	80	27
Allyl chloride	120	49	Butyl alcohol	200	93
Alum	x	x	Butyl phthalate	210	99
Aluminum acetate	210	99	Butyric acid	180	82
Aluminum chloride, aqueous	x	x	Calcium bisulfite[d]	300	149
Aluminum chloride, dry	150	66	Calcium carbonate	210	99
Aluminum fluoride	x	x	Calcium chlorate 10%	210	99
Aluminum hydroxide	80	27	Calcium chloride[b,c]	80	27
Aluminum nitrate	80	27	Calcium hydroxide 10%	210	99
Aluminum sulfate[b]	210	99	Calcium hydroxide, sat.	200	93
Ammonia gas	90	32	Calcium hypochlorite	x	x
Ammonium carbonate	200	93	Calcium nitrate	90	32
Ammonium chloride 10%	230	110	Calcium oxide	90	32
Ammonium chloride 50%	x	x	Calcium sulfate	210	99
Ammonium chloride, sat.	x	x	Caprylic acid[a]	210	99
Ammonium fluoride 10%	x	x	Carbon bisulfide	210	99
Ammonium fluoride 25%	x	x	Carbon dioxide, dry	210	99
Ammonium hydroxide 25%	230	110	Carbon dioxide, wet	200	93
Ammonium hydroxide, sat.	210	99	Carbon disulfide	210	99
Ammonium nitrate[c]	210	99	Carbon monoxide	570	299
Ammonium persulfate	x	x	Carbon tetrachloride	210	99
Ammonium phosphate 40%	130	54	Carbonic acid	210	99
Ammonium sulfate 10–40%	x	x	Cellosolve	210	99
Ammonium sulfide	210	99	Chloracetic acid, 50% water	x	x
Ammonium sulfite	210	99	Chloracetic acid	x	x
Amyl acetate	300	149	Chlorine gas, dry	x	x
Amyl alcohol	80	27	Chlorine gas, wet	x	x
Amyl chloride	150	66	Chlorine, liquid[b]	110	43
Aniline	500	260	Chlorobenzene	210	99
Antimony trichloride	x	x	Chloroform[c]	210	99
Aqua regia 3:1	x	x	Chlorosulfonic acid	x	x
Barium carbonate	80	27	Chromic acid 10%	200	93
Barium chloride	x	x	Chromic acid 50%	90	32
Barium hydroxide	230	110	Chromyl chloride	210	99
Barium sulfate	210	99	Citric acid 15%	210	99
Barium sulfide	210	99	Citric acid, concentrated	80	27

Table A.14 Continued

Chemical	Maximum temp. °F	Maximum temp. °C	Chemical	Maximum temp. °F	Maximum temp. °C
Copper acetate	210	99	Nitric acid 5%	210	99
Copper carbonate 10%	80	27	Nitric acid 20%	190	88
Copper chloride	x	x	Nitric acid 70%	170	77
Copper cyanide	210	99	Nitric acid, anhydrous	80	27
Copper sulfate[d]	210	99	Nitrous acid, concentrated	80	27
Cresol	160	71	Oleum	100	38
Cupric chloride 5%	x	x	Perchloric acid 10%	x	x
Cupric chloride 50%	x	x	Perchloric acid 70%	x	x
Cyclohexane	100	38	Phenol[b]	560	293
Cyclohexanol	80	27	Phosphoric acid 50–80%[d]	120	49
Dichloroethane (ethylene dichloride)	210	99	Picric acid[b]	300	149
Ethylene glycol	210	99	Potassium bromide 30%	210	99
Ferric chloride	x	x	Salicylic acid	210	99
Ferric chloride 50% in water	x	x	Silver bromide 10%	x	x
Ferric nitrate 10–50%	210	99	Sodium carbonate 30%	210	99
Ferrous chloride	x	x	Sodium chloride to 30%[b]	210	99
Fluorine gas, dry	470	243	Sodium hydroxide 10%	210	99
Fluorine gas, moist	x	x	Sodium hydroxide 50%	210	99
Hydrobromic acid, dilute	x	x	Sodium hydroxide, concentrated	90	32
Hydrobromic acid 20%	x	x	Sodium hypochlorite 20%	x	x
Hydrobromic acid 50%	x	x	Sodium hypochlorite, concentrated	x	x
Hydrochloric acid 20%	x	x	Sodium sulfide to 50%[b]	210	99
Hydrochloric acid 38%	x	x	Stannic chloride	x	x
Hydrocyanic acid 10%	210	99	Stannous chloride	x	x
Hydrofluoric acid 30%	x	x	Sulfuric acid 10%	x	x
Hydrofluoric acid 70%	x	x	Sulfuric acid 50%	x	x
Hydrofluoric acid 100%	x	x	Sulfuric acid 70%	x	x
Hypochlorous acid	x	x	Sulfuric acid 90%[d]	80	27
Iodine solution 10%	x	x	Sulfuric acid 98%[d]	80	27
Ketones, general	200	93	Sulfuric acid 100%[d]	80	27
Lactic acid 25%[b,d]	120	49	Sulfuric acid, fuming	90	32
Lactic acid, concentrated[b,d]	80	27	Sulfurous acid	x	x
Magnesium chloride	x	x	Thionyl chloride	x	x
Malic acid 50%	120	49	Toluene	210	99
Manganese chloride	x	x	Trichloroacetic acid	x	x
Methyl chloride[b]	210	99	White liquor	100	38
Methyl ethyl ketone	200	93	Zinc chloride	x	x
Methyl isobutyl ketone	200	93			
Muriatic acid	x	x			

[a]The chemicals listed are in the pure state or in a saturated solution unless otherwise indicated. Compatibility is shown to the maximum allowable temperature for which data are available. Incompatibility is shown by an x. A blank space indicates that data are unavailable. When compatible, the corrosion rate is < 20 mpy.
[b]Subject to pitting.
[c]Subject to stress cracking.
[d]Subject to intergranular attack (type 304).

Source: PA Schweitzer. *Corrosion Resistance Tables*. 4th ed. Vols. 1–3. New York: Marcel Dekker, 1995.

Table A.15 Mechanical and Physical Properties of Type 316 and 316L Stainless Steels

	Alloy type	
Property	316	316L
Modulus of elasticity × 10^6, psi	28	28
Tensile strength × 10^3, psi	75	70
Yield strength 0.2% offset × 10^3, psi	30	25
Elongation in 2 in., %	50	50
Hardness, Rockwell	B-80	B-80
Density, lb/in.^{3}	0.286	0.286
Specific gravity	7.95	7.95
Specific heat (32–212°F) Btu/lb°F	0.12	0.12
Thermal conductivity, Btu/h ft^2 °F		
at 70°F	9.3	9.3
at 1500°F	12.4	12.4
Thermal expansion coefficient (32–212°F) × 10^{-6}, in./in.°F	8.9	8.9
Izod impact, ft-lb	110	110

additional advantage of preventing intergranular precipitation of chromium carbide during welding and stress relieving.

The newest of the high molybdenum containing type 300 stainless steels is type 317LN also known as type 317MN or 317LXN. In addition to the molybdenum it also contains nitrogen as a critical alloying ingredient. Overall improvement in its corrosion resistance has not yet been proven, but pitting resistance appears to be better than for the other 317 alloys.

In general these alloys can be used wherever the type 316 series are suitable and in many instances with improved performance. Table A-17 gives the mechanical and physical properties, while Table A-18 provides the compatibility of types 317 and 317L stainless steels with selected corrodents.

Type 321

By alloying austenitic alloys with a small amount of an element having a higher affinity for carbon than does chromium, carbon is restrained from diffusing to the grain boundaries, and any carbon that reaches the boundary reacts with the element instead of with the chromium. These are known as stabilized grades. Type 321 is such an alloy. It has titanium added to stabilize it. Table A-19 gives the mechanical and physical properties.

This alloy is particularly useful in high temperature service in the carbide precipitation range. Even though it has improved resistance to carbide precipitation, it may still be susceptible to chloride stress corrosion cracking. Refer to Table A-20 for the compatibility of type 321 with selected corrodents.

Type 347

This is a columbium stabilized stainless steel. Physical and mechanical properties are shown in Table A-21. The intergranular corrosion resistance has been improved by the

Table A.16 Compatibility of Types 316, 316L Stainless Steels with Selected Corrodents[a]

Chemical	Maximum temp. °F	°C	Chemical	Maximum temp. °F	°C
Acetaldehyde	210	99	Barium hydroxide	400	204
Acetamide	340	171	Barium sulfate	210	99
Acetic acid 10%	420	216	Barium sulfide	210	99
Acetic acid 50%	400	204	Benzaldehyde	400	204
Acetic acid 80%	230	110	Benzene	400	204
Acetic acid, glacial	400	204	Benzene sulfonic acid 10%	210	99
Acetic anhydride	380	193	Benzoic acid	400	204
Acetone	400	204	Benzyl alcohol	400	204
Acetyl chloride	400	204	Benzyl chloride	210	99
Acrylic acid	120	49	Borax	400	204
Acrylonitrile	210	99	Boric acid	400	204
Adipic acid	210	99	Bromine gas, dry	x	x
Allyl alcohol	400	204	Bromine gas, moist	x	x
Allyl chloride	100	38	Bromine liquid	x	x
Alum	200	93	Butadiene	400	204
Aluminum acetate	200	93	Butyl acetate	380	193
Aluminum chloride, aqueous	x	x	Butyl alcohol	400	204
Aluminum chloride, dry	150	66	*n*-Butylamine	400	204
Aluminum fluoride	90	32	Butyl phthalate	210	99
Aluminum hydroxide	400	204	Butyric acid	400	204
Aluminum nitrate	200	93	Calcium bisulfide	60	16
Aluminum sulfate[b]	210	99	Calcium bisulfite	350	177
Ammonia gas	90	32	Calcium carbonate	205	96
Ammonium bifluoride 10%	90	32	Calcium chloride[b]	210	99
Ammonium carbonate	400	204	Calcium hydroxide 10%	210	99
Ammonium chloride 10%	230	110	Calcium hypochlorite	80	27
Ammonium chloride 50%	x	x	Calcium nitrate	350	177
Ammonium chloride, sat.	x	x	Calcium oxide	80	27
Ammonium fluoride 10%	90	32	Calcium sulfate	210	99
Ammonium fluoride 25%	x	x	Caprylic acid	400	204
Ammonium hydroxide 25%	230	110	Carbon bisulfide	400	204
Ammonium hydroxide, sat.	210	99	Carbon dioxide, dry	570	299
Ammonium nitrate[b]	300	149	Carbon dioxide, wet	200	93
Ammonium persulfate 10%	360	182	Carbon disulfide	400	204
Ammonium phosphate 40%	130	54	Carbon monoxide	570	299
Ammonium sulfate 10–40%	400	204	Carbon tetrachloride[b,c]	400	204
Ammonium sulfide	390	171	Carbonic acid	350	177
Ammonium sulfite	210	99	Cellosolve	400	204
Amyl acetate	300	149	Chloracetic acid, 50% water	x	x
Amyl alcohol	400	204	Chloracetic acid	x	x
Amyl chloride	150	66	Chlorine gas, dry	400	204
Aniline	500	260	Chlorine gas, wet	x	x
Antimony trichloride	x	x	Chlorine, liquid dry	120	49
Aqua regia 3:1	x	x	Chlorobenzene, ELC only	260	127
Barium carbonate	80	27	Chloroform[b]	210	99
Barium chloride[c]	210	99	Chlorosulfonic acid	x	x

Table A.16 Continued

Chemical	Maximum temp. °F	Maximum temp. °C	Chemical	Maximum temp. °F	Maximum temp. °C
Chromic acid 10%[d]	400	204	Methyl ethyl ketone	330	166
Chromic acid 50%[d]	150	49	Methyl isobutyl ketone	350	177
Chromyl chloride	210	99	Muriatic acid	x	x
Citric acid 15%[c]	200	93	Nitric acid 5%[e]	210	99
Citric acid, concentrated[c]	380	193	Nitric acid 20%[e]	270	132
Copper acetate	210	99	Nitric acid 70%[e]	400	204
Copper carbonate 10%	80	27	Nitric acid, anhydrous[e]	110	43
Copper chloride	x	x	Nitrous acid, concentrated	80	27
Copper cyanide	210	99	Oleum	80	27
Copper sulfate	400	204	Perchloric acid 10%	x	x
Cresol	100	38	Perchloric acid 70%	x	x
Cupric chloride 5%	x	x	Phenol	570	299
Cupric chloride 50%	x	x	Phosphoric acid 50–80%[e]	400	204
Cyclohexane	400	204	Picric acid	400	204
Cyclohexanol	80	27	Potassium bromide 30%[c]	350	177
Dichloroethane (ethylene dichloride)	400	204	Salicylic acid	350	177
Ethylene glycol	340	171	Silver bromide 10%	x	x
Ferric chloride	x	x	Sodium carbonate	350	177
Ferric chloride 50% in water	x	x	Sodium chloride to 30%[b]	350	177
Ferric nitrate 10–50%	350	177	Sodium hydroxide 10%	350	177
Ferrous chloride	x	x	Sodium hydroxide 50%[a]	350	177
Fluorine gas, dry	420	216	Sodium hydroxide, concentrated	350	177
Fluorine gas, moist	x	x	Sodium hypochlorite 20%	x	x
Hydrobromic acid, dilute	x	x	Sodium hypochlorite, concentrated	x	x
Hydrobromic acid 20%	x	x	Sodium sulfide to 50%	190	88
Hydrobromic acid 50%	x	x	Stannic chloride	x	x
Hydrochloric acid 20%	x	x	Stannous chloride 10%	210	99
Hydrochloric acid 38%	x	x	Sulfuric acid 10%	x	x
Hydrocyanic acid 10%	210	99	Sulfuric acid 50%	x	x
Hydrofluoric acid 30%	x	x	Sulfuric acid 70%	x	x
Hydrofluoric acid 70%	x	x	Sulfuric acid 90%[e]	80	27
Hydrofluoric acid 100%	80	27	Sulfuric acid 98%[e]	210	99
Hypochlorous acid	x	x	Sulfuric acid 100%[e]	210	99
Iodine solution 10%	x	x	Sulfuric acid, fuming	210	99
Ketones, general	250	121	Sulfurous acid[e]	150	66
Lactic acid 25%	210	99	Thionyl chloride	x	x
Lactic acid, concentrated[c,e]	300	149	Toluene	350	177
Magnesium chloride 50%[b,c]	210	99	Trichloroacetic acid	x	x
Malic acid	250	121	White liquor	100	38
Manganese chloride 30%	210	99	Zinc chloride	200	93
Methyl chloride, dry	350	177			

[a]The chemicals listed are in the pure state or in a saturated solution unless otherwise indicated. Compatibility is shown to the maximum allowable temperature for which data are available. Incompatibility is shown by an x. A blank space indicates that data are unavailable. When compatible, the corrosion rate is < 20 mpy.

[b]Subject to stress cracking.

[c]Subject to pitting.

[d]Subject to crevice attack.

[e]Subject to intergranular corrosion.

Source: PA Schweitzer. *Corrosion Resistance Tables.* 4th ed. Vols. 1–3. New York: Marcel Dekker, 1995.

Table A.17 Mechanical and Physical Properties of Type 317 and 317-L Stainless Steels

	Alloy type	
Property	317	317-L
Modulus of elasticity × 10^6, psi	28.0	28.0
Tensile strength × 10^3, psi	75	75
Yield strength 0.2% offset × 10^3, psi	30	30
Elongation in 2 in., %	35	35
Hardness, Rockwell	B-85	B-85
Density, lb/in.^{3}	0.286	0.286
Specific gravity		
Specific heat (32–212°F) Btu/lb°F	0.12	0.12
Thermal conductivity, Btu/h ft^2 °F		
at 70°F	9.3	9.3
at 1500°F	12.4	12.4
Thermal expansion coefficient (32–212°F) × 10^{-6}, in./in.°F	9.2	9.2
Izod impact, ft-lb	110	110

addition of columbium. Basically this alloy is equivalent to type 304 stainless steel, with added protection against carbide precipitation, and can be used wherever type 304 stainless steel is suitable. Refer to Table A-14.

Type 304L also supplies protection from carbide precipitation but has a maximum operating temperature of 800°F (427°C), while type 347 can be operated as high as 1000°F (538°C).

Type 348

This is a tantalum stabilized stainless steel to prevent carbide precipitation.

20Cb3 Stainless Steel

This alloy was originally developed to provide improved corrosion resistance to sulfuric acid. However it has found wide application throughout the chemical processing industry. The alloy is stabilized with columbium and tantalum and has a high nickel content, approximately 33%. It exhibits excellent resistance to chloride stress corrosion cracking, with minimum carbide precipitation due to welding. Refer to Table A-22 for mechanical and physical properties and Table A-23 for elevated temperature tensile properties.

In high concentrations of chlorides, alloy 20 is vulnerable to pitting and crevice attack. For improved resistance to these localized types of corrosion, the 2% molybdenum content must be increased to 4 or 6% as has been done in alloy 20Mo-4 and alloy 20Mo-6.

Alloy 20 finds wide application in the manufacture of synthetic rubber, high octane gasoline, solvents. explosives, plastics, synthetic fibers, heavy chemicals, organic chemicals, pharmaceuticals, and food processing equipment. Refer to Table A-24 for the compatibility of alloy 20Cb3 with selected corrodents.

20Mo-4 Stainless Steel

This alloy is similar to alloy 20Cb3 but with a 4% molybdenum content in place of the 2%, providing improved pitting and crevice corrosion resistance over alloy 20Cb3. Other corrosion resistant properties remain the same as those of alloy 20Cb3.

Table A.18 Compatibility of Types 317 and 317-L Stainless Steel with Selected Corrodents[a]

Chemical	Maximum temp. (°F/°C)	Chemical	Maximum temp. (°F/°C)
Acetaldehyde	150/66	Copper sulfate	70/21
Acetic acid 10%	232/111	Ferric chloride	70/21
Acetic acid 50%	232/111	Hydrochloric acid 20%	x
Acetic acid 80%	240/116	Hydrochloric acid 38%	x
Acetic acid, glacial	240/116	Hydrofluoric acid 30%	x
Acetic anhydride	70/21	Hydrofluoric acid 70%	x
Acetone	70/21	Hydrofluoric acid 100%	x
Aluminum chloride, aqueous	x	Iodine solution 10%	70/21
Aluminum chloride, dry	x	Lactic acid 25%	70/21
Aluminum sulfate 50–55%	225/107	Lactic acid, concentrated	330/166
Ammonium nitrate 66%	70/21	Magnesium chloride	70/21
Ammonium phosphate	80/27	Nitric acid 5%	70/21
Ammonium sulfate 10–40%	100/38	Nitric acid 20%	210/99
Benzene	100/38	Nitric acid 70%	210/99
Boric acid	210/99	Phenol	70/21
Bromine gas, dry	x	Phosphoric acid 50–80%	140/60
Bromine gas, moist	x	Sodium carbonate	210/99
Bromine liquid	x	Sodium chloride	x
Butyl alcohol 5%	195/91	Sodium hydroxide 10%	210/99
Calcium chloride	210/99	Sodium hydroxide 50%	70/21
Calcium hypochlorite	70/21	Sodium hydochlorite 20%	70/21
Carbon tetrachloride	70/21	Sodium hypochlorite, concentrated	70/21
Carbonic acid	70/21	Sodium sulfide to 50%	210/99
Chloracetic acid 78%	122/50	Sulfuric acid 10%	120/49
Chlorine, liquid	x	Sulfuric acid 50%	x
Chlorobenzene	265/129	Sulfuric acid 70%	x
Chromic acid 10%	x	Sulfuric acid 90%	x
Chromic acid 50%	x	Sulfuric acid 98%	x
Citric acid 15%	210/99	Sulfuric acid 100%	x
Citric acic, concentrated	210/99	Sulfurous acid	x

[a]The chemicals listed are in the pure state or in a saturated solution unless otherwise indicated. Compatibility is shown to the maximum allowable temperature for which data are available. Incompatibility is shown by an x. A blank space indicates that data are unavailable. When compatible the corrosion rate is < 20 mpy.
Source: PA Schweitzer. *Corrosion Resistance Tables*. 4th ed. Vols. 1–3. New York: Marcel Dekker, 1995.

20Mo-6 Stainless Steel

Of the three grades of alloy 20 this one offers the highest level of pitting and crevice corrosion resistance. It also possesses the same overall corrosion resistant properties as alloy 20Cb3. Physical and mechanical properties are shown in Table A-25.

Alloy 25-6Mo

This is an austenitic alloy produced by Inco International containing nickel, chromium, molybdenum, copper, nitrogen, carbon, manganese, phosphorus, sulfur, and silicon as alloying ingredients.

Table A.19 Mechanical and Physical Properties of Type 321 Stainless Steel

Modulus of elasticity × 10^6, psi	29
Tensile strength × 10^3, psi	75
Yield strength 0.2% offset × 10^3, psi	30
Elongation in 2 in., %	35
Hardness, Rockwell	B-85
Density, lb/in.^{3}	0.286
Specific gravity	7.92
Specific heat (32–212°F) Btu/lb°F	0.12
Thermal conductivity, Btu/h ft^2 °F	
at 70°F	9.3
at 1500°F	12.8
Thermal expansion coefficient (312–212°F) × 10^{-6}, in./in.°F	9.3
Izod impact, ft-lb	110

Table A.20 Compatibility of Type 321 Stainless Steel with Selected Corrodents[a]

Chemical	Maximum temp. (°F/°C)	Chemical	Maximum temp. (°F/°C)
Acetic acid 10%	x	Copper sulfate	70/21
Acetic acid 50%	x	Ferric chloride	x
Acetic acid 80%	x	Hydrochloric acid 20%	x
Acetic acid, glacial	x	Hydrochloric acid 38%	x
Acetic anhydride	70/21	Hydrofluoric acid 30%	x
Alum	x	Hydrofluoric acid 70%	x
Aluminum chloride, aqueous	x	Hydrofluoric acid 100%	x
Aluminum chloride, dry	x	Iodine solution 10%	x
Aluminum sulfate	70/21	Lactic acid 25%	70/21
Ammonium phosphate	70/21	Lactic acid, concentrated	70/21
Ammonium sulfate 10–40%	70/21	Magnesium chloride	x
Benzene	100/38	Nitric acid 5%	70/21
Boric acid	210/99	Nitric acid 20%	210/99
Bromine gas, dry	x	Nitric acid 70%	210/99
Bromine gas, moist	x	Phenol	x
Bromine liquid	x	Phosphoric acid 50–80%	70/21
Calcium chloride	x	Sodium carbonate	70/21
Calcium hypochlorite	x	Sodium chloride	x
Carbon tetrachloride	x	Sodium hydroxide 10%	70/21
Carbonic acid	70/21	Sodium hydroxide 50%	70/21
Chloracetic acid	x	Sodium hypochlorite 20%	x
Chlorine, liquid	x	Sodium hypochlorite, concentrated	x
Chromic acid 10%	x	Sodium sulfide to 50%	70/21
Chromic acid 50%	x	Sulfuric acid 98%	x
Citric acid 15%	70/21	Sulfuric acid 100%	x
Citric acid, concentrated	70/21	Sulfurous acid	x

[a]The chemicals listed are in the pure state or in a saturated solution unless otherwise indicated. Compatibility is shown to the maximum allowable temperature for which data are available. Incompatibility is shown by an x. A blank space indicates that data are unavailable. When compatible the corrosion rate is < 20 mpy.

Table A.21 Mechanical and Physical Properties of Type 347 Stainless Steel

Modulus of elasticity × 10^6, psi	29.0
Tensile strength × 10^3, psi	75
Yield strength 0.2% offset × 10^3, psi	30
Elongation in 2 in., %	35
Hardness, Rockwell	B-85
Density, lb/in.^{3}	0.285
Specific gravity	7.92
Specific heat (32–212°F) Btu/lb°F	
Thermal conductivity, Btu/h ft^2 °F	
at 70–212°F	9.3
at 1500°F	12.8
Thermal expansion coefficient (32–212°F) × 10^{-6}, in./in.°F	9.3
Izod impact, ft-lb	110

The mechanical properties of alloy 25-6Mo are superior to the other austenitic stainless steels such as 316L. Typical mechanical properties are

Tensile strength: 1000,000 psi
Yield strength (0.2% offset): 48,000 psi
Elongation: 42%

Physical properties of the annealed material at room temperature are

Density: 0.296 lb/in^3

Table A.22 Mechanical and Physical Properties of 20Cb-3 Stainless Steel

Modulus of elasticity × 10^6, psi	28
Tensile strength × 10^3, psi	91
Yield strength 0.2% offset × 10^3, psi	45
Elongation in 2 in., %	45
Hardness, Rockwell	B-86
Density, lb/in.^{3}	0.292
Specific gravity	8.08
Specific heat, Btu/lb°F	0.12
Thermal conductivity, Btu/h ft^2 °F	
at 122°F	7.05
at 212°F	7.57
at 392°F	8.50
at 572°F	9.53
at 752°F	10.50
Thermal expansion coefficient, in./in.°F	
at 77–212°F	8.16×10^{-6}
at 77–842°F	8.84×10^{-6}
at 77–1652°F	9.53×10^{-6}
Charpy V-notch impact strength, ft-lb	200

Table A.23 Elevated Temperature Tensile Properties of 20Cb-3 Stainless Steel

Temperature (°F/°C)	0.2% yield strength (ksi)	Ultimate tensile strength (ksi)
Room	45	91
200/93	40	86
400/204	35	83
600/316	33	80
800/427	30	79
1000/538	28	77
1400/760	26	45
1600/871	19	29

Specific heat: 0.12 Btu/lb°F
Thermal conductivity: 116 Btu-in/ft^2 hr°F
Coefficient of expansion between 68 and 212°F (20 and 100°C): 8.4 in/in°F x 10^{-6}
Young's modulus: 28.1 x 10^6 psi

One of the outstanding attributes of alloy 25-6Mo is its resistance to environments containing chlorides or other halides. It is especially suited for application in high chloride environments such as brackish water, sea water, caustic chlorides, and pulp mill bleach systems.

The alloy offers excellent resistance to pitting and crevice corrosion. Performance in these areas is often measured using critical pitting temperatures (CPT), critical crevice temperatures (CCT), and pitting resistance equivalent number (PREN).

As a general rule, the higher the PREN, the better the resistance. Alloys having similar values may differ considerably in actual service. Those alloys with values greater than 38 on the PREN scale offer more corrosion resistance than the austenitic stainless steels. Alloy 25-6Mo has a PREN of 47. The pitting resistance number is determined by the chromium, molybdenum, and nitrogen contents as

$$\text{PREN} = \%\text{Cr} + 3.3 \times \%\text{Mo} + 30 \times \%\text{N}$$

For 25-6Mo,

$$\text{PREN} = 20 + 3.3(6.5) + 30(0.2) = 47.45$$

The critical pitting temperature for alloy 25-6Mo is 140°F (60°C) or higher, while the critical crevice temperature is 90°F (32.5°C).

In brackish and wastewater systems, microbially influenced corrosion can occur, especially in systems where equipment has been idle for extended periods. A 6% molybdenum alloy offers protection from manganese-bearing, sulfur-bearing, and generally reducing types of bacteria. Because of its resistance to microbially influenced corrosion, alloy 25-6Mo is being used in wastewater piping systems of power plants.

In saturated sodium chloride environments and pH values of 6–8, alloy 25-6Mo exhibits a corrosion rate of less than 1 mpy. Even under more aggressive oxidizing conditions involving sodium chlorate, alloy 25-6Mo maintains a corrosion rate of less than 1 mpy and shows no pitting even at temperatures up to boiling.

Table A.24 Compatibility of Type 20Cb3 Stainless Steel with Selected Corrodents[a]

Chemical	Maximum temp. °F	Maximum temp. °C	Chemical	Maximum temp. °F	Maximum temp. °C
Acetaldehyde	200	93	Barium sulfate	210	99
Acetamide	60	16	Barium sulfide	210	99
Acetic acid 10%	220	104	Benzaldehyde	210	99
Acetic acid 50%	300	149	Benzene	230	110
Acetic acid 80%	300	149	Benzene sulfonic acid 10%	210	99
Acetic acid, glacial	300	149	Benzoic acid	400	204
Acetic anhydride	180	82	Benzyl alcohol	210	99
Acetone	220	104	Benzyl chloride	230	110
Acetyl chloride	210	99	Borax	100	38
Acrylonitrile	210	99	Boric acid	130	54
Adipic acid	210	99	Bromine gas, dry	80	27
Allyl alcohol	300	149	Bromine gas, moist	x	x
Allyl chloride	200	93	Butadiene	180	82
Alum	200	93	Butyl acetate	300	149
Aluminum acetate	60	16	Butyl alcohol	90	32
Aluminum chloride, aqueous	120	43	Butyl phthalate	210	99
Aluminum chloride, dry	120	43	Butyric acid	300	149
Aluminum fluoride	x	x	Calcium bisulfite	300	149
Aluminum hydroxide	80	27	Calcium carbonate	210	99
Aluminum nitrate	80	27	Calcium chlorate	90	32
Aluminum sulfate	210	99	Calcium chloride	210	99
Ammonia gas	90	32	Calcium hydroxide 10%	210	99
Ammonium bifluoride	90	32	Calcium hydroxide, sat.	210	99
Ammonium carbonate	310	154	Calcium hypochlorite	90	32
Ammonium chloride 10%	230	110	Calcium oxide	80	27
Ammonium chloride 50%	170	77	Calcium sulfate	210	99
Ammonium chloride, sat.[b]	210	99	Caprylic acid	400	204
Ammonium fluoride 10%	90	32	Carbon bisulfide	210	99
Ammonium fluoride 25%	90	32	Carbon dioxide, dry	570	299
Ammonium hydroxide 25%	90	32	Carbon dioxide, wet	400	204
Ammonium hydroxide, sat.	210	99	Carbon disulfide	210	99
Ammonium nitrate[b]	210	99	Carbon monoxide	570	299
Ammonium persulfate	210	99	Carbon tetrachloride	210	99
Ammonium phosphate	210	99	Carbonic acid	570	299
Ammonium sulfate 10–40%	210	99	Cellosolve	210	99
Ammonium sulfide	210	99	Chloracetic acid	80	27
Ammonium sulfite	210	99	Chlorine gas, dry	400	204
Amyl acetate	310	154	Chlorine gas, wet	x	x
Amyl alcohol	160	71	Chlorobenzene, dry	100	38
Amyl chloride	130	54	Chloroform	210	99
Aniline	500	260	Chlorosulfonic acid	130	54
Antimony trichloride	200	93	Chromic acid 10%	130	54
Aqua regia 3:1	x	x	Chromic acid 50%	140	60
Barium carbonate	90	32	Chromyl chloride	210	99
Barium chloride 40%	210	99	Citric acid 15%	210	99
Barium hydroxide 50%	230	110	Citric acid, concentrated	210	99

Table A.24 Continued

Chemical	Maximum temp. °F	Maximum temp. °C	Chemical	Maximum temp. °F	Maximum temp. °C
Copper acetate	100	38	Muriatic acid	x	x
Copper carbonate	90	32	Nitric acid 5%	210	99
Copper chloride	x	x	Nitric acid 20%	210	99
Copper cyanide	210	99	Nitric acid 70%	210	99
Copper sulfate	210	99	Nitric acid, anhydrous	80	27
Cupric chloride 5%	60	16	Nitrous acid, concentrated	90	32
Cupric chloride 50%	x	x	Oleum	110	43
Cyclohexane	200	93	Perchloric acid 10%	100	38
Cyclohexanol	80	27	Perchloric acid 70%	110	43
Dichloroethane (ethylene dichloride)	210	99	Phenol	570	299
Ethylene glycol	210	99	Phosphoric acid 50–80%	210	99
Ferric chloride	x	x	Picric acid	300	149
Ferric chloride 50% in water	x	x	Potassium bromide 30%	210	99
Ferric nitrate 10–50%	210	99	Salicylic acid	210	99
Ferrous chloride	x	x	Silver bromide 10%	90	32
Fluorine gas, dry	570	299	Sodium carbonate	570	299
Fluorine gas, moist	x	x	Sodium chloride to 30%[b]	210	99
Hydrobromic acid, dilute	x	x	Sodium hydroxide 10%	300	149
Hydrobromic acid 20%	x	x	Sodium hydroxide 50%[c]	300	149
Hydrobromic acid 50%	x	x	Sodium hydroxide, concentrated	200	93
Hydrochloric acid 20%	x	x	Sodium hypochlorite 30%	90	32
Hydrochloric acid 38%	x	x	Sodium sulfide to 50%	200	93
Hydrocyanic acid 10%	210	99	Stannic chloride	x	x
Hydrofluoric acid 30%	190	88	Stannous chloride 10%	90	32
Hydrofluoric acid 70%	x	x	Sulfuric acid 10%	200	93
Hydrofluoric acid 100%	80	27	Sulfuric acid 50%	110	43
Iodine solution 10%	x	x	Sulfuric acid 70%	120	49
Ketones, general	100	38	Sulfuric acid 90%	100	38
Lactic acid 25%[b]	210	49	Sulfuric acid 98%	300	149
Lactic acid, concentrated, air free	300	149	Sulfuric acid 100%	300	149
Magnesium chloride	200	93	Sulfuric acid, fuming	210	99
Malic acid 50%	160	71	Sulfurous acid	360	182
Manganese chloride 40%	210	99	Toluene	210	99
Methyl chloride	210	99	White liquor	100	38
Methyl ethyl ketone	200	93	Zinc chloride	210	99
Methyl isobutyl ketone	210	99			

[a]The chemicals listed are in the pure state or in a saturated solution unless otherwise indicated. Compatibility is shown to the maximum allowable temperature for which data are available. Incompatibility is shown by an x. A blank space indicates that data are unavailable. When compatible the corrosion rate is < 20 mpy.

[b]Material subject to intergranular corrosion.

[c]Material subject to stress cracking.

Source: PA Schweitzer. *Corrosion Resistance Tables*. 4th ed. Vols. 1–3. New York: Marcel Dekker, 1995.

Table A.25 Mechanical and Physical Properties of 20Mo-6 Stainless Steel

Modulus of elasticity × 10^6, psi	27.0
Tensile strength × 10^3, psi	88
Yield strength 0.2% offset × 10^3, psi	45
Elongation in 2 in., %	50
Hardness, Rockwell	B-85
Density, lb/in.^{3}	0.294
Specific gravity	8.133
Specific heat, Btu/lb°F (32–12°F)	0.11
Thermal conductivity, Btu/h ft^2 °F	
at 122°F	6.99
at 212°F	7.51
at 392°F	8.55
at 572°F	9.53
at 752°F	10.52
Thermal expansion coefficient, in./in.°F × 10^{-6}	
at 77–212°F	8.22
at 77–392°F	8.29
at 77–752°F	8.73
at 77–1112°F	9.29
at 77–1652°F	9.86

904L Stainless Steel

The high nickel and chromium contents make alloy 904L resistant to corrosion in a wide variety of both oxidizing and reducing environments. Molybdenum and copper are included in the alloy for increased resistance to pitting and crevice corrosion, and to general corrosion in reducing acids. Other advantages of the alloy's composition are sufficient nickel for resistance to chloride-ion stress corrosion cracking and a low carbon content for resistance to intergranular corrosion. Mechanical and physical properties are shown in Table A-26.

The alloy's outstanding attributes are resistance to nonoxidizing acids along with resistance to pitting, crevice corrosion, and stress corrosion cracking in such media as stack gas condensate and brackish water. Alloy 904L is especially suited to handle hot solutions of sulfuric acid at moderate concentrations. It also has excellent resistance to phosphoric acid.

AL-6XN Stainless Steel

This alloy is a low carbon, high purity superaustenitic stainless steel containing chromium, nickel, molybdenum, and nitrogen. Mechanical and physical properties are given in Table A-27. It is manufactured by Allegheny Ludlum Corporation. The presence of nitrogen provides the alloy with improved pitting and crevice corrosion resistance and greater resistance to localized corrosion in oxidizing chlorides and reducing solutions.

The corrosion resistant properties of A1-6XN show exceptional resistance to pitting,

Table A.26 Mechanical and Physical Properties of 904-L Stainless Steel

Modulus of elasticity $\times 10^6$, psi	28.5
Tensile strength $\times 10^3$, psi	75
Yield strength 0.2% offset $\times 10^3$, psi	32
Elongation in 2 in., %	35
Hardness, Rockwell	B-90
Density, lb/in.^{3}	0.29
Specific gravity	7.9
Specific heat, Btu/lb°F	0.105
Thermal conductivity, Btu-in./ft^2 hr °F	
at 68°F	90
at 200°F	94
at 400°F	97
at 600°F	105
at 800°F	113
at 1000°F	121
Thermal expansion coefficient, in./in. °F $\times 10^{-6}$	
at 200°F	8.4
at 600°F	8.8
at 800°F	9.0
at 1000°F	9.2
at 1500°F	10.0
at 1800°F	10.3
Charpy V-notch impact, ft-lb	125

crevice attack, and stress cracking in high chlorides, and general resistance in various acid, alkaline, and salt solutions. It also exhibits excellent corrosion resistance to oxidizing chlorides, reducing solutions, and seawater corrosion. A1-6XN can also handle various concentrations at a variety of temperatures of sulfuric, nitric, phosphoric, acetic, and formic acids. Refer to Table A-28.

Alloy 31

With a 6.5% molybdenum content, alloy 31 exhibits excellent resistance to pitting and crevice corrosion in neutral and aqueous acid solutions. The high chromium content of 27% imparts superior resistance to corrosive attack by oxidizing media.

Alloy 654

This alloy has better resistance to localized corrosion than other superaustenitics and is on the same level as nickel based alloys. Indications are that alloy 654 is as corrosion resistant as alloy C-276, based on tests in filtered seawater, bleach plants, and other aggressive chloride environments. It is intended to compete with titanium in the handling of high chloride environments.

Table A.27 Mechanical and Physical Properties of AL-6XN[a] Stainless Steels

Modulus of elasticity × 10^6, psi	27
Tensile strength × 10^3, psi	107.5
Yield strength 0.2% offset × 10^3, psi	54.7
Elongation in 2 in., %	50
Density, lb/in.^{3}	0.291
Specific gravity	8.06
Hardness, Rockwell	B-90
Thermal conductivity at 68–212°F, Btu/hr °F	7.9
Thermal expansion coefficient, in./in. °F × 10^{-6}	
at 68–212°F	8.5
at 68–932°F	8.9
at 68–1472°F	10.0
Izod impact ft-lb	140

[a]Registered trademark of Allegheny Ludlum Corporation.

Pitting Resistance Summary

Although the austenitic stainless steels have a wide range of corrosion resistance and are the most widely used family of alloys in the corrosion resistance field, their single largest drawback has been their tendency to pit. As a result, new alloys have been developed for the purpose of improving their pitting and crevice corrosion resistance. As explained

Table A.28 Compatibility of AL-6XN Stainless Steel with Selected Corrodents[a]

Chemical	Maximum temperature (°F/°C)
Acetic acid 20%	210/99
Acetic acid 80%	217/103
Formic acid 45%	220/104
Formic acid 50%	220/104
Nitric acid 10%	194/90
Nitric acid 65%	241/116
Oxalic acid 10%	210/99
Phosphoric acid 20%	210/99
Phosphoric acid 85%	158/76
Sulfamic acid 10%	210/99
Sulfuric acid 10%	x/x
Sulfuric acid 60%	122/50
Sulfuric acid 95%	86/30
Sodium bisulfate 10%	210/99
Sodium hydroxide 50%	210/99

aCompatibility is shown to the maximum allowable temperature for which data are available. Incompatibility is shown by an x. When compatible, the corrosion rate is < 20 mpy previously, performance in these areas is often measured using pitting resistance equivalent number (PREN). As a general rule, the higher the PREN the better the resistance. The PREN for various austenitic stainless steel alloys is listed in the table.

Alloy	PREN	Alloy	PREN
654	63.09	316LN	31.80
31	54.45	316	27.90
25-6Mo	47.45	20Cb3	27.26
A1-6XN	46.96	348	25.60
20Mo-6	42.81	347	19.0
317LM	39.6		
904L	36.51	331	19.0
20Mo-4	36.20	304N	18.3
317	33.0	304	18.0

Sec Refs. 3, 10, 14, and 18.

REFERENCES

1. WL Sheppard Jr. Chemically Resistant Masonry. 2nd ed. New York: Marcel Dekker, 1982.
2. PA Schweitzer. Corrosion Resistance of Elastomers. New York: Marcel Dekker, 1990.
3. PA Schweitzer. Corrosion Resistance Tables. 4th ed. Vols. 1–3. New York: Marcel Dekker, 1995.
4. N Sridhar, G. Hodge. Nickel and high nickel alloys. In: PA Schweitzer, ed. Corrosion and Corrosion Protection Handbook. New York: Marcel Dekker, 1989, pp 96–124.
5. EH Hollingsworth, HY Hunsicher. Aluminum alloys. In: PA Schweitzer, ed. Corrosion and Corrosion Protection Handbook. 2nd ed. New York: Marcel Dekker, 1989, pp 153–187.
6. C Leygraf. Atmospheric corrosion. In: P Marchus and J Oudar, eds. Corrosion Mechanisms in Theory and Practice. New York: Marcel Dekker, 1995, pp 439–440.
7. BW Lifka. Aluminum and aluminum alloys. In: PA Schweitzer, ed. Corrosion Engineering Handbook. New York: Marcel Dekker, 1996, pp 99–156.
8. LJ Judson. General properties of materials. In: T Baumeister, ed. Mark's Standard Handbook for Mechanical Engineers. 7th ed. New York: McGraw-Hill, 1967, pp 23–27, 69–70.
9. I Suzuki. Corrosion Resistant Coatings Technology. New York: Marcel Dekker, 1989.
10. CP Dillon. Corrosion Control in the Chemical Process Industries 2nd ed. St. Louis: Material Technology Institute of the Chemical Process Industries, 1994.
11. V Kucera, E Mattsson. Atmospheric corrosion. In: M Florian, ed. Corrosion Mechanisms. New York: Marcel Dekker, 1987, pp 211–284.
12. HH Ulhig. Corrosion and Corrosion Control. New York: John Wiley, 1963.
13. PA Schweitzer. Atmosperic corrosion. In: PA Schweitzer, ed. Corrosion and Corrosion Protection Handbook. 2nd ed. New York: Marcel Dekker, 1989, pp 23–32.
14. CP Dillon. Corrosion Resistance of Stainless Steels. New York: Marcel Dekker, 1995.
15. FC Porter. Corrosion Resistance of Zinc and Zinc Alloys. New York: Marcel Dekker, 1994.
16. GW George, PG Breig. Cast alloys. In: PA Schweitzer, ed. Corrosion and Corrosion Protection Handbook. 2nd ed. New York: Marcel Dekker, 1989, pp 289–290.
17. JL Gossett. Corrosion resistance of cast alloys. In: PA Schweitzer, ed. Corrosion Engineering Handbook. New York: Marcel Dekker, 1996, pp 260–261.
18. PK Whitcraft. Corrosion of stainless steels. In: PA Schweitzer, ed. Corrosion Engineering Handbook. New York: Marcel Dekker, 1996, pp 53–77.

B

BACTERIAL CORROSION

When certain bacteria produce substances such as sulfuric acid, ammonia, etc., the resulting corrosion is known as bacterial corrosion. Also see "Biological Corrosion."

BARRIER COATINGS

Barrier coatings are used to keep moisture and/or corrosive materials away from a metallic (usually steel) substrate. These protective barriers may vary in thickness from a thin paint film of only a few mils to a mastic coating applied about 1/4 to 1/2 inch thick, to acid proof brick linings several inches thick. Barrier coatings are effective because they keep moisture, oxygen, and corrosives away from the metallic substrate. The lower the moisture vapor transmission of the polymer the more effective it is as a vehicle for protective coatings. Protective coatings vary greatly in composition, cost, and performance. Refer to "Liquid Applied Linings" and "Paint Coatings."

BASE

A compound of a metal or metallike group, with hydrogen and oxygen in the proportion to form an OH radical, which ionizes in aqueous solution to yield hydroxyl ions. A base is formed when a metallic oxide reacts with water.

BAUME SCALE

This is a special hydrometer used to determine the specific gravities of liquids. Its scale utilizes Baume degrees, which can be converted to true specific gravities or concentrations by reference to conversion tables.

BEARING CORROSION

During operation the lubricating oil or grease contained within the bearing may be subject to chemical deterioration and produce a corrosive material. When this corrodent attacks one of the metals in the bearing alloy the action is referred to as bearing corrosion.

BIOLOGICAL CORROSION

Corrosive conditions can be developed by living organisms as a result of their influence on anodic and cathodic reactions. This metabolic activity can directly or indirectly cause deterioration of a metal by the corrosion process. This activity can

1. Produce a corrosive environment
2. Create electrolytic concentration cells on the metal surface
3. Alter the resistance of surface films
4. Have an influence on the rate of anodic or cathodic reaction
5. Alter the environment composition

Because this form of corrosion gives the appearance of pitting, it is first necessary to diagnose the presence of bacteria. Once established, prevention can be accomplished by the use of biocides or by the selection of a more resistant material of construction. For some species of bacteria a change in pH will provide control. Refer to "Microbial Corrosion."

See Refs. 1 and 2.

BISPHENOL POLYESTERS

Also see "Polymers and Thermoset Polymers."

The bisphenol polyesters are superior in their corrosion resistant properties to the isophthalic polyesters. They show good performance with moderate alkaline solutions and excellent resistance to the various categories of bleaching agents. The bisphenol polyesters will break down under highly concentrated acids or alkalies. These resins can be used in the handling of the following materials:

Acids (to 200°F/93°C)

acetic	fatty acids	stearic
benzoic	hydrochloric (10%)	sulfonic (50%)
boric	lactic	tannic
butyric	maleic	tartaric
chloroacetic (15%)	oleic	trichloroacetic (50%)
chromic (5%)	oxalic	rayon spin bath
citric	phosphoric (80%)	

Salts (solution to 200°F/93°C)

all aluminum salts	copper salts
most ammonium salts	iron salts
calcium salts	zinc salts
	most plating solutions

Solvents (all solvents shown are for the isophthalic resins)

sour crude oil	linseed oil
alcohols at ambient temperature	glycerine

Alkalies

ammonium hydroxide 5%	potassium hydroxide 25%
calcium hydroxide 25%	sodium hydroxide 25%
calcium hypochlorite 20%	chlorite
chlorine dioxide 15%	hydrosulfite

Solvents such as benzene, carbon disulfide, ether, methyl ethyl ketone, tolulene, xylene, trichloroethylene, and trichloroethane will attack the resin. Sulfuric acid above 70% concentration, 73% sodium hydroxide, and 30% chromic acid will also attack the resin. Refer to Table B-1 for the compatibility of bisphenol A–fumarate polyester with selected corrodents and Table B-2 for hydrogenated bisphenol A–Bisphenol A resin with selected corrodents. Refer to Ref.3 for the compatibility of the bisphenol esters with a wider range of selected corrodents.

See also Refs. 4–6.

Table B.1 Compatibility of Bisphenol A-Fumarate Polyester with Selected Corrodents[a]

Chemical	Maximum temp. °F	Maximum temp. °C	Chemical	Maximum temp. °F	Maximum temp. °C
Acetaldehyde	x	x	Ammonium persulfate	180	82
Acetic acid 10%	220	104	Ammonium phosphate	80	27
Acetic acid 50%	160	171	Ammonium sulfate 10–40%	220	104
Acetic acid 80%	160	171	Ammonium sulfide	110	43
Acetic acid, glacial	x	x	Ammonium sulfite	80	27
Acetic amhydride	110	43	Amyl acetate	80	27
Acetone	x	x	Amyl alcohol	200	93
Acetyl chloride	x	x	Amyl chloride	x	x
Acrylic acid	100	38	Aniline	x	x
Acrylonitrile	x	x	Antimony trichloride	220	104
Adipic acid	220	104	Aqua regia 3:1	x	x
Allyl alcohol	x	x	Barium carbonate	200	93
Allyl chloride	x	x	Barium chloride	220	104
Alum	220	104	Barium hydroxide	150	66
Aluminum chloride, aqueous	200	93	Barium sulfate	220	104
Aluminum fluoride 10%	90	32	Barium sulfide	140	60
Aluminum hydroxide	160	71	Benzaldehyde	x	x
Aluminum nitrate	200	93	Benzene	x	x
Aluminum sulfate	200	93	Benzene sulfonic acid 10%	200	93
Ammonia gas	200	93	Benzoic acid	180	82
Ammonium carbonate	90	32	Benzyl alcohol	x	x
Ammonium chloride 10%	200	93	Benzyl chloride	x	x
Ammonium chloride 50%	220	104	Borax	220	104
Ammonium chloride, sat.	220	104	Boric acid	220	104
Ammonium fluoride 10%	180	82	Bromine gas, dry	90	32
Ammonium fluoride 25%	120	49	Bromine gas, moist	100	38
Ammonium hydroxide 25%	100	38	Bromine liquid	x	x
Ammonium hydroxide, 20%	140	60	Butyl acetate	80	27
Ammonium nitrate	220	104	Butyl alcohol	80	27

Table B.1 Continued

Chemical	Maximum temp. °F	Maximum temp. °C	Chemical	Maximum temp. °F	Maximum temp. °C
n-Butylamine	x	x	Ferrous chloride	220	104
Butyric acid	220	93	Ferrous nitrate	220	104
Calcium bisulfite	180	82	Fluorine gas, moist		
Calcium carbonate	210	99	Hydrobromic acid, dilute	220	104
Calcium chlorate	200	93	Hydrobromic acid 20%	220	104
Calcium chloride	220	104	Hydrobromic acid 50%	160	71
Calcium hydroxide 10%	180	82	Hydrochloric acid 20%	190	88
Calcium hydroxide, sat.	160	71	Hydrochloric acid 38%	x	x
Calcium hypochlorite 10%	80	27	Hydrocyanic acid 10%	200	93
Calcium nitrate	220	93	Hydrofluoric acid 30%	90	32
Calcium sulfate	220	93	Hypochlorous acid 20%	90	32
Caprylic acid	160	71	Iodine solution 10%	200	104
Carbon bisulfide	x	x	Lactic acid 25%	210	99
Carbon dioxide, dry	350	177	Lactic acid, concentrated	220	104
Carbon dioxide, wet	210	99	Magnesium chloride	220	104
Carbon disulfide	x	x	Malic acid	160	71
Carbon monoxide	350	177	Methyl ethyl ketone	x	x
Carbon tetrachloride	110	43	Methyl isobutyl ketone	x	x
Carbonic acid	90	32	Muriatic acid	130	54
Cellosolve	140	60	Nitric acid 5%	160	71
Chloracetic acid, 50% water	140	60	Nitric acid 20%	100	38
Chloracetic acid to 25%	80	27	Nitric acid 70%	x	x
Chlorine gas, dry	200	93	Nitric acid, anhydrous	x	x
Chlorine gas, wet	200	93	Oleum	x	x
Chlorine, liquid	x	x	Phenol	x	x
Chlorobenzene	x	x	Phosphoric acid 50–80%	220	104
Chloroform	x	x	Picric acid	110	43
Chlorosulfonic acid	x	x	Potassium bromide 30%	200	93
Chromic acid 10%	x	x	Salicylic acid	150	66
Chromic acid 50%	x	x	Sodium carbonate	160	71
Chromyl chloride	150	66	Sodium chloride	220	104
Citric acid 15%	220	104	Sodium hydroxide 10%	130	54
Citric acid, concentrated	220	104	Sodium hydroxide 50%	220	104
Copper acetate	180	82	Sodium hydroxide, concentrated	200	93
Copper chloride	220	104	Sodium hypochlorite 20%	x	x
Copper cyanide	220	104	Sodium sulfide to 50%	210	99
Copper sulfate	220	104	Stannic chloride	200	93
Cresol	x	x	Stannous chloride	220	104
Cyclohexane	x	x	Sulfuric acid 10%	220	104
Dichloroacetic acid	100	38	Sulfuric acid 50%	220	104
Dichloroethane (ethylene dichloride)	x	x	Sulfuric acid 70%	160	71
Ethylene glycol	220	104	Sulfuric acid 90%	x	x
Ferric chloride	220	104	Sulfuric acid 98%	x	x
Ferric chloride 50% in water	220	104	Sulfuric acid, 100%	x	x
Ferric nitrate 10–50%	220	104	Sulfuric acid, fuming	x	x

Table B.1 Continued

Chemical	Maximum temp. °F	°C	Chemical	Maximum temp. °F	°C
Sulfurous acid	110	43	Trichloroacetic acid 50%	180	82
Thionyl chloride	x	x	White liquor	180	82
Toluene	x	x	Zinc chloride	250	121

[a]The chemicals listed are in the pure state or in a saturated solution unless otherwise indicated. Compatibility is shown to the maximum allowable temperature for which data are available. Incompatibility is shown by an x. A blank space indicates that data are unavailable.
Source: PA Schweitzer. *Corrosion Resistance Tables.* 4th ed. Vols. 1–3. New York: Marcel Dekker, 1995.

Table B.2 Compatibility of Hydrogenated Bisphenol A–Bisphenol A Polyester with Selected Corrodents[a]

Chemical	Maximum temp. °F	°C	Chemical	Maximum temp. °F	°C
Acetic acid 10%	200	93	Butyl acetate	x	x
Acetic acid 50%	160	71	*n*-Butylamine	x	x
Acetic anhydride	x	x	Butyric acid	x	x
Acetone	x	x	Calcium bisulfide	120	49
Acetyl chloride	x	x	Calcium chlorate	210	99
Acrylonitrile	x	x	Calcium chloride	210	99
Aluminum acetate			Calcium hypochlorite 10%	180	82
Aluminum chloride, aqueous	200	93	Carbon bisulfide	x	x
Aluminum fluoride	x	x	Carbon disulfide	x	x
Aluminum sulfate	200	93	Carbon tetrachloride	x	x
Ammonium chloride, sat.	200	93	Chloracetic acid, 50% water	90	32
Ammonium nitrate	200	93	Chlorine gas, dry	210	99
Ammonium persulfate	200	93	Chlorine gas, wet	210	99
Ammonium sulfide	100	38	Chloroform	x	x
Amyl acetate	x	x	Chromic acid 50%	x	x
Amyl alcohol	200	93	Citric acid 15%	200	93
Amyl chloride	90	32	Citric acid, concentrated	210	99
Aniline	x	x	Copper acetate	210	99
Antimony trichloride	80	27	Copper chloride	210	99
Aqua regia 3:1	x	x	Copper cyanide	210	99
Barium carbonate	180	82	Copper sulfate	210	99
Barium chloride	200	93	Cresol	x	x
Benzaldehyde	x	x	Cyclohexane	210	99
Benzene	x	x	Dichloroethane (ethylene dichloride)	x	x
Benzoic acid	210	99	Ferric chloride	210	99
Benzyl alcohol	x	x	Ferric chloride 50% in water	200	93
Benzyl chloride	x	x	Ferric nitrate 10–50%	200	93
Boric acid	210	99	Ferrous chloride	210	99
Bromine liquid	x	x	Ferrous nitrate	210	99

Table B.2 Continued

Chemical	Maximum temp. °F	°C	Chemical	Maximum temp. °F	°C
Hydrobromic acid 20%	90	32	Phenol	x	x
Hydrobromic acid 50%	90	32	Phosphoric acid 50–80%	210	99
Hydrochloric acid 20%	180	82	Sodium carbonate 10%	100	38
Hydrochloric acid 38%	190	88	Sodium chloride	210	99
Hydrocyanic acid 10%	x	x	Sodium hydroxide 10%	100	38
Hydrofluoric acid 30%	x	x	Sodium hydroxide 50%	x	x
Hydrofluoric acid 70%	x	x	Sodium hydroxide, concentrated	x	x
Hydrofluoric acid 100%	x	x	Sodium hypochlorite 10%	160	71
Hypochlorous acid 50%	210	99	Sulfuric acid 10%	210	99
Lactic acid 25%	210	99	Sulfuric acid 50%	210	99
Lactic acid, concentrated	210	99	Sulfuric acid 70%	90	32
Magnesium chloride	210	99	Sulfuric acid 90%	x	x
Methyl ethyl ketone	x	x	Sulfuric acid 98%	x	x
Methyl isobutyl ketone	x	x	Sulfuric acid 100%	x	x
Muriatic acid	190	88	Sulfuric acid, fuming	x	x
Nitric acid 5%	90	32	Sulfurous acid 25%	210	99
Oleum	x	x	Toluene	90	32
Perchloric acid 10%	x	x	Trichloroacetic acid	90	32
Perchloric acid 70%	x	x	Zinc chloride	200	93

[a]The chemicals listed are in the pure state or in a saturated solution unless otherwise indicated. Compatibility is shown to the maximum allowable temperature for which data are available. Incompatibility is shown by an x. A blank space indicates that data are unavailable.
Source: PA Schweitzer. *Corrosion Resistance Tables*. 4th ed. Vols. 1–3. New York: Marcel Dekker, 1995.

BLISTER CRACKING

Blister cracking is a hydrogen-induced failure in steels containing internal flaws by nonmetallic inclusions due to superficial corrosion of the steel by an acid hydrogen sulfide environment liberating atomic hydrogen which diffuses into the metal and is released at the inclusion metal interface as molecular hydrogen under high pressure.

BLISTERING

Early stages of corrosion can be recognized as blistering. Frequently blistering occurs without external evidence of rusting or corrosion. Blistering is mainly the result of volume expansion due to swelling, gas inclusion, gas formation, soluble impurities at the film/support interface from osmotic processes, or electroosmotic effects. Water and chemical gases pass through the film, dissolve ionic material either from the film or from the substrate material, causing an osmotic pressure greater than that of the external face of the coating. This establishes a solute concentration gradient, with water building up at these sites until the film eventually blisters.

Blistering is also the effect of hydrogen damage, particularly to low strength alloys.

This occurs when atomic hydrogen diffuses to internal defects and then precipitates as molecular hydrogen. See "Blister Cracking."

BORON CARBIDE

Boron carbide is used as a high strength reinforcing material for thermosetting resins. See "Thermoset Reinforcing Materials."

BOROSILICATE GLASS

Of the many glass compositions available, the one most commonly used for corrosive applications is borosilicate glass. This particular composition has been selected because of its wide range of corrosion resistance, relatively high operating temperature, good heat resistance due to low thermal expansion, transparency to ultraviolet light, and ability to be prestressed.

The chemical stability of borosilicate glass is one of the most comprehensive of any known construction material. It is highly resistant to water, acids, salt solutions, organic substances, and even halogens like chlorine and bromine.

Only hydrofluoric acid, phosphoric acid with fluorides, or strong alkalis at temperatures above 102°F (49°C) can visibly affect the glass surface. Refer to Table B-3 for the compatibility of borosilicate glass with selected corrodents.

BRASS

See "Copper-Zinc Alloys."

BUTADIENE-STYRENE RUBBER (SBR, Buna-S, GR-S)

During World War II a shortage of natural rubber was created when Japan occupied the Far Eastern nations from which natural rubber was obtained. Because of the great need for rubber, the U.S. government developed what was originally known as Government Rubber Styrene-Type (GR-S) because it was the most practical to put into rapid production on a wartime scale. It was later designated GR-S.

The rubber is produced by copolymerizing butadiene and styrene. As with natural rubber and the other synthetic elastomers, compounding with other ingredients will improve certain properties. Continued development since World War II has improved its properties considerably over what was initially produced by either Germany or the Unites States.

Physical and Mechanical Properties

In general, Buna-S is very similar to natural rubber, although some of its physical and mechanical properties are inferior. It is lacking in tensile strength, elongation, resilience, hot tear, and hysteresis. These disadvantages are offset somewhat by its low cost,

Table B.3 Compatibility of Borosilicate Glass with Selected Corrodents[a]

Chemical	Maximum temp. °F	Maximum temp. °C	Chemical	Maximum temp. °F	Maximum temp. °C
Acetaldehyde	450	232	Benzoic acid	200	93
Acetamide	270	132	Benzyl alcohol	200	93
Acetic acid 10%	400	204	Benzyl chloride	200	93
Acetic acid 50%	400	204	Borax	250	121
Acetic acid 80%	400	204	Boric acid	300	149
Acetic acid, glacial	400	204	Bromine gas, moist	250	121
Acetic anhydride	250	121	Bromine liquid	90	32
Acetone	250	121	Butadiene	90	32
Adipic acid	210	99	Butyl acetate	250	121
Allyl alcohol	120	49	Butyl alcohol	200	93
Allyl chloride	250	121	Butyric acid	200	93
Alum	250	121	Calcium bisulfite	250	121
Aluminum chloride, aqueous	250	121	Calcium carbonate	250	121
Aluminum chloride, dry	180	82	Calcium chlorate	200	93
Aluminum fluoride	x	x	Calcium chloride	200	93
Aluminum hydroxide	250	121	Calcium hydroxide 10%	250	121
Aluminum nitrate	100	38	Calcium hydroxide, sat.	x	x
Aluminum oxychloride	190	88	Calcium hypochlorite	200	93
Aluminum sulfate	250	121	Calcium nitrate	100	38
Ammonium bifluoride	x	x	Carbon bisulfide	250	121
Ammonium carbonate	250	121	Carbon dioxide, dry	160	71
Ammonium chloride 10%	250	121	Carbon dioxide, wet	160	71
Ammonium chloride 50%	250	121	Carbon disulfide	250	121
Ammonium chloride, sat.	250	121	Carbon monoxide	450	232
Ammonium fluoride 10%	x	x	Carbon tetrachloride	200	93
Ammonium fluoride 25%	x	x	Carbonic acid	200	93
Ammonium hydroxide 25%	250	121	Cellosolve	160	71
Ammonium hydroxide, sat.	250	121	Chloracetic acid, 50% water	250	121
Ammonium nitrate	200	93	Chloracetic acid	250	121
Ammonium persulfate	200	93	Chlorine gas, dry	450	232
Ammonium phosphate	90	32	Chlorine gas, wet	400	204
Ammonium sulfate 10–40%	200	93	Chlorine, liquid	140	60
Amyl acetate	200	93	Chlorobenzene	200	93
Amyl alcohol	250	121	Chloroform	200	93
Amyl chloride	250	121	Chlorosulfonic acid	200	93
Aniline	200	93	Chromic acid 10%	200	93
Antimony trichloride	250	121	Chromic acid 50%	200	93
Aqua regia 3:1	200	93	Citric acid 15%	200	93
Barium carbonate	250	121	Citric acid, concentrated	200	93
Barium chloride	250	121	Copper chloride	250	121
Barium hydroxide	250	121	Copper sulfate	200	93
Barium sulfate	250	121	Cresol	200	93
Barium sulfide	250	121	Cupric chloride 5%	160	71
Benzaldehyde	200	93	Cupric chloride 50%	160	71
Benzene	200	93	Cyclohexane	200	93
Benzene sulfonic acid 10%	200	93	Cyclohexanol		

Table B.3 Continued

Chemical	Maximum temp. °F	°C	Chemical	Maximum temp. °F	°C
Dichloroacetic acid	310	154	Nitric acid, anhydrous	250	121
Dichloroethane (ethylene dichloride)	250	121	Oleum	400	204
Ethylene glycol	210	99	Perchloric acid 10%	200	93
Ferric chloride	290	143	Perchloric acid 70%	200	93
Ferric chloride 50% in water	280	138	Phenol	200	93
Ferric nitrate 10–50%	180	82	Phosphoric acid 50–80%	300	149
Ferrous chloride	200	93	Picric acid	200	93
Fluorine gas, dry	300	149	Potassium bromide 30%	250	121
Fluorine gas, moist	x	x	Silver bromide 10%		
Hydrobromic acid, dilute	200	93	Sodium carbonate	250	121
Hydrobromic acid 20%	200	93	Sodium chloride	250	121
Hydrobromic acid 50%	200	93	Sodium hydroxide 10%	x	x
Hydrochloric acid 20%	200	93	Sodium hydroxide 50%	x	x
Hydrochloric acid 38%	200	93	Sodium hydroxide, concentrated	x	x
Hydrocyanic acid 10%	200	93	Sodium hypochlorite 20%	150	66
Hydrofluoric acid 30%	x	x	Sodium hypochlorite, concentrated	150	66
Hydrofluoric acid 70%	x	x	Sodium sulfide to 50%	x	x
Hydrofluoric acid 100%	x	x	Stannic chloride	210	99
Hypochlorous acid	190	88	Stannous chloride	210	99
Iodine solution 10%	200	93	Sulfuric acid 10%	400	204
Ketones, general	200	93	Sulfuric acid 50%	400	204
Lactic acid 25%	200	93	Sulfuric acid 70%	400	204
Lactic acid, concentrated	200	93	Sulfuric acid 90%	400	204
Magnesium chloride	250	121	Sulfuric acid 98%	400	204
Malic acid	160	72	Sulfuric acid 100%	400	204
Methyl chloride	200	93	Sulfurous acid	210	99
Methyl ethyl ketone	200	93	Thionyl chloride	210	99
Methyl isobutyl ketone	200	93	Toluene	250	121
Nitric acid 5%	400	204	Trichloroacetic acid	210	99
Nitric acid 20%	400	204	White liquor	210	99
Nitric acid 70%	400	204	Zinc chloride	210	99

[a]The chemicals listed are in the pure state or in a saturated solution unless otherwise indicated. Compatibility is shown to the maximum allowable temperature for which data are available. Incompatibility is shown by an x. A blank space indicates that data are unavailable.
Source: PA Schweitzer. *Corrosion Resistance Tables*. 4th ed. Vols. 1–3. New York: Marcel Dekker, 1995.

cleanliness, slightly better heat-aging properties, slightly better wear than natural rubber for passenger tires, and availability at a stable price. The electrical properties of SBR are generally good but are not outstanding in any one area.

Buna-S has a maximum operating temperature of 175°F (80°C), which is not exceptional. At reduced temperatures, below 0°F, Buna-S products are more flexible than those produced from natural rubber.

Butadiene-styrene rubber has poor flame resistance and will support combustion.

Table B-4 lists the physical and mechanical properties of Buna-S.

Table B.4 Physical and Mechanical Properties of Butadiene-Styrene Rubber (SBR, Buna-S, GR-S)[a]

Specific gravity	0.94
Refractive index	1.53
Specific heat, cal/g	0.454
Brittle point	–76°F (–60°C)
Insulation resistance, ohms/cm	10^{15}
Dielectric constant at 50 Hz	2.9
Swelling, % by volume	
in kerosene at 77°F (25°C)	100
in benzene at 77°F (25°C)	200
in acetone at 77°F (25°C)	30
in mineral oil at 100°F (38°C)	150
Tear resistance, psi	550
Creep at 70°C	14.6
Tensile strength, psi	1600–3700
Elongation, % at break	650
Hardness, Shore A	35–90
Abrasion resistance	Excellent
Maximum temperature, continuous use	175°F (80°C)
Resistance to compression set	Poor
Machining qualities	Can be ground
Resistance to sunlight	Deteriorates
Effect of aging	Little effect
Resistance to heat	Stiffens

[a]These are representative values since they may be altered by compounding.

Resistance to Sun, Weather, and Ozone

Butadiene-styrene rubber has poor weathering and aging properties. Sunlight will cause it to deteriorate. However, it does have better water resistance than natural rubber.

Chemical Resistance

The chemical resistance of Buna-S is similar to that of natural rubber. It is resistant to water and exhibits fair to good resistance to dilute acids, alkalies, and alcohols. It is not resistant to oils, gasoline, hydrocarbons, or oxidizing agents.

Applications

The major use of Buna-S is in the manufacture of automobile tires, although Buna-S materials are also used to manufacture conveyor belts, hose, gaskets, and seals against air, moisture, sound, and dirt.

See Ref. 7.

BUTYL RUBBER (IIR) AND CHLOROBUTYL RUBBER (IIR)

Butyl rubber contains isobutylene

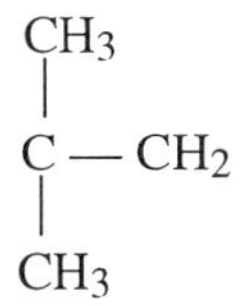

as its parent material, with small proportions of butadiene or isoprene added. Commercial butyl rubber may contain 5% butadiene as a copolymer. It is a general-purpose synthetic rubber whose outstanding physical properties are low permeability to air (approximately one-fifth that of natural rubber) and high energy absorption.

Chlorobutyl rubber is chlorinated isobutylene-isoprene. It has the same general properties as butyl rubber but with slightly higher allowable operating temperatures.

Physical and Mechanical Properties

The single outstanding physical property of butyl rubber is its impermeability. It does not permit gases like hydrogen or air to diffuse through it nearly as rapidly as ordinary rubber does, and it has excellent resistance to the aging action of air. These properties make butyl rubber valuable in the production of life jackets (inflatable type), life rafts, and inner tubes for tires.

At room temperature the resiliency of butyl rubber is poor, but as the temperature increases the resiliency increases. At elevated temperatures butyl rubber exhibits good resiliency. Its abrasion resistance, tear resistance, tensile strength, and adhesion to fabrics and metals is good. Butyl rubber has a maximum continuous service temperature of 250–350°F (120–150°C), with good resistance to heat aging. Its electrical properties are generally good but not outstanding in any one category. The flame resistance of butyl rubber is poor.

Table B-5 lists the physical and mechanical properties of butyl rubber.

Chlorobutyl (CIIR) rubbers have a maximum operating temperature of 300°F (150°C) and can be operated as low as -30°F (-34°C). The other physical and mechanical properties are similar to those of butyl rubber.

Table B.5 Physical and Mechanical Properties of Butyl Rubber (IIR)[a]

Specific gravity	0.91
Dielectric strength, V/mm	25,000
Tensile strength, psi	500–3000
Hardness, Shore A	15–90
Abrasion resistance	Excellent
Maximum temperature, continuous use	250–300°F (120–148°C)
Machining qualities	Can be ground
Resistance to sunlight	Excellent
Effect of aging	Highly resistant
Resistance to heat	Stiffens slightly

[a]These are representative values since they may be altered by compounding.

Resistance to Sun, Weather, and Ozone

Butyl rubber has excellent resistance to sun, weather, and ozone. Its weathering qualities are outstanding, as is its resistance to water absorption.

Chemical Resistance

Butyl rubber is very nonpolar. It has exceptional resistance to dilute mineral acids, alkalies, phosphate ester oils, acetone, ethylene, ethylene glycol, and water. Resistance to concentrated acids, except nitric and sulfuric, is good. Unlike natural rubber, it is very resistant to swelling by vegetable and animal oils. It has poor resistance to petroleum oils, gasoline, and most solvents (except oxygenated solvents).

CIIR has the same general resistance as natural rubber but can be used at higher temperatures. Unlike butyl rubber, CIIR cannot be used with hydrochloric acid. Refer to Table B-6 for the compatibility of butyl rubber with selected corrodents and Table B-7 for the compatibility of chlorobutyl rubber with selected corrodents.

Applications

Because of its impermeability, butyl rubber finds many uses in the manufacture of inflatable items such as life jackets, lifeboats, balloons, and inner tubes. The excellent resistance it exhibits in the presence of water and steam makes it suitable for hoses and diaphragms. Applications are also found as flexible electrical insulation, shock and vibration absorbers, curing bags for tire vulcanization, and molding.

See Refs. 3 and 7.

Table B.6 Compatibility of Butyl Rubber with Selected Corrodents[a]

Chemical	Maximum temp. °F	°C	Chemical	Maximum temp. °F	°C
Acetaldehyde	80	27	Aluminum nitrate	100	38
Acetic acid 10%	150	66	Aluminum sulfate	200	93
Acetic acid 50%	110	43	Ammonium bifluoride	x	x
Acetic acid 80%	110	43	Ammonium carbonate	190	88
Acetic acid, glacial	x	x	Ammonium chloride 10%	200	93
Acetic anhydride	x	x	Ammonium chloride 50%	200	93
Acetone	100	38	Ammonium chloride, sat.	200	93
Acrylonitrile	x	x	Ammonium fluoride 10%	150	66
Adipic acid	x	x	Ammonium fluoride 25%	150	66
Allyl alcohol	190	88	Ammonium hydroxide 25%	190	88
Allyl chloride	x	x	Ammonium hydroxide, sat.	190	88
Alum	200	93	Ammonium nitrate	200	93
Aluminum acetate	200	93	Ammonium persulfate	190	88
Aluminum chloride, aqueous	200	93	Ammonium phosphate	150	66
Aluminum chloride, dry	200	93	Ammonium sulfate 10–40%	150	66
Aluminum fluoride	180	82	Amyl acetate	x	x
Aluminum hydroxide	100	38	Amyl alcohol	150	66

Table B.6 Continued

Chemical	Maximum temp. °F	Maximum temp. °C	Chemical	Maximum temp. °F	Maximum temp. °C
Aniline	150	66	Cupric chloride 5%	150	66
Antimony trichloride	150	66	Cupric chloride 50%	150	66
Barium chloride	150	66	Cyclohexane	x	x
Barium hydroxide	190	88	Dichloroethane (ethylene dichloride)	x	x
Barium sulfide	190	88	Ethylene glycol	200	93
Benzaldehyde	90	32	Ferric chloride	175	79
Benzene	x	x	Ferric chloride 50% in water	160	71
Benzene sulfonic acid 10%	90	32	Ferric nitrate 10–50%	190	88
Benzoic acid	150	66	Ferrous chloride	175	79
Benzyl alcohol	190	88	Ferrous nitrate	190	88
Benzyl chloride	x	x	Fluorine gas, dry	x	x
Borax	190	88	Hydrobromic acid, dilute	125	52
Boric acid	150	66	Hydrobromic acid 20%	125	52
Butyl acetate	x	x	Hydrobromic acid 50%	125	52
Butyl alcohol	140	60	Hydrochloric acid 20%	125	52
Butyric acid	x	x	Hydrochloric acid 38%	125	52
Calcium bisulfite	120	49	Hydrocyanic acid 10%	140	60
Calcium carbonate	150	66	Hydrofluoric acid 30%	150	66
Calcium chlorate	190	88	Hydrofluoric acid 70%	150	66
Calcium chloride	190	88	Hydrofluoric acid 100%	150	66
Calcium hydroxide 10%	190	88	Hypochlorous acid	x	x
Calcium hydroxide, sat.	190	88	Lactic acid 25%	125	52
Calcium hypochlorite	x	x	Lactic acid, concentrated	125	52
Calcium nitrate	190	88	Magnesium chloride	200	93
Calcium sulfate	100	38	Malic acid	x	x
Carbon dioxide, dry	190	88	Methyl chloride	90	32
Carbon dioxide, wet	190	88	Methyl ethyl ketone	100	38
Carbon disulfide	190	88	Methyl isobutyl ketone	80	27
Carbon monoxide	x	x	Muriatic acid	x	x
Carbon tetrachloride	90	32	Nitric acid 5%	200	93
Carbonic acid	150	66	Nitric acid 20%	150	66
Cellosolve	150	66	Nitric acid 70%	x	x
Chloracetic acid, 50% water	150	66	Nitric acid, anhydrous	x	x
Chloracetic acid	100	38	Nitrous acid, concentrated	125	52
Chlorine gas, dry	x	x	Oleum	x	x
Chlorine, liquid	x	x	Perchloric acid 10%	150	66
Chlorobenzene	x	x	Phenol	150	66
Chloroform	x	x	Phosphoric acid 50–80%	150	66
Chlorosulfonic acid	x	x	Salicylic acid	80	27
Chromic acid 10%	x	x	Sodium chloride	200	93
Chromic acid 50%	x	x	Sodium hydroxide 10%	150	66
Citric acid 15%	190	88	Sodium hydroxide 50%	150	66
Citric acid, concentrated	190	88	Sodium hydroxide, concentrated	150	66
Copper chloride	150	66	Sodium hypochlorite 20%	x	x
Copper sulfate	190	88	Sodium hypochlorite, concentrated	x	x
Cresol	x	x	Sodium sulfide to 50%	150	66

Table B.6 Continued

Chemical	Maximum temp. °F	Maximum temp. °C	Chemical	Maximum temp. °F	Maximum temp. °C
Stannic chloride	150	66	Sulfuric acid 100%	x	x
Stannous chloride	150	66	Sulfuric acid, fuming	x	x
Sulfuric acid 10%	200	93	Sulfurous acid	200	93
Sulfuric acid 50%	150	66	Thionyl chloride	x	x
Sulfuric acid 70%	x	x	Toluene	x	x
Sulfuric acid 90%	x	x	Trichloroacetic acid	x	x
Sulfuric acid 98%	x	x	Zinc chloride	200	93

[a]The chemicals listed are in the pure state or in a saturated solution unless otherwise indicated. Compatibility is shown to the maximum allowable temperature for which data are available. Incompatibility is shown by an x. A blank space indicates that data are unavailable.

Source: PA Schweitzer. *Corrosion Resistance Tables.* 4th ed. Vols. 1–3. New York: Marcel Dekker, 1995.

Table B.7 Compatibility of Chlorobutyl Rubber with Selected Corrodents[a]

Chemical	Maximum temp. °F	Maximum temp. °C	Chemical	Maximum temp. °F	Maximum temp. °C
Acetic acid 10%	150	60	Carbon monoxide	100	38
Acetic acid 50%	150	60	Carbonic acid	150	66
Acetic acid 80%	150	60	Chloracetic acid	100	38
Acetic acid, glacial	x	x	Chromic acid 10%	x	x
Acetic anhydride	x	x	Chromic acid 50%	x	x
Acetone	100	38	Citric acid 15%	90	32
Alum	200	93	Copper chloride	150	66
Aluminum chloride, aqueous	200	93	Copper cyanide	160	71
Aluminum nitrate	190	88	Copper sulfate	160	71
Aluminum sulfate	200	93	Cupric chloride 5%	150	66
Ammonium carbonate	200	93	Cupric chloride 50%	150	66
Ammonium chloride 10%	200	93	Ethylene glycol	200	93
Ammonium chloride 50%	200	93	Ferric chloride	175	79
Ammonium chloride, sat.	200	93	Ferric chloride 50% in water	100	38
Ammonium nitrate	200	93	Ferric nitrate 10–50%	160	71
Ammonium phosphate	150	66	Ferrous chloride	175	79
Ammonium sulfate 10–40%	150	66	Hydrobromic acid, dilute	125	52
Amyl alcohol	150	66	Hydrobromic acid 20%	125	52
Aniline	150	66	Hydrobromic acid 50%	125	52
Antimony trichloride	150	66	Hydrochloric acid 20%	x	x
Barium chloride	150	66	Hydrochloric acid 38%	x	x
Benzoic acid	150	66	Hydrofluoric acid 70%	x	x
Boric acid	150	66	Hydrofluoric acid 100%	x	x
Calcium chloride	160	71	Lactic acid 25%	125	52
Calcium nitrate	160	71	Lactic acid, concentrated	125	52
Calcium sulfate	160	71	Magnesium chloride	200	93

Table B.7 Continued

Chemical	Maximum temp. °F	Maximum temp. °C	Chemical	Maximum temp. °F	Maximum temp. °C
Nitric acid 5%	200	93	Sodium sulfide to 50%	150	66
Nitric acid 20%	150	66	Sulfuric acid 10%	200	93
Nitric acid 70%	x	x	Sulfuric acid 70%	x	x
Nitric acid, anhydrous	x	x	Sulfuric acid 90%	x	x
Nitrous acid, concentrated	125	52	Sulfuric acid 98%	x	x
Phenol	150	66	Sulfuric acid 100%	x	x
Phosphoric acid 50–80%	150	66	Sulfuric acid, fuming	x	x
Sodium chloride	200	93	Sulfurous acid	200	93
Sodium hydroxide 10%	150	66	Zinc chloride	200	93

[a]The chemicals listed are in the pure state or in a saturated solution unless otherwise indicated. Compatibility is shown to the maximum allowable temperature for which data are available. Incompatibility is shown by an x. A blank space indicates that data are unavailable.

Source: PA Schweitzer. *Corrosion Resistance Tables.* 4th ed. Vols. 1–3. New York: Marcel Dekker, 1995.

REFERENCES

1. D Thierry, W Sand. Microbially influenced corrosion. In: P Marcus and J Oudar, eds. Corrosion Mechanisms in Theory and Practice. New York: Marcel Dekker, 1995, pp. 457–500.
2. HH Ulhig. Corrosion and Corrosion Control. New York: John Wiley, 1963.
3. PA Schweitzer. Corrosion Resistance Tables. 4th ed. New York: Marcel Dekker, 1995.
4. GT Murray. Introduction to Engineering Materials. New York: Marcel Dekker, 1993.
5. JH Mallinson. Corrosion-Resistant Plastic Composites in Chemical Plant Design. New York: Marcel Dekker, 1988.
6. PA Schweitzer. Corrosion Resistant Piping Systems. New York: Marcel Dekker, 1994.
7. PA Schweitzer. Corrosion Resistance of Elastomers. New York: Marcel Dekker, 1990.

C

CADMIUM COATINGS

These coatings are produced almost exclusively by electro deposition. A cadmium coating on steel does not provide as much protection to the steel as does a zinc coating, since the potential between cadmium and iron is not as great as that between zinc and iron. Therefore it becomes important to minimize defects in the cadmium coating.

Unlike zinc, a cadmium coating will retain a bright metallic appearance. It is more resistant to attack by salt spray and atmospheric condensation. In aqueous solutions cadmium will resist attack by strong alkalies but will be corroded by dilute acids and aqueous ammonia.

Since cadmium salts are toxic, these coatings should not be allowed to come into contact with food products. This coating is commonly used on nuts and bolts.

See Refs. 1 and 2.

CAPPED STEEL

See "Killed Carbon Steel."

CARBIDE PRECIPITATION

Carbon is added to stainless steels as an alloying ingredient to increase strength. During melting and high temperature working operations, such as welding, the carbon content in stainless steel is generally in solid solution. As the steel cools from a temperature of approximately 1600°F (872°C) there is a preference for the formation of a chromium carbide compound, which precipitates preferentially at grain boundaries.

The chromium carbides in themselves do not suffer from poor corrosion resistance. The problem lies in the fact that in the formation of these chromium carbides the chromium has been depleted from the surrounding matrix. This depletion can be to such an extent that the chromium content locally can be below 11%, which is considered the minimum value for stainless steel, leaving this area open to corrosion.

The problem of carbide precipitation can be alleviated by the addition of titanium or niobium (columbium) as an alloying ingredient. These elements tie up the carbon, preventing the precipitation of chromium carbide. Another approach is to reduce the carbon content in the alloy from the usual 0.08% to below 0.035%. This prevents the precipitation of harmful levels of chromium carbide precipitate. The latter approach results in stainless steels known as low carbon, carrying the suffix L after the grade, e.g., 304L, 316L.

See Ref. 3.

CARBON

Carbon has extremely good chemical resistance. It is produced from carbon particles bonded with materials that carbonize during subsequent heating. The operation is usually carried out below 2250°F (1230°C).

In an oxidizing atmosphere it may be used to 660°F (350°C) while in an inert or reducing environment it can be used to 5000°F (2760°C).

Table C.1 lists the compatibility of carbon in contact with selected corrodents. For a more complete listing see Ref. 4.

See also Ref. 5.

CARBON FIBERS

Carbon fibers are used to reinforce FRP laminates and to impart conductivity. See "Thermoset Reinforcing Materials."

CARBON FIBER REINFORCED THERMOPLASTICS

Also See "ZYMAXX." There are many composite thermoplastic materials having carbon filler for reinforcement. Some typical examples of such thermoplastics are as in the table.

Nylon 6	acetal
Nylon 6/6	PBT
Nylon 6/10	PPS
Nylon 6/12	PEEK
ABS	polycarbonate
Polyetherimide	polysulfone
PES	ETFE
FEP	PFA
PVDF	

These basic resins are available with various degrees of carbon reinforcement and may also contain a secondary reinforcing material or a lubricant additive. In all cases the mechanical properties of the base resin are improved by the addition of the reinforcement.

CARBON/GRAPHITE YARNS

For many years the predominent sealing material and mechanical material has been asbestos. However the elimination of asbestos as an environmentally unsafe material has

Table C.1 Compatibility of Carbon with Selected Corrodents[a]

Chemical	Max. Temp. °F	°C
Acetaldehyde	340	171
Acetamide	340	171
Acetic acid 10%	340	171
Acetic acid 50%	340	171
Acetic acid 80%	340	171
Acetic acid, glacial	340	171
Acetic anhydride	340	171
Acetone	340	171
Acetyl chloride	340	171
Acrylonitrile	340	171
Adipic acid	340	171
Allyl alcohol	340	171
Allyl chloride	100	38
Alum	340	171
Aluminum chloride, aqueous	340	171
Aluminum chloride, dry	340	171
Aluminum fluoride	340	171
Aluminum hydroxide	340	171
Aluminum nitrate	340	171
Ammonia gas	340	171
Ammonium bifluoride	390	199
Ammonium carbonate	340	171
Ammonium chloride 10%	340	171
Ammonium chloride 50%	340	171
Ammonium chloride, sat.	340	171
Ammonium fluoride 10%	330	166
Ammonium fluoride 25%	340	171
Ammonium hydroxide 25%	200	93
Ammonium hydroxide, sat.	220	104
Ammonium nitrate	340	171
Ammonium persulfate	340	171
Ammonium phosphate	340	171
Ammonium sulfate 10–40%	340	171
Ammonium sulfide	340	171
Amyl acetate	340	171
Amyl alcohol	200	93
Amyl chloride	210	99
Aniline	340	171
Barium carbonate	250	121
Barium chloride	250	121
Barium hydroxide	250	121
Barium sulfate	250	121
Barium sulfide	250	121
Benzaldehyde	340	171

Table C.1 Continued

Chemical	Max. Temp. °F	Max. Temp. °C
Benzene	200	93
Benzene sulfonic acid 10%	340	171
Benzoic acid	350	177
Borax	250	121
Boric acid	210	99
Bromine gas, dry	x	x
Bromine gas, moist	x	x
Bromine liquid	x	x
Butadiene	340	171
Butyl acetate	340	171
Butyl alcohol	210	99
n-Butylamine	100	38
Butyl phthalate	90	32
Butyric acid	340	171
Calcium bisulfide	340	171
Calcium bisulfite	340	171
Calcium carbonate	340	171
Calcium chlorate 10%	140	60
Calcium chloride	340	171
Calcium hydroxide 10%	200	93
Calcium hydroxide, sat.	250	121
Calcium hypochlorite	170	77
Calcium nitrate	340	171
Calcium oxide	340	171
Calcium sulfate	340	171
Caprylic acid	340	171
Carbon bisulfide	340	171
Carbon dioxide, dry	340	171
Carbon dioxide, wet	340	171
Carbon disulfide	340	171
Carbon monoxide	340	171
Carbon tetrachloride	250	121
Carbonic acid	340	171
Cellosolve	200	93
Chloracetic acid, 50% water	340	171
Chloracetic acid	340	171
Chlorine gas, dry	180	82
Chlorine gas, wet	80	27
Chlorobenzene	340	171
Chloroform	340	171
Chlorosulfonic acid	340	171
Chromic acid 10%	x	x
Chromic acid 50%	x	x
Citric acid 15%	340	171
Citric acid, conc.	340	171
Copper carbonate	340	171
Copper chloride	340	171

Table C.1 Continued

Chemical	Max. Temp. °F	°C
Copper cyanide	340	171
Copper sulfate	340	171
Cresol	400	238
Cupric chloride 5%	340	171
Cupric chloride 50%	340	171
Cyclohexane	340	171
Ethylene glycol	340	171
Ferric chloride	340	171
Ferric chloride 50% in water	340	171
Ferric nitrate 10%–50%	340	171
Ferrous chloride	340	171
Ferrous nitrate	340	171
Fluorine gas, dry	x	x
Hydrobromic acid, dilute	340	171
Hydrobromic acid 20%	340	171
Hydrobromic acid 50%	340	171
Hydrochloric acid 20%	340	171
Hydrochloric acid 38%	340	171
Hydrocyanic acid 10%	340	171
Hydrofluoric acid 30%	340	171
Hydrofluoric acid 70%	x	x
Hydrofluoric acid 100%	x	x
Hypochlorous acid	100	38
Ketones, general	340	171
Lactic acid 25%	340	171
Lactic acid, concentrated	340	171
Magnesium chloride	170	77
Malic acid	100	38
Manganese chloride	400	227
Methyl chloride	340	171
Methyl ethyl ketone	340	171
Methyl isobutyl ketone	340	171
Muriatic acid	340	171
Nitric acid 5%	180	82
Nitric acid 20%	140	60
Nitric acid 70%	x	x
Nitric acid, anhydrous	x	x
Nitrous acid, concentrated	x	x
Perchloric acid 10%	340	171
Perchloric acid 70%	340	171
Phenol	340	171
Phosphoric acid 50–80%	200	93
Picric acid	100	38
Potassium bromide 30%	340	171
Salicylic acid	340	171
Sodium bromide	340	171
Sodium carbonate	340	171

Table C.1 Continued

Chemical	Max. Temp. °F	°C
Sodium chloride	340	171
Sodium hydroxide 10%	240	116
Sodium hydroxide 50%	270	132
Sodium hydroxide, concentrated	260	127
Sodium hypochlorite 20%	x	x
Sodium hypochlorite, concentrated	x	x
Sodium sulfide to 50%	120	49
Stannic chloride	340	171
Sulfuric acid 10%	340	171
Sulfuric acid 50%	340	171
Sulfuric acid 70%	340	171
Sulfuric acid 90%	180	82
Sulfuric acid 98%	x	x
Sulfuric acid 100%	x	x
Sulfurous acid	340	171
Toluene	340	171
Trichloroacetic acid	340	171
White liquor	100	38
Zinc chloride	340	171

[a]The chemicals listed are in the pure state or in a saturated solution unless otherwise indicated. Compatibility is shown to the maximum allowable temperature for which data are available. Incompatibility is shown by an x.
Source: Extracted from PA Schweitzer. *Corrosion Resistance Tables,* 4th ed., Marcel Dekker, New York, 1995.

led to the acceptance of carbon/graphite as a reliable substitute. Carbon/graphite possesses the properties of strength, density, modulus, thermal conductivity, thermal stability, and corrosion resistance.

Its strong chemical resistance makes it ideally suited to the packing and sealing industries for use with acids, caustics, alkalies, and high temperature applications. It is available in many styles and weights.

In general it is inert to most chemicals in the pH range of 2–12. Typical examples are as in the table.

See Ref. 4.

	Concentration	Temperature
Inorganic acids		
Hydrochloric acid	all	boiling point
Hydrofluoric acid	all	boiling point
Phosphoric acid	all	boiling point
Sulfuric acid	0–70%	boiling point
Chromic acid	0–10%	392°F (200°C)
Nitric acid	0–10%	185°F (85°C)
Nitric acid	0–20%	140°F (60°C)
Nitric acid	over 20%	104°F (40°C)

	Concentration	Temperature
Organic acids		
Phenylsulfonic acid	60%	boiling point
Acetic acid	all	boiling point
Acetic anhydride	100%	boiling point
Chloracetic acid	all	boiling point
Amino acid	all	boiling point
Alkalies		
Caustic soda	all	boiling point
Sodium hydroxide	solid	melting point
Solvents		
Benzene	0–100%	boiling point
Ethers	0–100%	boiling point
Alcohols	0–100%	boiling point
Esters	0–100%	boiling point
Ketones	0–100%	boiling point
Halogenated hydrocarbons	0–100%	boiling point
Vinyl chloride	0–100%	boiling point
Mineral oils	0–100%	boiling point

CARBURIZATION

Carburization is the absorption of carbon atoms into a metal surface at high temperature, which reduces the effectiveness of a prior oxide film by the formation of chromium carbides. This depletes the matrix of chromium.

CAST AUSTENITIC STAINLESS STEEL

The various grades of the cast austenitic alloys are the equivalents to the 300 series of wrought stainless steels. The 300 series of stainless steels in the wrought form are fully austenitic. The cast compositions of these alloys can contain anywhere from 5 to 40% ferrite. Because of this, these alloys can display a magnetic response from none to quite strong. The presence of the ferrite in these alloys offers a number of advantages. The ferrite

1. Prevents hot cracking of the castings
2. Enhances the weldability
3. Increases the mechanical strength in proportion to the amount of ferrite present
4. Increases the resistance to stress corrosion cracking as compared to the wrought alloys.

Alloy CF-8 is the equivalent of wrought 304. It has excellent resistance to boiling nitric acid and other strong oxidizing media. Good resistance is also exhibited in contact with copper sulfate, adipic acid, fatty acids, sewage, organic liquids and acids, sodium carbonate, sodium sulfite, vinegar, and white liquor.

Alloy CF-8M is the cast equivalent of 316 stainless steel. Overall corrosion resistance is equal to or better than that of wrought 316. It has good resistance to reducing media, acetone, acetic acid, chloride solutions, black liquor, fatty acids, hot dyes, diluted and concentrated sulfuric acid, phosphoric acid, sulfate and sulfite liquors, sulfurous acid, and vinyl alcohol.

See Refs. 6 and 7.

CAST FERRITIC STAINLESS STEELS

The two major grades of this group of alloys are CB-30 and CC-50. Both are all ferritic and nonhardenable. The major drawback to CB-30 is its poor impact toughness. This alloy has been used to handle food products, nitric acid, steam, sulfur atmospheres, and other oxidizing atmospheres, at temperatures up to 1400°F (760°C).

CC-50, with its higher chromium content, has improved corrosion resistance in oxidizing media. This alloy is used to handle sulfuric and nitric acid mixtures, sulfurous liquors, alkaline liquors, and acid mine waters.

See Refs. 6 and 7.

CAST INCONEL 600

Inconel is the trademark of Inco Alloys. CY-40 is the cast equivalent of wrought Inconel 600. The primary use of this alloy is in the applications requiring strength retention at elevated temperatures and for oxidation resistance. This alloy resists chloride stress corrosion cracking and is often used in caustic soda which contains halogens. Because it resists chloride stress corrosion cracking and corrosion by high purity water it is widely used in nuclear reactor services. It also finds some application in food processing.

See Refs. 4, 6, and 7.

CAST IRON

The term cast iron does not describe a specific material but rather a family of materials. These materials contain over 2% carbon, which can take several different forms. It may be in the form of graphite flakes, graphite spheres, irregular graphite nodules, iron carbides, or cementite, and combinations of these. Included in this group of materials are gray iron, white iron, malleable iron, nodular or ductile iron, and high alloy irons. Use of the term cast iron by itself is discouraged because it does not truly describe a specific material.

CAST MARTENSITIC STAINLESS STEEL

There is a range of martensitic stainless steels of varying grades. The different grades may be heat treated similar to carbon steels to produce hardnesses varying by grade. Their low cost and their ability to be hardened are their main advantages.

The martensitic grades exhibit good corrosion resistance in mildly corrosive organic services. They are also resistant to atmospheric corrosion. These materials have been used to handle alkaline liquors, ammonia water, boiler feed water, pulp, steam, and food products.

When used at high hardness levels they are subject to stress corrosion cracking. They are attacked by most inorganic acids and will rust quickly in marine or humid atmospheres. In general the martensitic grades are less corrosion resistant than the austenitic grades.

See Refs. 6 and 7.

CAST MONEL

Monel is the trademark of the International Nickel Company for their nickel–copper alloy. The most common cast grade is M35-1. The various monel castings find widespread use in the handling of seawater and brackish water. Excellent resistance is also exhibited in the presence of reducing mineral acids, organic acids, salt solutions, and chlorinated solvents. They can also be used in sulfuric acid where reducing conditions exist. Oxidizing conditions or contaminants accelerate the rate of corrosion in all applications.

See Refs. 4, 6, and 7.

CAST NICKEL

Cast nickel, CZ-100, is the equivalent of commercially pure wrought nickel. CZ-100 is used predominantly in alkaline service. Except for ammonium hydroxide it has excellent resistance to all bases. Concentrations greater than 1% of ammonium hydroxide will cause rapid attack. CZ-100 is resistant to all concentrations and temperatures of sodium and potassium hydroxide. If the caustic solutions contain chlorates or oxidizable sulfur compounds the corrosion rates will be accelerated. Applications requiring product purity, such as food processing, also make use of CZ-100.

See Refs. 4, 6, and 7.

CAST SUPERAUSTENITIC STAINLESS STEELS

These castings are particularly suitable for high temperature, chloride-containing environments where pitting and stress corrosion cracking are common causes of failure with other stainless steels. These alloys also exhibit excellent resistance to sulfide stress cracking. Alloy CK3MCuN illustrates the improved resistance to pitting of these alloys. In the presence of 6% $FeCl_3$, conventional stainless steels will pit at ambient temperature, while alloy CK3MCuN will resist pitting at 104°F (40°C).

Listed here are several casting alloys with their equivalent wrought alloy.

Casting	Wrought
CE3MN	AL6XN
CN7M	alloy 20Cb3
CU5MCuC	Incoloy 825
CK3MCuN	254SMO

In general the corrosion resistance of these alloys is equivalent to that of their wrought counterpart with improved resistance in the areas of pitting and SCC.

See Refs. 6 and 7.

CATHODE

A cathode is a negatively charged electrode where reduction is the principal reaction.

CATHODIC CORROSION

Cathodic corrosion is an unusual condition in which metal loss is accelerated at the cathode as a result of the alkaline condition there being corrosive to certain amphoteric metals, primarily aluminum, zinc, and lead.

CATHODIC DELAMINATION

Cathodic delamination is the loss of adhesion of a paint film adjacent to defects, when cathodic protection is applied to a coated metal.

CATHODIC PROTECTION

When dissimilar metals are in physical or electrical contact (the latter via a conductive electrolyte) such as a process fluid or soil, galvanic corrosion can take place. The galvanic corrosion process is similar to the action of a simple DC cell in which the more active metal becomes an anode, and corrodes, while the less active metal becomes a cathode, and is protected. It is possible to predict which metals will corrode when in contact with others based on the galvanic series shown in the table.

Galvanic Series

Anodic end	
Magnesium	Hastelloy C (active)
Magnesium alloys	Brasses
Zinc	Copper
Aluminum 5052	Bronzes
Aluminum 6061	Cupronickel alloys
Cadmium	Monel
Aluminum AA2017	Silver solder
Iron and carbon steel	Nickel (passive)
Copper steel	Inconel (passive)
4–6% chromium steel	Ferritic stainless (passive)
Ferritic stainless (active) 400 series	Austenitic stainless (passive)
Austenitic stainless (active) 18-8 series	Titanium

Galvanic Series

Lead-tin solder	Hastelloy C (passive)
Lead	Silver
Tin	Graphite
Nickel (active)	Gold
Inconel (active)	Platinum
	Cathodic end

All metals and alloys have certain inherent properties that cause them to react as anodes or cathodes when in contact with dissimilar metals or alloys. Whether a particular material will react as a cathode or an anode can be determined from its relative position in the galvanic series. The further apart two materials are from each other in the galvanic series, all other factors being equal, the greater the rate of corrosion. The material closest to the anodic end will be the one to corrode. For example, if tin and zinc were in contact the zinc would corrode, whereas if tin and copper were in contact the tin would corrode.

The rate of attack is also affected by the relative size of the material and the specific electrolyte present. A small anode area in contact with a large cathode area will result in a rapid severe attack. Conversely, a large anode area in contact with a small cathode area will lessen the rate of galvanic attack since the same total emf driving force of corrosion will be spread out over a larger area. Also the higher the degree of ionization of the electrolyte the greater the rate of attack.

Galvanic corrosion can also take place when metals having the same analysis have different surface conditions and an electrolyte is present. In general, the formation of a corrosion cell is induced by the nonuniformity of the surface condition, such as with defects in the surface oxide film, localized distribution of elements, and difference in crystal face or phase. These nonuniformities of surface cause potential difference between portions of the surface and thereby promote the formation of a corrosion cell.

Galvanic corrosion can be stopped by means of cathodic protection, which is an electrochemical technique. It can be applied to metals immersed in water, buried in soil, or in contact with electrolytes in a process application. Cathodic protection consists of a cathodic current flowing through the metal electrolyte interface favoring the reduction reaction over the anodic metal dissolution. The entire structure works as a cathode.

This electrochemical technique was developed by Sir Humphrey Davy in 1824. The British Admiralty had blocks of iron attached to the hulls of copper-sheathed vessels to provide cathodic protection. Unfortunately cathodically protected copper is subject to fouling by marine life, which reduced the speed of vessels under sail and forced the Admiralty to discontinue the practice. Unprotected copper provides a sufficient number of copper ions to poison fouling organisms. However the corrosion rate of the copper had been appreciably reduced.

In 1829 Edmund Davy was successful in protecting the iron portions of buoys by using zinc blocks, and in 1840 Robert Mallet produced a zinc alloy that was particularly suited as a sacrificial anode. As steel hulls replaced wooden hulls, the fitting of zinc slabs to the steel hulls, to provide cathodic protection, became standard practice.

In 1950 the Canadian Navy determined that the proper use of antifouling paints in conjunction with corrosion resisting paints made cathodic protection of ships feasible and could reduce maintenance costs.

Cathodic protection is achieved by applying electrochemical principles to metallic components buried in soil or immersed in water. It is accomplished by flowing a cathodic

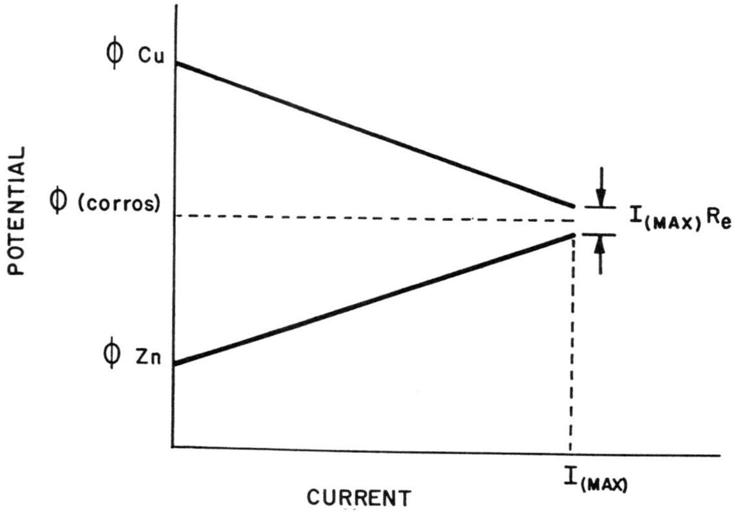

Figure C.1 Polarization for copper-zinc cell.

current through a metal–electrolyte interface favoring the reduction reaction over the anodic metal dissolution. This enables the entire structure to work as a cathode.

The basis of cathodic protection is shown in the polarization diagram for a copper–zinc cell as in Fig. C.1. If polarization of the cathode is continued by use of an external current beyond the corrosion potential to the open-circuit potential of the anode, both electrodes reach the same potential and no corrosion of the zinc can take place.

Cathodic protection is accomplished by supplying an external current to the corroding metal on the surface of which local action cells operate, as shown in Fig. C.2. Current flows from the auxiliary anode and enters the anodic and cathodic areas of the corrosion cells, returning to the source of the DC current B. Local action current will cease to flow when all the metal surface is at the same potential as a result of the cathodic areas being polarized by an external current to the open-circuit potential of the anodes. As long as this external current is maintained the metal cannot corrode.

There are two methods by which cathodic protection can be accomplished. One is by coupling the structure with a more active metal, such as zinc or magnesium. This produces a galvanic cell in which the active metal works as an anode and provides a flux of electrons to the structure. The structure then becomes the cathode and is protected while the anode is destroyed progressively and is called a sacrificial anode.

The second method is to impress a direct current between an inert anode and the structure. The structure receives the excess of electrons which protects it. About 1910–1912 the first application of cathodic protection by means of an impressed electric current was undertaken in England and the United States. Since that time the general use of cathodic protection has been widespread. There are thousands of miles of buried pipelines and cables that are protected in this manner. This form of protection is also used for water tanks, submarines, canal gates, marine piping, condensers, and chemical equipment.

Sacrificial Anodes

In cathodic protection, the structure to be protected must receive a cathodic current flow so that it operates as a cathode. The need for an external DC current to accomplish this

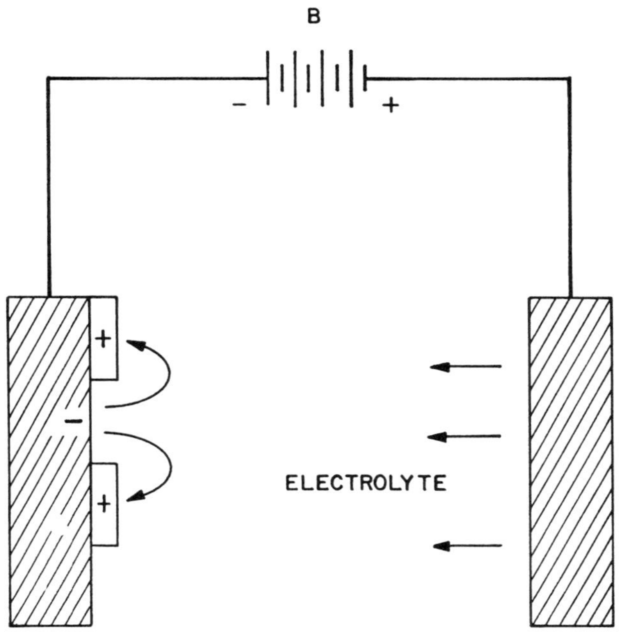

Figure C.2 Cathodic protection using impressed current on local action cell.

can be eliminated by selecting an anode constructed of a metal that is more active in the galvanic series than the metal to be protected. A galvanic cell will be established with the current direction as required. These sacrificial anodes are usually composed of magnesium or magnesium based alloys. On occasion zinc or aluminum have been used.

Magnesium is more active than steel, it has a greater tendency to ionize, and its potential is more active than iron. The open-circuit potential difference between magnesium and steel is about 1 volt. This means that one anode can protect only a limited length of pipeline. This low voltage can have an advantage over higher impressed voltages in that the danger of overprotection to some portions of the structure is less and because the total current per anode is limited; the danger of stray-current damage to adjoining metal structures is reduced.

Magnesium rods have also been placed in steel hot water tanks to increase their life. The greatest degree of protection is afforded in hard waters, since the degree of conductivity is greater than in soft waters.

Sacrificial Anode Requirements

To provide cathodic protection, a current density of a few milliamps (mA) is required. In order to determine the anodic requirement it is necessary to know the energy content of the anode and its efficiency. With this information it is possible to determine the size of the anode required, its expected life, and the number of anodes required.

The three most common metals used as sacrificial anodes are magnesium, zinc, and aluminum. The energy content and efficiency of these metals are shown in the table.

Metal	Theoretical energy content (A h/lb)	Anodic efficiency %	Practical energy content (A h/lb) (PE)
Magnesium	1000	50	500
Zinc	370	90	333
Aluminum	1345	60	810

Zinc is more economical to use than magnesium, but because of the relatively small cell voltage it produces, it is primarily useful under special circumstances, such as to protect ships in seawater or to prevent corrosion of systems with low current requirements. Although magnesium is more expensive than zinc, and although it is consumed faster than zinc or aluminum, it does provide the largest cell voltage and the largest current. Care must be taken not to use aluminum in environments having a pH of 8 or greater, since alkaline conditions will produce a rapid self-corrosion of aluminum.

In determining anodic requirements to provide cathodic protection, several calculations are required. The number of pounds of metal required to provide a current of 1 A for one year is calculated as

$$\text{lb metal/A–yr} = \frac{8760\ \text{h/yr}}{PE}$$

For magnesium this would be:

$$\text{lb Mg/A–yr} = \frac{8760}{500} = 17.52$$

The number of years (*YN*) for which 1 lb of metal can produce a current of 1 mA is determined from

$$YN = \frac{PE}{10^{-3}\ \text{A}\ 8760\ \text{h/yr}}$$

For magnesium this would be

$$YN = \frac{500}{10^{-3}\,(8760)} = 60\ \text{yr}$$

The current density requirements for cathodic protection is on the order of a few mA. The life expectancy (*L*) of an anode of *W* lb, delivering a current of 1 mA is calculated as

$$\text{L} = \frac{YN\,(W)}{i}$$

For magnesium this would be

$$\text{L} = \frac{60W}{i}$$

which is based on 50% anodic efficiency. Since actual efficiencies tend to be somewhat less, it is advisable to a apply a safety factor and multiply the result by 0.75.

The current required to secure protection of a structure and the available cell voltage between the metal structure and the sacrificial anode determine the number of anodes required. This can be illustrated by the following example.

Assume that an underground pipeline has an external area of 200 sq ft and a soil resistivity of 600 ohm cm. Field tests indicate that 6 mA/sq ft is required for protection. To provide protection for the entire pipeline

$$(6\text{mA/ft}^2)\ (200\text{ft}^2) = 1200\ \text{mA}$$

is required. Magnesium anodes used in this particular soil have a voltage of –1.65 V or a galvanic cell voltage of

$$\begin{aligned} E_{\text{calc}} &= E_{\text{c}} - E_{\text{a}} \\ &= -0.85 - (-1.65) \\ &= +0.8 + V \end{aligned}$$

Therefore the resistance is

$$R = \frac{V}{I} = \frac{0.8}{1.2} = 0.67\text{ohm}$$

As the number of anodes is increased, the total resistance of the system decreases. Each anode that is added provides a new path for current flow, parallel to the existing system. The relationship between the resistance of the system and the number of anodes is shown in the Suade equation:

$$R = \frac{0.00521P}{NL}\left(2.3 \log \frac{8L}{d-1} + \frac{2L}{S}\ 2.3 \log 0.656\ N\right)$$

where

R = resistance in ohms
P = soil resistivity in ohm-cm
N = number of anodes
L = anode length ft
d = diameter of anode ft
S = distance between anodes ft

Figure C.3 shows the typical plotting of the results of this equation. Different anodic shapes will have different curves.

Impressed Current Systems

For these systems the source of electricity is external. A rectifier converts high voltage to a low voltage DC current. This direct current is impressed between buried anodes and the structure to be protected.

It is preferred to use inert anodes, which will last for the longest possible times. Typical materials used for these anodes are graphite, silicon, titanium, and niobium plated with platinum.

For a given applied voltage, the current is limited by electrolyte resistivity and by the anodic and cathodic polarization. With the impressed current system it is possible to impose whatever potential is necessary to obtain the current density required by means of the rectifier.

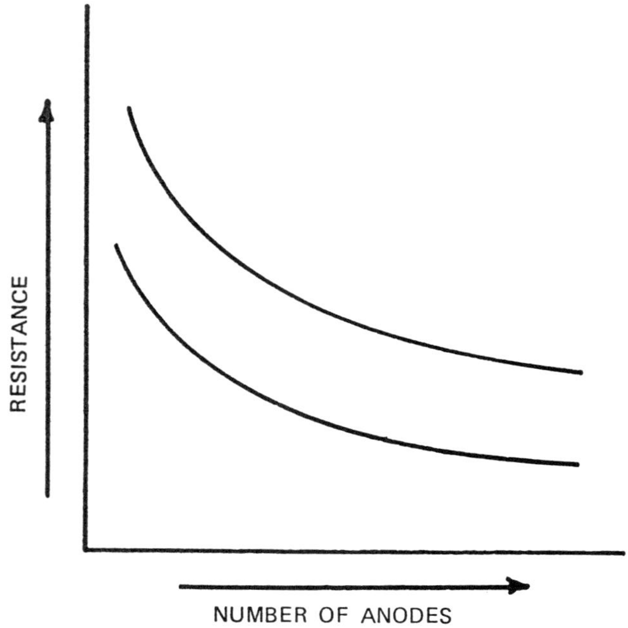

Figure C.3 Plot of Sunde equation.

Electric current flows in the soil from the buried anode to the underground structure to be protected. Therefore the anode must be connected to the positive pole of the rectifier and the structure to the negative pole. All cables from the rectifier to the anode, and the structure must be electrically insulated. If not, those from the rectifier to the anode will act as an anode and deteriorate rapidly, while those from the rectifier to the structure may pick up some of the electric current, which would then be lost for protection.

Current Requirements

The specific metal and the environment will determine the current density required for complete protection. The applied current density must always exceed the current density equivalent to the measured corrosion rate under the same conditions. Therefore as the corrosion rate increases, the impressed current density must be increased to provide protection.

Factors that affect current requirements are

1. The nature of the electrolyte
2. The soil resistivity
3. The degree of aeration

The more acid the electrolyte, the greater will be the potential for corrosion and the greater will be the current requirement. Soils that exhibit a high resistance require a lower cathodic current to provide protection. In the area of violent agitation or high aeration, an increase in current will be required. The required current to provide cathodic protection can vary from 0.5 to 20 mA/sq ft of bare surface.

Field testing may be required to determine the necessary current density to provide cathodic protection in a specific area. These testing techniques will only provide an approximation. After completion of the installation, it will be necessary to conduct a potential survey and make the necessary adjustments to provide the desired degree of protection.

Anode Materials and Backfill

Although it is generally preferred to use inert anodes, it is possible to use scrap iron. Scrap iron is consumed at a considerably faster rate than graphite or other inert anode materials. The advantage of scrap iron is its lower initial cost and lower operating cost, since its power requirements are less. In areas where replacement poses a problem the cost of the use of the more inert anodes outweighs the reduced cost of the scrap iron.

Platinum clad or 2% silver–lead electrodes have been used for protection of structures in seawater and are estimated to last 10 years, whereas sacrificial magnesium anodes have a life of 2 years.

Since the effective resistivity of soil surrounding an anode is limited to the immediate area of the electrode, this local resistance is usually reduced by using backfill. The anode is usually surrounded by a thick bed of coke mixed with 3 or 4 parts of gypsum to one part of sodium chloride. The consumption of the anode itself is reduced somewhat, since the coke backfill carries a part of the current. Backfill is not required when the anode is immersed in a river bed, lake, or ocean.

Testing for Completeness of Protection

Once the system has been installed it must be tested for completeness of protection. The preferred method is to take potential measurements. By measuring the potential of the protected structure, the degree of protection including overprotection can be determined. The basis for this determination is the fundamental concept that cathodic protection is complete when the protected structure is polarized to the open-circuit anodic potential of the local action cells.

The reference electrode is placed as closely as possible to the protected structure to avoid and to minimize an error caused by internal resistance (IR) drop through the soil. For buried pipelines a compromise location is directly over the buried pipe at the soil surface because cathodic protection currents flow mostly to the lower surface and are minimum at the upper surface of the pipe buried a few feet below the surface.

The potential for steel is equal to –0.85 V versus the copper saturated copper sulfate half-cell, or 0.53 V on the standard hydrogen scale. The theoretical open-circuit anodic potential for other metals may be calculated using the Nernst equation. Several typical calculated values are shown in the table.

Metal	E° (V)	Solubility product $M(OH)_2$	OH_2 scale (V)	OVs Cu – $CuSO_4$ reference (V)
Iron	0.440	1.8×10^{-15}	–0.59	–0.91
Copper	–0.337	1.6×10^{-19}	0.16	–0.16
Zinc	0.763	4.5×10^{-17}	–0.93	–1.25
Lead	0.126	4.2×10^{-15}	–0.27	–0.59

Overprotection of steel structures, to a moderate degree, does not cause any problems. The primary disadvantages are waste of electric power and increased consumption of auxiliary anodes. When overprotection is excessive, hydrogen can be generated at the protected structure in sufficient quantities to cause blistering of organic coatings, hydrogen embrittlement of the steel, or hydrogen cracking.

Overprotection of systems with amphoteric metals (e.g. aluminum, zinc, lead, tin) will damage the metal by causing increased attack instead of reduction of corrosion. This stresses the need for making potential measurements of protected structures.

Use with Coatings

It is advantageous to use insulating coatings with sacrificial anodes or impressed current systems when supplying cathodic protection. These coatings need not be pore free, because the protective current flows preferentially to the exposed metal areas, which require the protection. Coatings are useful in distributing the protective current, in reducing total current requirements, and in extending the life of the anode. Compared to a bare pipeline, the current distribution in a coated pipeline is greatly improved, the total number of anodes required is reduced, and the total current required is less. In addition, one anode can protect a much longer section of pipeline. For example, one magnesium anode is capable of protecting approximately 100 feet (30 m) of a bare pipeline, whereas the same anode can provide protection for approximately 5 miles of a coated pipeline.

In a hot water tank coated with glass or an organic coating, the life of the magnesium anode is increased and more uniform protection is supplied to the tank. Without the coating the tendency is for excess current to flow to the side and insufficient current flows to the top and bottom.

Because of these factors cathodic protection is usually provided with coated surfaces.

Economics

The installation of cathodic protection systems has made it economically feasible to transport oil and high pressure natural gas across the American continent by

1. Guaranteeing there will be no corrosion on the soil side of the pipe
2. Permitting the use of thinner-walled pipe
3. Eliminating the need for an external corrosion allowance
4. Reducing maintenance costs
5. Permitting longer operating periods between routine inspections and maintenance periods

The cost of the cathodic protection system is more than recovered as a result of the above savings. Similar savings and advantages have been realized on other types of installations where cathodic protection systems have been installed.

See Refs. 1, 8–10.

CAUSTIC EMBRITTLEMENT

Caustic embrittlement is a form of stress corrosion cracking occuring in metals in contact with caustic under certain conditions. The developing cracks result from the combined actions of tensile stress and corrosion. The cracking can be intergranular or transgranular.

CAVITATION CORROSION

This form of corrosion is similar to erosion corrosion. It is caused by the formation and collapse of tiny vapor bubbles near a metallic surface in the presence of a corrodent. The protective film is damaged by the high pressure caused by the collapse of the bubbles. This form of corrosion is found quite frequently on pump impellers and condensers.

CELL POTENTIALS

A reaction will only occur if there is a negative free energy change. (ΔG). For electrochemical reactions the free energy change is calculated by

$$\Delta G = -nFE$$

where n is the number of electrons, F is Faraday's constant and E is the cell potential.

For a given reaction to take place the cell potential must be positive. The cell potential is the difference between the two half-cell reactions, the one at the cathode minus the one at the anode.

The cell potential for iron corroding freely in acid is calculated to be

$$E = \text{cathode half-cell} - \text{anode half-cell}$$
$$E = E(H^+/H_2) - E(Fe/Fe^{2+})$$
$$E = (0) - (-0.440) = +0.44$$

The reaction can take place because the cell potential is positive. The larger the potential difference the greater the driving force for the reaction. In order for corrosion to occur, there must be a current flow and a completed circuit, which is then governed by Ohm's law.

$$I = \frac{E}{R}$$

The cell potential calculated here represents the peak value for the case of the two independent reactions. If the resistance were infinite, the cell potential would remain as calculated, but there would be no corrosion at all. If the resistance of the circuit were zero, the potentials of each half-cell would approach the other while the rate of corrosion would be infinite.

See Refs. 10, 11.

CERAMIC MATERIALS

Ceramic materials are various hard, brittle, heat resistant, and corrosion resistant materials produced by firing (heat treating) clay, other minerals, or synthetic inorganic compositions. They usually consist of one or more metals in combination with a nonmetal, usually oxygen.

Ceramics are subject to many of the forms of corrosion that metals are subject to, including uniform corrosion, crevice corrosion, pitting, cavitation corrosion, erosion corrosion, galvanic corrosion, intergranular corrosion, and corrosion assisted cracking.

Corrosion data reported on ceramic materials presents two problems, the wide variety of units used and the use of units that make it difficult to compare results from different investigators. From an engineering and practical aspect for ceramics the two main criteria for corrosion performance are loss of physical dimension and loss of mechanical properties.

The preferred units for measuring loss of physical dimension are penetration rates such as mils per year (mpy). This is comparable to the measurement used for metallic corrosion.

Corrosion of ceramic materials is a complicated process. It can take place by any one or a combination of mechanisms. In general, the environment will attack the ceramic, forming a reaction product that may be a gas, a liquid, a solid, or a combination. When the reaction product formed is a solid, it may form a protective layer preventing further corrosion. If the reaction product formed is a combination of a solid and a liquid, the protective layer formed may be removed by the process of erosion.

Some of the fundamental concepts of chemistry will help to permit understanding of the corrosion of a ceramic section. A ceramic with acidic character tends to be attacked by an environment with a basic character, and vice-versa; ionic materials tend to be soluble in polar solvents while covalent materials tend to be soluble in nonpolar solvents; the solubility of solids in liquids generally increases with increasing temperatures.

Crystalline Materials

Polycrystalline materials are made up of several components. Corrosive attack of these materials starts with the least corrosion resistant component, which normally is the ingredient used for bonding, or more generally the minor component of the material.

The corrosion of a solid crystalline material by a liquid can result from either indirect dissolution or direct dissolution. In the former, an interface or reaction product is formed between the solid crystalline material and the solvent. This reaction product, being less soluble than the bulk solid, may or may not form an attached surface layer. In direct dissolution the solid crystalline material dissolves directly into the solvent.

When a silicate is leached by an aqueous solution, an ion is removed from a site within the crystal structure and is placed into the aqueous phase. Whether or not leaching occurs will depend upon the ease with which the ions can be removed from the crystal structure.

The corrosion of polycrystalline ceramic by a vapor can be more serious than attack by either liquids or solids. Porosity or permeability of the ceramic is one of the most important properties related to its corrosion by a vapor or gas. If the vapor can penetrate the material, the surface area exposed to attack is greatly increased and corrosion proceeds rapidly. A combined attack of vapor may also take place. In this situation the vapor may penetrate the material under a thermal gradient to a lower temperature, condense, and then dissolve material by liquid solution. The liquid solution can then penetrate further along temperature gradients until it freezes. If the thermal gradient is changed, it is possible for the solid reaction products to melt, causing excessive corrosion and spalling at the point of melting.

If two dissimilar solid materials react when in contact with each other, corrosion can take place. Common types of reactions involve the formation of a solid, a liquid, or a gas. Solid–solid reactions are predominantly reactions involving diffusion.

Porosity plays an important role in the corrosion resistance of ceramics. The greater the porosity, the greater will be the corrosion. The fact that one material may yield a better corrosion resistance than another does not necessarily make it a better material, if the two materials have different porosities.

Ceramics that have an acid or base characteristic similar to the corrodent will tend to resist corrosion the best. In some cases the minor components of a ceramic, such as the bonding agent, may have a different acid/base character than the major component. In this instance the acid/base character of the corrodent will determine which phase corrodes first.

In the area of corrosion resistance of ceramics we will be dealing with the so-called traditional ceramics, which include brick type products.

Common bricks, though hard to the touch, have a high water adsorption (anywhere from 8% to 15%) and are leached or destroyed by exposure to strong acid or alkali. Acid brick, whether of shale or fireclay body, are made from selected clays containing little acid-soluble components. These bricks are fired for a longer period of time at higher temperatures than the same clay when used to make common brick. This firing eliminates any organics that may be present and produces a brick with a much lower absorption rate, under 1% for best quality red shale and under 5% for high quality fireclay. See "Acid Brick."

Zirconia containing materials will be attacked by alkali solutions containing lithium, potassium, or sodium hydroxide, and potassium carbonate.

The transition metal carbides and nitrides are chemically stable at room temperature but exhibit some attack by concentrated acid solutions. The normally protective layer of silicon oxide that forms on the surface of silicon carbide and silicon nitride can exhibit accelerated corrosion when various molten salts are present. None of the carbides or nitrides are stable in oxygen containing environments. Under certain conditions some carbides and nitrides form a protective metal oxide layer that allows them to exhibit reasonably good oxidation resistance. Silicon carbide and nitride are reasonably inert to most silicate liquors as long as they do not contain significant amounts of iron oxide.

Quartz (silica) is not attacked by hydrochloric, nitric, or sulfuric acids at room temperature but will be slowly attacked by alkaline solutions. At elevated temperatures, quartz is readily attacked by sodium hydroxide, potassium hydroxide, sodium carbonate, and sodium borate. The presence of organics dissolved in the water increases the solubility of silica. Fused silica is attacked by molten sodium sulfate.

Table C.2 shows the compatibility of various ceramic materials with selected corrodents.

Glassy Materials

Typical glassy materials consist of silicate glasses, borosilicate glasses, lead containing glasses, phosphor containing glasses, fluoride glasses, and chalcogenide–halide glasses.

Glassy materials corrode primarily through the action of aqueous media. In general the very high silica glasses (96% SiO_2) such as aluminosilicate and borosilicate compositions have excellent resistance to a variety of corrodents.

Borosilicate glass is the primary composition used in the corrosion resistance field. It is resistant to all chemicals except hydrofluoric acid, fluorides, and such strong caustics as sodium or potassium hydroxide. However, caustics of even up to 50% concentration at room temperature will not be detrimental to borosilicate glass.

Table C.2A Chemical Resistance of Various Ceramic Materials

	Resistance to							
	Slags		Molten metals					
Material	Acid	Basic	Al	Fe	Na	Pb	Zn	Mg
Zircon	G	F	G	G	P	G	G	F
Bonded 99% alumina	G	G	G	G	P	G	G	G
Fused cast alumina	EX	G	G	G	F	G	G	EX
Zirconia stabilized	G	P	G	G	P	G	G	F
Silicon-carbide	G	F	G	P	P		EX	G
Silicon-nitride bonded silicon-carbide	EX	F	EX	P	P		EX	G
Magnesite	P	G	F	G	P	G	F	F
Chrome	P	F	P	G	P	G	F	F
Fosterite	P	F	F	G	P	G	F	F
Synthetic mullite	G	G	F	G	P	G	G	F
Converted mullite	G	F	F	G	P	G	G	F
Silica	G	F	P	G	P	F	G	P
Fireclay	G	F	F	G	P	F	G	P

EX = Excellent, G = Good, F = Fair, P = Poor; A = No reaction, material stable; B = Slight reaction, material suitable; C = Reaction, material suitable under certain conditions; D = Reaction, material not suited unless tested under operating conditions; E = Rapid reaction, material not suitable.

1. Chlorine attacks silicates above 1300°F (704°C).
2. Nascent or atomic hydrogen attacks silica and iron.
3. Sulfur in strong concentrations reacts with silica above 1700°F (927°C).

Table C.2B Chemical Resistance of Various Ceramic Materials: Resistance to Gases

Material	CO_2	CO	Steam	Cl_2[1]	H_2[2]	HCl	NH_3	SO_2	S[3]
Zircon	A	A	A	B–D	E	A	A	A	B–D
Bonded 99% alumina	A	A	A	A	A	A	A	A	A
Fused cast alumina	A	A	A	A–D	A	A	A	A	A–C
Zirconia stabilized	A	A	B–C	B	E	A	E	A	A–B
Silicon-carbide	B	B	B	D		A	A	A	D
Silicon-nitride bonded silicon-carbide	B	B	B	D		A	A	A	A–C
Magnesite	A	B–C	A	D		D	A	C–D	B–D
Chrome	A	B–C	A	C–D		C–D	A	C–D	B–C
Fosterite	A	A	A	C–D		C–D	A	C–D	B–C
Synthetic mullite	A	A	A	B–D		A	A–C	A	A
Converted mullite	A	B–C	A	B–E		A	A–C	A–C	A–C
Silica	A	A	A	C–E		A	A–C	B–C	B–C
Fireclay	A	C	A	C–E		A	A–C	B–C	B–C

EX = Excellent, G = Good, F = Fair, P = Poor; A = No reaction, material stable; B = Slight reaction, material suitable; C = Reaction, material suitable under certain conditions; D = Reaction, material not suited unless tested under operating conditions; E = Rapid reaction, material not suitable.

1. Chlorine attacks silicates above 1300°F (704°C).
2. Nascent or atomic hydrogen attacks silica and iron.
3. Sulfur in strong concentrations reacts with silica above 1700°F (927°C).

Table C.2C Chemical Resistance of Various Ceramic Materials: Resistance to Heated Acids

Material	Nitric	Sulfuric	Hydrochloric	Hydrofluoric	Phosphoric	Hydrocarbons
Zircon	A	A	A	B–C	A–B	A–D
Bonded 99% alumina	A	A	A	A–C	A	A
Fused cast alumina	A	A	A	A–C	A	A
Zirconia stabilized	A	A	A	B–C	A	A
Silicon-carbide	A	A	A	B–D	A–B	A–D
Silicon-nitride bonded silicon-carbide	A	A	A	C–D	A	A–C
Magnesite	D	D	D	D	D	B–C
Chrome	B–C	B–D	B–D	B–E	B–D	B–C
Fosterite	B–D	B–D	B–D	B–E	B–E	B–C
Synthetic mullite	A	A	A	C–E	A–B	A–D
Converted mullite	A	A	A	C–E	A–B	A–D
Silica	A–C	A–C	A–C	E	A–C	B–D
Fireclay	A–C	A–C	A–C	E	A–C	B–D

EX = Excellent, G = Good, F = Fair, P = Poor; A = No reaction, material stable; B = Slight reaction, material suitable; C = Reaction, material suitable under certain conditions; D = Reaction, material not suited unless tested under operating conditions; E = Rapid reaction, material not suitable.
1. Chlorine attacks silicates above 1300°F (704°C).
2. Nascent or atomic hydrogen attacks silica and iron.
3. Sulfur in strong concentrations reacts with silica above 1700°F (927°C).

Because of its inertness, borosilicate glass has found wide usage in contact with high purity products. The glass will not impart contamination to the material it comes into contact with. See "Borosilicate Glass."

There are many glass compositions. A list of about 30 glass compositions with their resistance to weathering, water, and acid may be found on page 572 of *Encyclopedia of Glass Technology*, 2nd ed., vol. 10, published by Wiley in New York.

See Refs. 12 and 13.

C-GLASS

This is a calciumaluminosilicate glass used for reinforcing thermosetting resins. See "Thermoset Reinforcing Materials."

CHECKING

Checking is the development of slight breaks in a coating film that do not penetrate to the underlying surface. It is also cracking in a crosshatch manner resembling mud cracking. It usually forms as the coating ages and becomes harder and more brittle as a result of shrinkage of the film.

CHEMICAL SYNONYMS

Chemical	Synonym
Acetic acid, crude	Pyroligneous acid
Acetic acid amide	Acetamide
Acetic ether	Ethyl acetate
Acetol	Diacetone alcohol
Acetylbenzene	Acetophenone
Acetylene tetrachloride	Tetrachloroethane
Almond oil	Benzaldehyde
Aluminum hydrate	Aluminum hydroxide
Aluminum potassium chrome	Chrome alum
Alum potash	Aluminum potassium sulfate
Amino benzene	Aniline
Ammonium fluoride, acid	Ammonium bifluoride
Baking soda	Sodium carbonate
Benzene carbonal	Benzaldehyde
Benzene carboxylic acid	Benzoic acid
Benzol	Benzene
Boracic acid	Boric acid
Bromomethane	Methyl bromide
Butanoic acid	Butyric acid
Butanol-1	Butyl alcohol
Butanone	Methyl ethyl ketone
Butter of antimony	Antimony trichloride
Butyl phthalate	Dibutyl phthalate
Calcium sulfide	Lime sulfur
Carbamide	Urea
Carbolic acid	Phenol
Carbonyl chloride	Phosgene
Caustic potash	Potassium hydroxide
Caustic soda	Sodium hydroxide
Chlorobenzene	Monochlorobenzene
1-Chlorobutane	Butyl chloride
Chloroethane	Ethyl chloride
Chloroethanoic acid	Chloroacetic acid
Chloromethane	Methyl chloride
Chloropentane	Amyl chloride
3-Chloropropene-1	Allyl chloride
Chlorotoluene	Benzyl chloride
Chromium trioxide	Chromic acid
Cupric acetate	Copper acetate
Cupric carbonate	Copper carbonate
Cupric fluoride	Copper fluoride
Cupric nitrate	Copper nitrate
Cupric sulfate	Copper sulfate
Cuprous chloride	Copper chloride
Diacetone	Diacetone alcohol
Dibromomethane	Ethylene bromide
Dibutyl ether	Butyl ether

Chemical	Synonym
Dichloroethane	Ethylene dichloride
Dichloroethane	Dichloroethylene
Dichloromethane	Methylene chloride
Diethyl	Butane
Diethylene dioxide	Dioxane
Diethylenimide oxide	Morpholine
Dihydroxy ethane	Ethylene glycol
Dimethylbenzene	Xylene
Dimethyl polysiloxane	Silicone oil
Dipropyl	Hexane
Dipropyl ether	Isopropyl ether
Dowtherm	Diphenyl
Epsom salt	Magnesium sulfate
Ethanal	Acetaldehyde
Ethanamide	Acetamide
Ethanoic acid	Acetic acid
Ethnoic anhydride	Acetic anhydride
Ethanol	Ethyl alcohol
Ethanonitrile	Acetonitrile
Ethanoyl chloride	Acetyl chloride
Ethanoxy ethanol	Cellosolve
Ethylene chloride	Ethylene dichloride
Ethyl ether	Diethyl ether
Formalin	Formaldehyde
Furfuraldehyde	Furfural
Furfurol	Furfural
Glucose	Dextrose
Glycerol	Glycerine
Glycol	Ethylene glycol
Glycol ether	Diethylene glycol
Glycol methyl ether	Methyl Cellosolve
Hexamethylene	Cyclohexane
Hexandioic acid	Adipic acid
Hexose	Dextrose
Hexyl alcohol	Hexanol
Hydroxybenzoic acid	Salicylic acid
Hypo photographic solution	Sodum bisulfate
Lime	Calcium oxide
Marsh gas	Methane
Methanal	Formaldehyde
Methaoic acid	Formic acid
Methanol	Methyl alcohol
Methylbenzene	Toluene
Methyl chloroform	Trichloroethane
Methyl cyanide	Acetonitrile
Methyl phenol	Cresol
Methyl phenol ketone	Acetophenone
Methyl phthalate	Dimethyl phthalate
Methyl propane-2	Butyl alcohol, tertiary
Muriatic acid	Hydrochloric acid

Chemical	Synonym
Nitrogen trioxide	Nitrous acid
Oil of mirbane	Nitrobenzene
Oil of wintergreen	Methyl salicylate
Oxalic nitrile	Cyanogen
Phenylamine	Aniline
Phenyl bromide	Bromobenzene
Phenyl carbinol	Benzyl alcohol
Phenol chloride	Chlorobenzene
Phenyl ethane	Ethylbenzene
Pimelic ketone	Cyclohexanone
Propanoic acid	Propionic acid
Propanol	Propyl alcohol
Propanone	Acetone
Propenyl alcohol	Allyl alcohol
Propenoic acid	Acrylic acid
Propyl acetate	Isopropyl acetate
Prussic acid	Hydrocyanic acid
Pyrogallol	Progallic acid
Red oil	Oleic acid
Sal ammoniac	Ammonium chloride
Sodium borate, tetra	Borax
Sodium phosphate, diabasic	Disodium phosphate
Starch gum	Dextrin
Sugar of lead	Lead acetate
Sulfuric chlorohydrin	Chlorosulfonic acid
Tannin	Tannic acid
Tetrachloroethylene	Perchloroethylene
Tetrachloromethane	Carbon tetrachloride
Trichloromethane	Chloroform
Trihydroxybenzene	Pyrogallic acid
Trihydroxybenzoic acid	Gallic acid
Trinitrophenol	Picric acid
Vinyl cyanide	Acrylonitrile
Water glass	Sodium silicate

CHLORINATED POLYVINYL CHLORIDE (CPVC)

See also "Polymers." When acetylene and hydrochloric acid are reacted to produce polyvinyl chloride, the chlorination is approximately 56.8%. Further chlorination of the PVC to approximately 67% produces CPVC whose chemical structure is as follows:

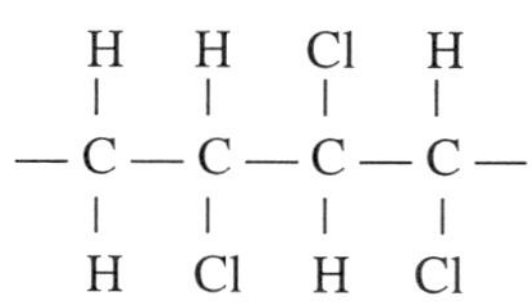

Table C.3 Physical and Mechanical Properties of CPVC

Specific gravity	1.55
Water absorption 24 h at 73°F (23°C), %	0.03
Tensile strength at 73°F (23°C), psi	8000
Modulus of elasticity in tension at 73°F (23°C) $\times 10^5$	4.15
Compressive strength at 73°F (23°C), psi	9000
Flexural strength, psi	15,100
Izod impact strength at 73°F (23°C)	1.5
Coefficient of thermal expansion	
in./in.–°F $\times 10^{-5}$	3.4
in./10°F/100 ft	0.034
Thermal conductivity Btu/h/sq ft/°F/in.	0.95
Heat distortion temperature, °F/°C	
at 66 psi	238/114
at 264 psi	217/102
Resistance to heat at continuous drainage, °F/°C	200/93
Limiting oxygen index, %	60
Flame spread	15
Underwriters lab rating (U.L. 94)	VO;5VA;5VB

Source: Courtesy of B. F. Goodrich, Specialty Polymers and Chemical Division.

The additional chlorine increases the heat deflection temperature and permits a higher allowable operating temperature. While PVC is limited to a maximum operating temperature of 140°F (60°C), CPVC has a maximum operating temperature of 180°F (82°C). Because of the higher operating temperature, CPVC finds application as piping for condensate return lines in areas having corrosive external conditions. It has also found application for hot water piping. The physical and mechanical properties are given in Table C.3.

The corrosion resistance of CPVC is similar to that of PVC but not identical. CPVC can be used to handle most acids, alkalies, salts, halogens, and many corrosive wastes. In general it cannot be used in contact with most polar organic materials including chlorinated or aromatic hydrocarbons, esters, and ketones. Refer to Table C.4 for the compatibility of CPVC with selected corrodents. Reference 4 provides a more comprehensive listing of the compatibility of CPVC with selected corrodents.

See Refs. 4 and 14,

CHLOROBUTYL RUBBER

See "Butyl Rubber."

CHLOROSULFONATED POLYETHYLENE RUBBER (HYPALON)

Chlorosulfonated polyethylene rubber (CSM) is manufactured by DuPont under the trade name Hypalon. In many respects it is similar to neoprene, but it does possess some

Table C.4 Compatibility of CPVC with Selected Corrodents[a]

Chemical	Maximum temp. °F	Maximum temp. °C	Chemical	Maximum temp. °F	Maximum temp. °C
Acetaldehyde	x	x	Barium hydroxide	180	82
Acetic acid 10%	90	32	Barium sulfate	180	82
Acetic acid 50%	x	x	Barium sulfide	180	82
Acetic acid 80%	x	x	Benzaldehyde	x	x
Acetic acid, glacial	x	x	Benzene	x	x
Acetic anhydride	x	x	Benzene sulfonic acid 10%	180	82
Acetone	x	x	Benzoic acid	200	93
Acetyl chloride	x	x	Benzyl alcohol	x	x
Acrylic acid	x	x	Benzyl chloride	x	x
Acrylonitrile	x	x	Borax	200	93
Adipic acid	200	93	Boric acid	210	99
Allyl alcohol 96%	200	93	Bromine gas, dry	x	x
Allyl chloride	x	x	Bromine gas, moist	x	x
Alum	200	93	Bromine liquid	x	x
Aluminum acetate	100	38	Butadiene	150	66
Aluminum chloride, aqueous	200	93	Butyl acetate	x	x
Aluminum chloride, dry	180	82	Butyl alcohol	140	60
Aluminum fluoride	200	93	*n*-Butylamine	x	x
Aluminum hydroxide	200	93	Butyric acid	140	60
Aluminum nitrate	200	93	Calcium bisulfide	180	82
Aluminum oxychloride	200	93	Calcium bisulfite	210	99
Aluminum sulfate	200	93	Calcium carbonate	210	99
Ammonia gas, dry	200	93	Calcium chlorate	180	82
Ammonium bifluoride	140	60	Calcium chloride	180	82
Ammonium carbonate	200	93	Calcium hydroxide 10%	170	77
Ammonium chloride 10%	180	82	Calcium hydroxide, sat.	210	99
Ammonium chloride 50%	180	82	Calcium hypochlorite	200	93
Ammonium chloride, sat.	200	93	Calcium nitrate	180	82
Ammonium fluoride 10%	200	93	Calcium oxide	180	82
Ammonium fluoride 25%	200	93	Calcium sulfate	180	82
Ammonium hydroxide 25%	x	x	Caprylic acid	180	82
Ammonium hydroxide, sat.	x	x	Carbon bisulfide	x	x
Ammonium nitrate	200	93	Carbon dioxide, dry	210	99
Ammonium persulfate	200	93	Carbon dioxide, wet	160	71
Ammonium phosphate	200	93	Carbon disulfide	x	x
Ammonium sulfate 10–40%	200	93	Carbon monoxide	210	99
Ammonium sulfide	200	93	Carbon tetrachloride	x	x
Ammonium sulfite	160	71	Carbonic acid	180	82
Amyl acetate	x	x	Cellosolve	180	82
Amyl alcohol	130	54	Chloracetic acid, 50% water	100	38
Amyl chloride	x	x	Chloracetic acid	x	x
Aniline	x	x	Chlorine gas, dry	140	60
Antimony trichloride	200	93	Chlorine gas, wet	x	x
Aqua regia 3:1	80	27	Chlorine, liquid	x	x
Barium carbonate	200	93	Chlorobenzene	x	x
Barium chloride	180	82	Chloroform	x	x

Table C.4 Continued

Chemical	Maximum temp. °F	Maximum temp. °C	Chemical	Maximum temp. °F	Maximum temp. °C
Chlorosulfonic acid	x	x	Methyl chloride	x	x
Chromic acid 10%	210	99	Methyl ethyl ketone	x	x
Chromic acid 50%	210	99	Methyl isobutyl ketone	x	x
Chromyl chloride	180	82	Muriatic acid	170	77
Citric acid 15%	180	82	Nitric acid 5%	180	82
Citric acid, conc.	180	82	Nitric acid 20%	160	71
Copper acetate	80	27	Nitric acid 70%	180	82
Copper carbonate	180	82	Nitric acid, anhydrous	x	x
Copper chloride	210	99	Nitrous acid, concentrated	80	27
Copper cyanide	180	82	Oleum	x	x
Copper sulfate	210	99	Perchloric acid 10%	180	82
Cresol	x	x	Perchloric acid 70%	180	82
Cupric chloride 5%	180	82	Phenol	140	60
Cupric chloride 50%	180	82	Phosphoric acid 50–80%	180	82
Cyclohexane	x	x	Picric acid	x	x
Cyclohexanol	x	x	Potassium bromide 30%	180	82
Dichloroacetic acid, 20%	100	38	Salicylic acid	x	x
Dichloroethane (ethylene dichloride)	x	x	Silver bromide 10%	170	77
Ethylene glycol	210	99	Sodium carbonate	210	99
Ferric chloride	210	99	Sodium chloride	210	99
Ferric chloride 50% in water	180	82	Sodium hydroxide 10%	190	88
Ferric nitrate 10–50%	180	82	Sodium hydroxide 50%	180	82
Ferrous chloride	210	99	Sodium hydroxide, concentrated	190	88
Ferrous nitrate	180	82	Sodium hypochlorite 20%	190	88
Fluorine gas, dry	x	x	Sodium hypochlorite, concentrated	180	82
Fluorine gas, moist	80	27	Sodium sulfide to 50%	180	82
Hydrobromic acid, dilute	130	54	Stannic chloride	180	82
Hydrobromic acid 20%	180	82	Stannous chloride	180	82
Hydrobromic acid 50%	190	88	Sulfuric acid 10%	180	82
Hydrochloric acid 20%	180	82	Sulfuric acid 50%	180	82
Hydrochloric acid 38%	170	77	Sulfuric acid 70%	200	93
Hydrocyanic acid 10%	80	27	Sulfuric acid 90%	x	x
Hydrofluoric acid 30%	x	x	Sulfuric acid 98%	x	x
Hydrofluoric acid 70%	90	32	Sulfuric acid 100%	x	x
Hydrofluoric acid 100%	x	x	Sulfuric acid, fuming	x	x
Hypochlorous acid	180	82	Sulfurous acid	180	82
Ketones, general	x	x	Thionyl chloride	x	x
Lactic acid 25%	180	82	Toluene	x	x
Lactic acid, concentrated	100	38	Trichloroacetic acid, 20%	140	60
Magnesium chloride	230	110	White liquor	180	82
Malic acid	180	82	Zinc chloride	180	82
Manganese chloride	180	82			

[a]The chemicals listed are in the pure state or in a saturated solution unless otherwise indicated. Compatibility is shown to the maximum allowable temperature for which data are available. Incompatibility is shown by an x.

Source: PA Schweitzer. *Corrosion Resistance Tables*, 4th ed. New York: Vols. 1–3. Marcel Dekker, 1995.

advantages over neoprene in certain types of service. It has better heat and ozone resistance, better electrical properties, better color stability, and better chemical resistance.

Hypalon, when properly compounded, also exhibits good resistance to wear and abrasion, good flex life, high impact resistance, and good resistance to permanent deformation under heavy loading.

Physical and Mechanical Properties

The ability of Hypalon to retain its electrical properties after long-term exposure to heat, water immersion, and weathering is outstanding. These properties make the elastomer useful as insulation for low voltage applications (less than 600 V), particularly as a covering for power and control cable, mine trailing cable, locomotive wire, nuclear power station cable, and motor lead wire. Because of Hypalon's outstanding weathering resistance, it is used as an outer protective jacket in high voltage applications. The elastomer also exhibits excellent resistance to corona discharge.

Another property of Hypalon that is important in electrical applications is its ability to be colored and not discolor or fade when exposed to sunlight and ultraviolet light for long periods of time. The white raw polymer will accept any color, including light pastels, without impairing the true brilliance or hue. Because of the polymer's natural ozone resistance, it is not necessary to add antiozonates during compounding. The antizonates are strong discoloring agents and when added to elastomers will cause colors to fade and become unstable.

When coloring agents are added to most elastomers it is usually necessary to sacrifice some physical properties. This is not the case with Hypalon. Except in cases where the elastomer is being specially compounded for exceptionally high heat resistance or set characteristics, its physical properties will be unaffected by the addition of coloring agents. In these special cases a black material must be used if the maximum performance is to be gotten from the elastomer.

Hypalon will burn in an actual fire situation but is classified as self-extinguishing. If the flame is removed the elastomer will stop burning. This phenomenon is due to its chlorine content, which makes it more resistant to burning than exclusively hydrocarbon polymers.

Hypalon's resistance to abrasion is superior to that of natural rubber and many other elastomers by as much as 2 to 1. It also possesses high resistance to fatigue cracking and cut growth from constant flexing. These latter properties make Hypalon suitable for products intended for dynamic operation. Good resistance to impact, crushing, cutting, gouging, and other types of physical abuse are also present in rubber parts produced from this elastomer.

The chlorine content of the elastomer protects it against the attack of microorganisms, and it will not promote the growth of mold, mildew, fungus, or bacteria. This feature is important when the elastomer is to be used in coating fabrics. To maintain this property it is important that proper compounding procedures be followed. The addition of wax and those plasticizers that provide food for microorganisms should be avoided if the maximum resistance to mold, mildew, and fungus is to be maintained.

On the low temperature side, conventional compounds can be used continuously down to 0 to -20°F (-18 to -28°C). Special compounds can be produced that will retain

Table C.5 Physical and Mechanical Properties of Chlorosulfonated Polyethylene (Hypalon; CSM)[a]

Specific gravity	1.08–1.28
Brittle point	–40°F to –80°F (–40°C to –62°C)
Dielectric strength, V/mil	500
Dielectric constant at 1000 Hz	8^{-10}
Dissipation factor at 1000 Hz	0.05–0.07
Tensile strengh, psi	2500
Elongation, % at break	430–540
Hardness, Shore A	60
Abrasion resistance	Excellent
Maximum temperature, continuous use	250°F (121°C)
Impact resistance	Good
Compression set, %	
at 158°D (70°C)	16
at 212°F (100°C)	25
at 250°F (121°C)	44
Resistance to sunlight	Excellent
Effect of aging	None
Resistance to heat	Good

[a]These are representative values since they may be altered by compounding.

their flexibility down to –40°F (–40°C), but to produce such a compound it is necessary to sacrifice performance of some of the other properties.

Heat aging does not have any effect on the tensile strength of Hypalon, since it acts as additional heat curing. However, the elongation at break does not decrease as the temperature increases.

Hypalon exhibits good recovery from deformation after being subjected to a heavy load or a prolonged deflection. Refer to Table C.5 for compression set values.

Resistance to Sun, Weather, and Ozone

Hypalon is one of the most weather resistant elastomers available. Oxidation takes place at a very slow rate. Sunlight and ultraviolet light have little if any adverse effect on its physical properties. It is also inherently resistant to ozone attack without the need for the addition of special antioxidants or antizonants to the formulation.

Many elastomers are degraded by ozone concentrations of less than 1 part per million parts of air. Hypalon, however, is unaffected by concentrations as high as 1 part per 100 parts of air.

Chemical Resistance

When properly compounded, Hypalon is highly resistant to attack by hydrocarbon oils and fuels, even at elevated temperatures. It is also resistant to such oxidizing chemicals as sodium hypochlorite, sodium peroxide, ferric chloride, and sulfuric, chromic, and hydrofluoric acids. Concentrated hydrochloric acid (37%) at elevated temperatures above 158°F (70°C) will attack Hypalon, but it can be handled without adverse effect at all concentrations below this temperature. Nitric acid at room temperature and up to 60% concentration can also be handled without adverse effects.

Hypalon is also resistant to salt solutions, alcohols, and both weak and concentrated alkalies and is generally unaffected by soil chemicals, moisture, and other deteriorating factors associated with burial in the earth. Long-term contact with water has little or no effect on Hypalon. It is also resistant to radiation.

Hypalon has poor resistance to aliphatic, aromatic, and chlorinated hydrocarbons, aldehydes, and ketones.

Fabrics coated with Hypalon are highly resistant to soiling and staining from atmospheric deposits and from abrasive contact with soiling agents. Most deposits left on the elastomeric surface can be removed by the application of soap and water. Stubborn deposits can be removed when necessary with detergents, dry cleaning fluids, bleaches, and other cleaning agents without causing damage to the elastomers.

Table C.5 lists the physical and mechanical properties of Hypalon.

Hypalon has a broad range of service temperatures with excellent thermal properties. General-purpose compounds can operate continuously at temperatures of 248–275°F (120–135°C). Special compounds can be formulated that can be used intermittently up to 302°F (150°C).

Refer to Table C.6 for the compatibility of Hypalon with selected corrodents.

Applications

Hypalon finds useful applications in many industries and many fields. Because of its outstanding resistance to oxidizing acids, it has found widespread use as acid transfer hose. For the same reason it is used to line railroad tank cars and other tanks containing acids and other oxidizing chemicals. Its physical and mechanical properties make it suitable for use in hoses undergoing continuous flexing and/or those carrying hot water or steam.

The electrical industry makes use of Hypalon to cover automotive ignition and primary wire, nuclear power station cable, control cable, and welding cable. As an added protection from storms at sea, power and lighting cable on off-shore oil platforms are sheathed with Hypalon. Because of its heat and radiation resistance it is also used as a jacketing material on heating cable imbedded in roadways to melt ice and on x-ray machine cable leads. It is also used in appliance cord, insulating hoods and blankets, and many other electrical accessories.

In the automotive industry, advantage is taken of Hypalon's color stability and good weathering properties by using the elastomer for exterior parts on cars, trucks, and other commercial vehicles. Its resistance to heat, ozone, oil, and grease makes it useful for application under the hood for such components as emission control hose, tubing, ignition wire jacketing, spark plug boots, and air-conditioning and power steering hoses. The ability to remain soil-free and easily cleanable makes it suitable for tire whitewalls.

When combined with cork, Hypalon provides a compressible set resistant gasket suitable for automobile crankcase and rocker pans. The Hypalon protects the cork from oxidation at elevated temperatures and also provides excellent resistance to oil, grease, and fuels.

Table C.6 Compatibility of Hypalon with Selected Corrodents[a]

Chemical	Maximum temp. °F	°C	Chemical	Maximum temp. °F	°C
Acetaldehyde	60	16	Boric acid	200	93
Acetamide	x	x	Bromine gas, dry	60	16
Acetic acid 10%	200	93	Bromine gas, moist	60	16
Acetic acid 50%	200	93	Bromine liquid	60	16
Acetic acid 80%	200	93	Butadiene	x	x
Acetic acid, glacial	x	x	Butyl acetate	60	16
Acetic anhydride	200	93	Butyl alcohol	200	93
Acetone	x	x	Butyric acid	x	x
Acetyl chloride	x	x	Calcium bisulfite	200	93
Acrylonitrile	140	60	Calcium carbonate	90	32
Adipic acid	140	60	Calcium chlorate	90	32
Allyl alcohol	200	93	Calcium chloride	200	93
Aluminum fluoride	200	93	Calcium hydroxide 10%	200	93
Aluminum hydroxide	200	93	Calcium hydroxide, sat.	200	93
Aluminum nitrate	200	93	Calcium hypochlorite	200	93
Aluminum sulfate	180	82	Calcium nitrate	100	38
Ammonia gas	90	32	Calcium oxide	200	93
Ammonia carbonate	140	60	Calcium sulfate	200	93
Ammonium chloride 10%	190	88	Caprylic acid	x	x
Ammonium chloride 50%	190	88	Carbon dioxide, dry	200	93
Ammonium chloride, sat.	190	88	Carbon dioxide, wet	200	93
Ammonium fluoride 10%	200	93	Carbon disulfide	200	93
Ammonium hydroxide 25%	200	93	Carbon monoxide	x	x
Ammonium hydroxide, sat.	200	93	Carbon tetrachloride	200	93
Ammonium nitrate	200	93	Carbonic acid	x	x
Ammonium persulfate	80	27	Chloracetic acid	x	x
Ammonium phosphate	140	60	Chlorine gas, dry	x	x
Ammonium sulfate 10–40%	200	93	Chlorine gas, wet	90	32
Ammonium sulfide	200	93	Chlorobenzene	x	x
Amyl acetate	60	16	Chloroform	x	x
Amyl alcohol	200	93	Chlorosulfonic acid	x	x
Amyl chloride	x	x	Chromic acid 10%	150	66
Aniline	140	60	Chromic acid 50%	150	66
Antimony trichloride	140	60	Chromyl chloride		
Barium carbonate	200	93	Citric acid 15%	200	93
Barium chloride	200	93	Citric acid, concentrated	200	93
Barium hydroxide	200	93	Copper acetate	x	x
Barium sulfate	200	93	Copper chloride	200	93
Barium sulfide	200	93	Copper cyanide	200	93
Benzaldehyde	x	x	Copper sulfate	200	93
Benzene	x	x	Cresol	x	x
Benzene sulfonic acid 10%	x	x	Cupric chloride 5%	200	93
Benzoic acid	200	93	Cupric chloride 50%	200	93
Benzyl alcohol	140	60	Cyclohexane	x	x
Benzyl chloride	x	x	Cyclohexanol	x	x
Borax	200	93	Dichloroethane (ethylene dichloride)	x	x

Table C.6 Continued

Chemical	Maximum temp. °F	Maximum temp. °C	Chemical	Maximum temp. °F	Maximum temp. °C
Ethylene glycol	200	93	Nitric acid, anhydrous	x	x
Ferric chloride	200	93	Oleum	x	x
Ferric chloride 50% in water	200	93	Perchloric acid 10%	100	38
Ferric nitrate 10–50%	200	93	Perchloric acid 70%	90	32
Ferrous chloride	200	93	Phenol	x	x
Fluorine gas, dry	140	60	Phosphoric acid 50–80%	200	93
Hydrobromic acid, dilute	90	32	Picric acid	80	27
Hydrobromic acid 20%	100	38	Potassium bromide 30%	200	93
Hydrobromic acid 50%	100	38	Sodium carbonate	200	93
Hydrochloric acid 20%	160	71	Sodium chloride	200	93
Hydrochloric acid 38%	140	60	Sodum hydroxide 10%	200	93
Hydrocyanic acid 10%	90	32	Sodium hydroxide 50%	200	93
Hydrofluoric acid 30%	90	32	Sodium hydroxide, concentrated	200	93
Hydrofluoric acid 70%	90	32	Sodium hypochlorite 20%	200	93
Hydrofluoric acid 100%	90	32	Sodium hypochlorite, concentrated		
Hypochlorous acid	x	x	Sodium sulfide to 50%	200	93
Ketones, general	x	x	Stannic chloride	90	32
Lactic acid 25%	140	60	Stannous chloride	200	93
Lactic acid, concentrated	80	27	Sulfuric acid 10%	200	93
Magnesium chloride	200	93	Sulfuric acid 50%	200	93
Manganese chloride	180	82	Sulfuric acid 70%	160	71
Methyl chloride	x	x	Sulfuric acid 90%	x	x
Methyl ethyl ketone	x	x	Sulfuric acid 98%	x	x
Methyl isobutyl ketone	x	x	Sulfuric acid 100%	x	x
Muriatic acid	140	60	Sulfurous acid	160	71
Nitric acid 5%	100	38	Toluene	x	x
Nitric acid 20%	100	38	Zinc chloride	200	93
Nitric acid 70%	x	x			

[a]The chemicals listed are in the pure state or in a saturated solution unless otherwise indicated. Compatibility is shown to the maximum allowable temperature for which data are available. Incompatibility is shown by an x.
Source: PA Schweitzer. *Corrosion Resistance Tables*, 4th ed. Vols. 1–3. New York: Marcel Dekker, 1995.

The construction industry has made use of Hypalon for sheet roofing, pond liners, reservoir covers, curtain wall gaskets, floor tiles, escalator rails, and decorative and maintenance coatings. In these applications the properties of color stability, excellent weatherability, abrasion resistance, useful temperature range, light weight, flexibility, and good aging characteristics are of importance.

Application is also found in the coating of fabrics that are used for inflatable structures, flexible fuel tanks, tarpaulins, and hatch and boat covers. These products offer the advantages of being lightweight and colorful. Consumer items such as awnings, boating garb, convertible tops, and other products also make use of fabrics coated with this elastomer.

See Refs. 4, 15.

CHROMATING

Chromating is a process for producing a conversion coating containing chromium compounds on metal surfaces. The chromate conversion coatings are applied to the protective top coats of metallic products and are the bases for organic coatings. The coating processes are dipping, electrolysis, and roll coating. The metals usually chromated include aluminum, copper, magnesium, cadmium, silver, zinc, and their alloys.

See Ref. 1.

CHROMIUM COATINGS

Chromium coatings are for decorative purposes, they are used mostly on duplex nickel or copper strike undercoat. Decorative chromium coatings are used for bicycle parts and electric components.

See Ref. 1.

CLAD STEELS

A clad steel plate is a composite plate made of carbon steel with a cladding of corrosion resistant or heat resistant metal on one or both sides. The clad steels are used in place of solid corrosion resistant or heat resistant materials, particularly when relatively thick sections are required because of high pressure applications in processing vessels, in order to reduce the cost. They also find application where corrosion is a minor problem but where freedom from contamination of the materials handled is essential. In addition to the savings in material costs, the clad steels are frequently easier to fabricate than solid plates of the cladding material resulting in reduced labor costs. Their high heat conductivity is another reason for their selection for many applications.

Various grades of stainless steels, nickel, Monel, Inconel, cupronickel, titanium, or silver may be used as a cladding material. The thickness of the cladding material is normally held to 10 to 20% of the thickness of the clad plate, but it may vary from 5 to 50%. Clad steels are available in the form of sheet, plate, and strip and may be obtained also as wire.

Clad steels are used for processing equipment in the chemical, food, beverage, drug, paper, textile, oil, and associated industries.

Cladding can be applied in any one of six methods from the insertion of a loose liner to explosion cladding.

The least expensive form of cladding is the installation of a thin corrosion resistant liner inside a process vessel constructed of a less expensive base metal. With this type of cladding the liner is normally relatively thin, anywhere from 0.3 to 2 mm thick, and is used only for corrosion resistance. The base metal provides all of the structural strength. The advantages of this method are

1. Relatively low cost.
2. Availability. When the base metal and the liner material are available, the finished piece of equipment can be produced in a short time.
3. Repairs can be made relatively easily.
4. Base metal and liner material do not have to be metallurgically compatible.

The disadvantages of a loose liner are

1. Heat transfer is reduced.
2. The liner is easily damaged.
3. A vacuum will cause the liner to collapse. This disadvantage can be overcome by periodically attaching the liner to the base metal, or by increasing its thickness, but this does increase the cost.

Roll Cladding

Roll cladding produces full-sized sheets of clad material that a fabricator then forms into a finished product. These sheets are rolled at the mill. The bond formed is partly mechanical and partly metallurgical, and consequently metallurgically incompatible materials normally cannot be produced. One exception is the production of titanium-clad sheets being produced by Nippon Kokan KK of Tokyo. Cladding thicknesses range from 5 to 50% of the compatible thickness.

Composites are produced with cladding designed to resist wear, abrasion, or corrosion.

Explosion Cladding

Explosion cladding produces full-sized sheets of clad material, as roll cladding does, that a fabricator then forms and welds into a finished product.

The explosion bonding technique was originally developed by DuPont. In this process, detonation of an explosive presses the plates together with such force that the lower elastic limit of the metals is exceeded, and the unmelted surface metal is jetted through the rapidly closing space between the plates, destroying interfering layers of metal and resulting in a metallurgical bond with a relatively smooth surface.

Metallurgically incompatible metals can be coupled by use of an intermediate material.

Thick plates up to 510 mm may be produced by this method, but unless relatively thick sections are required this process is not economical.

Weld Overlaying

Weld overlaying is used with cladding materials and base metals that are metallurgically compatible and is best restricted for use on small, complex parts. Readily available commercial alloys, such as stainless steels and nickel and copper based alloys can be used to provide a corrosion resistant overlay.

Because of the heat required for welding, care must be taken not to distort the member being clad.

Thermal Spraying

Thermal spraying can also produce a clad material. It is accomplished by heating the metal cladding (or nonmetallic) particles to a molten state and spraying them on the prepared surface of the base metal. As the molten particles impact on the surface of the base metal, they form an overlapping multilayered cladding, ranging in thickness from 0.2 to 2.5 mm. One of the advantages of this process is that the temperature of the base metal normally does not exceed 300 to 410°F (150 to 200°C), which minimizes thermal distortion of the base metal. This also does not permit the coating to become diluted, which is essential when the cladding material and the base metal are not metallurgically compatible. Combinations of certain metals form mixtures that are affected adversely by the intermixing or interaction of the components. For example, nickel and copper are metallurgically compatible, whereas steel and zirconium are not. This process is relatively inexpensive and is a common operation performed by many shops.

Certain disadvantages are prevalent with these spraying processes. The claddings or coatings have a tendency to be porous. This can be overcome somewhat by increasing the density of the coating. In addition, the bond between the coating/cladding is mechanical, and any leakage of corrodent to the base metal through a porous section of the coating can cause the coating to spall off. It is also very difficult, and at times impossible, to clad complex shapes.

Resistance Cladding

Resistance cladding provides a means of applying a lining to a base metal regardless of whether or not the base metal and the liner material are metallurgically compatible. In this process, use is made of resistance welding and proprietary intermediate materials to bond thin-guage corrosion resistant materials to heavier less expensive base metals.

This type of cladding can be applied to completed fabrications of equipment or to components during fabrication. It is also possible to apply this cladding to existing used equipment. Completed fabrications are relatively inexpensive.

Because of the relatively thin layer of cladding, the corrosion rate of the process must be low. The process also does not lend itself to providing a finished surface. Economics do not favor this process unless the cladding material is expensive.

COATINGS

The development of new and improved coatings has been increasing over the past several years. New technologies have evolved that have expanded the usage of these materials. By incorporating these coatings with a substrate having the required physical and mechanical properties, it is possible to obtain the desired strength and the optimum corrosion resistance at an economical cost.

The available coatings can be categorized as metallic, inorganic, or organic in nature.

Each category has its own specific area of application, with a range of properties dependent on the specific material.

Metallic Coatings

There are several methods by which metallic coatings may be applied.

1. Brief immersion in a molten bath of metal, called hot dipping
2. Electroplating from an aqueous electrolyte
3. Spraying, in which a gun is used that simultaneously melts and propels small droplets of metal onto the surface to be coated, as with spray painting
4. Cementation, in which the material to be coated is tumbled in a mixture of metal powder and an appropriate flux at elevated temperatures, which allows the metal to diffuse into the base metal
5. Gas phase reaction
6. Chemical reduction of metal–salt solutions, the precipated metal forming an overlay on the base metal (nickel coatings of this type are referred to as "electroless" nickel plate)

Coatings from a corrosion viewpoint are classified as either noble or sacrificial. All metal coatings contain some degree of porosity. Coating performance is therefore determined by the degree of galvanic action that takes place at the base of a pore, scratch, or other imperfection in the coating.

Noble coatings, consisting of nickel, silver, copper, lead, or chromium on steel, are noble in the galvanic series with respect to steel, resulting in galvanic current attack at the base of the pores of the base metal and eventually undermining the coating. See Fig. C.4.

In order to reduce this rate of attack it is important that this type of coating be prepared with a minimum number of pores and that any pores present be as small as possible. This can be accomplished by increased coating thickness.

In sacrificial coatings consisting of zinc, cadmium, and in certain environments aluminum and tin on steel, the base metal is noble in the galvanic series to the coating material, resulting in cathodic protection to the base metal and attack on the coating material. See Fig. C.5.

As long as sufficient current flows and the coating remains in electrical contact, the

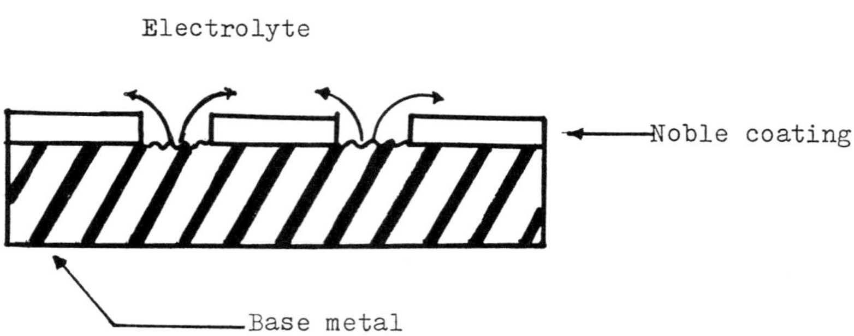

Figure C.4 Galvanic action with a noble coating.

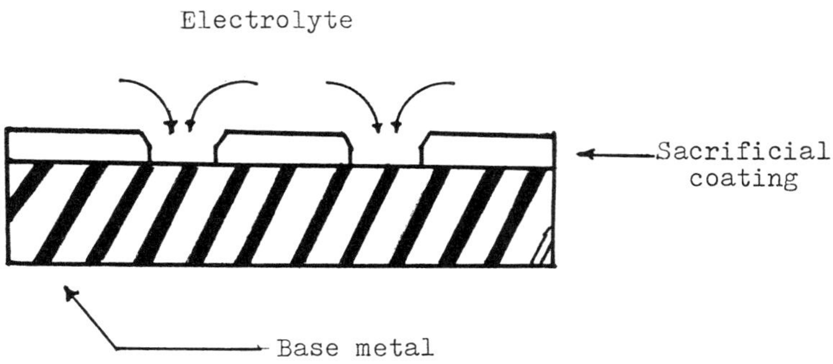

Figure C.5 Galvanic action with a sacrificial coating.

base metal will be protected from corrosion. Contrary to noble coatings, the degree of porosity of sacrificial coatings is relatively insignificant. However, the thicker the coating the longer cathodic protection will be provided to the base metal.

Nickel Coatings

The most common method of applying nickel coatings is by electroplating, either directly on steel or over an intermediate coating of copper. Copper is used as an underlayer to facilitate buffing, since it is softer than steel, and to increase the required coating thickness with a material less expensive than nickel.

Zinc Coatings

Galvanized or zinc coating is probably one of the most common coatings. It is applied by either hot dipping or electrodeposition. Electrodeposited coatings tend to be more ductile than hot dipped coatings, but otherwise they are comparable in corrosion resistance with one exception. Hot dipped coatings tend to pit less in hot or cold water and soils than coatings applied by other methods.

Zinc coatings stand up extremely well in rural atmospheres and in marine atmospheres except when salt water spray comes into direct contact with the coating.

In aqueous environments at room temperatures within a pH range of 7 to 12, good corrosion resistance will be obtained.

Any welding or forming should, if possible, be performed prior to the galvanizing operation.

Cadmium Coatings

These coatings are produced almost exclusively by electrodeposition. A cadmium coating on steel does not provide as much cathodic protection to the steel as does a zinc coating, since the potential between cadmium and iron is not as great as between zinc and iron. Therefore it becomes important to minimize defects in the cadmium coating.

Unlike zinc, a cadmium coating will retain a bright metallic appearance. It is more resistant to attack by salt spray and atmospheric condensation. In aqueous solutions cadmium will resist attack by strong alkalies but will be corroded by dilute acids and aqueous ammonia.

Since cadmium salts are toxic, these coatings should not be allowed to come into contact with food products. This coating is commonly used on nuts and bolts.

Tin Coatings

Most of the tinplate (tin coating on steel) is used for the manufacture of food containers (tin cans). The nontoxic nature of tin salts makes tinplate an ideal material for the handling of foods and beverages.

An inspection of the galvanic series will indicate that tin is more noble than steel and consequently the steel corrodes at the base of the pores. On the outside of the tinned container this is what happens—the tin is cathodic to the steel. However, on the inside of the container there is a reversal of the polarity due to the complexing of the stannous ions by many food products. This greatly reduces the activity of the stannous ion, resulting in a change in the potential of tin in the active direction.

This change in polarity is absolutely necessary, since most tin coatings are thin and therefore porous. In order to avoid perforation of the can, the tin must act as a sacrificial coating. Figure C.6 illustrates this reversal of activity between the outside and inside of the can.

Tin will react with both acids and alkalies but is relatively resistant to neutral or near neutral media.

It is extremely resistant to soft waters and consequently has found wide usage as a piping material for distilled water. Only its cost and availability have precluded it from monopolizing this market.

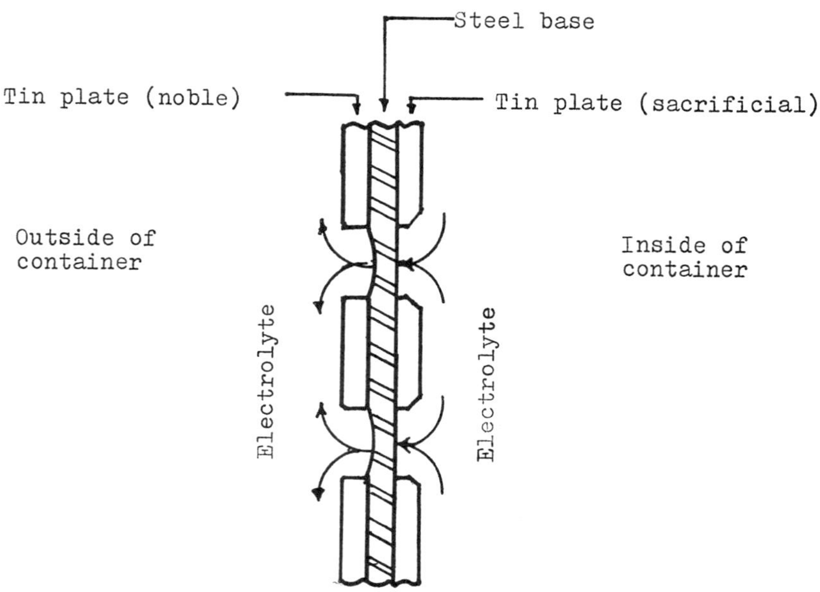

Figure C.6 Tin acting as both a noble and sacrificial coating.

Aluminum Coatings

Aluminum coatings on steel are applied primarily by hot dipping or by spraying. Silicon is usually added to the molten bath so as to retard the formation of a brittle alloy layer. Organic lacquers or paints are used as sealers over sprayed coatings.

Hot-dipped coatings are used mostly to provide oxidation resistance at moderately elevated temperatures, e.g., oven construction. They find limited applications as protection against atmospheric corrosion because they are more expensive than zinc and have a variable performance.

Vitreous Enamels

Vitreous enamels, glass linings, or porcelain enamels are all essentially glass coatings that have been fused on metals. Powdered glass is applied to a pickled or otherwise prepared metal surface and heated in a furnace at a temperature that softens the glass and permits it to bond to the metal. Several thin coats are applied to provide the required final thickness. These coatings are normally applied to steel, but some coatings can be applied to brass, aluminum, and copper.

There are many glass formulations, but those with very high silica ($< 96\%$ S_1O_2), aluminosilicate, and borosilicate compositions have the highest corrosion resistance to a wide range of corrosive environments. Glass is assumed to be inert to most liquids, but in reality it slowly dissolves.

The greatest danger of failure of a glass coating comes from mechanical damage or from cracking as a result of thermal shock. Thus care must be taken in handling glassed equipment so as not to damage the lining, and sudden temperature changes in the operation must be avoided, particularly cold shock, which poses a greater danger of failure than hot shock.

Cold shock is the sudden introduction of a cold material onto a hot glassed surface; hot shock is the reverse. Manufactureres of this type of equipment will specify the maximum allowable thermal shock. These precautions must be followed.

Masonry

Monolithic corrosion resistant masonry linings are normally applied by means of pneumatic gunning (guniting), although they may be trowled on. Thicknesses vary from 1/2 in. up to several inches. When a thickness greater than 1 in. is to be applied, it is necessary to anchor the lining in place using wire mesh and studs. There should be a minimum of 1/2 in. cover over the highest point.

This type of lining has three main advantages:

1. Curved or irregular surfaces can be covered uniformly.
2. Monolithic linings bond to steel, brick, and concrete.
3. Monolithic linings can be gunited horizontally, vertically, or overhead without the need for complex forms, supports, or scaffolds.

Sodium Silicate Base Monolithics

This material is supplied as two separate components, a powder and a liquid. The two components are mixed and applied for use. Hardening occurs as a result of a chemical reaction. Application can be by means of casting, pouring into forms, or guniting. Its resistance to acid is excellent over a pH range of 0.0 to 7.0.

Modified Silicate Base Monolithics

There are two types of modified silicate base materials available, both of which are supplied in powder form and must be mixed with water prior to application. Application is by guniting.

The first type is unaffected by acids (except hydrofluoric), mild alkalies, water, and solvents and can be operated up to a temperature of 1740°F (950°C) through a pH range of 0.0 to 9.0. It weighs approximately 135 lbs/ft^3.

The second type is lighter in weight, weighing approximately 98 lbs/ft^3 and is thermally insulating. It has a K factor of 2.25 to 2.50. This type can be operated up to a temperature of 1695°F (925°C) through a pH range of 0.0 to 9.0.

Calcium Aluminate Base Monolithics

This type consists of a calcium aluminate base cement to which various inert aggregates have been added. It is supplied in powder form and mixed with water when used, and may be cast, poured, or gunited into place. These monolithics are similar to portland cement in that they are hydraulic in nature and consume water in their reaction mechanism to form hydrated phases. In contrast to portland cement their rates of hardening are very rapid, and full strength is usually attained within 24 hours at room temperatures of 73°F (23°C).

They are not useful in contact with acids below a pH of 5, although they do exhibit better mild acid resistance than portland cement. These types should not be used for halogen service or for alkali service above a pH of 12.

Portland Cement Coatings

Although portland cement does not have the corrosion resistance of other monolithics, it does have the advantage of low cost and ease of repair when accessible. The coatings can be applied by centrifugal casting, troweling, or guniting. Thicknesses usually range from 1/4 to more than 1 in. The thicker coatings are reinforced with wire mesh.

Primary applications for portland cement linings are the protection of hot or cold water tanks, oil tanks, and chemical storage tanks. The disadvantage of portland cement coatings is their sensitivity to damage by mechanical shock.

Since cement compositions may vary, care should be taken that the proper selection be made for the specific application.

Chemical Conversion Coatings

These are protective coatings formed by a chemical reaction taking place on the surface of the metal. Included in this category are phosphate coatings on steel (sometimes referred to as "parkerizing" or "bonderizing"), oxide coatings on steel and aluminum, and chromate coatings on zinc.

Phosphate Coatings

These coatings are not used to provide corrosion protection since they offer little. They are used to provide a base for the application of paints, by providing good adherence of the paint to the steel and decreasing the tendency for corrosion to undercut the paint film at scratches or defects.

Oxide Coatings

Oxide coatings are not applied for providing increased corrosion resistance since they do not appreciably improve the resistance of the base metal. Oxide coatings produced on aluminum result in a product known as anodized aluminum. During this process the oxide coating can be dyed various colors, for aesthetic purposes.

As with phosphate coatings, the main advantage lies in providing an improved base for paints.

Chromate Coatings

These coatings are produced on zinc, imparting a slight yellow color and protecting the metal against spotting or staining by condensed moisture. The coating will extend the life of zinc somewhat, when exposed to the atmosphere.

Organic Coatings

Organic coatings are widely used to protect metallic surfaces from corrosion. The effectiveness of such coatings is dependent not only on the properties of the coatings, which are related to the polymeric network and possible flaws in this network, but also on the character of the metal substrate, the surface pretreatment, and the application procedures. Therefore when considering the application of a coating it is necessary to take into account the properties of the entire system.

Organic coatings provide protection either by the formation of a barrier action from the layer or from active corrosion inhibition provided by pigments in the coating. In actual practice the barrier properties are limited since all organic coatings are permeable to water and oxygen to some extent. The average transmission rate of water through a coating is about 10 to 100 times larger than the water consumption rate of a freely corroding surface, and in normal outdoor conditions an organic coating is saturated with water at least half its service life. For the remainder of the time it contains a quantity of water comparable in its behavior to an atmosphere of high humidity. It has also been determined that in most cases the diffusion of oxygen through the coating is large enough to allow unlimited corrosion. Taking these factors into account indicates that the physical barrier properties alone do not account for the protective action of coatings.

Additional protection may be supplied by resistance inhibition, which is also a part of the barrier mechanism. Retardation of the corrosion action is accomplished by inhibiting the charge transport between cathodic and anodic sites. The reaction rate may be reduced by an increase in the electronic resistance and/or the ionic resistance in the corrosion cycle. Applying an organic coating on a metal surface increases the ionic resistance. The electronic resistance may be increased by the formation of an oxide film on the metal. This is the case for aluminum substrates.

Organic coatings are relatively easily damaged under mechanical and thermal load, which may lead to corrosion under the paint film at or near the site. Under these conditions the otherwise adequate barrier properties of the coating will no longer provide adequate protection. In an attempt to compensate for this, active pigments are incorporated in the matrix of the primer (first coating layer). These pigments provide protection through an active inhibitive mechanism immediately when water and some corrosive agent reach the metal surface. The protection provided is of a passivating, blocking, or galvanic action.

In order for the coating to provide protection, the adhesion of the coating must be good. The quality of the coating is determined to some extent by the mechanical properties of the polymer that determine the formability of coated substances and also the sensitivity to external damage.

Water Permeation and Underfilm Corrosion Initiation

In order for corrosion to take place under a coating it is necessary for an electrochemical double layer to be established. In order for this to occur it is necessary for the adhesion between the coating and the substrate to be broken. When this happens a thin water layer at the interface can be formed when the water permeates the coating. All organic coatings are permeable to water to some degree. The permeability of a coating is often given in terms of the permeation coefficient P. This is defined as the product of the solubility of water in the coating (S Kg/cm^3), the diffusion coefficient of water in the coating (D m^2/s), and the specific mass of water (p Kg/m^2). Therefore different coatings can have the same permeation coefficient even though the solubility and the diffusion coefficient, both being material constants, are very different. This limits the usefulness of the permeation coefficient.

Water permeation takes place under the influence of several driving forces:

1. A concentration gradient during immersion or during exposure to a humid atmosphere, resulting in true diffusion through the polymer
2. Capillary forces in the coating resulting from poor curing, improper solvent evaporation, bad interaction between binder and additives, or entrapment of air during application
3. Osmosis due to impurities, or corrosion products at the interface between the metal and the coating

Given sufficient time a coating system that is exposed to an aqueous solution or a humid atmosphere will be permeated. Water molecules will eventually reach the coatings substrate interface. Saturation will occur after a relatively short period of time (of the order of one hour) depending on the values for D and S and the thickness of the layer. Typical values for D and S are 10^{-13} m^2/s and 3%. Periods of saturation under atmospheric exposure are determined by the actual cyclic behavior of the temperature and the humidity. In any case, situations will develop in which water molecules reach the coating–metal interface, where they can interfere with the bonding between the coating and the substrate, eventually resulting in loss of adhesion and corrosion initiation, providing that a cathodic reaction can take place. A constant supply of water or oxygen is required for the corrosion reaction to proceed. Water permeation may also result in the build up of high osmotic pressures resulting in blistering and delamination.

Wet Adhesion

Adhesion between the coating and the substrate may be affected when water molecules have reached the substrate–coating interface. The degree to which permeated water may change the adhesion properties of a coated system is referred to as wet adhesion. Two different theories have been proposed for the mechanism for the loss of adhesion due to water.

1. Chemical disbondment resulting from the chemical interaction of water molecules with covalent hydrogen, or polar bonds between polymer and metal (oxide).
2. Mechanical or hydrodynamic disbondment as a result of forces caused by accumulation of water and osmotic pressure.

For chemical disbondment to take place it is not necessary that there be any sites of poorly bonded coating. This is not the case for mechanical disbonding where water is supposed to condense at existing sites of bad adhesion. Water volume at the interface may subsequently increase due to osmosis. As the water volume increases under the coating, hydrodynamic stresses develop. These stresses eventually result in an increase of the nonadherent surface area.

Osmosis

Osmotic pressure may result from one or more of the following conditions:

1. Presence of soluble salts as contaminants at the original metal surface.
2. Inhomogenities in the metal surface such as precipitates, grain boundaries, or particles from blasting pretreatment.
3. Surface roughness due to abrasion. Once corrosion has started at the interface, the corrosion products produced can be responsible for the increase in osmotic pressure.

Blistering

Various phenomena may be responsible for the formation of blisters and the start of underfilm corrosion. These include the presence of voids, wet adhesion problems, swelling of the coating during water uptake, gas inclusions, impurity ions in the coating, poor general adhesion properties, and defects in the coating.

When a coating is exposed to an aqueous solution, water vapor molecules and some oxygen diffuse into the film and end up at the substrate interface. Eventually a thin film of water may develop at sites of poor adhesion or at sites where wet adhesion problems arise. A corrosion reaction can start with the presence of an aqueous electrolyte with an electrochemical double layer, oxygen and the metal. This reaction will cause the formation of macroscopic blisters. Depending on the specific materials and circumstances, the blisters may grow out because of the hydrodynamic pressure, in combination with one of the chemical propagation mechanisms such as cathodic delamination and anodic undermining.

Cathodic Delamination

Loss of adhesion of the paint film adjacent to defects on a coated metal to which cathodic protection is applied is known as cathodic delamination. It derives its name since the driving force is the cathodic reaction taking place at the interface. As a result of the high pH values resulting from the cathodic reactions, delamination occurs. In the case of cathodic overprotection, blistering, due to the evolution of hydrogen gas, can take place.

Anodic Undermining

This is a class of corrosion reactions underneath an organic coating in which loss of adhesion is caused by anodic dissolution of the substrate metal or its oxide in contrast to cathodic delamination. In this case the metal is anodic at the blister edges. Anodic undermining usually is initiated at a corrosion-sensitive site underneath the coating such as an enclosed particle, or a section of the metal with potentially increased activity caused by scratches. These sites become active once corrodents have penetrated to the metal surfaces. The corrosion rates start out low, but as corrosion products are formed osmotic pressure develops, which stimulate blister growth. Once formed the blister will grow due to a type of anodic crevice corrosion at the edge of the blister.

In general coated aluminum tends to be susceptible to anodic undermining, while coated steel is more susceptible to cathodic delamination.

Filiform Corrosion

Filiform corrosion is a thread like undermining of the coating and is sometimes referred to as worm track corrosion. It generally occurs in humid chloride containing environments and is common under organic coatings on steel, aluminum, magnesium, and galvanized steel. The majority of problems occur on coated aluminum and is a special form of anodic undermining.

Filiform corrosion takes place when the following conditions are present:

1. The coating has defects.
2. The coating is permeable to water.
3. High relative humidity, specifically in the range 80–95%.
4. Contaminants are present on or in the coating or at the coating–substrate interface.

Early Rusting

When a latex paint is applied to a cold steel substrate under high moisture conditions a measles like rusty appearance may develop immediately when the coating is touch dry. This corrosion takes place when the following conditions are met:

1. The air humidity is high.

2. The substrate temperature is low.
3. A thin (up to 40 μm) latex coating has been applied.

Flash Rusting

Flash rusting is the appearance of brown rust stains on a blasted steel surface immediately after applying a water based primer. Contaminants remaining on the metal surface after blast cleaning are responsible for this corrosion. The grit on the surface provides crevices or local galvanic cells that activate the corrosion process as soon as the surface is wetted by the water based primer.

Stages of Corrosion

To prevent excessive corrosion, good inspection procedures and preventative maintenance practices are required. Proper design considerations are also necessary as well as selection of the proper coating system. Regular inspections of coatings should be conducted. Since corrosion of substrates under coatings takes place in stages, early detection will permit correction of the problem, thereby preventing ultimate failure.

First Stages of Corrosion

The first stages of corrosion are indicated by rust spotting or the appearance of a few small blisters. Rust spotting is the very earliest stage of corrosion and in many cases is left unattended. Standards have been established for evaluating the degree of rust spotting and may be found in ASTM D610-68 or "Steel Structures Painting Council Vis-2." One rust spot in 1 square foot may provide a 9+ rating but three or four rust spots drop the rating to 8. If the rust spots go unattended, a mechanism for further corrosion is provided.

Blistering is another form of early corrosion. Frequently blistering occurs without external evidence of rusting or corrosion. The mechanism of blistering is attributed to osmotic attack or a dilution of the coating film at the interface with the steel under the influence of moisture. Water and gases pass through the film and dissolve ionic material from either the film or the substrate, causing an osmotic pressure greater than that of the external face of the coating. This produces a solution concentration gradient, with water building up at these sites until the film eventually blisters. Visual blistering standards are found in ASTM D-714-56.

Electrochemical reactions also assist in the formation of blisters. Water diffuses through a coating also by an electroendesmotic gradient. Once corrosion has started, moisture is pulled through the coating by an electrical potential gradient between the corroding areas and the protected areas that are in electrical contact. Therefore osmosis starts the blistering, and once corrosion begins, electro-endesmotic reactions accelerate the corrosion process. The addition of heat and acidic chemicals increases the rate of breakdown. Temperatures of 150°F to 200°F (66°C to 93°C) accelerate the chemical reaction. Under these conditions steel will literally dissolve in a chemical environment. Moisture is always present and often condenses on the surface behind the blister. This condensation offers a solute for gaseous penetrants to dissolve. When the environment is acidic the pH of the water behind the blister can be as low as 1.0 to 2.0, subjecting the steel to severe attack.

Second Stage of Corrosion

After the initial one or two rust spots have been observed, or after a few blisters are found, a general rusting in the form of multiple rest spots develops. This rusting is predominately Fe_2O_3, a red rust. In atmospheres lacking sufficient oxygen, such as in sulfur dioxide scrubbers, a black FeO rust develops. Once the unit has been shut down and more oxygen becomes available, the FeO will eventually convert to Fe_2O_3.

Third Stage of Corrosion

This advanced stage of corrosion is the total disbondment of the coating from the substrate exposing the substrate directly to the corrodents. Corrosion can occur at an uninhibited rate since the coating is no longer protecting the steel.

Fourth Stage of Corrosion

Attack of the metal substrate after the removal of the coating is not usually of a uniform nature, but rather that of a localized attack, resulting in pitting.

Fifth Stage of Corrosion

Deep pits formed in the substrate during the fourth stage of attack may eventually penetrate completely to cause holes. Within the corrosion cell, pitting has occurred to such a degree that undercutting, flaking, and delamination of the substrate takes place. As the small hole develops, the electrolyte has access to the reverse side and corrosion now takes place on both sides of the substrate.

Final Stage of Corrosion

Corrosion is now taking place at its most rapid and aggressive rate. Large gaping holes are formed causing severe structural damage.

Composition of Coatings

The most commonly used organic coating is paint. When applied for corrosion protection, paints are referred to as coatings. Paints consist of binders, pigments, fillers, additives, and solvents.

Binder

The binder forms the continuous polymeric phase in which all of the other ingredients are incorporated. Its density and composition determine the permeability, chemical resistance, and ultraviolet resistance of the coating. The protective film is formed through physical curing, chemical curing, or a combination of these.

Pigments

Pigments are added to the coating for two reasons: to provide color and to improve the corrosion resistance of the coating. The improvement in corrosion resistance may be accomplished by one or more of the following:

1. Anticorrosion pigments dissolve slowly in the coating and provide protection by covering corrosion sensitive sites under the coating, or by sacrificially corroding themselves, thereby protecting the substrate metal, or by passivating the surface.
2. Blocking pigments absorb at the active metal surface, reducing the active area for corrosion, and form a transport barrier for ionic species to and from the substrate.
3. Galvanic pigments are nonnoble metal particles (relative to the substrate). These particles when exposed corrode preferentially, while at the original metal surface only the cathodic reaction occurs.
4. Passivating pigments stabilize the oxide film on the exposed metal substrate. Chromates with limited water solubility are generally used.

Fillers

Fillers are used to increase the volume of the coating. They are also used to improve such properties as impact, abrasion resistance, and water permeability.

Additives

Additives are made up of numerous materials that are added in small amounts to enhance certain specific properties. They consist of thickeners, antifugal agents, dispersing agents, antifoam agents, anticoalescence agents, UV absorbers and fire retarding agents.

Solvents

Solvents have two different roles to perform in a coating. Prior to application, the solvent has to function to reduce viscosity of the binder and other components to permit their homogeneous mixing. The reduced viscosity also enables the coating to be applied in a thin, smooth, continuous film. Prior to application, the liquid mixture should be a solution or stable dispersion or emulsion of binder, pigments, and additives in the solvent. After the paint has been applied a major attractive force between the components is necessary for the formation of a continuous film. There should be no interaction between the solvent and other components, so that the solvent is free to evaporate from the curing film.

Two component epoxy coatings do not require the use of a solvent. These coatings have a low viscosity. The two components are mixed, usually at elevated temperatures, to reduce the viscosity as much as possible.

Complex Coating Systems

Because of poor adhesion, the corrosion protection supplied by organic coatings is not always satisfactory. To compensate for this, conversion layers are applied to the substrate metal. These layers provide ions, which become part of the protective coating after (electro) chemical reaction of the substrate with a reactive medium. Common conversion layers are phosphate layers on steel and zinc, chromate layers on zinc and aluminum, and anodized layers on aluminum, the latter without an organic topcoat. Anodizing is an electrochemical treatment of a metal (mainly aluminum) while the metal itself is the anode. This produces a reasonably thick oxide layer, which is passive.

Pretreatment layers are used for a variety of reasons:

1. To provide a uniform grease free surface
2. To obtain electrically insulating barrier layers
3. To improve the adherence of the organic layer
4. To provide active corrosion inhibition by passivating the metallic substrate or by reducing the rate of oxygen reduction reaction.

There are two types of resin systems, thermoplastic and thermosetting. Thermoplastic solvent-deposited coatings do not undergo any chemical change from the time of application until the attainment of final properties as a protective film. Thermosetting resins differ in that a chemical change takes place after application and solvent evaporation. The coating is said to cure as the chemical reaction is taking place. This curing can take place at room temperature or, in the case of baked coatings, at elevated temperatures. The reaction is irreversible and unlike with thermoplastic coatings, high temperatures or exposure to solvents do not cause the coating to melt or soften. The more commonly used industrial paints are discussed.

Vinyls (Thermoplastic)

These are polyvinyls dissolved in aromatics, ketones, or ester.

1. Resistance: insoluble in oils, greases, aliphatic hydrocarbons, and alcohols; resistant to water and aqueous salt solutions; at room temperature resistant to inorganic acids and alkalies; fire resistant
2. Temperature resistance: 180°F (82°C) dry; 140°F (60°C) wet
3. Limitations: dissolved by ketones, aromatics, and ester solvents
4. Applications: used on surfaces exposed to potable water and on sanitary equipment

Chlorinated rubbers (Thermoplastic)

These are resins dissolved in hydrocarbon solvents.

1. Resistance: chemically resistant to acids and alkalies; low permeability to water vapor; abrasion resistant; fire resistant
2. Temperature resistance: 200°F (93°C) dry; 120°F (49°C) wet
3. Limitations: degraded by ultraviolet light; attacked by hydrocarbons
4. Applications: used on structures exposed to water and marine atmospheres (swimming pools, etc.); excellent adherence to concrete and masonry

Epoxy (Thermoset)

These are a series of various epoxy paints, all of which are of the thermoset variety.

Epoxy (Thermoset)

This is a polamide plus epoxy resin (amine epoxy).

1. Resistance: resistant to acids, acid salts, alkalies, and organic solvents
2. Temperature resistance: 225°F (107°C) dry; 190°F (88°C) wet
3. Limitations: harder and less flexible than other epoxies; less tolerant of moisture during application
4. Applications: widest range of chemical and solvent resistance of epoxies; used for piping and vessels

Epoxy (Thermoset)

This is polyamide plus epoxy resin.

1. Resistance: partially resistant to acids, acid salts, alkalies, and organic solvents; resistant to moisture
2. Temperature resistance: 150°F (66°C) dry; 225°F (108°C) wet
3. Limitations: chemical resistance inferior to that of the polyamide epoxies
4. Applications: used on wet surfaces or under water, as in tidal zone areas of pilings, oil rigs, etc.

Epoxy (Thermoset)

This is aliphatic polyamine plus partially prepolymerized epoxy.

1. Resistance: partially resistant to acids, acid salts, and organic solvents
2. Temperature resistance: 225°F (108°C) dry; 150°F (66°C) wet
3. Limitations: film formed has greater permeability than the other amine epoxies
4. Applications: used for protection against mild atmospheric corrosion

Epoxy (Thermoset)

Esters of epoxies and fatty acids are modified (epoxy ester).

1. Resistance: resistant to weathering; attacked by alkalies
2. Temperature resistance: 225°F (107°C) dry; 150°F (66°C) wet
3. Limitations: chemical resistance generally poor
4. Applications: used where properties of a high quality oil base paint is required

Epoxy (Thermoset)

This is a cool tar plus epoxy resin (amine or polyamide cured).

1. Resistance: excellent resistance to freshwater, saltwater, and inorganic acids
2. Temperature resistance: 225°F (107°C) dry; 150°F (66°C) wet

3. Limitations: attacked by organic solvents
4. Application: used on steel for immersion or below grade service

Oil Base

This comprises coating formulations with vehicles (alkyd), epoxy (urethane), combined with drying oils.

1. Resistance: resistant to weathering
2. Temperature resistance: 225°F (107°C) dry; 150°F (66°C) wet
3. Limitations: chemical resistance generally poor
4. Applications: used on wood exterior surfaces because of its penetrating power

Urethanes

This is a moisture cured isocyanate prepolymer reacting with atmospheric moisture.

1. Resistance: abrasion resistant; if cross-linked, resistant to chemicals and solvents
2. Temperature resistance: 250°F (121°C) dry; 150°F (66°C) wet
3. Limitations: may yellow under ultraviolet light; poor chemical resistance
4. Applications: used on furniture and floors

Urethanes

These are catalyzed aliphatic or aromatic isocyanate reacted with polyesters, epoxy, or acrylic polyhydroxyls.

1. Resistance: good chemical resistance; similar to polyamide epoxy
2. Temperature resistance: 225°F (107°C) dry; 150°F (66°C) wet
3. Limitations: not recommended for exposure to or immersion in strong acids or alkalies
4. Applications: used as a decorative coating of tank cars and steel in highly corrosive atmospheres

Silicones

Consider the high temperature type.

1. Resistance: water repellent
2. Temperature resistance: 1200°F (649°C) dry and wet
3. Requires baking for good cure; not chemically resistant
4. Applications: water solvent formulations used on limestone, cement, and nonsilaceous materials; solvent formulations used on bricks and non-calcaceous masonry

Water Base

These are aqueous emulsions of polyvinyl acetate, acrylic, or styrene-butadiene latex.

1. Resistance: poor chemical resistance; resistant to weather
2. Temperature resistance 150°F (66°C) dry and wet

3. Limitations: not suitable for immersion service
4. Applications: used in general decorative applications primarily on wood

Polyesters

These are organic acids condensed with polybasic alcohols. Styrene is a reactive diluent.

1. Resistance: excellent resistance to acids and aliphatic solvents; good resistance to weathering
2. Temperature resistance 180°F (82°C) dry and wet
3. Limitations: not suitable for use with alkalies and most aromatic solvents, since they swell and soften these coatings
4. Applications: lining materials for tanks and chemical process equipment

Coal Tar

This is a distilled coking by-product in aromatic solvent.

1. Resistance: excellent resistance to moisture; good resistance to weak acids, weak alkalies, petroleum oils, and salts.
2. Temperature resistance: 100°F (38°C) dry and wet
3. Limitations: ultraviolet light and weathering will degrade
4. Applications: used on submerged or buried steel and concrete

Asphalt

These are solids from crude oil refining in aliphatic solvents.

1. Resistance: good resistance to weak acids, alkalies, and salts
2. Temperature resistance: 230°F (110°C) dry and wet
3. Applications: used in above-ground weathering environments and chemical fume atmospheres

Zinc Rich

This is metallic zinc in a vehicle of organic or inorganic type.

1. Resistance: highly resistant to galvanic and pitting type corrosion
2. Temperature resistance: 700°F (371°C) dry and wet
3. Limitations: must have top coat in severe environments or when pH is below 6 or above 10.5
4. Applications: jet fuel storage tanks; petroleum products

See Refs. 1, 8, 16, 2, 17–20.

COBALT ALLOYS

Cobalt alloys are primarily used for hard face applications such as in valve seats. The purpose of hard facing is to improve resistance to abrasion, friction, galling, and/or impact. The cobalt hard face alloys usually contain 30 to 60% cobalt with additives of carbon, nickel, chromium, tungsten, and/or molybdenum. Application is made by either welding or thermal spray processes. Their corrosion resistance is approximately that of the 300 series stainless steels. A typical alloy is Stellite.

See Ref. 4.

COLD WATER PITTING

Cold water pitting is the electrochemical pitting of copper tubes and fittings in domestic water systems that transport ground waters containing free carbon dioxide in conjunction with dissolved oxygen. The action may be accelerated by the presence of chlorides and sulfates in the water.

COMPOSITES

For the purpose of corrosion resisting materials a composite is defined as a mixture of two or more materials that are distinct in composition and form, each being present in significant quantities (e.g., greater than 5 volume percent). By this definition conventional alloyed steel would not be considered a composite since the alloying ingredients are present in quantities of much less than 1%, by weight or volume and most often less than 0.1%.

The object of composite materials is to achieve properties in composite form that exceed those of their individual components alone. In forming composites at times it is necessary to accept a trade-off. For example combining a strong but brittle ceramic fiber in a ductile and weaker metal matrix results in a composite whose strength lies somewhere between the strength of the ceramic fiber and that of the metal matrix, but which is not as brittle as the ceramic alone.

The possibilities of forming composites are quite extensive. The duplex stainless steels, which contain approximately equal amounts of ferrite and austenite, are examples of metallic composites (see "Duplex Stainless Steels").

Glass fiber reinforced polyesters are considered the first engineered composites. In addition to glass fiber, other types of fibers used to produce composites include boron, carbon, silicon carbide, and aramid fibers. For more detailed information on composites refer to the following topics in this book:

Composite laminates
Duplex stainless steels
Thermoplastic composites
Thermoset reinforcing materials
Thermoplastic laminates

See Ref. 21.

COMPOSITE LAMINATES

Composite laminates are combinations of two composite materials joined to form a dual laminate, one material being on the exterior while the other is on the interior. Typical combinations include ABS and polyester, bisphenol and isophthalic–fibrous glass systems, vinyl ester and polyester, epoxy and polyester, glass and reinforced polyester, polypropylene lined reinforced polyester, and PVC and polyester composite.

When applying a composite laminate it is necessary to evaluate each member of the composite as to its compatibility to the corrosive environment.

Also see "Composites."

CONCENTRATED

Concentrated refers to a solution in which the maximum possible amount of solute has been dissolved in the solvent.

CONCENTRATION

Concentration refers to the amount of solute, usually by weight, that has been dissolved in the solvent.

CONCENTRATION CELLS

These are the cause of pitting and crevice corrosion. The primary factors causing pitting are electrical contact between dissimilar materials or areas of the same metal where oxygen or conductive salts in water differ. Such a couple is sufficient to cause a difference of potential, causing an electric current to flow through the water, or across moist steel, from the metallic anode to a nearby cathode. The cathode may be mill scale or any other portion of the metal surface that is cathodic to the more active metal areas. Mill scale is cathodic to steel and is one of the more common causes of pitting. If the cathodic area is relatively large compared to the anodic area, the damage is spread out and usually negligible. When the anode area is relatively small the metal loss is concentrated and may be serious.

Concentration cells are capable of causing severe corrosion, leading to pitting, when differences in dissolved oxygen concentrations occur. That portion of the metal that is in contact with water relatively low in dissolved oxygen concentration is anodic to adjoining areas with water higher in dissolved oxygen concentration. This lack of oxygen may be caused by exhaustion of dissolved oxygen in a crevice. The low oxygen area is always anodic. This type of cell is responsible for corrosion at crevices that are formed at the interface of two coupled pipes, or at threaded connections, since the oxygen concentration is lower within the crevice or at the threads than elsewhere. It also is responsible for pitting damage under rust or at the water line (air–water interface). These differential aeration cells are responsible for initiating pits in stainless steel, aluminum, nickel, and other so-called passive metals when they are exposed to aqueous environments such as seawater.

See Ref. 22.

CONVERSION COATINGS

Conversion coatings are coatings either formed on metal naturally by reaction with the environment or man-made using chemical or electrochemical treatment. The films or coatings formed by these treatments serve two purposes. They not only improve the corrosion resistance of the metal but also increase the adhesive bonding of paint coatings.

Typical examples of this type of coating include anodizing of aluminum, magnesium, and titanium alloys; phosphate coatings on iron and steel, aluminum, and zinc; chromate coatings on zinc, aluminum, and cadmium; oxide bluing of iron and steel; oxide coatings on cadmium, iron, steel, copper, and zinc alloys; and pack cementation to form diffusion coatings on various metals.

See Refs. 1 and 2.

COPOLYMER

A copolymer is a polymer produced from two or more types of different monomers.

COPPER

Copper is a very useful material. It has excellent electrical and thermal conductivity properties, is malleable and machinable but has low mechanical properties. The mechanical and physical properties are given in Table C.7. In order to obtain strength the metal must be cold worked or alloyed. As a result there are hundreds of copper alloys. The Copper Development Association, together with the American Society of Testing and Materials and the Society of Automotive Engineers, has developed a five digit system to identify these alloys. This system is part of the unified numbering system for metals and alloys. The numbers C-10000 through C-79999 denote the wrought alloys while the cast copper and copper alloys are numbered C-80000 through C-99999.

To be classified as copper the compound must contain a minimum of 99.3% copper. Elements such as silver, arsenic, phosphorous, antimony, tellurium, nickel, cadmium, lead, sulfur, zirconium, magnesium, boron and bismuth may be present singly or in combination.

Copper is noble to hydrogen in the emf series and thermodynamicaly stable with no tendency to corrode in water and in nonoxidizing acids free of dissolved oxygen. With copper and its alloys the predominant cathodic reaction is the reduction of oxygen to form hydroxide ions. Therefore oxygen or other oxidizing agents are necessary for corrosion to take place. In oxidizing acids or in aerated solutions of ions that form Cu complexes, e.g., CN^-, $MH^-{}_A$, corrosion can be severe. Copper is also subject to attack by turbulently flowing solutions, even though the metal may be resistant to the solution in a stagnant condition. Most of the corrosion products formed on copper and copper alloys produce adherent, relatively impervious films with low solubility that provide the corrosion protection.

Copper finds many applications in the handling of seawater and/or freshwater. Copper pipe was used by the Egyptian pharaoh Cheops to transport water to the royal bath.

Table C.7 Mechanical and Physical Properties of Copper

Property	Annealed	Hard-drawn
Modulus of elasticity $\times 10^6$, psi	17	17
Tensile strength $\times 10^3$, psi	33	45
Yield strength 0.2% offset $\times 10^3$, psi	10	40
Elongation in 2 in., %	45	10
Hardness, Rockwell	F-45	B-40
Density, lb/in.^{3}	0.323	0.323
Specific gravity	8.91	8.91
Specific heat, Btu/hr °F	.092	.092
Thermal conductivity at 68°F, Btu/hr ft^2 °F	2364	2364
Coefficient of thermal expansion at 77–572 °F, in./in. °F $\times 10^{-6}$	9.8	9.8

Several years ago a remnant of this pipe was unearthed, still in usable condition, a testimony to copper's durability and resistance to corrosion.

The corrosion resistance of copper, when in contact with freshwater or seawater, is dependent upon the surface oxide film that forms. In order for corrosion to continue, oxygen must diffuse through this film. High velocity water will disturb the film, while carbonic acid or organic acids, which are present in some freshwaters and soils, will dissolve the film. Either situation leads to an appreciably high corrosion rate. If the water velocity is limited to a maximum of 4–5 ft/sec the film will not be disturbed.

Sodium and potassium hydroxide solutions can be handled at room temperature by copper, in all concentrations. Copper is not corroded by perfectly dry ammonia, but it may be rapidly corroded by moist ammonia and ammonium hydroxide solutions. Alkaline salts, such as sodium carbonate, sodium phosphate, or sodium silicate act like the hydroxides but are less corrosive.

When exposed to the atmosphere over long periods of time the protective film that forms is initially dark in color, gradually turning green. This corrosion product is known as patina. Since the coloration is given by copper hydroxide products, the length of time required to form this coloration is dependent upon the atmosphere. In a marine atmosphere, the compound is a mixture of copper/hydroxide/chloride and in an urban or industrial atmosphere a mixture of copper/hydroxide/sulfate.

Pure copper is immune to stress corrosion cracking. However, alloys of copper containing more than 15% zinc are particularly subject to this type of corrosion.

The coppers are resistant to urban, marine, and industrial atmospheres. For this reason copper is used in many architectural applications such as building fronts, downspouts, flashing, gutters, roofing, and screening. In addition to their corrosion resistance, their good thermal conductivity properties make the coppers ideal for use in solar panels and related tubing and piping used in solar energy conversion. These same properties plus their resistance to engine coolants have made the coppers suitable for use as radiators.

Large amounts of copper are also used in the beverage industry, particularly in the brewing and distilling operations.

In general the coppers are generally resistant to

1. Seawater
2. Freshwaters, hot or cold
3. Deaerated; hot or cold, dilute sulfuric acid, phosphoric acid, acetic acid, and other nonoxidizing acids
4. atmospheric exposure

The coppers are not resistant to

1. Oxidizing acids such as nitric, hot concentrated sulfuric, and aerated nonoxidizing acids (including carbonic acid)
2. Ammonium hydroxide (plus oxygen). A complex ion $Cu(NH_3)_4^{++}$ forms. Substituted ammonia compounds (amines) are also corrosive
3. High velocity aerated waters and aqueous solutions
4. Oxidizing heavy metal salts (ferric chloride, ferric sulfate, etc.)
5. Hydrogen sulfide, sulfur, and some sulfur compounds

The compatibility of copper with selected corrodents is shown in Table C.8. Reference 4 provides a comprehensive listing of the compatibility of copper with a variety of corrodents.

See Refs. 4, 23–25.

COPPER ALLOYS (HIGH COPPER)

These alloys contain a minimum of 95% copper if wrought and 94% copper when cast. The corrosion resistance of these alloys is approximately the same as that of copper. They find their main area of application in the electrical and electronics field. Application in corrosion resistant service is found when mechanical strength greater than that of the coppers is required. Refer to Table C.8 for the compatibility with selected corrodents.

COPPER–ALUMINUM ALLOYS

These alloys are known as aluminum bronzes and find application where combinations of strength, corrosion resistance, and wear resistance are of importance. They are available in both wrought and cast forms.

Alkalies such as sodium and potassium hydroxides can be handled. When constantly immersed in concentrations of from 1 to 45% of sodium hydroxide at temperatures from room to 125˚F (52˚C), the corrosion rate will be less than 5 mpy for alloys C-61400 and C-96300. At a temperature of 175˚F (80˚C) the corrosion rate will range from 0.1 to 8.1 mpy in concentrations of 5 to 10%.

These alloys are resistant to nonoxidizing mineral acids such as sulfuric and phosphoric. Resistance is controlled by the presence of oxygen or an oxidizing agent. At room temperature up to 90˚F (32˚C) the corrosion rate of alloy C-95400 in sulfuric acid concentrations ranging from 0.5 to 50% will be less than 5 mpy.

Aluminum bronzes are also resistant to many organic acids such as acetic, citric, formic, and lactic. When in contact with these materials it is possible to experience copper pickup in the finished product. Although the concentration is very low, discoloration of the

Table C.8 Compatibility of Copper, Aluminum Bronze, and Red Brass with Selected Corrodents

	Maximum temperature (°F/°C)		
Chemical	Copper	Aluminum bronze	Red brass
Acetaldehyde	x	x	x
Acetamide		60/16	
Acetic acid 10%	100/38		x
Acetic acid 50%	x	x	x
Acetic acid 80%	x	x	x
Acetic acid, glacial	x	x	x
Acetic anhydride	80/27	90/32	x
Acetone	140/60	90/32	220/104
Acetyl chloride	x	60/16	x
Acrylonitrile	80/27	90/32	210/99
Adipic acid	80/27		
Allyl alcohol	90/32	90/32	90/32
Alum	90/32	60/16	80/27
Aluminum acetate	60/16		
Aluminum chloride, aqueous	x	x	x
Aluminum chloride, dry	60/16		
Aluminum fluoride	x	90/32	
Aluminum hydroxide	90/32	x	80/27
Aluminum nitrate		x	x
Aluminum oxychloride			
Aluminum sulfate	80/27	x	x
Ammonia gas	x	90/32	x
Ammonium bifluoride	x		x
Ammonium carbonate	x		x
Ammonium chloride 10%	x	x	
Ammonium chloride 50%	x	x	
Ammonium chloride, sat.	x	x	x
Ammonium fluoride 10%	x		x
Ammonium fluoride 25%	x	x	x
Ammonium hydroxide 25%	x	x	x
Ammonium hydroxide, sat.	x	x	x
Ammonium nitrate	x	x	x
Ammonium persulfate	90/32	x	x
Ammonium phosphate	x	90/32	x
Ammonium sulfate 10–40%	x	x	x
Ammonium sulfide	x	x	x
Ammonium sulfite	x		x
Amyl acetate	90/32	x	400/204
Amyl alcohol	80/27	90/32	90/32
Amyl chloride	80/27	90/32	80/27
Aniline	x	90/32	x
Antimony trichloride	80/27	x	x
Aqua regia 3:1	x	x	x
Barium carbonate	80/27	90/32	90/32
Barium chloride	80/27	80/27	80/27

Table C.8 Continued

Chemical	Maximum temperature (°F/°C) Copper	Aluminum bronze	Red brass
Barium hydroxide	80/27	x	80/27
Barium sulfate	80/27	60/16	210/99
Barium sulfide	x	x	x
Benzaldehyde	80/27	90/32	210/99
Benzene	100/38	80/27	210/99
Benzene sulfonic acid 10%			90/32
Benzoic acid 10%	80/27	90/32	210/99
Benzyl alcohol	80/27	90/32	210/99
Benzyl chloride	x		x
Borax	80/27	90/32	80/27
Boric acid	100/38	90/32	x
Bromine gas, dry	60/16	x	
Bromine gas, moist	x	x	
Bromine liquid		x	
Butadiene	80/27		80/27
Butyl acetate	80/27		300/149
Butyl alcohol	80/27	90/32	90/32
Butyl phthalate	80/27		210/99
Butyric acid	60/16	90/32	80/27
Calcium bisulfite	80/27	x	x
Calcium carbonate	80/27	x	80/27
Calcium chlorate	x	x	x
Calcium chloride	210/99	x	80/27
Calcium hydroxide 10%	210/99	80/27	
Calcium hydroxide, sat.	210/99	60/16	210/99
Calcium hypochlorite	x	x	x
Calcium nitrate		x	x
Calcium sulfate	80/27	x	80/27
Caprylic acid	x		x
Carbon bisulfide	80/27		x
Carbon dioxide, dry	90/32	90/32	570/299
Carbon dioxide, wet	90/32	90/32	x
Carbon disulfide	80/27		x
Carbon monoxide		60/16	570/299
Carbon tetrachloride	210/99	90/32	180/82
Carbonic acid	80/27	x	210/99
Cellosolve	80/27	60/16	210/99
Chloracetic acid, 50% water	x		
Chloracetic acid	x	80/27	x
Chlorine gas, dry	210/99	90/32	570/299
Chlorine gas, wet	x	x	x
Chlorine, liquid			
Chlorobenzene	90/32	60/16	210/99
Chloroform	80/27	90/32	80/27
Chlorosulfonic acid	x	x	x
Chromic acid 10%	x	x	x

Table C.8 Continued

Chemical	Maximum temperature (°F/°C) Copper	Aluminum bronze	Red brass
Chromic acid, 50%	x	x	x
Citric acid 15%	210/99	90/32	x
Citric acid, concentrated	x	x	x
Copper acetate	90/32	x	x
Copper carbonate	90/32		
Copper chloride	x	x	x
Copper cyanide	x	x	x
Copper sulfate	x	x	x
Cupric chloride 5%	x		
Cupric chloride 50%			
Cyclohexane	80/27	80/27	80/27
Cyclohexanol	80/27		80/27
Dichloroethane			210/99
Ethylene glycol	100/38	80/27	80/27
Ferric chloride	80/27	x	x
Ferric chloride 50% in water	x	x	x
Ferric nitrate 10–50%	x	x	x
Ferrous chloride		x	x
Ferrous nitrate			
Fluorine gas, dry	x	x	x
Fluorine gas, moist	x		
Hydrobromic acid, dilute	x	x	x
Hydrobromic acid 20%	x	x	x
Hydrobromic acid 50%	x	x	x
Hydrochloric acid 20%	x	x	x
Hydrochloric acid 38%	x	x	x
Hydrocyanic acid 10%	x	x	x
Hydrofluoric acid 30%	x	x	x
Hydrofluoric acid 70%	x	x	x
Hydrofluoric acid 100%	x	x	x
Hypochlorous acid	x		x
Iodine solution 10%			
Ketones, general		90/32	100/38
Lactic acid 25%		x	90/32
Lactic acid, concentrated	90/32	90/32	90/32
Magnesium chloride	300/149	90/32	x
Malic acid	x		
Manganese chloride	x		
Methyl chloride	90/32	x	210/99
Methyl ethyl ketone	80/27	60/16	210/99
Methyl isobutyl ketone	90/32		210/99
Muriatic acid	x		
Nitric acid 5%	x	x	x
Nitric acid 20%	x	x	x
Nitric acid 70%	x	x	x
Nitric acid, anhydrous	x	x	x

Table C.8 Continued

Chemical	Maximum temperature (°F/°C) Copper	Aluminum bronze	Red brass
Nitrous acid, concentrated	80/27	x	
Oleum		x	x
Perchloric acid 10%			x
Perchloric acid 70%			x
Phenol	x	x	570/299
Phosphoric acid 50–80%	x	x	x
Picric acid	x	x	x
Potassium bromide 30%	80/27		
Salicylic acid	90/32		210/99
Silver bromide 10%	x		
Sodium carbonate	120/49	60/16	90/32
Sodium chloride to 30%	210/99	60/16	210/99
Sodium hydroxide 10%	210/99	60/16	210/99
Sodium hydroxide 50%	x	x	x
Sodium hydroxide, concentrated	x	x	x
Sodium hypochlorite 20%	x	x	80/27
Sodium hypochlorite, concentrated	x	x	x
Sodium sulfide to 50%	x	x	x
Stannic chloride	x	x	x
Stannous chloride	x	x	x
Sulfuric acid 10%	x	x	200/93
Sulfuric acid 50%	x	x	x
Sulfuric acid 70%	x	x	x
Sulfuric acid 90%	x	x	x
Sulfuric acid 98%	x	x	x
Sulfuric acid 100%	x	x	x
Sulfuric acid, fuming	x	x	x
Sulfurous acid	x	x	90/32
Toluene	210/99	90/32	210/99
Trichloroacetic acid	80/27	x	80/27
Zinc chloride	x	x	x

The chemicals listed are in the pure state or in a saturated solution unless otherwise indicated. Compatibility is shown to the maximum allowable temperature for which data are available. Incompatibility is shown by an x. A blank space indicates that data are unavailable. When compatible, corrosion rate is <20 mpy.
Source: PA Schweitzer. *Corrosion Resistance Tables,* 4th ed. Vols. 1–3. New York: Marcel Dekker, 1995.

product may occur. The color contaminant can easily be removed by a carbon filtration polishing step.

This alloy also finds many applications in the handling of seawater for such items as valves, fittings, and pump casings, particularly in desalination plants aboard naval and commercial vessels.

Aluminum bronzes are not affected by pitting, crevice corrosion, or stress corrosion.

Refer to Table C.8 for the compatibility of aluminum bronze with selected corrodents.

See Refs. 4 and 25.

COPPER–NICKEL ALLOYS

These are referred to as cupronickels and are useful in waters ranging from fresh to brackish and sea. Their biofouling resistance is excellent. Corrosion rates for alloys C-70600 and C-71500 in seawater are approximately 1 mpy. They find their greatest use in saltwater service where they are used as piping, fittings, condenser tubes and plates, and pump casings.

Of all the copper alloys, the copper–nickel are the most resistant to stress-corrosion cracking in the presence of ammonia and amoniacal solutions and are highly resistant to stress-cracking in general.

Of the several copper–nickel alloys commercially available, alloy 706 offers the best combination of properties for marine application and has the broadest application in seawater service. Mechanical and physical properties are given in Table C.9.

Alloy 706 has been used aboard ships for seawater distribution and shipboard fire protection. It is also used in many desalting plants. Exposed to seawater, alloy 706 forms a thin but tightly adhering oxide film on its surface. To the extent that this film forms, copper–nickel does in fact corrode in marine environments. However, the copper–nickel oxide film is firmly bonded to the underlying metal and is nearly insoluble in seawater. It therefore protects the alloy against any further attack once it has formed. Initial corrosion rates may be in the range of less than 1.0 to about 2.5 mils per year. In the absence of turbulence, as would be the case in a properly designed piping system, the copper–nickel's corrosion rate will decrease with time, eventually dropping to as low as 0.05 mpy after several years of service.

Another important advantage of the oxide film developed on alloy 706 is that it is an extremely poor medium for adherence and growth of marine life. Algae and barnacles,

Table C.9 Mechanical and Physical Properties of 90-10 Copper–Nickel Alloy 706

Modulus of elasticity × 10^6, psi	18
Tensile strength × 10^3, psi	
4½ in. O.D.	40
5½ in. O.D.	38
Yield strength at 0.5% extension under load × 10^3, psi	
4½ in. O.D.	15
5½ in. O.D.	13
Elongation in 2 in., %	25
Density, lb/in.^{3}	0.323
Specific heat, Btu/lb °F	0.09
Thermal conductivity, Btu/h/ft/at ft.2/°F	26
Coefficient of thermal expansion at 68–572°F, in./in. °F × 10^{-6}	9.5

the two most common forms of marine biofouling, simply will not grow on alloy 706. Alloy 706 piping therefore remains clean and smooth, neither corroding appreciably nor becoming encrusted with growth.

Flow velocities play an important role in seawater corrosion. Velocity-dependent erosion corrosion is often associated with turbulence and entrained particles. Recommended velocities for continuous service are as follows:

Pipe size (in.)	Velocity (fps)
≤ 3	5
4–8 with short radius bends	6.5
≥ 4 with long radius bends	11
≥ 8	11

See Ref. 14.

COPPER–TIN ALLOYS

These alloys are known as tin bronzes or phosphor bronzes. Although tin is the principal alloying ingredient, phosphorus is always present in small amounts, usually less than 0.5%, because of its use as an oxidizer.

These alloys are probably the oldest alloys known, having been the bronzes of the Bronze Age. Even today many of the artifacts produced during that age are still in existence. Items such as statues, vases, bells, and swords have survived hundreds of years of exposure to a wide variety of environments testifying to the corrosion resistance of these materials.

Alloys that contain more than 5% tin are especially resistant to impingement attack. In general the tin bronzes are noted for their high strength. Their main application is in water service for such items as valves, valve components, pump casings, and similar items. Because of their corrosion resistance in stagnant waters they also find wide application as components of fire protection systems.

COPPER–ZINC ALLOYS (BRASSES)

Brasses contain zinc as their principal alloying ingredient. Other alloying additions are lead, tin, and aluminum. Lead is added to improve machineability and does not improve the corrosion resistance. The addition of approximately 1% tin increases the dealloying resistance of the alloys. Aluminum is added to stabilize the protective surface film. Alloys containing in excess of 15% zinc are susceptible to dealloying in environments such as acids, both organic and inorganic, dilute and concentrated alkalies, neutral solutions of chlorides and sulfates, and mild oxidizing agents.

This is a type of corrosion in which the brass dissolves as an alloy and the copper constituent redeposits from the solution onto the surface of the brass as a metal in porous form. The zinc constituent may be deposited in place of an insoluble compound or carried away from the brass as a soluble salt. The corrosion can take place uniformly or be local. Uniform corrosion is more apt to take place in acid environments while local corrosion is more apt to take place in alkaline, neutral, or slightly acid environments. The addition of tin or arsenic will inhibit this form of corrosion.

Conditions of the environment that favor dezincification are high temperature, stagnant solutions, especially acid, and porous inorganic scale formation. Other factors that stimulate the process are increasing zinc concentrations and the presence of both cuprous and chloride ions.

As the dealloying procedes, a porous layer of pure or almost pure copper is left behind. This reaction layer is of poor mechanical strength. The dezincification process on copper–zinc alloys is therefore very detrimental. These alloys are also subject to stress corrosion cracking. Moist ammonia in the presence of air will cause this form of corrosion. The quantity of ammonia present need not be great, as long as the other factors are present.

Relative resistance of the brasses to stress corrosion cracking is as in the table.

Low Resistance
brasses containing more than 15% zinc
brasses containing more than 15% zinc and small amounts of lead, tin, or aluminum
Intermediate Resistance
brasses containing less than 15% zinc

If the metal is cold formed, residual stresses may be present that can also cause stress corrosion cracking. By heating the metal to a temperature high enough to permit recrystallization the stresses will be removed. It is also possible to provide a stress relieving anneal at a lower temperature without substantially changing the mechanical properties of the cold-worked metal.

As stated earlier, the addition of small amounts of tin improves the dezincification resistance of these alloys. The brasses most commonly used for corrosion resistant applications are

Copper Alloy UNS No.
C-27000
C-28000
C-44300
C-44400
C-44500
C-46400
C-46500
C-46600
C-46700
C-68700

Copper alloys UNS C-44300 through C-44500 are known as admiralty brasses. They are resistant to dealloying as a result of the presence of tin in the alloy.

Admiralty brass finds application mainly in the handling of seawater and/or freshwater, particularly in condensers. These brasses are also resistant to hydrogen sulfide and therefore find application in petroleum refineries.

The high zinc brasses, such as C-27000, C-28000, C-44300, and C-46400 resist sulfides better than do the low zinc brasses. Dry hydrogen sulfide is well resisted.

Alloys containing 15% or less of zinc resist dealloying and are generally more corrosion resistant than the high zinc-bearing alloys. These alloys are resistant to many acids, alkalies, and salt solutions that cause dealloying in the high zinc brasses. Dissolved air, oxidizing materials such as chlorine and ferric salts, compounds that form soluble copper complexes (e.g., ammonia) and compounds that react directly with copper (e.g., sulfur and mercury) are corrosive to the low zinc brasses. These alloys are more resistant to stress corrosion cracking than the high zinc containing alloys.

Red brass is a typical alloy in this group containing 15% zinc. It has basically the same corrosion resistance as copper. The mechanical and physical properties of red brass are shown in Table C.10. Refer to the corrosion resistance Table C.8 under the-heading of "Copper" above for the compatibility of red brass with selected corrodents.

Reference 4 provides a comprehensive listing of the compatibility of various brasses with a variety of corrodents.

See Refs. 4 and 25.

CORROSION ALLOWANCE

Corrosion allowance is an extra thickness of material above what is needed for mechanical strength. This extra thickness is added to the design thickness in an amount that will compensate for lifetime corrosion loss.

CORROSION COUPONS

See "Corrosion Testing."

CORROSION FATIGUE

When a metal cracks upon being stressed alternately or repeatedly it is said to fail by fatigue. The greater the applied stresses at each cycle, the shorter is the time to failure. In a corrosive environment, failure at a given stress level usually occurs within fewer cycles, and a true fatigue limit is no longer observed. Cracking of a metal resulting from the combined action of a corrosive environment and repeated or alternate stress is called corrosion fatigue.

Table C.10 Mechanical and Physical Properties of Red Brass

Modulus of elasticity × 10^6, psi	17
Tensile strength × 10^3, psi	40
Yield strength 0.2% offset × 10^3, psi	15
Elongation in 2 in., %	50
Hardness, Brinell	50
Density, lb/in.^{3}	0.316
Specific gravity	8.75
Specific heat, Btu/lb °F	0.09
Thermal conductivity at 32–212°F, Btu/ft^2/h/°F/in.	1100
Coefficient of thermal expansion at 31–212°F, in./in. °F × 10^{-6}	9.8

See Refs. 26 and 10.

CORROSION INHIBITORS

Corrosion of metalic surfaces can be reduced or controlled by the addition of chemical compounds to the corrodent. This form of corrosion control is called inhibition, and the compounds added are known as inhibitors. These inhibitors will reduce the rate of either anodic oxidation or cathodic reduction or both processes. The inhibitors themselves form a protective film on the surface of the metal. It has been postulated that the inhibitors are absorbed into the metal surface either by physical (electrostatic) adsorption or chemisorption.

Physical adsorption is the result of electrostatic attractive forces between the organic ions and the electrically charged metal surface. Chemisorption is the transfer of, or sharing of, the inhibitor molecule's charge to the metal surface, forming a coordinate bond. The adsorbed inhibitor reduces the corrosion rate of the metal surface by either retarding the anodic dissolution reaction of the metal or by cathodic evolution of hydrogen or both. Inhibitors can be used at pH values of from acid to near neutral to alkaline.

Inhibitors can be classified in many different ways, according to

1. Their chemical nature (organic or inorganic substances)
2. Their characteristics (oxidizing or nonoxidizing compounds)
3. Their technical field of application (pickling, descaling, acid cleaning, cooling water systems, and the like)

The most common and widely known use of inhibitors are their application in automobile cooling systems and boiler feed waters.

Inhibitor Evaluation

Since there may be more than one inhibitor suitable for a specific application, it is necessary to have a means of comparing the performance of each. This can be done by determining the inhibitor efficiency according to the correlation

$$I_{\text{eff}} = \frac{R_0 - R_i}{R_0} \times 100$$

where

I_{eff} = efficiency of inhibitor, %

R_0 = corrosion rate of metal without inhibitor present

R_i = corrosion rate of metal with inhibitor present

R_0 and R_i can be determined by any of the standard corrosion testing techniques. The corrosion rate can be measured in any units, such as weight loss (mpy), as long as consistent units are used for both tests.

Classification of Inhibitors

Inhibitors can be classified in several ways as indicated previously. We will classify and discuss inhibitors under the headings

1. Passivation inhibitors
2. Organic inhibitors
3. Precipitation inhibitors

Passivation Inhibitors

Passivation inhibitors are chemical oxidizing materials such as chromate ($Cr_2O_4^{2-}$) and nitrite (NO_2^-) or substances such as Na_3PO_4 or $NaBrO_7$. These materials favor the adsorption on the metal surface of dissolved oxygen.

This type of inhibitor is the most effective and consequently widely used. Chromatics are the least expensive inhibitors for use in water systems and are widely used in the recirculation—cooling—systems of internal combustion engines, rectifiers, and cooling towers. Sodium chromate in concentrations of 0.04 to 0.1% is used for this purpose. At higher temperatures or in freshwater that has chloride concentrations above 10 ppm, higher concentrations are required. If necessary sodium hydroxide is added to adjust the pH to a range of 7.5 to 9.5. If the concentration of chromate falls below a concentration of 0.016%, corrosion will be accelerated. Therefore it is essential that periodic colorimetric analysis be conducted to prevent this from occuring.

Recent environmental regulations have been imposed on the use of chromates. They are toxic and on prolonged contact with the skin can cause a rash. It is usually required that the Cr^{+6} ion be converted to Cr^{+3} before discharge. The Cr^{+3} ion is water soluble and toxic. Even so the Cr^{+3} sludge is classified as a hazardous waste and must be constantly monitored. Because of the cost of the conversion of the chromate ions, the constant monitoring required, and the disposal of the hazardous wastes, the economics of the use of these inhibitors are not as attractive as they formerly were.

Since most antifreeze solutions contain methanol or ethylene glycol, the chromates cannot be used in this application since the chromates have a tendency to react with organic compounds. In these applications borax ($Na_2B_4O_7 \cdot 10H_2O$), to which has been added sulfonated oils to produce an oily coating, and mercaptobenzothiazole are used. The latter material is a specific inhibitor for the corrosion of copper.

Nitrites are also used in antifreeze type cooling water systems since they have little tendency to react with alcohols or ethylene glycol. Since they are gradually decomposed by bacteria they are not recommended for use in cooling tower waters. Nitrites are the corrosion inhibitors of the internal surfaces of pipelines used to transport petroleum products or gasoline, which is accomplished by continuously injecting a 5 to 30% sodium nitrite solution into the line.

At lower temperatures, such as in underground storage tanks, gasoline can be corrosive to steel as dissolved water is released. This water, in contact with the large quantities of oxygen dissolved in the gasoline, corrodes the steel and forms large quantities of rust. The sodium nitrite enters the water phase and effectively inhibits corrosion.

The nitrites are also used to inhibit corrosion by cutting oil–water emulsions used in the machining of metals.

Passivation inhibitors can actually cause pitting and accelerate corrosion when concentrations fall below minimum limits. For this reason it is essential that constant monitoring of the inhibitor concentration be performed.

Organic Inhibitors

These materials build up a protective film of adsorbed molecules on the metal surface, which provides a barrier to the dissolution of the metal in the electrolyte. Since the metal surface covered is proportional to the inhibitor concentrates, the concentration of the inhibitor in the medium is critical. For any specific inhibitor in any given medium there is an optimal concentration. For example, a concentration of 0.05% sodium benzoate, or 0.2% sodium cinnamate, is effective in water that has a pH of 7.5 and contains 17 ppm sodium chloride or 0.5% by weight of ethyl octanol.

The corrosion due to ethylene glycol cooling water systems can be controlled by the use of ethanolamine as an inhibitor.

Precipitation Inhibitors

These are compounds that cause the formation of precipitates on the surface of the metal, thereby providing a protective film. Hard water, which is high in calcium and magnesium, is less corrosive than soft water because of the tendency of the salts in the hard water to precipitate on the surface of the metal and form a protective film.

If the water pH is adjusted in the range of 5 to 6, a concentration of 10 to 100 ppm of sodium pyrophosphate will cause a precipitate of calcium or magnesium orthophosphate to form on the metal surface providing a protective film. The inhibition can be improved by the addition of zinc salts.

Inhibition of Acid Solution

The inhibition of corrosion in acid solutions can be accomplished by the use of a variety of organic compounds. Among those used for this purpose are triple-bonded hydrocarbons, acetylenic alcohols, sulfoxides and mercaptans, aliphatic, aromatic, and heterocyclic compounds containing nitrogen, and many other families of simple organic compounds of condensation products formed by the reaction between two different species such as amines and aldehydes.

Incorrect choice or use of organic inhibitors in acid solutions can lead to corrosion stimulation and/or hydrogen penetration into the metal. In general stimulation of corrosion is not related to the type and structure of the organic molecule. Stimulation of acid corrosion of iron has been found with mercaptans, sulfoxides, azole and treazole deratives, nitrites, and quinoline. This adverse action depends on the type of acid. For example, bis(4-dimethylaminophenyl) antipyrilcarbinol and its derivatives at a 10^{-4} M concentration inhibited attack of steel in hydrochloric acid solutions but stimulated attack in sulfuric acid solutions. Much work has been done studying the inhibiting and stimulating phenomena of organic compounds on ferrous as well as nonferrous metals. Organic inhibitors have a critical concentration value, below which inhibition decreases and stimulation begins. Therefore it is essential that when organic inhibitors are used, constant monitoring of the solution should take place to insure that the inhibitor concentration does not fall below the critical value.

Inhibition of Near Neutral Solutions

Because of the differences in the mechanisms of the corrosion process between acid and near neutral solutions, the inhibitors used in acid solutions usually have little or no inhibition effect in near neutral solutions. In acid solutions the inhibitor action is due to adsorption on oxide-free metal surfaces. In these media the main cathodic process is hydrogen evolution.

In almost neutral solutions the corrosion process of metals results in the formation of sparingly soluble surface products such as oxides, hydroxides, or salts. The cathodic partial reaction is oxygen reduction.

Inorganic or organic compounds as well as chelating agents are used as inhibitors in near neutral aqueous solutions. Inorganic inhibitors can be classified according to their mechanism of action:

1. Formation and maintenance of protective films can be accomplished by the addition of inorganic anions such as polyphosphates, phosphates, silicates, and borates.
2. Oxidizing inhibitors such as chromates and nitrites cause self-passivation of the metallic material. It is essential that the concentration of these inhibitors be maintained above a "safe" level. If not, severe corrosion can occur as a result of pitting or localized attack caused by the oxidizer.
3. Precipitation of carbonates on the metal surfaces forming a protective film. This usually occurs due to the presence of Ca^{2+} and Mg^{2+} ions usually present in industrial waters.
4. Modification of surface film protective properties is accomplished by the addition of Ni^{2+}, Co^{2+}, Zn^{2+}, or Fe^{2+}.

The sodium salts of organic acids such as benzoate, salicylate, cinnamate, tartrate, and azelate can be used as alternatives to the inorganic inhibitors, particularly in ferrous solutions. When using these particular compounds in solutions containing certain anions such as chlorides or sulfates, the inhibitor concentration necessary for effective protection will depend on the concentration of the aggressive anions. Therefore the critical pH value for inhibition must be considered rather than the critical concentration. Other formulations for organic inhibition of near neutral solutions are given in Table C.11.

Chelating agents of the surface active variety also act as efficient corrosion inhibitors when insoluble surface chelates are formed. Various surface acting chelating agents recommended for corrosion inhibition of different metals are given in Table C.12.

Inhibition of Alkaline Solutions

All metals whose hydroxides are amphoteric and metals covered by protective oxides that are broken in the presence of alkalies are subject to caustic attack. Localized attack may also occur as a result of pitting and crevice formation.

Organic substances such as tannions, gelatin, saponin, and agar-agar are often used

Table C.11 Organic Inhibitors for Use in Near-Neutral Solutions

Inhibitor	Type of metal protected
Organic phosphorus-containing compounds, salts of aminomethylenephosphonic acid, hydroxyethylenediphosphonic acid, phosphenocarboxylic acid, polyacrylate, polymethacrylate	Ferrous
Borate or nitrocinnimate anions (dissolving oxygen in solution required)	Zinc, zinc alloys
Acetate or benzoate anions	Aluminum
Heterocylic compounds such as benzotriazole and its derivatives, 2-mercaptobenzothiazole, 2-mercaptobenzimidazole	Copper and copper based alloys

as inhibitors for the protection of aluminum, zinc, copper, and iron. Other materials that have also been found to be effective are thiourea, substituted phenols, and naphthols, beta-diketones, 8-hydroxyquinoline, and quinalizarin.

Temporary Protection with Inhibitors

Occasions arise when temporary protection of metallic surfaces against atmospheric corrosion is required. Typical instances are in the case of finished metallic materials or of machinery parts during transportation and/or storage prior to use. When ready to be used the surface treatment or protective layer can be easily removed.

It is also possible to provide protection by controlling the aggressive environment either by eliminating the moisture and the aggressive gases or by introducing a vapor phase inhibitor. This latter procedure can only be accomplished in a closed environment such as a sealed container, a museum showcase, or a similar enclosure.

Organic substances used as contact inhibitors or vapor inhibitors are compounds belonging to the following classes:

1. Aliphatic, cycloaliphatic, aromatic, and heterocyclic amines

Table C.12 Chelating Agents Used as Corrosion—Inhibitors in Near-Neutral Solutions

Chelating agent	Type of metal protected
Alkyl-catechol derivatives, sarcosine derivatives, carboxymethylated fatty amines, and mercaptocarboxylic acids	Steel in industrial cooling systems
Azo compounds, cupferron, and rubeanic acid	Aluminum alloys
Azole derivatives and alkyl esters of thioglycolic acid	Zinc and galvinized steel
Oximes and quinoline derivatives	Copper
Cresol phthalexon and thymol phthalexon derivatives	Titamium in sulfuric acid solutions

2. Amine salts with carbonic, carbamic, acetic, benzoic, nitrous, and chromic acids
3. Organic esters
4. Nitro derivatives
5. Acetylenic alcohols

Summary

Corrosion inhibitors are usually able to prevent general or uniform corrosion. However they are very limited in their ability to prevent localized corrosion such as pitting, crevice corrosion, galvanic corrosion, dezincification, or stress corrosion cracking. Additional research is being undertaken in the use of inhibitors to prevent these types of corrosion.

See Refs. 27, 10 and 28.

CORROSION MEASUREMENT

See "Monitoring Corrosion" and "Corrosion Testing."

CORROSION MECHANISMS

There are many forms of corrosion but the mechanisms whereby corrosion takes place are fewer in number. Mechanisms covering the main corrosion phenomena are responsible for all of the forms of corrosion. The basic mechanisms are anodic dissolution, adsorption, passivation, passivity breakdown, localized corrosion, and stress corrosion cracking. Each of these is discussed under its own heading.

See Refs. 29 and 23.

CORROSION MONITORING

See "Monitoring Corrosion."

CORROSION TESTING

When a corrosion test is designed and conducted properly, reliable corrosion data may be obtained. There are a number of testing techniques that may be employed. The simplest of these involve the determination of a change in weight or dimension and observation of the corroded surface. Other more complex methods involve the measurement of hydrogen diffusion or electrical resistance or determining electrochemical characteristics.

Weight Change

Corrosion testing utilizing weight change involves the use of corrosion coupons. Coupons can be made in any size or shape. They are carefully weighed and measured before assembly on a test rack. They may be mounted in different configurations to study different types of corrosion mechanisms, such as galvanic attack, crevice corrosion, and stress corrosion. By mounting different materials of construction on the same test rack it is possible to evaluate the difference in resistance to corrosion of various materials. When assembling the coupons on the test rack it is necessary that they be insulated both from the test rack itself and from each other. This is to avoid galvanic corrosion taking place between the test pieces. See Fig. C.7.

When the test is completed the test rack is disassembled and the coupons are cleaned, weighed, and measured. The formula for calculating the uniform corrosion rate is

$$\text{mpy} = \frac{534W}{DAT}$$

where

W = weight loss, mg
D = density of specimen, g/cm^3
A = area of specimen, in.^{2}
T = exposure time, h

The low cost, the ability to evaluate several materials at one time, and closer resemblence to actual conditions of the equipment are the main advantages to using corrosion coupons. The time involved in preparing and evaluating the coupons to determine the corrosion rates after sufficient exposure time, and the limited locations for placing the test coupons, are the primary disadvantages. The use of test coupons does not permit the engineer to evaluate the results of changing process conditions, which can also be a disadvantage.

Whenever possible, actual field conditions should be used to determine which metal is going to provide the best service. Coupons should be exposed to all conditions to which the metal will be subjected. Different locations in the process may have different corrosive effects. For example the effects in the vapor space may vary from those in the liquid phase or in the condensing area. Because of this it is necessary to install test racks in each of the various locations.

If field testing is not possible, then laboratory tests should be planned to duplicate as closely as possible actual field conditions. Remember that very small changes in the

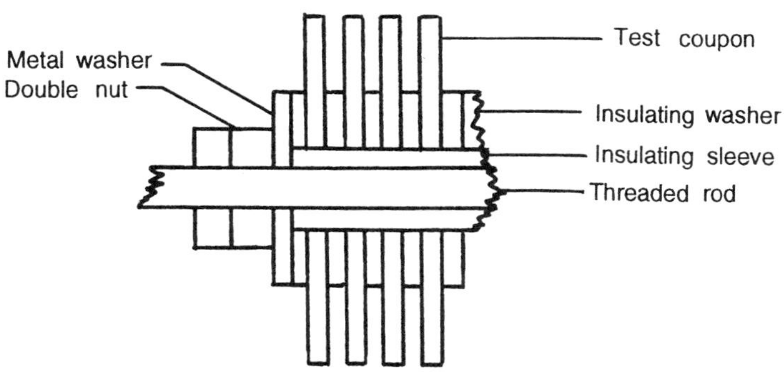

Figure C.7 Corrosion coupons insulated from each other and the coupon rack.

environment can produce large changes in the corrosion rates of metals. Test apparatus must be used that will simulate the different conditions that will be faced in the actual equipment. For example, condensate is often more corrosive than the bulk of the liquid; also heat transfer surfaces can exhibit different corrosive effects from other surfaces.

Manufacturers of metals and new products usually have fairly large amounts of data available. These data are a good guide as to which materials should be considered, but testing should still be conducted to insure that the material will be satisfactory for the specific application.

It is important that the proper data be recorded. This will vary depending on the purpose of the test. The following are guidelines as to what data should be recorded.

1. For corrosive media, the overall concentration and variation in concentration during the test, also any contaminants that may be present
2. For test metals, the trade name, chemical composition, product type (plate, sheet, rod, etc.), metallurgical condition (cold rolled, hot rolled, quenched and tempered, solution heat treated, stabilized, cast, etc.), and the size and shape of the coupon
3. Volume of test solution, for laboratory tests
4. The temperature: average, variation, and whether it was a heat transfer test
5. For aeration, the technique or conditions for the laboratory test, process exposed to atmosphere for field test
6. The apparatus and test rack type
7. The test time
8. The exposure location
9. The cleaning technique
10. The weight loss
11. The type and nature of localized corrosion: stress corrosion cracking, intergranular corrosion, pitting (maximum and average depth), crevice corrosion, etc.
12. The agitation: velocity for field tests, and technique for laboratory tests
13. The corrosion rate (which if localized corrosion is present may be misleading)

Dimension Change

Changes in dimensions can be measured using ultrasonic measuring techniques, microscopic examination, or eddy current. The most straightforward is the ultrasonic measurement of parts.

1. Ultrasonic thickness measurement. This technique can be used to measure thickness either while the part is in service or after it has been removed from service. Before testing starts, the thickness of the part is measured and thereafter at regular time intervals. After each measurement the thickness is plotted against time as shown in Fig. C.8. The difference between thickness measurements can be divided by the time interval to obtain the corrosion rate. This information is particuarly useful in determining the remaining service life of a vessel.

   Assume that the pressure vessel represented in Fig. C.8 had a corrosion allowance of 0.125 inches. During the initial plant start-up (A) the corrosion rate was 15 mpy based on the ultrasonic measurement. The corrosion rate during the

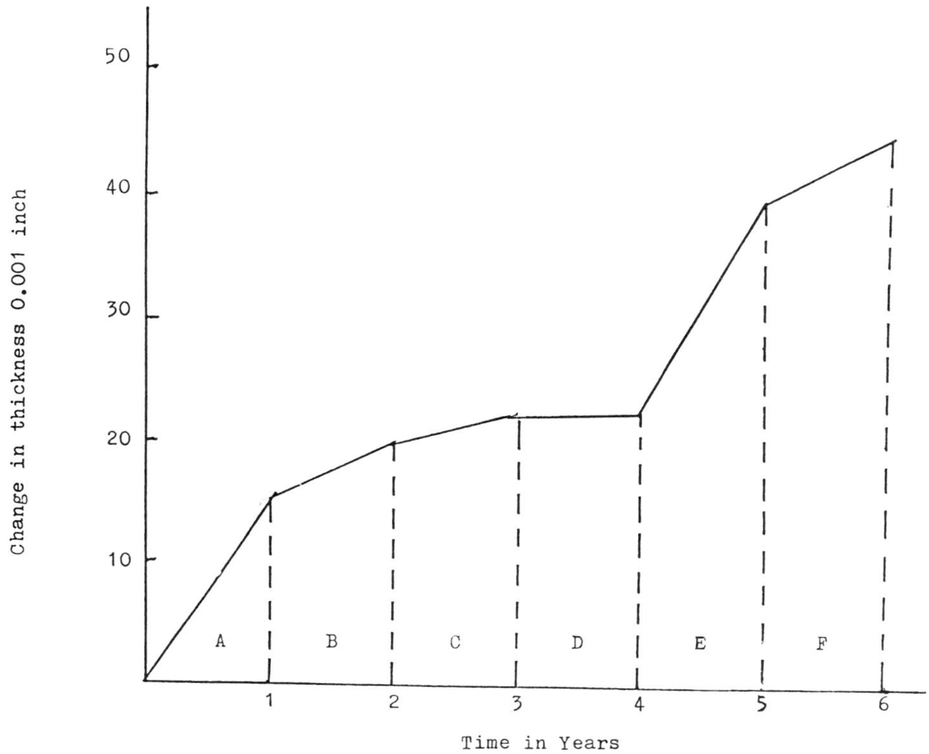

Figure C.8 Change in wall thickness plotted against time.

second year (B) equals 5 mpy. The overall rate equals 10 mpy. The process changed during the third year (C) with a corrosion rate equal to 2 mpy, making for an average corrosion rate of 7 mpy. During the fourth year (D) the plant was shut down. A new process was introduced in the fifth year (E) with an average corrosion rate equal to 18 mpy. The overall corrosion rate equals 8 mpy. The vessel continues in service during the sixth year with a corrosion rate equaling 6 mpy. Of the original corrosion allowance of 0.125 inches there is now 0.080 inch remaining, or an expected life of 13 years based on the overall corrosion rate or 16 years based on the last year's corrosion rate.

This measurement technique is not accurate enough for most laboratory testing, but it is the most accurate technique that can be used to measure the thickness of parts while in service. It also has the advantage of being able to detect changes in corrosion rates when there are process changes, or if inhibitors are added. If there is a possibility of nonuniform corrosion, an instrument with a cathode ray tube should be used, since digital instruments can give misleading readings.

2. Microscopic examination. When testing for high temperature corrosion or dealloying (dezincification), it is usually necessary to examine a polished cross-section to determine how much unaffected metal remains. In these types of corrosion, considerable damage can take place because of inward diffusion

of a corrodent such as oxygen or sulfur or because of the removal of some of the elements from the solid alloy. A change in external dimensions or a loss of weight is not an indication of the amount of damage that may have taken place. To determine accurately how much metal is left, it is necessary to prepare a cross-section of corroded metal for a metallographic examination and determine microscopically how much metal is left.

3. Eddy current. An eddy current instrument and a probe are used to measure the wall thickness of nonferromagnetic tubes. In this manner tubes in a heat exchanger may be inspected for corrosion while still in place. The instrument must first be calibrated on a tube of known thickness of the same metal as the tube to be inspected.

   Changes in thickness can be measured with an accuracy of $\pm 2\%$. It is also possible to test for nonuniform corrosion.

Electrochemical Techniques

The three most often used electrochemical techniques are zero-resistance ammeters, polarization curves, and linear polarization resistance curves.

1. Zero-resistance ammeter. A zero-resistance ammeter is a potentiostat that has been programmed to zero potential difference between the reference electrode and the test electrode. The current lead to the counterelectrode is connected to the reference electrode. This permits measurement of the amount and direction of the current that flows between the two electrodes that are electrically short-circuited. Corrosion rates of the anode cannot be calculated directly from the galvanic current because it is only a measure of how much faster the anode is corroding than the cathode.
2. Polarization curves. This is primarily a laboratory technique to study corrosion, particularly pitting. A variety of methods and equipment are available to conduct these studies. The following types are generally used:

Potentiostatic
    Potential held constant
Galvanostatic
    Current held constant
Potentiodynamic
    (a) Potential changed constantly at a specified rate
    (b) Potential changed in steps and held constant at each step
Galvanodynamic
    (a) Current changed continuously at a specified rate
    (b) Current changed in steps and held constant at each step

   As with all electrochemical techniques, they can only be used with sufficiently conductive media and when the area of the wetted electrode is known. Because of the high polarization potentials required, the estimation of corrosion rates is less precise than with linear polarization resistance methods.

3. Linear polarization resistance. The linear polarization technique permits measurement of the corrosion rate of a metal at any instant. In order to be utilized the electrodes must be exposed to an electrolyte that has a continuous

path between them. Laboratory or field testing can be conducted using either manual or automatic linear polarization equipment.

Hydrogen Diffusion

Some corrosive reactions produce atomic hydrogen at the cathode, which can diffuse through steel and most other metals if it does not combine to form hydrogen molecules. When sulfides are present, the atomic hydrogen produced by the corrosion reaction readily diffuses through steel. If hydrogen diffusion is detected, corrosion is taking place. Hydrogen diffusion can be measured using either a hydrogen probe (pressure measurement) or a hydrogen monitoring system (electrochemical).

Electrical Resistance

As the products of corrosion build up, small changes in electrical resistance occur. Low corrosion rates can be measured in this manner by not removing the products of corrosion. Probes are available with test elements made from all of the common alloys used to fabricate process equipment. Temperature changes can result in erroneous readings, since resistance changes with temperature. Although there have been modifications in probe design, it is still not possible to measure small changes in corrosion rate with a single reading unless you are absolutely sure that the temperature remained constant.

Corrosion is measured by first taking a reading on the test and check element. The probe is then inserted into the test environment and allowed to come to the test temperature. Another reading is then taken on the test and check element. Corrosion is allowed to take place for a few hours, after which a new set of readings is taken. The corrosion rate is calculated using the equation

$$\text{Corrosion rate (mpy)} = \frac{CR \times PM \times 0.365}{CT}$$

where

CR = the current reading minus the previous reading. If the reading is negative, the results are not corrosion. They are either the temperature of a conductive film or the test element.

PM = probe multiplier supplied by the manufacturer

CT = change of time, in days

The overall corrosion rate (rate over the total exposed time) and the corrosion rate between readings should both be calculated to determine whether the corrosion rate is changing with time.

See Refs. 30 and 31.

CORROSION TESTING FOR ENVIRONMENTALLY ASSISTED CRACKING (EAC)

There is no single testing technique that will take into account all the factors that come into play for a particular material and environment for the evaluation of EAC. The testing program undertaken will take into account as many of the factors as possible. This may require

1. Different alternative configurations of the same specimen
2. More than one type of test specimen
3. Several test techniques with the same specimen

It is also important that the laboratory and field or inplant test data be correlated with service experience. There are three basic general types of tests that can be performed. They are constant load/deflection techniques, slow strain rate tests, and fracture mechanics tests.

Constant Load/Deflection Tests

Types of tests included in this category include

1. Tension tests per ASTM G-49
2. Bent beam per ASTM G-38
3. C-Ring tests per ASTM G-38
4. U-Bend tests per ASTM G-30

Each type of test provides data relating to the specific type of stress or strain to which the specimen will be subjected under constant load.

Slow Strain Rate Tests

This test replaces the constant load with a slow extension of the sample until failure. A detailed description of this test method is given in ASTM G-129. The advantage of SSR testing is that it produces a result in a relatively short period of time. The primary benefit of SSR testing is that it permits the evaluation of the effect of alloy composition heat treatment and/or environmental changes such as aeration, concentration, and inhibition.

Fracture Mechanics Tests

These tests are conducted to determine the effects of metallurgical or environmental changes on EAC when the specimen contains a sharp crack. There are several fracture mechanics techniques that can be used to evaluate EAC. Regardless of which technique is used, the specimen is usually one of constant load or deflection.

When a constant load specimen is employed, a load is applied to a fracture mechanics specimen using a directly applied dead weight or through a pulley or lever system to magnify the dead weight load.

Alternatively, constant deflection specimens may be used. In this situation either constant tension or double cantilever beam specimens are loaded to an initial level of crack tip stress intensity by deflection of the arms of the specimen. The deflection is obtained either by tightening a bolt arrangement that deflects the arms of the specimen, or by inserting a wedge into the specimen. The initial stress intensity must be above the threshold stress for EAC that permits cracking to start. Once started, cracking proceeds while the stress intensity decreases as the crack proceeds through the specimen.

See Ref. 32.

CORROSION UNDER INSULATION

Insulation is applied to vessels and piping as a means of conserving heat or of providing personnel protection from hot surfaces. As a result, the selection of a particular insulating material is normally based on installed cost versus energy saved. However, there are other costs associated with insulation that are generally overlooked, namely the cost of corrosion and maintenance.

Corrosion that takes place under the insulation can be caused by the insulation itself or by improper application. If after a period it is necessary to remove sections of insulation to make repairs on the equipment, these costs and the cost of repairing the insulation should be considered during the selection process.

Thermal insulation, when exposed to water, can hold a reservoir of available moisture which together with the permeability of air causes severe attack, up to 200 to 300 mpy. This is particularly true on warm steel surfaces. Severe corrosion may occur on cold surfaces where structural members abut the insulated vessel or pipe, permitting rime ice to form. Depending upon the specific containment, insulation can cause stress corrosion cracking of high strength copper alloys and external stress corrosion cracking of type 300 series stainless steels. Chlorides or alkaline containments will rapidly attack aluminum. The use of an appropriate coating system, such as a catalyzed epoxy-phenolic and modified silicone, will help to prevent such corrosion in the event of ingress of water. Zinc and chloride free coatings should be used for stainless steels.

Types of Corrosion Under Insulation

There are three types of corrosion that can take place under insulation:

1. Alkaline or acidic corrosion
2. Chloride corrosion
3. Galvanic corrosion

Alkaline or acidic corrosion takes place as a result of either acid or alkali being present in contact with moisture in the insulating material. When these materials are applied to hot surfaces at temperatures above 250°F (121°C), the water is driven off, but it may condense at the edge of the insulation, dissolving any alkaline or acidic materials present. This results in corrosion of aluminum with certain alkaline waters.

Some insulating cements, before drying, contain alkaline chemicals and water. During the drying operation these alkaline chemicals will attack such metals as stainless steel, copper, brass, and aluminum. If the vessel being insulated is constructed of one of these metals, it may be subject to corrosion. Under normal circumstances steel would not be affected during the time required for the cement to dry.

Foam insulations containing fire retardant chemicals, such as brominated or chlorinated compounds, can produce acidic solutions. This has been found to be true with polyurethane and phenolic foams.

Steps can be taken to prevent this type of corrosion. When external corrosion of the jacket is a problem, a good all-weather plastic jacket should be considered. If a metal jacket is to be used, a moisture barrier should be installed on the inside of the insulation. It goes without saying that all joints must be tightly sealed to prevent external water from entering the insulation.

Care should be taken in selection of the insulating material. Insulating cements are best mixed with clean potable water. The use of distilled water will increase the aggressiveness of the attack.

Chloride corrosion results when the insulating material contains leachable chlorides, temperatures are above 140°F (60°C), and the substrate surface is a 300 series stainless steel. The attack will usually be a typical stress corrosion cracking. Corrosion is propagated when water enters the insulation and diffuses inward toward the hot surface, eventually finding a "dry" spot. Adjacent to this dry spot will be found an area in which the pores of the insulation are filled with a saturated salt solution in which chlorides may be present. As the vessel wall cools down, this saturated area "moves" into the metal wall. When the wall is reheated it will be temporarily in contact with the saturated salt solution and any chlorides present. This will initiate stress corrosion cracking.

The most efficient means of preventing chloride corrosion under insulation is to use the proper insulating materials. Insulation that meets ASTM C-795 or MIL-I-24244 specifications can safely be used over 300 series stainless steels. Care must be taken, however, to prevent chlorides in the atmosphere from impregnating the insulation.

As with acidic or alkaline corrosion, care must be taken when installing the outer insulation jacket and barrier, making sure that all joints are properly sealed.

Galvanic corrosion under insulation occurs when wet insulation, that has an electrolyte salt present, allows a current to flow between dissimilar metals such as the insulated metal surface and the outer metal jacket or other metallic accessories. This can result in the corroding of the metal jacket or of the vessel, depending upon which is the less noble metal.

Galvanic corrosion can be prevented by using a cellular insulation and applying a synthetic rubber or plastic jacket over the insulation. Hypalon performs well in this application.

Summary

Corrosion under insulation can be prevented by taking into account

Insulation selection
Equipment design
Weather barriers

The type of insulation to be used will be dependent upon the application, keeping in mind the types of corrosion that can occur. Equipment design also enters into the picture. Adequate insulation supports should be provided and additional protection should be considered where leakage or mechanical damage is possible. Additional flashing should also be installed where spills or hosing down are prevalent.

Weather barriers are a must because all corrosion under insulation requires moisture of some kind. Therefore it becomes a necessity to make sure that all joints are properly sealed and that an adequate weather barrier is installed.

CRACK INDUCING AGENTS

Crack inducing agents can be either active or passive. Most crack inducing agents require the presence of an electrolyte to become active. Passive agents, though present, cause no harm. For example hydrogen sulfide requires the presence of liquid water or some other electrolyte to become an active agent. On the other hand ammonia does not require the presence of an electrolyte to attack copper.

Hydrogen Sulfide

Wet hydrogen sulfide can cause several forms of hydrogen cracking including sulfide stress corrosion cracking (SSCC), hydrogen induced cracking (HIC), and stress oriented hydrogen induced cracking (SOHIC). SSCC occurs in many steels and alloys. HIC occurs in "dirty" steels. It is not necessary for stress to be present for this cracking to initiate. SOHIC is a stress assisted form of HIC. It usually occurs in the heat affected zones of restrained welds where residual stresses probably assist the cracking operation.

Wet hydrogen sulfide corrosion is the beginning of the hydrogen cracking. As the sulfide ion combines with iron to form iron sulfide, hydrogen is released as a corrosion by-product. The sulfide ion being a cathodic poison encourages two phenomena:

1. The nascent hydrogen tends to dissolve into the metal rather than combining with another hydrogen atom to form hydrogen gas.
2. Under normal circumstances this type of corrosion is rapidly slowed by the formation of a polarizing layer of hydrogen gas at the anode. However, the sulfide ion prevents such polarization. As a result, large amounts of nascent hydrogen are produced, as corrosion continues, until a thick film of dense iron sulfide forms, stopping further corrosion.

Sulfide Stress Cracking

Sulfide stress cracking is a form of hydrogen stress cracking. It can develop in areas of excessive metal hardness. The low to medium strength carbon steels are most resistant to SSCC. Microalloyed carbon steels should not be used in welded construction subject to SCC since they have a tendency to produce excessively hard weld heat affected zones. Such heat affected zones are difficult to temper by post weld heat treatment. It is difficult to generate weld metal or heat affected zone hardnesses to cause SSCC in low to medium strength carbon steels. When welding ordinary carbon steels, preheat and proper welding procedures will provide the necessary control to prevent excessive hardness. Post weld heat treatment is usually not necessary unless other crack inducing agents (such as amines) or other cathodic poisons (such as cyanide) are also present. In these situations reduction of resident stresses is necessary to reduce the susceptibility to stress cracking.

Hydrogen Induced Cracking

Hydrogen induced cracking is not technically a form of stress corrosion cracking but it is related. HIC occurs primarily in steel plates containing excessive amounts of nonmetallic inclusions (primarily manganese sulfides) which have been flattened by the rolling process. Hydrogen induced cracking is essentially a crack initiation mechanism. Nascent hydrogen diffuses into the steel as a result of hydrogen sulfide corroding the surface of the steel. Catalyst sites are established by nonmetallic inclusions, causing the diffusing nascent hydrogen to recombine into hydrogen gas. Pressure builds up adjacent to the inclusions as a result of the accumulating hydrogen gas, causing the inclusion matrix to split, initiating an HIC crack. A split can develop parallel to the surfaces of the steel plate when multiple initiation sites are present and the plate is relatively thin, less than ½ inch. On relatively thick plates staggered internal HIC cracks can link up and in extreme cases can cause through-thickness cracks.

Hydrogen induced cracking takes place in the temperature range of 32 to 130°F (0 to 55°C). HIC damage proceeds slowly above 130°F (55°C).

Stress Oriented Hydrogen Induced Cracking

Stress oriented hydrogen induced cracking (SOHIC) usually occurs in heat affected zones associated with the residual stresses of welds. The mechanics involves two components:

1. HTC cracks form in a stacked manner, producing a crack plane perpendicular to the surface of the plane. The cracks are generally short but closely spaced.
2. By shearing the ligaments between the stacked HIC cracks a through thickness crack develops. These cracks can be minimized by post weld heat treatment and normalizing.

Mercury

Intergranular cracking of copper alloys can be caused by liquid mercury, as well as intergranular cracking and pitting corrosion of aluminum. This effect is known as liquid metal embrittlement.

Zinc

Liquid metal embrittlement in iron and aluminum alloys can be caused by zinc. For this reason galvanized carbon steel should not be used at temperatures exceeding 390°F (200°C). Intergranular penetration of the steel substrate by zinc from the galvanized coating is possible. At temperatures exceeding 1380°F (750°C) molten zinc will rapidly attack the grain boundaries of austenitic stainless steel at a rate of inches per second.

The following conditions can cause failures:

1. Welding or cutting stainless steel components that have been coated with a zinc-rich product such as zinc paint.
2. Welding a galvanized steel part to an austenitic steel component without first removing the galvanizing adjacent to the weld preparation.

Higher alloys such as alloy 20Cb-3 and alloy 276 are not as susceptible as the conventional austenitic stainless steels.

Cyanides

Cyanides by themselves do not cause stress corrosion cracking. However, in combination with wet hydrogen sulfide they can increase the rate of sulfide stress corrosion cracking of carbon and low alloy steels. The hydrogen sulfide need only be present at any concentration greater than 20 ppm.

Cyanides can also accelerate the rate of wet hydrogen sulfide corrosion. Sulfide films are usually stable and limit corrosion, but the presence of cyanides converts the iron sulfide scale deposits into soluble salt complexes. This subjects the underlying carbon steel to rapid corrosion.

Chlorides

Chlorides are capable of causing stress corrosion cracking of austenitic stainless steels, providing the exposed surface is in tension. Residual tensile stresses can be the result of welding or of cold work. Such stresses can be relieved by solution annealing, post weld heat treatment, stress relief heat treatment, and shot peening which ensures that the exposed surface is in compression.

Austenitic stainless steels may be exposed to external chloride stress corrosion cracking as well as internal exposure from the process. External chloride attack can be the result of exposure to wet chlorides in atmospheric marine environments or to chlorides deposited extenally by wind, dust, or water.

Caustics

Caustics can cause stress corrosion cracking of carbon steel and low alloy steels. Under severe conditions caustics can also cause cracking of stainless steels and nickel based alloys. For moderate temperatures and concentrations carbon steel is the recommended material of construction. It is advisable to stress relieve or post weld heat treat carbon steel equipment to be used in caustic service.

Carbonates and Bicarbonates

Carbon steel is subject to stress corrosion cracking when combinations of carbonates and bicarbonates (either singly or in combination) exceed 1 weight percent. Post weld heat treatment does not prevent cracking since cracking generally occurs in the parent metal. A better choice is to upgrade to duplex or austenitic stainless steels, or for low pressure applications to use a nonmetallic material. Coatings are not recommended.

Amines

Amines can cause alkaline stress corrosion cracking in carbon steel. If the amine concentration exceeds two weight percent, all carbon steel components should be post weld heat treated regardless of service temperature. If the amine is fresh and uncontaminated, the heat treating may be eliminated. Some amines are rich in hydrogen sulfide and can cause various types of hydrogen related cracking.

Ammonia

Copper and copper alloys are subject to stress cracking in the presence of anhydrous liquid ammonia if there is less than 0.1 weight percent of water present.

See Refs. 33–35, 10, and 36.

CREVICE CORROSION

Crevice corrosion is a localized type of corrosion occurring within or adjacent to narrow gaps or openings formed by metal to metal or metal to nonmetal contact. When this occurs, attack is usually more severe than on surrounding areas of the same surface. This form of corrosion can develop as the result of a deficiency of oxygen in the crevice, or depletion of an inhibitor.

Prevention can be accomplished by proper design and operating procedures. Nonabsorbant gasketing material should be used at flanged joints, while fully penetrated butt welded joints are preferable to threaded joints. In the design of tankage, butt welded joints are preferable to lap joints. If lap joints are used, the laps should be filled with fillet welding or a suitable caulking compound designed to prevent crevice corrosion.

See Refs. 10 and 37.

CRITICAL CREVICE CORROSION TEMPERATURE

The critical crevice corrosion temperature of an alloy is that temperature at which crevice corrosion is first observed when immersed in a ferric chloride solution. Listed below are the critical crevice corrosion temperatures of several alloys in 10% ferric chloride solution.

Alloy	Temperature °F/°C
Type 316	27/–3
Alloy 825	27/–3
Type 317	36/2
Alloy 904L	59/15
Alloy 220S	68/20
E-Brite	70/21
Alloy G	86/30
Alloy 625	100/38
AL-6XN	100/43
Alloy 276	130/55

Table C.13 Range of Physical and Mechanical Properties of Cycoloy Based on the Specific Formulation

Property	Value	Units
Specific gravity	1.12–1.8	
Water absorption, 24 h at 73°F (23°C)	0.07–0.20	%
Tensile strength, type 1		
0.125 in. (3.2 mm) yield	6.5–9.1	psi × 10^3
break	7.25	psi × 10^3
Tensile elongation, type 1		
0.125 in. (3.2 mm) yield	4.0–5.0	%
break		%
Tensile modulus type 1		
0.125 in. (3.2 mm)	2.8–3.9	psi × 10^5
Flexural strength		
0.125 in. (3.2 mm) yield	11.2–14.8	psi
break	12–13.2	psi
Flexural modulus		
0.125 in. (3.2 mm)	3.0–4.0	psi
0.250 in. (6.4 mm)	3.85	psi
Hardness, Rockwell R	115	

CRITICAL PITTING TEMPERATURE

The critical pitting temperature of an alloy is the temperature of a solution at which pitting is first observed. These temperatures are usually determined in ferric chloride (10% $FeCl_3 - 6H_2O$) and in an acidic mixture of chlorides and sulfates.

CYCOLOY

Cycoloy is G. E. Plastics' trademark for their polycarbonate/acrylonitrile-butadiene-styrene thermoplastic alloy. It is available in several blends. The alloy can be formulated to maintain its impact and ductility below –40°F (–40°C). See Table C.13 for the range of physical and mechanical properties based on the specific formulation.

REFERENCES

1. I Suzuki. Corrosion Resistant Coatings Technology. New York: Marcel Dekker, 1989.
2. H Leidheiser Jr. Coatings. In: F Mansfield, ed. Corrosion Mechanisms. New York: Marcel Dekker, 1987, pp. 165–209.
3. PK Whitcraft. Corrosion of stainless steels. In: PA Schweitzer, ed. Corrosion Engineering Handbook. New York: Marcel Dekker, 1996, pp. 53–77.
4. PA Schweitzer. Corrosion Resistance Tables. 4th ed. Vols. 1–3. New York: Marcel Dekker, 1995.
5. PF Lafyates. Carbon and graphite. In: BJ Moniz and WL Pollock, eds. Process Industries Corrosion—Theory and Practice. Houston: NACE International, 1986, pp. 703–770.

6. JL Gossett. Corrosion resistance of cast alloys. In: PA Schweitzer, ed. Corrosion Engineering Handbook. New York: Marcel Dekker, 1996, pp. 268–273.
7. GW George and PG Breig. Cast alloys. In: PA Schweitzer, ed. Corrosion and Corrosion Protection Handbook. 2nd ed. New York: Marcel Dekker, 1989, pp. 296-301.
8. FC Porter. Corrosion Resistance of Zinc and Zinc Alloys. New York: Marcel Dekker, 1994.
9. PA Schweitzer. Cathodic protection. In: PA Schweitzer, ed. Corrosion and Corrosion Protection Handbook. 2nd ed. New York: Marcel Dekker, 1989, pp. 33–45.
10. HH Uhlig. Corrosion and Corrosion Control. New York: John Wiley 1963.
11. PK Whitcraft. Fundamental of metallic corrosion. In: PA Schweitzer, ed. Corrosion Engineering Handbook. New York: Marcel Dekker, 1996, pp. 11–12.
12. RA McCauley. Corrosion of Ceramics. New York: Marcel Dekker, 1995.
13. EL Liening and JM Macki. Aqueous corrosion of advanced ceramics. In: PA Schweitzer, ed. Corrosion Engineering Handbook. New York: Marcel Dekker, 1996, pp. 419–458.
14. PA Schweitzer. Corrosion Resistant Piping Systems. New York: Marcel Dekker, 1994.
15. PA Schweitzer. Corrosion Resistance of Elastomers. New York: Marcel Dekker, 1990.
16. KB Tator. Coatings. In: PA Schweitzer, ed. Corrosion, and Corrosion Protection Handbook. New York: Marcel Dekker, 1989, pp. 453–490.
17. W Funk. Prog Org Coating 9:29, 1981.
18. KR Gowers and D Scautlebury. Corros Sci 23:935, 1983.
19. W Funk. Ind Eng Chem Prod Res Dev 24:343, 1985.
20. JHW de Wit. Inorganic and organic coatings. In: P Marcus and J Oudar, eds. Corrosion Mechanisms in Theory and Practice. New York: Marcel Dekker, 1995, pp. 581–628.
21. GT Murray. Introduction to Engineering Materials. New York: Marcel Dekker, 1993.
22. DM Berger. Fundamentals and prevention of metallic corrosion. In: PA Schweitzer, ed. Corrosion and Corrosion Protection Handbook. 2nd ed. New York: Marcel Dekker, 1989, pp. 1–22.
23. P Marcus and J Oudar. Corrosion Mechanisms in Theory and Practice. New York: Marcel Dekker, 1995.
24. V Kucera and E Mattsson. Atmospheric corrosion. In: F Mansfield, ed. Corrosion Mechanisms. New York: Marcel Dekker, 1987.
25. JM Ciesiweicz. Copper and copper alloys. In: PA Schweitzer, ed. Corrosion and Corrosion Protection Handbook. 2nd ed. New York: Marcel Dekker, 1989, pp. 125–152.
26. DJ Duquette. Corrosion fatigue. In: M Florian, ed. Corrosion Mechanisms. New York: Marcel Dekker, 1987, pp. 367–397.
27. PA Schweitzer. Corrosion inhibitors. In: PA Schweitzer, ed. Corrosion and Corrosion Protection Handbook. 2nd ed. New York: Marcel Dekker, 1989, pp. 47–50.
28. G Trabanelli. Corrosion inhibitors. In: F Mansfield, ed. Corrosion Mechanisms. New York: Marcel Dekker, 1987, pp. 119–163.
29. F Mansfield. Corrosion Mechanisms. New York: Marcel Dekker, 1987.
30. CG Arnold and PA Schweitzer. Corrosion-testing techniques. In: PA Schweitzer, ed. Corrosion and Corrosion Protection Handbook. 2nd ed. New York: Marcel Dekker, 1989, pp. 587–618.
31. A Perkins. Corrosion monitoring. In: PA Schweitzer, ed. Corrosion Engineering Handbook. New York: Marcel Dekker, 1996, pp. 623–652.
32. RD Kane. Corrosion testing. In: PA Schweitzer, ed. Corrosion Engineering Handbook. New York: Marcel Dekker, 1996; pp. 607–621.
33. RC Newman. Stress corrosion cracking mechanisms. In: P Marcus and J Oudar, eds. Corrosion Mechanisms in Theory and Practice. New York: Marcel Dekker, 1995, pp. 311–372.
34. CP Dillon. Corrosion Control in the Chemical Process Industries. 2nd ed. St. Louis: Materials Technology Institute of the Chemical Process Industries, 1994.
35. CP Dillon. Corrosion Resistance of Stainless Steels. New York: Marcel Dekker, 1995.
36. DA Hansen and RB Puyear. Materials Selection for Hydrocarbon and Chemical Plants. New York: Marcel Dekker, 1996.
37. H Bohni. Localized corrosion. In: F Mansfield, ed. Corrosion Mechanisms. New York: Marcel Dekker, 1987, p. 285.

D

DEALLOYING

See "Dezincification."

DECARBURIZATION

See "Hydrogen Attack."

DEPOSIT ATTACK

See "Poultice Corrosion."

DEPOSIT CORROSION

See "Poultice Corrosion."

DEW POINT CORROSION

Dew point corrosion is a form of attack that occurs when the temperature of the metal surface is below the dew point of the atmosphere. This can occur outdoors during the night when the surface temperature may decrease by radiant heat transfer between the metal structure and the sky. It is also possible to have dew formation in the early morning when the temperature of the air increases faster than the temperature of the metal surface. Dew may also form when metal products are brought into warm storage after cold transport.

Dew point corrosion can also take place in the low temperature sections of fossil fuel power plant combustion equipment resulting from acidic flue gas vapors that condense and cause corrosion damage.

DEZINCIFICATION (DEALLOYING)

When one element in a solid alloy is removed by corrosion, the process is known as dezincification or dealloying. The most common example is the removal of zinc from brass alloys. When the zinc corrodes preferentially, a porous residue of copper and corrosion products remain. The corroded part often retains its original shape and may appear undamaged except for surface tarnish. However, its tensile strength and particularly its ductility have been seriously reduced.

See Refs. 1–3.

DIFFERENTIAL AERATION CELL

See "Oxygen Concentration Cell."

DISSIMILAR METAL CORROSION

See "Galvanic Corrosion."

DRY CORROSION

See "Gaseous Corrosion."

DURALUMIN

Duralumin is a heat treatable aluminum–copper alloy developed in Germany in 1919. It is produced in the United States as alloy 2017, containing approximately 4% copper and usually lesser amounts of magnesium, manganese, and on occasion silicon. Recently lithium has also been added.

See Ref. 4.

DUCTILE (NODULAR) IRON

Ductile iron not only retains all of the attractive qualities of gray iron, such as machineability and corrosion resistance, but also provides additional strength, toughness, and ductility. Ductile iron differs from gray iron in that its graphite form is spheroidal, or nodular, instead of the flake form found in gray iron. Due to its nodular graphite form, ductile iron has approximately twice the strength of gray iron as determined by tensile, beam, ring bending, and bursting tests. Its tensile and impact strength and elongation are many times greater than that of gray iron.

The corrosion resistance of ductile iron is essentially the same as that of gray iron. It exhibits good resistance to alkaline solutions, such as sodium hydroxide and molten caustic soda. It is also resistant to alkaline salt solutions, such as cyanides, carbonates, sulfides, and silicates. Acids and oxidizing salts rapidly attack ductile iron.

See Refs. 5 and 6.

DUPLEX STAINLESS STEELS

While the austenitic stainless steels have a wide range of corrosion resistance, they do exhibit some shortcomings, such as in the area of chloride stress corrosion cracking, pitting, and maintaining corrosion resistance after welding. The duplex stainless steels were produced to improve upon these shortcomings.

The duplex stainless steels are alloys whose micro-structures are a mixture of austenite and ferrite. The original duplex stainless steels did not have nitrogen added specifically as an alloying ingredient. By adding 0.15–0.25% nitrogen, the chromium partitioning between the two phases is reduced, resulting in an improvement in the pitting and crevice corrosion resistance of the austenite. This nitrogen addition also improves the weldability of the stainless steel without losing any of its corrosion resistance. It is not necessary to heat treat these stainless steels after welding.

The high chromium and molybdenum contents of the duplex stainless steels are particularly important in providing resistance in oxidizing environments and are also responsible for exceptionally good pitting and crevice corrosion resistance, especially in chloride environments. In general, these stainless steels have greater pitting resistance than alloy 904L.

The resistance to crevice corrosion of the duplexes is superior to the resistance of the 300 series austenitics. They also provide a greater resistance to crevice corrosion cracking. The duplexes are resistant to chloride stress corrosion cracking in chloride-containing environments. However, under very severe conditions, such as boiling magnesium chloride, the duplexes will crack.

The duplexes have less resistance in reducing environments than the austenitics as a result of the lower nickel content. This reduction in nickel content is necessary in order to achieve the desired microstructure. The high chromium and molybdenum contents partially offset this loss, and consequently they can be used in some reducing environments of dilute concentrations and lower temperatures.

Type 329

This is one of the original duplexes. Because this alloy has a high carbon content there is a considerable loss in corrosion resistance as a result of welding. Postweld heat treatment is required to maintain corrosion resistance.

Alloy 2205

This duplex resists oxidizing mineral acids and most organic acids in addition to reducing acids and chloride environments. Refer to Table D.1 for the physical and mechanical properties.

7Mo Plus*

The general corrosion resistance of 7Mo Plus is superior to those of stainless steel such as type 304 type and 316 in many environments. The high chromium content provides good corrosion resistance in strongly oxidizing media such as nitric acid. The corrosion resistance of this alloy is extended into the less oxidizing environments as a result of the molybdenum content. The combination of the chromium and molybdenum content imparts a high level of resistance to pitting and crevice corrosion. This alloy also provides excellent chloride stress corrosion cracking resistance in applications where type 304 and type 316

*Registered trademark of Carpenter Technology Corp.

Table D.1 Mechanical and Physical Properties of Alloy 2205 Duplex Stainless Steel

Modulus of elasticity × 10^6, psi	29.0
Tensile strength × 10^3, psi	90
Yield strength 0.2% offset × 10^3, psi	65
Elongation in 2 in., %	25
Hardness, Rockwell	C30.5
Density, lb/in.^{3}	0.283
Specific gravity	7.83
Thermal conductivity at 70°F (20°C), Btu/h °F	10
Thermal expansion coefficient at 68–212°F, in./in. °F × 10^{-6}	7.5

stainless steels would be used for general corrosion resistance. Refer to Table D.2 for physical and mechanical properties.

See Refs. 7–10.

DURIRON

See "High Silicon Iron."

Table D.2 Mechanical and Physical Properties of Type 7 Mo Plus[a] Stainless Steel

Modulus of elasticity × 10^6, psi	29.0
Tensile strength × 10^3, psi	90
Yield strength 0.2% offset × 10^3, psi	70
Elongation in 2 in., %	20
Hardness, Rockwell	C30.5
Density, lb/in.^{3}	0.280
Specific gravity	7.74
Specific heat, (75–212°F), Btu/lb °F	0.114
Thermal conductivity, Btu/h °F	
at 70°F (20°C)	8.8
at 1500°F (815°C)	12.5
Thermal expansion coefficient, in./in. °F × 10^{-6}	
at 75–400°F	6.39
at 75–600°F	6.94
at 75–800°F	7.49
at 75–1000°F	7.38
Charpy V-notch impact at 75°F (20°C), ft-lb	101

[a]Registered trademark of Carpenter Technology Corp.

REFERENCES

1. FC Porter. Corrosion Resistance of Zinc and Zinc Alloys. New York: Marcel Dekker, 1994.
2. PA Schweitzer. Corrosion Resistant Piping Systems. New York: Marcel Dekker, 1994.
3. HH Uhlig. Corrosion and Corrosion Control. New York: John Wiley, 1963.
4. BW Lifka. Corrosion of aluminum and aluminum alloys. In PE Schweitzer, ed. Corrosion Engineering Handbook. New York: Marcel Dekker, 1996, pp. 121–122.
5. GW George and PG Breig. Cast alloys. In: PA Schweitzer, ed. Corrosion and Corrosion Protection Handbook. 2nd ed. New York: Marcel Dekker, 1989, pp. 285–289.
6. JL Gossett. Corrosion resistance of cast alloys. In: PA Schweitzer, ed. Corrosion Engineering Handbook. New York: Marcel Dekker, 1996, pp. 258–259.
7. PA Schweitzer. Stainless steels. In: PA Schweitzer, ed. Corrosion and Corrosion Protection Handbook. 2nd ed. New York: Marcel Dekker, 1989, pp. 82–85.
8. CP Dillon. Corrosion Resistance of Stainless Steels, New York: Marcel Dekker, 1995.
9. RC Newman. Stress-corrosion cracking mechanisms. In: Corrosion Mechanisms in Theory and Practice. P Marcus and J Oudar, eds. New York: Marcel Dekker, 1995, pp. 331–332.
10. CP Dillon. Corrosion Control in the Chemical Process Industries. 2nd ed. St. Louis: Materials Technology Institute of the Chemical Process Industries, 1994.

E

E-GLASS

This is a boroalumiosilicate glass used for reinforcing thermosetting resins. See "Thermoset Reinforcing Materials."

ELASTOMERS

Also see "Permeation," "Adsorption," and "Polymers." An elastomer is generally considered to be any material, either natural or synthetic, that is elastic or resilient and in general resembles natural rubber in feeling and appearance. A more detailed technical definition is provided by ASTM, which states

> An elastomer is a polymeric material which at room temperature can be stretched to at least twice its original length and upon immediate release of the stress will return quickly to its original length.

These materials are sometimes referred to as rubbers.

Natural rubber is a polymerized hydrocarbon whose commercial synthesis proved to be difficult. Synthetic rubbers now produced are similar to but not identical with natural rubber. Natural rubber has the hydrocarbon butadiene as the simplest unit. Butadiene, $CH_2 = CH — CH = CH_2$, has two unsaturated linkages and is easily polymerized. It is produced commercially by cracking petroleum and also from ethyl alcohol. Natural rubber is a polymer of methyl butadiene (isoprene):

$$\begin{array}{l} \quad\quad\; CH_3 \\ \quad\quad\quad | \\ CH_2{=}C{-}C{=}CH_2 \end{array}$$

When butadiene or its derivatives become polymerized, the units link together to form long chains that each contain over 1000 units. In early attempts to develop a synthetic rubber it was found that simple butadiene does not yield a good grade of rubber, apparently because the chains are too smooth and do not interlock sufficiently strongly. Better results are obtained by introducing side groups into the chain either by modifying butadiene or by making a copolymer of butadiene and some other compound.

As development work continued in the production of synthetic rubbers, other compounds were used as the parent material in place of butadiene. Two of them were isobutylene and ethylene.

Elastomers are primarily composed of large molecules that tend to form spiral threads, similar to coiled springs, that are attached to each other at frequent intervals. These coils tend to stretch or compress when a small stress is applied but exert an increasing resistance to the application of additional stress. This phenonema is illustrated by the reaction of rubber to the application of additional stress.

In the raw state elastomers tend to be soft and sticky when hot, and hard and brittle when cold. Compounding increases the utility of rubber and synthetic elastomers. Vulcanization extends the temperature range within which they are flexible and elastic. In addition to vulcanizing agents, ingredients are added to make elastomers stronger, tougher, or harder, to make them age better, to color them, and in general to improve specific properties to meet specific application needs. The following examples illustrate some of the important properties that are required of elastomers and the typical services that require these properties:

Resistance to abrasive wear: automobile tire treads, conveyor belt covers, soles and heels, cables, hose covers

Resistance to tearing: tire treads, footwear, hot water bags, hose covers, belt covers, O-rings

Resistance to flexing: auto tires, transmission belts, V-belts, mountings, footwear

Resistance to high temperature: auto tires, belts conveying hot materials, steam hose, steam packing, O-rings

Resistance to cold: airplane parts, automotive parts, auto tires, refrigeration hose, O-rings

Minimum heat buildup: auto tires, transmission belts, V-belts, mountings

High resilience: sponge rubber, mountings, elastic bands, thread, sandblast hose, jar rings, O-rings

High rigidity: packing, soles and heels, valve cups, suction hose, battery boxes

Long life: fire hose, transmission belts, tubing

Electrical resistivity: electrician's tape, switchboard mats, electrician's gloves, wire insulation

Electrical conductivity: hospital flooring, nonstatic hose, matting

Impermeability to gases: balloons, life rafts, gasoline hose, special diaphragms

Resistance to ozone: ignition distributor gaskets, ignition cables, windshield wipers

Resistance to sunlight: wearing apparel, hose covers, bathing caps, windshield wipers

Resistance to chemicals: tank linings, hose for chemicals

Resistance to oils: oil-suction hose, paint hose, creamery hose, packing house hose, special belts, tank linings, special footwear

Stickiness: cements, electrician's tape, adhesive tapes, pressure sensitive tapes

Low specific gravity: airplane parts, forestry hose, balloons

Lack of odor or taste: milk tubing, brewery and winery hose, nipples, jar rings

Acceptance of color pigments: ponchos, life rafts, welding hose

Table E.1 provides a comparison of the important properties of the most common elastomers. Specific values of each property will be found in the section dealing with each elastomer. Tensile strength and elongation as applied to elastomers are defined by the American Society for Testing and Materials as follows:

Tensile strength is the force per unit of the original cross-sectional area which is applied at the time of the rupture of the specimen.

Elongation or strain is the extension between benchmarks produced by a tensile force applied to a specimen and is expressed as a percentage of the original distance between the marks. Ultimate elongation is the elongation at the time of rupture.

The procedure for conducting tensile tests is standardized and described in ASTM D412. Dumbbell-shaped specimens 4 or 5 inches long are die-cut from flat sheets and marked in the narrow section with benchmarks 1 or 2 in. apart. The ends of a specimen are placed in the grips of a vertical testing machine. The lower grip is power driven at 20 in./min and stretches the specimen until it breaks. As the distance between benchmarks widens, measurements are taken between their centers to determine elongation.

Tension tests are frequently conducted before and after an exposure test to determine the relative resistance of a group of compounds to deterioration by such things as oil, sunlight, weathering, ozone, heat, oxygen, and chemicals. Even a small amount of deterioration results in appreciable changes in tension properties.

Hardness, as applied to elastomeric products, is defined as relative resistance of the surface to indentation by a Shore A durometer. In this device the indenter point projects upward from the flat bottom of the case, held in the zero position by a spring. When pressed against a sample, the indenter point is pushed back into the case against the spring: This motion is translated through a rack and pinion mechanism into movement of the pointer on the durometer dial. Hardness numbers from a durometer scale for typical products are as follows:

Faucet washer, flooring, typewriter platen	90 ± 5
Shoe sole	80 ± 5
Solid tire, heel	70 ± 5
Tire tread, hose cover, conveyor belt cover	60 ± 5
Inner tube, bathing cap	50 ± 5
Rubber band	40 ± 5

Erasers and printing rolls usually have hardness values below 30.

Compression set is permanent deformation that remains in the elastomer after a compression force has been removed. The area of the elastomer that has compression set is not only permanently deformed but also less resilient than normal. The possibility of compression set occurring increases with increasing temperature, compression force, and length of time that the force is applied. Each type of elastomer has a different resistance to compression set. As with any material, each elastomer also has a limiting temperature range within which it may be used. Table E.2 shows the allowable operating temperature range of each of the common elastomers.

Causes of Failure

Elastomeric materials can fail as the result of chemical action and/or mechanical damage. Chemical deterioration occurs as the result of a chemical reaction between the elastomer and the medium or by the absorption of the medium into the elastomer. This attack results in the swelling of the elastomer and a reduction in its tensile strength.

The degree of deterioration is a function of the temperature and concentration of the corrodent. In general the higher the temperature and the higher the concentration of the corrodent, the greater will be the chemical attack. Elastomers, unlike metals, absorb varying quantities of the material they are in contact with, especially organic liquids. This

Table E.1 Comparative Properties of Elastomers[a]

Property	Natural rubber (NR)	Isoprene (IR)	Neoprene (CR)	Butadiene-styrene (Buna S)	Nitrile-NBR (Buna N)	Butyl (IIR)	Chlorobutyl (CIIR)	Hypalon (CSM)	Polybutadiene (BR)
Abrasion resistance	E	E	G	G	G	G	G	G	E
Acid resistance	P	P	GE	P	F	G	P	E	G
Chemical resistance									
Aliphatic hydrocarbons	P	P	G	P	E	P	E	G	P
Aromatic hydrocarbons	P	P	F	P	G	P	P	F	P
Oxygentaed (ketones, etc.)	G	G	P	G	P	G	P	P	G
Oil and gasoline	P	P	FG	P	E	P	E	G	P
Animal and vegetable oils	PG	PG	G	PG	E	E	G	G	PG
Resistance to									
Water absorption	E	E	G	GE	FG	G	G	GE	G
Ozone	P	P	GE	P	P	E	E	E	P
Sunlight aging	P	P	E	P	P	G	G	E	P
Heat aging	P	G	G	F	G	G	G	E	G
Flame	P	P	G	P	P	P	P	G	P
Electrical properties	G	G	F	E	F	E	PF	G	E
Impermeability	G	G	G	GE	G	E	G	E	G
Compression set resistance	E	E	F	G	GE	P	F	G	G
Tear resistance	GE	GE	FG	P	FG	GE	G	G	G
Tensile strength	E	E	G	G	GE	G	G	GE	G
Water/steam resistance	E	E	F	E	FG	E	PF	G	E
Weather resistance	P	P	E	P	F	E	F	E	G
Adhesion to metals	E	E	E	E	E	G	G	E	E
Adhesion to fabrics	E	E	E	E	G	G	G	E	E
Rebound									
Cold	E	E	E	G	G	P	F	G	E
Hot	E	E	E	G	G	G	F	G	E

[a]E, excellent; G, good; F, fair; P, poor; PF, poor to fair; PG, poor to good; GE, good to excellent

Table E.1 Continued

Ethylene-acrylic (EA)	Acrylcate-butadiene (ABR)	Acrylic ester-acrylic halide (ACM)	Ethylene-propylene (EPDM)	Styrene-butadiene-styrene (SBS)	Styrene-ethylene-butylene-styrene (SEBS)	Polysulfide (ST)	Polysulfide (FA)	Urethane (AU)	Polyamides	Polyester (PE)	Thermoplastic (TPE)	Silicone (Si)	Fluorosilicone (FSI)	Vinylidene fluoride (HFP)	Fluoroelastomers (FKM)	Ethylene-tetrafluoroethylene (ETFE)	Ethylene-chlorotrifluoroethylene (ECTFE)	Perfluoroelastomers (FPM)
E	G	FG	GE	G	GE	PF	PF	E	E	E	G	P	P	G	G	G	E	G
PG	G	P	G	E	E	F	G	P	FG	F	G	FG	FG	E	E	E	E	E
G	E	E	P	P	P	E	E	E	G	E	P	G	E	G	E	E	E	E
F	P	P	P	P	P	E	E	P	G	G	P	P	FG	G	E	E	G	E
P	P	P	GE	P	P	G	G	P	G	F	P	P	PF	G	P	E	G	FG
G	E	E	P	F	FG	E	E	E	G	G	FG	P	G	E	E	E	E	E
G	G	G	G	P	G			E	FG	G	G	G	G	G	E	E	E	E
G	G	G	GE	E	E	G		G	G	G	E	G	G	G	GE	G	E	G
E	E	E	E	P	E	E	E	E	E	E	E	E	E	E	E	E	E	E
E	E	G	E	P	G	E	E	P	P	G	G	E	E	E	GE	E	E	E
G	E	G	E	G	E	P	G	F	G	E	G	E	E	E	E	E	E	E
P	P	P	P	P	PG	P	P	P	F	G	P	F	G	E	G	G		E
FG	G	PF	G	F	E	G	G	FG	G	FG	F	E	E	F	G	G	E	G
G	G	G	G	G	G	E	E	G	FG	G	P	P	G	G	E	G	G	E
G	G	FG	GE	G	G	P	P	F	G	F	G	GE	GE	GE	GE			G
G	G	G	GE	G	G	P	P	GE	G	E	G	P	P	F	F	G		P
G	G	G	GE	G	G	F	G	E	G	E	G	P	F	GE	GE	G	G	G
G	G	PF	E	FG	FG	E	G	P	G	FG	G	F	F	FG	G	G		G
G	E	F	E	F	G	E	E	G	E	G	E	E	E	E	E	E	E	E
G	G	G	GE	G	G	E	E	G	E	G	G	G	GE	G	GE	P		
G	G	G	G	G	G	G	G	G	E	G	G	E	E		GE	F		
P	G	F	G	G	G	P	P	E	G	G	G	G	G		G			
F	G	F	G	G	G	P	P	E	G	E	G	G	G		E			

Table E.2 Operating Temperature Range of Common Elastomers

	Temperature range			
	°F		°C	
Elastomer	Min	Max	Min	Max
NR, natural rubber	–59	175	–50	80
IR, isoprene rubber	–59	175	–50	80
CR, neoprene rubber	–13	203	–25	95
SBR, Buna-S	–66	175	–56	80
NBR, nitrile rubber, Buna-N	–40	250	–40	105
IIR, butyl rubber	–30	300	–34	149
CIIR, chlorobutyl rubber	–30	300	–34	149
CSM, Hypalon	–20	250	–30	105
BR, polybutadiene rubber	–150	200	–101	93
EA, ethylene-acrylic rubber	–40	340	–40	170
ABR, acrylate-butadiene rubber	–40	340	–40	170
EPDM, ethylene-propylene	–65	300	–54	149
SBS, styrene-butadiene-styrene		150		65
SEBS, styrene-ethylene-butylene-styrene	–102	220	–75	105
ST, polysulfide	–50	212	–45	100
FA, polysulfide	–30	250	–35	121
AU, polyurethane	–65	250	–54	121
Polyamides	–40	300	–40	149
PE, polyesters	–40	302	–40	150
TPE, thermoplastic elastomers	–40	277	–40	136
SI, silicone	–60	450	–51	232
FSI, fluorosilicone	–100	375	–73	190
HEP, vinylidene fluoride	–40	450	–40	232
FKM, fluoroelastomers	–10	400	–18	204
ETFE, ethylene-tetrafluoroethylene elastomer	–370	300	–223	149
ECTFE, ethylene-chlorotrifluoroethylene elastomer	–105	340	–76	171
FPM, perfluoroelastomers	–58	600	–50	316

can result in swelling, cracking, and penetration to the substrate in an elastomer lined vessel. Swelling can cause softening of the elastomer and in a lined vessel introduce high stresses and failure of the bond. If an elastomeric lining has high absorption, permeation will probably result. Some elastomers, such as the fluorocarbons, are easily permeated but have little absorption. An approximation of the expected permation and/or absorption of an elastomer can be based on the absorption of water.

Permeation is a factor closely related to absorption but is a function of other physical effects, such as diffusion and temperature. All materials are somewhat permeable to chemical molecules, but the permeability rate of elastomers tends to be an order of magnitude greater than that in metals. This permeation has been a factor in elastomer lined vessels where corrodents have permeated the rubber and formed bubbles between the rubber lining and the steel substrate. Permeation and absorption can result in

1. Bond failure and blistering. These are caused by an accumulation of fluids at the bond when the substrate is less permeable than the lining or from the formation of corrosion or reaction products if the substrate is attacked by the corrodent.
2. Failure of the substrate due to corrosive attack.
3. Loss of contents through lining and substrate as the result of eventual failure of the substrate.

Thickness of lining is a factor affecting permeation. For general corrosion resistance, thicknesses of 0.010–0.020 in. are usually satisfactory, depending upon the combination of elastomeric material and specific corrodent. When mechanical factors such as thinning due to cold flow, mechanical abuse, and permeation rates are a consideration, thicker linings may be required.

Increasing the lining thickness will normally decrease permeation by the square of the thickness. Although this would appear to be the approach to follow to control permeation, there are disadvantages. First, as the thickness increases, the thermal stresses on the boundary increase, which can result in bond failure. Temperature changes and large differences in coefficients of thermal expansion are the most common causes of bond failure. Thickness and modulus of elasticity of the elastomer are two of the factors that would influence these stresses. Second, as the thickness of the lining increases, installation becomes more difficult, with a resulting increase in labor costs.

The rate of permeation is also affected by temperature and temperature gradient in the lining. Lowering these will reduce the rate of permeation. Lined vessels that are used under ambient conditions, such as storage tanks, provide the best service.

In unbonded linings it is important that the space between the liner and the support member be vented to the atmosphere, not only to allow the escape of minute quantities of permeant vapors but also to prevent the expansion of entrapped air from collapsing the liner.

Although elastomers can be damaged by mechanical means alone, this is not usually the case. When in good physical condition, an elastomer will exhibit abrasion resistance superior to that of metal. The actual size, shape, and hardness of the particles and their velocity are the determining factors in how well a particular rubber resists mechanical damage from the medium. Hard, sharp objects, including those foreign to the normal medium, may cut or gouge the elastomer. Most mechanical damage occurs as a result of chemical deterioration of the elastomer. When the elastomer is in a deteriorated condition, the material is weakened, and consequently it is more suscepitble to mechanical damage from flowing or agitated media.

Elastomers in outdoor use can be subject to degradation as a result of the action of ozone, oxygen, and sunlight. These three weathering agents can greatly affect the properties and appearance of a large number of elastomeric materials. Surface cracking, discoloration of colored stocks, serious loss of tensile strength and elongation, and other rubber like properties are the result of this attack.

Selecting an Elastomer

Many factors must be taken into account when selecting an elastomer for a specific application. First and foremost is the compatibility of the elastomer with the medium at the temperature and concentration to which it will be exposed.

It should also be remembered that each of the materials can be formulated to improve certain of its properties. However, the improvement in one property may have an adverse effect on another property, such as corrosion resistance. Consequently, specifications of an elastomer should include the specific properties required for the application, such as resilience, hysteresis, static or dynamic shear and compression modulus, flex fatigue and cracking, creep resistance to oils and chemicals, permeability, and brittle point, all in the temperature range to be encountered in service. This must also be accompanied by a complete listing of the concentrations of all media to be encountered. Providing this information will permit a competent manufacturer to supply an elastomer that will give years of satisfactory service. Because of the ability to change the formulation of many of these elastomers, the wisest policy is to permit a competent manufacturer to make the selection of the elastomer to satisfy the application. In addition to being able to change the formulation of each elastomer, it is also a common practice to blend two or more elastomers to produce a compound having specific properties. By so doing the advantageous properties of each elastomer can be made use of.

Fabrics are very often used as reinforcing members in conjunction with elastomers. Cotton, because of its ease of processing, availability in a wide range of weaves, and high adhesive strength, is the most widely used. It is also priced relatively low in comparison with synthetic fibers. The disadvantages of cotton are its poor heat resistance and the need for bulk in order to obtain the proper strength.

When operating temperatures of reinforced elastomeric products are in the range of 200–250°F, (93–120°C), DuPont's Dacron polyester fiber is used to provide good service life. In addition to better heat resisting qualities than cotton, Dacron has strength comparable to that of cotton with considerably less bulk. On the negative side, Dacron is more difficult to process than cotton, has lower adhesive strength, and is initially more expensive. Table E.3 provides the corrosion resistance of selected elastomers and selected corrodents.

Applications

Elastomeric or rubber materials find a wide range of applications. One of the major areas of application is that of linings for vessels. Both natural and synthetic materials are used for this purpose. These linings have provided many years of service in the protection of steel vessels from corrosion. They are sheet applied and bonded to a steel substrate.

These materials are also used extensively as membranes in acid brick lined vessels to protect the steel shell from corrosive attack. The acid brick lining in turn protects the elastomer from abrasion and excessive temperature. Another major use is as an impermeable lining for settling ponds and basins. These materials are employed to prevent pond contaminants from seeping into the soil and causing pollution of groundwater and contamination of the soil.

Natural rubber and most of the synthetic elastomers are unsaturated compounds that oxidize and deteriorate rapidly when exposed to air in thin films. These materials can be saturated by reacting with chlorine under the proper conditions, producing compounds that are clear, odorless, nontoxic, and noninflammable. They may be dissolved and blended

Table E.3 Corrosion Resistance of Selected Elastomers

Chemical	Butyl		Hypalon		EPDM		EPT		Viton A		Kalrez		Natural rubber		Neoprene		Buna N	
	°F	°C	°F	°C	°F	°C	°F	°C	°F	°C	°F	°C	°F	°C	°F	°C	°F	°C
Acetaldehyde	80	27	60	16	200	93	210	99	x	x	x	x	x	x	200	93	x	x
Acetamide			x	x	200	93	200	93	210	99	x	x	x	x	200	93	180	82
Acetic acid 10%	150	66	200	93	140	60	x	x	190	88	200	93	150	66	160	71	200	93
Acetic acid 50%	110	43	200	93	140	60	x	x	180	82	200	93	x	x	160	71	200	93
Acetic acid 80%	110	43	200	93	140	60	x	x	180	82	90	32	x	x	160	71	210	99
Acetic acid, glacial	90	32	x	x	140	60	x	x	x	x	80	27	x	x	x	x	100	38
Acetic anhydride	150	66	200	93	x	x	x	x	x	x	210	99	x	x	90	32	200	93
Acetone	160	71	x	x	300	148	x	x	x	x	210	99	x	x	x	x	x	x
Acetyl chloride			x	x	x	x	x	x	190	88	210	99	x	x	x	x	x	x
Acrylic acid									x	x	210	99	x	x	x	x	x	x
Acrylonitrile	x	x	140	60	140	60	100	38	x	x	110	43	90	32	160	71	x	x
Adipic acid	x	x	140	60	200	93	140	60	180	82	210	99	80	27	160	71	180	82
Allyl alcohol	190	88	200	93	300	148	80	27	190	88			80	27	120	49	180	82
Allyl chloride	x	x			x	x	x	x	100	38			x	x	x	x	x	x
Alum	190	88	200	93	200	93	140	60	190	88	210	99	150	66	200	93	200	93
Aluminum acetate			x	x	200	93	180	82	180	82	210	99			x	x	200	93
Aluminum chloride, aqueous	150	66	250	121	210	99	180	82	190	88	210	99	140	60	200	93	200	93
Aluminum chloride, dry											190	88						
Aluminum fluoride	180	82	200	93	210	99	180	82	180	82	210	99	150	66	200	93	190	88
Aluminum hydroxide	100	38	250	121	210	99	140	60	190	88	210	99			180	82	180	82
Aluminum nitrate	190	88	250	121	210	99	180	82	190	88	210	99	150	66	200	93	200	93
Aluminum oxychloride									x	x								
Aluminum sulfate	190	88	200	93	210	99	210	99	190	88	210	99	160	71	200	93	210	99
Ammonia gas			140	60	140	60	140	60	x	x	210	99	x	x	140	60	190	88
Ammonium bifluoride	x	x			300	148	140	60	140	60	210	99	x	x	x	x	180	82
Ammonium carbonate	190	88	140	60	300	148	180	82	190	88	210	99	150	66	200	93	200	93
Ammonium chloride 10%	190	88	200	93	210	99	180	82	190	88	210	99	150	66	200	93	200	93
Ammonium chloride 50%	190	88	200	93	210	99	180	82	190	88	210	99	150	66	190	88	200	93
Ammonium chloride, sat.	190	88	200	93	300	148	180	82	190	88	210	99	150	66	200	93	200	93
Ammonium fluoride 10%	150	66	200	93	210	99	210	99	140	60	210	99	160	71	100	38	200	93

Table E.3 Continued

Chemical	Butyl °F	Butyl °C	Hypalon °F	Hypalon °C	EPDM °F	EPDM °C	EPT °F	EPT °C	Viton A °F	Viton A °C	Kalrez °F	Kalrez °C	Natural rubber °F	Natural rubber °C	Neoprene °F	Neoprene °C	Buna N °F	Buna N °C
Ammonium fluoride 25%	150	66			300	148	140	60	140	60	210	99	80	27	200	93	120	49
Ammonium hydroxide 25%	190	88	250	121	100	38	140	60	190	88	210	99	x	x	200	93	200	93
Ammonium hydroxide, sat.	190	88	250	121	100	38	140	60	190	88	210	99	90	32	210	99	200	93
Ammonium nitrate	180	82	200	93	250	121	180	82	x	x	300	148	170	77	200	93	180	82
Ammonium persulfate	190	88	80	27	300	148	210	99	140	60	210	99	150	66	200	93	200	93
Ammonium phosphate	180	82	140	60	300	148	180	82	180	82	210	99	150	66	200	93	200	93
Ammonium sulfate 10–40%	190	88	200	93	300	148	180	82	180	82	210	99	150	66	200	93	200	93
Ammonium sulfide			200	93	300	148	210	99	x	x	210	99			160	71	180	82
Ammonium sulfite											210	99					160	71
Amyl acetate	x	x	60	16	210	99	x	x	x	x	210	99	x	x	x	x	x	x
Amyl alcohol	180	82	200	93	210	99	180	82	200	93	210	99	150	66	200	93	180	82
Amyl chloride			x	x	x	x	x	x	190	88	210	99	x	x	x	x	x	x
Aniline	150	66	140	60	140	60	x	x	230	110	250	121	x	x	x	x	x	x
Antimony trichloride	150	66	140	60	300	148	x	x	190	88	210	99			140	60		
Aqua regia 3:1					x	x	x	x	190	88	210	99	x	x	x	x	x	x
Barium carbonate			200	93	300	148	180	82	250	121	210	99	180	82	160	71	180	82
Barium chloride	190	88	250	121	250	121	180	82	190	88			150	66	200	93	200	93
Barium hydroxide	190	88	250	121	250	121	180	82	190	88	210	99	150	66	200	93	200	93
Barium sulfate			200	93	300	148	180	82	190	88	210	99	180	82	160	71	180	82
Barium sulfide	190	88	200	93	140	60	140	60	190	88	210	99	150	66	200	93	200	93
Benzaldehyde	90	32	x	x	150	66	x	x	x	x	210	99	x	x	x	x	x	x
Benzene	x	x	x	x	x	x	x	x	190	88	210	99	x	x	x	x	x	x
Benzene sulfonic acid 10%	90	32	x	x	x	x	x	x	170	77	210	99	x	x	100	38	x	x
Benzoic acid	150	66	200	93	x	x	140	60	190	88	310	154	150	66	200	93	x	x
Benzyl alcohol	190	88	140	60	x	x	x	x	350	177	210	99	x	x	x	x	140	60
Benxyl chloride	x	x	x	x	x	x	x	x	110	43	210	99	x	x	x	x	x	x
Borax	190	88	200	93	300	148	210	99	190	88	210	99	150	66	200	93	180	82
Boric acid	190	88	290	143	190	88	140	60	190	88	210	99	150	66	200	93	180	82
Bromine gas, dry			60	16	x	x	x	x			210	99			x	x	x	x
Bromine gas, moist			60	16	x	x									x	x	x	x

Bromine liquid			60	16	x	x	x	x	350	177	140	60			x	x	x	x
Butadiene			x	x	x	x	x	x	190	88	210	99			140	60	200	93
Butyl acetate	x	x	60	16	140	60	x	x	x	x	210	99	x	x	60	16	x	x
Butyl alcohol	140	60	250	121	200	93	180	82	250	121	240	116	150	66	200	93	x	x
n-Butylamine							x	x	x	x	210	99					80	27
Butyl phthalate							x	x	80	27	210	99	x				x	x
Butyric acid	x	x	x	x	140	60	x	x	120	49	x	x	x	x	x	x	x	x
Calcium bisulfide							x	x	190	88	210	99					180	82
Calcium bisulfite	120	49	250	121	x	x	x	x	190	88	210	99	120	49	180	82	200	93
Calcium carbonate	150	66	90	32	210	99	140	60	190	88	200	93	180	82	60	16	180	82
Calcium chlorate	190	88	90	32	140	60	140	60	190	88	210	99	150	66	200	93	200	93
Calcium chloride	190	88	200	93	210	99	180	82	190	88	210	99	150	66	200	93	180	82
Calcium hydroxide 10%	190	88	200	93	210	99	180	82	190	88	210	99	200	93	220	104	180	82
Calcium hydroxide, sat.	190	88	250	121	220	104	180	82	190	88	210	99	200	93	220	104	180	82
Calcium hypochlorite	190	88	250	121	210	99	180	82	190	88	210	99	200	93	220	104	80	27
Calcium nitrate	190	88	100	38	300	148	180	82	190	88	210	99	150	66	200	93	200	93
Calcium oxide			200	93	210	99									160	71	180	82
Calcium sulfate	100	38	250	121	300	148	180	82	200	93	210	99	180	82	160	71	180	82
Caprylic acid											210	99						
Carbon bisulfide			x	x	x	x	x	x	190	88	210	99	x	x	x	x	x	x
Carbon dioxide, dry	190	88	200	93	250	121	180	82	x	x	210	99	150	66	200	93	200	93
Carbon dioxide, wet	190	88	200	93	250	121	180	82	x	x	210	99	150	66	200	93	200	93
Carbon disulfide	190	88	230	110	250	121	180	82	x	x	210	99	150	66	x	x	200	93
Carbon monoxide	x	x	x	x	x	x	x	x	190	88	210	99	x	x	x	x	x	x
Carbon tetrachloride	90	32	200	93	250	121	180	82	190	88	210	99	x	x	200	93	180	82
Carbonic acid	x	x	x	x	x	x	x	x	350	177	210	99	x	x	x	x	x	x
Cellosolve	150	66			300	148	x	x	x	x	210	99	x	x	x	x	x	x
Chloroacetic acid, 50% water	150	66					x	x	x	x	210	99	x	x	x	x	x	x
Chloroacetic acid	160	71	x	x	160	71	x	x	x	x	210	99	x	x	x	x	x	x
Chlorine gas, dry	x	x	x	x	x	x	x	x	190	88	210	99	x	x	x	x	x	x
Chlorine gas, wet			90	32	x	x	x	x	190	88					x	x	x	x
Chlorine liquid	x	x			x	x	x	x	190	88	210	99	x	x	x	x	x	x
Chlorobenzene	x	x	x	x	x	x	x	x	190	88	210	99	x	x	x	x	x	x
Chloroform	x	x	x	x	x	x	x	x	190	88	210	99	x	x	x	x	x	x
Chlorosulfonic acid	x	x	x	x	x	x	x	x	x	x	210	99	x	x	x	x	x	x
Chromic acid 10%	100	38	150	66			x	x	350	177	210	99	x	x	140	60	190	88
Chromic acid 50%	x	x	160	71	x	x	x	x	350	177	210	99	x	x	100	38	190	88

Table E.3 Continued

Chemical	Butyl		Hypalon		EPDM		EPT		Viton A		Kalrez		Natural rubber		Neoprene		Buna N	
	°F	°C	°F	°C	°F	°C	°F	°C	°F	°C	°F	°C	°F	°C	°F	°C	°F	°C
Chromyl chloride											210	99						
Citric acid 15%	190	88	250	121	210	99	180	82			210	99	110	43	200	93	180	82
Citric acid, conc.	190	88	250	121	210	99	180	82	190	88	210	99	150	66	200	93	180	82
Copper acetate			x	x	100	38	100	38	x	x	210	99			160	71	180	82
Copper carbonate					210	99	210	99	190	88	210	99					x	x
Copper chloride	190	88	200	93	210	99	180	82	190	88	210	99	150	66	200	93	200	93
Copper cyanide			250	121	210	99	210	99	190	88	210	99	160	71	160	71	180	82
Copper sulfate	190	88	250	121	210	99	180	82	190	88	210	99	150	66	200	93	200	93
Cresol	x	x	x	x	x	x	100	38	x	x	210	99	x	x	x	x	x	x
Cupric chloride 5%			200	93	210	99	210	99	180	82					210	99	210	99
Cupric chloride 50%			200	93	210	99	210	99	180	82					160	71	180	82
Cyclohexane	x	x	x	x	x	x	x	x	190	88	210	99	x	x	x	x	180	82
Cyclohexanol			x	x	x	x	x	x	190	88	210	99			x	x	x	x
Dichloroacetic acid											210	99			x	x		
Dichloroethane	x	x	x	x	x	x	x	x	190	88	210	99	x	x	x	x	x	x
Ethylene glycol	190	88	200	93	210	99	180	82	350	177	210	99	150	66	160	71	200	93
Ferric chloride	190	88	250	121	220	104	180	82	190	88	210	99	150	66	160	71	200	93
Ferric chloride, 50% water	160	71	250	121	210	99	180	82	180	82	210	99	150	66	160	71	180	82
Ferric nitrate 10–50%	190	88	250	121	210	99	180	82	190	88	210	99	150	66	200	93	200	93
Ferrous chloride	190	88	250	121	200	93	180	82	180	82	210	99	150	66	90	32	200	93
Ferrous nitrate	190	88			210	99	180	82	210	99			150	66	200	93	200	93
Fluorine gas, dry	x	x	140	60			x	x	x	x	x	x	x	x	x	x	x	x
Fluorine gas, moist					60	16	100	38	x	x	x	x	x		x	x	x	x
Hydrobromic acid, dil.	150	66	90	32	90	32	140	60	190	88	210	99	100	38	x	x	x	x
Hydrobromic acid 20%	160	71	100	38	140	60	140	60	190	88	210	99	110	43	x	x	x	x
Hydrobromic acid 50%	110	43	100	38	140	60	140	60	190	88	210	99	150	66	x	x	x	x
Hydrochloric acid 20%	x	x	160	71	100	38	x	x	350	177	210	99	150	66	90	32	130	54
Hydrochloric acid 38%	x	x	140	60	90	32	x	x	350	177	210	99	160	71	90	32	x	x
Hydrocyanic acid 10%	140	60	90	32	200	93	x	x	190	88	210	99	90	32	x	x	200	93
Hydrofluoric acid 30%	350	177	90	32	60	16	140	60	210	99	210	99	100	38	200	93	x	x

Hydrofluoric acid 70%	150	66	90	32	x	x	x	x	350	177	210	99	x	x	200	93	x	x
Hydrofluoric acid 100%	x	x	90	32	x	x	x	x	60	16	210	99	x	x	x	x	x	x
Hypochlorous acid	x	x	x	x	300	148	140	60	190	88	190	88	150	66	x	x	x	x
Iodine solution 10%					140	60	140	60	190	88					80	27	80	27
Ketones, general			x	x	x	x			x	x	210	99	x	x	x	x	x	x
Lactic acid 25%	120	49	140	60	140	60	210	99	190	88	210	99	x	x	140	60	x	x
Lactic acid, concentrated	120	49	80	27			210	99	150	66	210	99	80	27	90	32	x	x
Magnesium chloride	200	93	250	121	250	121	180	82	180	82	150	66	150	66	210	99	180	82
Malic acid	x	x			x	x	80	26	190	88	210	99	80	27			180	82
Manganese chloride			180	82			210	99	180	82	210	99			200	93	100	38
Methyl chloride	90	32	x	x	x	x	x	x	190	88			x	x	x	x	x	x
Methyl ethyl ketone	100	38	x	x	80	27	x	x	x	x	210	99	x	x	x	x	x	x
Methyl isobutyl ketone	80	27	x	x	60	16	x	x	x	x	210	99	x	x	x	x	x	x
Muriatic acid	x	x	140	60			x	x	350	177	210	99			x	x	x	x
Nitric acid 5%	160	71	100	38	60	16	x	x	190	88	210	99	x	x	x	x	x	x
Nitric acid 20%	160	71	100	38	60	16	x	x	190	88	210	99	x	x	x	x	x	x
Nitric acid 70%	90	32	x	x	x	x	x	x	190	88	160	71	x	x	x	x	x	x
Nitric acid, anhydrous	x	x	x	x	x	x	x	x	190	88	x	x	x	x	x	x	x	x
Nitrous acid, concentrated	120	49							100	38	210	99	x	x	x	x	x	x
Oleum	x	x	x	x	x	x	x	x	190	88	210	99	x	x	x	x	x	x
Perchloric acid 10%	150	66	100	38	140	60	190	88	190	88	210	99	150	66			x	x
Perchloric acid 70%			90	32			140	60	190	88					x	x	x	x
Phenol	150	66	x	x			80	27	210	99	210	99	x	x	x	x	x	x
Phosphoric acid 50–80%	150	66	200	93	140	60	180	82	190	88	210	99	110	43			x	x
Picric acid			80	27	300	148	140	60	190	88	210	99	x	x	200	93	130	54
Potassium bromide 30%			250	121	210	99	180	82	190	88	210	99	160	71	160	71	180	82
Salicylic acid	80	27			210	99	180	82	190	88	210	99						
Silver bromide 10%											210	99						
Sodium carbonate	180	82	250	121	300	148	180	82	190	88	210	99	180	82	200	93	200	93
Sodium chloride	180	82	240	116	140	60	180	82	190	88	210	99	130	54	200	93	180	82
Sodium hydroxide 10%	180	82	250	121	210	99	210	99	x	x	210	99	150	66	200	93	160	71
Sodium hydroxide 50%	190	88	250	121	180	82	200	93	x	x	210	99	150	66	200	93	150	66
Sodium hydroxide conc.	180	82	250	121	180	82	80	27	x	x	210	99	150	66	200	93	150	66
Sodium hypochlorite 20%	130	54	250	121	300	148	x	x	190	88			90	32	x	x	x	x
Sodium hypochlorite	90	32			300	148	x	x	190	88			90	32	x	x	x	x
Sodium sulfide to 50%	150	66	250	121	300	148	210	99	190	88	210	99	150	66	200	93	180	82
Stannic chloride	150	66	90	32	300	148	210	99	180	82	210	99	150	66	210	99	180	82

Table E.3 Continued

Chemical	Butyl		Hypalon		EPDM		EPT		Viton A		Kalrez		Natural rubber		Neoprene		Buna N	
	°F	°C	°F	°C	°F	°C	°F	°C	°F	°C	°F	°C	°F	°C	°F	°C	°F	°C
Stannous chloride	150	66	200	93	280	138	210	99	190	88	210	99	150	66	160	71	180	82
Sulfuric acid 10%	150	66	250	121	150	66	210	99	350	171	240	116	150	66	200	93	150	66
Sulfuric acid 50%	150	66	250	121	150	66	210	99	350	177	210	99	100	38	200	93	200	93
Sulfuric acid 70%	100	38	160	71	140	60	210	99	350	177	150	66	x	x	200	93	x	x
Sulfuric acid 90%	x	x	x	x	x	x	80	27	350	177	150	66	x	x	x	x	x	x
Sulfuric acid 98%	x	x	110	43	x	x	x	x	350	177			x	x	x	x	x	x
Sulfuric acid 100%	x	x	x	x	x	x	x	x	190	88			x	x	x	x	x	x
Sulfuric acid, fuming			x	x	x	x	x	x			210	99			x	x	x	x
Sulfurous acid	150	66	160	71	x	x	180	82	190	88	210	99	x	x	x	x	x	x
Thionyl chloride	x	x							x	x	210	99	x	x	x	x	x	x
Toluene	x	x	x	x	x	x	x	x	190	88	80	27	x	x	x	x	150	66
Trichloroacetic acid	x	x			80	27	x	x	190	88	210	99	x	x	x	x	x	x
White liquor					300	148	180	82	190	88	210	99	x	x	140	60	140	60
Zinc chloride	190	88	250	121	300	148	180	82	210	99	210	99	150	66	160	71	190	88

The chemicals listed are in the pure state or in a saturated solution unless otherwise indicated. Compatibility is shown to the maximum allowable temperature for which data are available. Incompatibility is indicated by an x. A blank space indicates that data are unavailable.

Source: Schweitzer, Philip A. *Corrosion Resistance Tables*, 4th ed. Vols. 1–3. New York: Marcel Dekker, 1995.

with varnishes to impart high resistance to moisture and to the action of alkalies. This makes these products particularly useful in paints for concrete, where the combination of moisture and alkali causes the disintegration of ordinary paints and varnishes. These materials also resist mildew and are used to impart flame resistance and waterproofing properties to canvas. Application of these paints to steel will provide a high degree of protection against corrosion.

Large quantities of elastomeric materials are used to produce a myriad of products such as hoses, cable insulation, O-rings, seals and gaskets, belting, vibration mounts, flexible couplings, expansion joints, automotive and airplane parts, electrical parts and accessories, and many other items. With such a wide variety of applications requiring very diverse properties, it is essential that an understanding of the properties of each elastomer be acquired so that proper choices can be made.

See Refs. 1–3.

ELASTOMER CROSS REFERENCE

Generic name	Designation	Manufacturers'[a] common or trade names
Natural rubber	NR	26–31
Isoprene	IR	
Polychloroprene	CR	26–31, neoprene (1), Bayprene (2)
Butadiene-styrene	SBR	26–30, Buna-S, GR-S
Butadiene-acrylonitrile	NBR	16, 26–31, nitrile rubber, Buna-N, Perbunan (2), Nytek (21)
Butyl rubber	IIR	Gr-1, 26–30, Kalar (19)
Chlorobutyl rubber	CIIR	26–30
Carboxylic-acrylonitrile-butadiene	NBR	16, 26–31
Chlorosulfonated polyethylene	CSM	26–28, 30, 31, Hypalon (1)
Polybutadiene	BR	26–28, 30, 31, Buna-85, Buna-DB (2)
Ethylene-acrylic	EA	13, 28 Vamac (1)
Acrylate-butadiene	ABR	13, 28
Acrylic ester-acryic halide	ACM	13, 28
Ethylene-propylene	EPDM	26–31
	EPT	Nordel (1), Royalene, EPDM (8), Dutral (9)
Styrene-butadiene styrene	SBS	Kraton G (3)
Styrene-ethylene-butylene-styrene	SEBS	Kraton G (3)
Polysulfide	ST	27, 28, 30, Thiokol (4)
	FA	Blak-Stretchy (14), Blak-Tufy (14), Gra-Tufy (14)
Urethane	AU	16, 27, 30, 38, 31, Adiprene (1), Baytec (2), Futrathane (11), Conathane (16), Texion (2), Urane (23), Pellethane (22), pure CMC (14)
Polyamides	Nylon	Nylon (1), Rilsan (12), Vydyne (18), Plaskin (25)
Polyester	PE	Hytrel (1), Kodar (20)

Generic name	Designation	Manufacturers'[a] common or trade names
Thermoplastic elastomers	TPE	Duracryn (1), Flexsorb (17), Geolast (18), Kodapak (20), Santoprene (18), Zurcon (24)
Silicone	SI	27–29, 32, Cohrplastic (15), Green-Sil (14), Parshield (13), Baysilone (2), Blue-Sil (14)
Fluorosilicone	FSI	Parshield (13)
Vinylidene fluoride	HFP	Kynar (7), Foraflon (5)
Fluoroelastomers	FKM	24,26, 28–31, Viton (1), Fluorel (6), Technoflon (9)
Ethylene-tetrafluoroethylene	ETFE	Tefzel (1), Halon ET (9)
Ethylene-chlorotrifluoroethylene	ECTFE	Halar (9)
Perfluoroelastomers	FPM	Kalrez (1), Chemraz (10), Kel-F (6)

[a]List of manufacturers:
1. E. I. du Pont
2. Mobay Corp.
3. Shell Chemical Co.
4. Morton Thiokol Inc.
5. Atochem Inc.
6. 3-M Corp
7. Pennwalt Corporation
8. Uniroyal
9. Ausimont
10. Greene, Tweed & Co., Inc.
11. Futura Coatings Inc.
12. Attochem Inc.
13. Parker Seal Group
14. The Perma-Flex Mold Co.
15. CHR Industries
16. Conap Inc.
17. Polymer Corp.
18. Monsanto Co.
19. Hardman Inc.
20. Eastman Chemical Products Inc.
21. Edmont Div. of Becton, Dickinsen & Co.
22. Dow Chemical USA
23. Krebs Engineers
24. W. S. Shamban & Co.
25. Allied Signal
26. General Rubber Co.
27. Hecht Rubber Co.
28. Minor Rubber Co.
29. Newco Holz Rubber Co.
30. Alvan Rubber Co.
31. Burke Rubber Co.
32. Unaflex

ELECTROCHEMICAL CORROSION

Corrosion of metals is caused by the flow of energy (electricity). This flow may be from one metal to another metal, or from one part of the surface of a metal to another part of the same metal, or from a metal to a recipient of some kind. This flow of electricity can take place in the atmosphere, underwater, or underground as long as a moist conductor or electrolyte such as water or especially saltwater is present.

The difference in potential that causes the electric currents is mainly due to contact between dissimilar metallic conductors, or differences in concentration of the solution, generally related to dissolved oxygen in natural waters. Any lack of homogenity on the metal surface may initiate attack by causing a difference in potentials that results in localized corrosion.

The flow of electricity (energy) may also be from a metal to a metal recipient of some kind, such as soil. Soils frequently contain dispersed metallic particles or bacterial pockets that provide a natural electrical pathway with buried metal. The electrical path will be from the metal to the soil, with corrosion resulting.

The presence of water is a key factor for corrosion to take place. For example, in dry air, such as a desert location, the corrosion of steel does not take place, and when the relative humidity of air is below 30% at normal or lower temperatures, corrosion is negligible.

See Refs. 4–6.

ELECTROLYSIS

Electrolysis is the process of the passage of an electric current through a solution with simultaneous chemical changes either in the electrodes or in the solutions in contact with the electrodes, or both. Corrosion of a buried or immersed structure due to stray DC from an external source, such as an electric motor or welding machine, is known as electrolysis. The first observance of this phenonema was around electric railways which were designed to furnish DC on an overhead wire and to have the current return to its source via the track or third rail. If the current found it easier to return to its source via an underground sewer system or water line, it would protect the pipe at the point of entry but cause severe corrosion damage at the point of discharge via the soil to the power source.

Similar problems have been experienced when underground austenitic stainless steel pipe is welded in place in the form of localized pitting.

ELECTROLYTE

An electrolyte is any substance that, when in solution or fused, forms a liquid that will conduct an electric current. Acids, bases, and salts are common electrolytes.

EMBEDDED IRON CORROSION

Embedded iron corrosion occurs when during fabrication of stainless steel equipment iron is embedded in the stainless steel surface. When exposed to moist air, or wetted, the iron corrodes, leaving rust streaks and possibly initiating crevice corrosion attack in the stainless steel.

EMBRITTLEMENT

This is the severe loss of ductility or toughness in a material, which may result in cracking. Some metals, when stressed, crack on exposure to corrosive environments, but corrosion is not necessarily a part of crack initiation or crack growth. This type of failure is not properly called stress corrosion cracking.

The most frequent occurrence of this form of attack is in steel equipment handling solutions containing hydrogen sulfide. Under these conditions corrosion of the steel generates atomic hydrogen, which penetrates the steel and at submicroscopic discontinuity of pressures high enough to cause cracking or blistering. Failures of this kind are called "Hydrogen Cracking" or hydrogen stress cracking.

See Refs. 7–10.

ENAMELING

See "Glass Linings."

ENGINEERING PLASTIC

This term is used interchangeably with the terms engineering polymers and high performance polymers. The *ASM Handbook* defines engineering plastics as synthetic polymers of a resin based material that have load-bearing characteristics and high performance properties, which permit them to be used in the same manner as metals and ceramics. In other words these materials are plastics and polymeric compositions having well-defined mechanical properties such that engineering rather than empirical methods can be used for the design and manufacture of products that require definite and predictable performance in structural applications over a substantial temperature range. Many engineering polymers are reinforced and/or alloy polymers, blends of different polymers.

Among the engineering plastics are poly (*p*-phenyleneterephthalamide) (aromatic polyamide or aramid), polyaromatic ester, polyetherketone, polyphenylene sulfide, polyamide-imidepolyether sulfone, polyether-imide, polysulfone, and polyimide (thermoplastic).

Elastomers are cross-linked linear thermoplastic polymers and many fall into the engineering category. However, the major products of the polymer industry, such as polyethylene, polyvinyl chloride, polypropylene, and polystyrene, are not considered engineering products because of their low strength.

EPOXY RESINS

See also "Polymers" and "Thermoset Polymers." The epoxy resins are also a family of resins as are the vinyl ester resins. They exhibit good resistance to alkalies, nonoxidizing acids, and many solvents. Specifically they are compatible to acids such as 10% acetic, benzoic, butyric, 10% hydrochloric, 20% sulfuric, oxalic, and fatty acids. On the alkaline side they are compatible with 50% sodium hydroxide, 10% sodium sulfite, calcium hydroxide, trisodium phosphate, magnesium hydroxide, aluminum, barium, calcium, iron, magnesium, potassium, and sodium. Solvents such as methanol, ethanol, isopropanol, benzene, ethyl acetate, naphtha, toluene, and xylene can also be handled safely.

Bromine water, chromic acid, bleaches, fluorine, methylene chloride, hydrogen peroxide, sulfuric acid above 80%, wet chlorine gas, and wet sulfur dioxide will attack the epoxies. Refer to Table E.4 for the compatibility of epoxies with selected corrodents.

Refer to Ref. 4 for the compatibility of the epoxies over a wider range of selected corrodents.

See Refs. 3, 2, 11, and 12.

EROSION CORROSION

This type of corrosion is also referred to as impingement attack and is caused by contact with high velocity liquids; it results in a pitting type of corrosion. It is most prevalent in condenser tubes and piping fittings such as elbows and tees. Prevention can be accomplished by one or more means:

1. Reduce the velocity.
2. Select a harder material.
3. Properly design the piping system or the condenser.

Erosion corrosion also occurs when abrasive materials destroy or damage the passive film on the surface of the metal, thereby permitting corrosion to take place.

ESTERS

Esters are organic compounds formed by reaction between alcohols and acids. When the organic radical is not specified in the name, ethyl is often understood; e.g., acetic ester is ethylacetate.

ETHYLENE-ACRYLIC (EA) RUBBER

Ethylene-acrylic rubber is produced from ethylene and acrylic acid. As with other synthetic elastomers, the properties of the EA rubbers can be altered by compounding. Basically EA is a cost-effective hot-oil resistant rubber with good low temperature properties.

Physical and Mechanical Properties

Ethylene-acrylic elastomers have good tear strength and tensile strength and high elongation at break. In addition, exceptionally low compression set values are an added advantage, making the product suitable for many hose, sealing, and cut gasket applications. A unique feature is its practically constant damping characteristics over broad ranges of temperature, frequency, and amplitude. Very little change in damping value takes place between –4 and 320°F (–20 and 160°C). This property, which shows up as a poor rebound in resiliency tests, is actually a design advantage. Combined with EA's heat and chemical resistance it allows the use of EA in vibration damping applications. This elastomer provides heat resistance surpassed by only the more expensive polymers such as the fluorocarbon or fluorosilicone elastomers. In measurements of dry heat resistance, EA outlasts other moderately priced oil resistant rubbers. Parts retain elasticity and remain functional after continuous air-oven exposures from 18 months at 250°F (121°C) to 7 days at 400°F (204°C). Parts fabricated of EA will perform at least as long as parts made of Hypalon or general-purpose nitrile rubber, but at exposure temperatures 50–100°F (27°C) higher.

The low temperature performance of EA is inherently superior to that of most other heat and oil resistant rubbers, including standard fluoroelastomers, chlorosulfonated polyethylene, polyacrylates, and polyepichlorhydrin. Typical unplasticized compounds are flexible to –20°F (–29°C) and have brittle points as low as –75°F (–60°C).

Table E.4 Compatiblity of Epoxy with Selected Corrodents[a]

Chemical	Maximum temp. °F	Maximum temp. °C	Chemical	Maximum temp. °F	Maximum temp. °C
Acetaldehyde	150	66	Barium sulfide	300	149
Acetamide	90	32	Benzaldehyde	x	x
Acetic acid 10%	190	88	Benzene	160	71
Acetic acid 50%	110	43	Benzene sulfonic acid 10%	160	71
Acetic acid 80%	110	43	Benzoic acid	200	93
Acetic anhydride	x	x	Benzyl alcohol	x	x
Acetone	110	43	Benzyl chloride	60	16
Acetyl chloride	x	x	Borax	250	121
Acrylic acid	x	x	Boric acid 4%	200	93
Acrylonitrile	90	32	Bromine gas, dry	x	x
Adipic acid	250	121	Bromine gas, moist	x	x
Allyl alcohol	x	x	Bromine liquid	x	x
Allyl chloride	140	60	Butadiene	100	38
Alum	300	149	Butyl acetate	170	77
Aluminum chloride,			Butyl alcohol	140	60
aqueous 1%	300	149	*n*-Butylamine	x	x
Aluminum chloride, dry	90	32	Butyric acid	210	99
Aluminum fluoride	180	82	Calcium bisulfide		
Aluminum hydroxide	180	82	Calcium bisulfite	200	93
Aluminum nitrate	250	121	Calcium carbonate	300	149
Aluminum sulfate	300	149	Calcium chlorate	200	93
Ammonia gas, dry	210	99	Calcium chloride 37.5%	190	88
Ammonium bifluoride	90	32	Calcium hydroxide, sat.	180	82
Ammonium carbonate	140	60	Calcium hypochlorite 70%	150	66
Ammonium chloride, sat.	180	82	Calcium nitrate	250	121
Ammonium fluoride 25%	150	66	Calcium sulfate	250	121
Ammonium hydroxide 25%	140	60	Caprylic acid	x	x
Ammonium hydroxide, sat.	150	66	Carbon bisulfide	100	38
Ammonium nitrate 25%	250	121	Carbon dioxide, dry	200	93
Ammonium persulfate	250	121	Carbon disulfide	100	38
Ammonium phosphate	140	60	Carbon monoxide	80	27
Ammonium sulfate 10–40%	300	149	Carbon tetrachloride	170	77
Ammonium sulfite	100	38	Carbonic acid	200	93
Amyl acetate	80	27	Cellosolve	140	60
Amyl alcohol	140	60	Chloracetic acid, 92% water	150	66
Amyl chloride	80	27	Chloracetic acid	x	x
Aniline	150	66	Chlorine gas, dry	150	66
Antimony trichloride	180	82	Chlorine gas, wet	x	x
Aqua regia 3:1	x	x	Chlorobenzene	150	66
Barium carbonate	240	116	Chloroform	110	43
Barium chloride	250	121	Chlorosulfonic acid	x	x
Barium hydroxide 10%	200	93	Chromic acid 10%	110	43
Barium sulfate	250	121	Chromic acid 50%	x	x

Table E.4 Continued

Chemical	Maximum temp. °F	Maximum temp. °C	Chemical	Maximum temp. °F	Maximum temp. °C
Citric acid 15%	190	88	Methyl isobutyl ketone	140	60
Citric acid, 32%	190	88	Muriatic acid	140	60
Copper acetate	200	93	Nitric acid 5%	160	71
Copper carbonate	150	66	Nitric acid 20%	100	38
Copper chloride	250	121	Nitric acid 70%	x	x
Copper cyanide	150	66	Nitric acid, anhydrous	x	x
Copper sulfate 17%	210	99	Nitrous acid, concentrated	x	x
Cresol	100	38	Oleum	x	x
Cupric chloride 5%	80	27	Perchloric acid 10%	90	32
Cupric chloride 50%	80	27	Perchloric acid 70%	80	27
Cyclohexane	90	32	Phenol	x	x
Cyclohexanol	80	27	Phosphoric acid 50–80%	110	43
Dichloroacetic acid	x	x	Picric acid	80	27
Dichloroethane (ethylene dichloride)	x	x	Potassium bromide 30%	200	93
			Salicylic acid	140	60
Ethylene glycol	300	149	Sodium carbonate	300	149
Ferric chloride	300	149	Sodium, chloride	210	99
Ferric chloride 50% in water	250	121	Sodium hydroxide 10%	190	88
Ferric nitrate 10–50%	250	121	Sodium hydroxide 50%	200	93
Ferrous chloride	250	121	Sodium hypochlorite 20%	x	x
Ferrous nitrate			Sodium hypochlorite, concentrated	x	x
Fluorine gas, dry	90	32			
Hydrobromic acid, dilute	180	82	Sodium sulfide to 10%	250	121
Hydrobromic acid 20%	180	82	Stannic chloride	200	93
Hydrobromic acid 50%	110	43	Stannous chloride	160	71
Hydrochloric acid 20%	200	93	Sulfuric acid 10%	140	60
Hydrochloric acid 38%	140	60	Sulfuric acid 50%	110	43
Hydrocyanic acid 10%	160	71	Sulfuric acid 70%	110	43
Hydrofluoric acid 30%	x	x	Sulfuric acid 90%	x	x
Hydrofluoric acid 70%	x	x	Sulfuric acid 98%	x	x
Hydrofluoric acid 100%	x	x	Sulfuric acid 100%	x	x
Hypochlorous acid	200	93	Sulfuric acid, fuming	x	x
Ketones, general	x	x	Sulfurous acid 20%	240	116
Lactic acid 25%	220	104	Thionyl chloride	x	x
Lactic acid, concentrated	200	93	Toluene	150	66
Magnesium chloride	190	88	Trichloroacetic acid	x	x
Methyl chloride	x	x	White liquor	90	32
Methyl ethyl ketone	90	32	Zinc chloride	250	121

[a]The chemicals listed are in the pure state or in a saturated solution unless otherwise indicated. Compatibility is shown to the maximum allowable temperature for which data are available. Incompatibility is shown by an x. A blank space indicates that the data are unavailable.

Source: Schweitzer, Philip A. *Corrosion Resistance Tables*, 4th ed. Vols 1–3. New York: Marcel Dekker, 1995.

Table E.5 Physical and Mechanical Properties of Ethylene-Acrylic (EA) Rubber[a]

Specific gravity	1.08–1.12
Hardness, Shore A	40–95
Tensile strength, psi	2500
Elongation, % at break	650
Compression set, %	Good
Tear resistance	Good
Maximum temperature, continuous use	340°F (170°C)
Brittle point	–75°F (–60°C)
Water absorption, %/24 hr	Very low
Abrasion resistance	Excellent
Resistance to sunlight	Excellent
Effect of aging	Nil
Resistance to heat	Excellent
Dielectric strength	Good
Electrical insulation	Fair to good
Permeability to gases	Very low

[a]These are representative values since they may be altered by compounding.

Compounding EA with ester plasticizers will extend its low temperature flexibility limits to –50°F (–46°C). When exposed to flame, EA has poor flame resistance but does have low smoke emission.

The physical and mechanical properties of ethylene-acrylic rubber are given in Table E.5.

Resistance to Sun, Weather, and Ozone

The EA elastomers have extremely good resistance to sun, weather, and ozone. Long-term exposures have no effect on these rubbers.

Chemical Resistance

The ethylene-acrylic elastomers exhibit very good resistance to hot oils, to hydrocarbon or glycol based proprietary lubricants, and to transmission and power steering fluids. The swelling characteristics of EA will be retained better than those of the silicone rubbers after oil immersion.

Ethylene-acrylic rubber also has outstanding resistance to hot water. Its resistance to water absorption is very good. Good resistance is also displayed to dilute acids, aliphatic hydrocarbons, gasoline, and animal and vegetable oils. Ethylene-acrylic rubber is not recommended for immersion in esters, ketones, highly aromatic hydrocarbons, or concentrated acids. Neither should it be used in applications calling for long-term exposure to high pressure steam.

Applications

Ethylene-acrylic rubber is used in such products as gaskets, hoses, seals, boots, damping components, low smoke floor tiling, and cable jackets for offshore oil platforms, ships, and building plenum installations. Ethylene-acrylic rubber in engine parts provides good resistance to heat, fluids, and wear as well as good low temperature sealing ability.

See Refs. 1 and 3.

ETHYLENE-CHLOROTRIFLUOROETHYLENE (ECTFE)

ECTFE is a 1:1 alternating copolymer of ethylene and chlorotrifluorethylene. The chemical structure of this thermoplast is

$$
\begin{array}{ccccccc}
 & \mathrm{H} & & \mathrm{H} & & \mathrm{F} & & \mathrm{F} \\
 & | & & | & & | & & | \\
- & \mathrm{C} & - & \mathrm{C} & - & \mathrm{C} & - & \mathrm{C} \\
 & | & & | & & | & & | \\
 & \mathrm{H} & & \mathrm{H} & & \mathrm{F} & & \mathrm{Cl}
\end{array}
$$

This chemical structure provides the polymer with a unique combination of properties. It possesses excellent chemical resistance, a broad use temperature range from cryogenic to 340°F (171°C) with continuous service to 300°F (149°C), and excellent abrasion resistance.

ECTFE exhibits excellent impact strength over its entire operating range, even in the cryogenic range. It also possesses good tensile, flexural, and wear related properties. ECTFE is also one of the most corrosion resistant polymers. Other important properties include a low coefficient of friction and the ability to be pigmented. Table E.6 lists the physical and mechanical properties of ECTFE.

ECTFE is resistant to strong mineral and oxidizing acids, alkalies, metal etchants, liquid oxygen, and practically all organic solvents except hot amines (aniline, dimethylamine, etc.). ECTFE is not subject to chemically induced stress cracking from strong acids, bases, or solvents. Some halogenated solvents can cause ECTFE to become slightly plasticized when it comes into contact with them. Under normal circumstances this does not affect the usefulness of the polymer since upon removal of the solvent from contact and upon drying its mechanical properties return to their original values, indicating that no chemical attack has taken place. Like other fluoropolymers, ECTFE will be attacked by metallic sodium and potassium. Table E.7 lists the compatibility of ECTFE with selected corrodents. Reference 3 provides a wide range of compatibility of ECTFE with selected corrodents.

See Refs. 3 and 2.

ETHYLENE-CHLOROTRIFLUOROETHYLENE (ECTFE) ELASTOMER

Ethylene-chlorotrifluoroethylene (ECTFE) elastomer is a 1:1 alternating copolymer of ethylene and chlorotrifluoroethylene. This chemical structure gives the polymer unique combination of properties. It possesses excellent chemical resistance, good electrical properties, and a broad use temperature range (from cryogenic to 340°F [171°C]) and meets the requirements of the UL-94V-0 vertical flame test in thicknesses as low as 7 mils. ECTFE is a tough material with excellent impact strength over its entire operating range. Of all the fluoropolymers, ECTFE ranks among the best for abrasion resistance.

Most techniques used for processing polyethylene can be used to process ECTFE. It can be extruded, injection molded, rotomolded, and applied by ordinary fluidized bed or electrostatic coating techniques.

Table E.6 Physical and Mechanical Properties of ECTFE

Specific gravity	1.68
Water absorption, 24 h at 73°F/23°C, %	<0.01
Tensile strength at 73°F/23°C, psi	4500
Modulus of elasticity in tension at 73°F/23°C × 10^5 psi	2.4
Flexural strength, psi	7000
Izod impact strength, notched at 73°F/23°C, ft-lb/in.	No break
Linear coefficient of thermal expansion, in./in. °F at:	
–22 to 122°F/–30 to 50°C	4.4×10^{-5}
122 to 185°F/50 to 80°C	5.6×10^{-5}
185 to 257°F/85 to 125°C	7.5×10^{-5}
257 to 356°F/125 to 180°C	9.2×10^{-5}
Thermal conductivity, Btu/h/ft^2/°F/in.	1.07
Heat distortion temperature, °F/°C	
at 66 psi	195/91
at 264 psi	151/66
Limiting oxygen index, %	60
Underwriters lab rating, Sub 94	V-O

Physical and Mechanical Properties

ECTFE possesses advantageous electrical properties. It has high resistivity and low loss. The dissipation factor varies somewhat with frequency, particularly above 1 KHz. The AC loss properties of ECTFE are superior to those of vinilydene fluoride and come close to those of PTFE. The dielectric constant is stable across a broad temperature and frequency range. Refer to Table E.8.

According to ASTM D-149 the dielectric strength of ECTFE has a value of 2000 V/mil in 1 mil thickness and 500V/mil in 1/8 in. thickness, which are similar to those obtained for PTFE or polyethylene.

The resistance to permeation by oxygen, carbon dioxide, chlorine gas, or hydrochloric acid is superior to that of PTFE and FEP, being 10–100 times better. Water absorption is less than 0.1%.

Other important physical properties include low coefficient of friction, excellent machineability, and the ability to be pigmented. In thicknesses as low as 7 mils, ECTFE has a UL-94-V-0 rating. The oxygen index (ASTM D2863) is 60 on a 1/16 in. thick sample and 48 on a 0.0005 in. filament yarn. ECTFE is a strong, highly impact resistant material that retains useful properties over a wide range of temperatures. Outstanding in this respect are properties related to impact at low temperatures. ECTFE can be applied at elevated temperatures in the range of 300–340°F (149–171°C). (Refer to Table E.8).

In addition to its excellent impact properties. ECTFE also possesses good tensile, flexural, and wear related properties.

The resistance of ECTFE to degradation by heat is excellent. It can resist temperatures of 300–340°F (149–171°C) for extended periods of time without degradation. It is one of the most radiation resistant polymers. Laboratory testing has determined that the following useful life can be expected at the temperatures indicated:

Temperature °F	Temperature °C	Useful life years
329	165	10
338	170	4.5
347	175	2
356	180	1.25

Resistance to Sun, Weather, and Ozone

ECTFE is extremely resistant to sun, weather, and ozone attack. Its physical properties undergo very little change after long exposures.

Chemical Resistance

The chemical resistance of ECTFE is outstanding. It is resistant to most of the common corrosive chemicals encountered in industry. Included in this list of chemicals are strong mineral and oxidizing acids, alkalies, metal etchants, liquid oxygen, and practically all organic solvents except hot amines (aniline, dimethylamine, etc.). No known solvent dissolves or stress cracks ECTFE at temperatures up to 250°F (120°C).

Some halogenated solvents can cause ECTFE to become slightly plasticized when it comes into contact with them. Under normal circumstances this does not impair the usefulness of the polymer. When the part is removed from contact with the solvent and allowed to dry, its mechanical properties return to their original values, indicating that no chemical attack has taken place.

As with other fluoropolymers, ECTFE will be attacked by metallic sodium and potassium.

The useful properties of ECTFE are maintained on exposure to cobalt-60 radiation of 200 Mrads. Refer to Table E.7 for the compatibility of ECTFE with selected corrodents.

Applications

This elastomer finds many applications in the electrical industry as wire and cable insulation, jacketing plenum cable insulation, oil well wire and cable insulation, logging wire jacketing, jacketing for cathodic protection, aircraft, mass transit, and automotive wire, connectors, coil forms, resistor sleeves, wire tie wraps, tapes, tubing, flexible printed circuitry, and flat cable.

Applications are also found in other industries as diaphragms, flexible tubing, closures, seals, gaskets, and convoluted tubing and hose, particularly in the chemical, cryogenic, and aerospace industries.

Materials of ECTFE are also used for lining vessels, pumps, and other equipment.

See Refs. 3 and 1.

Table E.7 Compatiblity of ECTFE with Selected Corrodents[a]

Chemical	Maximum temp. °F	Maximum temp. °C	Chemical	Maximum temp. °F	Maximum temp. °C
Acetic acid 10%	250	121	Barium sulfide	300	149
Acetic acid 50%	250	121	Benzaldehyde	150	66
Acetic acid 80%	150	66	Benzene	150	66
Acetic acid, glacial	200	93	Benzene sulfonic acid 10%	150	66
Acetic anhydride	100	38	Benzoic acid	250	121
Acetone	150	66	Benzyl alcohol	300	149
Acetyl chloride	150	66	Benzyl chloride	300	149
Acrylonitrile	150	66	Borax	300	149
Adipic acid	150	66	Boric acid	300	149
Allyl chloride	300	149	Bromine gas, dry	x	x
Alum	300	149	Bromine liquid	150	66
Aluminum chloride, aqueous	300	149	Butadiene	250	121
Aluminum chloride, dry			Butyl acetate	150	66
Aluminum fluoride	300	149	Butyl alcohol	300	149
Aluminum hydroxide	300	149	Butyric acid	250	121
Aluminum nitrate	300	149	Calcium bisulfide	300	149
Aluminum oxychloride	150	66	Calcium bisulfite	300	149
Aluminum sulfate	300	149	Calcium carbonate	300	149
Ammonia gas	300	149	Calcium chlorate	300	149
Ammonium bifluoride	300	149	Calcium chloride	300	149
Ammonium carbonate	300	149	Calcium hydroxide 10%	300	149
Ammonium chloride 10%	290	143	Calcium hydroxide, sat.	300	149
Ammonium chloride 50%	300	149	Calcium hypochlorite	300	149
Ammonium chloride, sat.	300	149	Calcium nitrate	300	149
Ammonium fluoride 10%	300	149	Calcium oxide	300	149
Ammonium fluoride 25%	300	149	Calcium sulfate	300	149
Ammonium hydroxide 25%	300	149	Caprylic acid	220	104
Ammonium hydroxide, sat.	300	149	Carbon bisulfide	80	27
Ammonium nitrate	300	149	Carbon dioxide, dry	300	149
Ammonium persulfate	150	66	Carbon dioxide, wet	300	149
Ammonium phosphate	300	149	Carbon disulfide	80	27
Ammonium sulfate 10–40%	300	149	Carbon monoxide	150	66
Ammonium sulfide	300	149	Carbon tetrachloride	300	149
Amyl acetate	160	71	Carbonic acid	300	149
Amyl alcohol	300	149	Cellosolve	300	149
Amyl chloride	300	149	Chloracetic acid, 50% water	250	121
Aniline	90	32	Chloracetic acid	250	121
Antimony trichloride	100	38	Chlorine gas, dry	150	66
Aqua regia 3:1	250	121	Chlorine gas, wet	250	121
Barium carbonate	300	149	Chlorine, liquid	250	121
Barium chloride	300	149	Chlorobenzene	150	66
Barium hydroxide	300	149	Chloroform	250	121
Barium sulfate	300	149	Chlorosulfonic acid	80	27

Table E.7 Continued

Chemical	Maximum temp. °F	°C	Chemical	Maximum temp. °F	°C
Chromic acid 10%	250	121	Muriatic acid	300	149
Chromic acid 50%	250	121	Nitric acid 5%	300	149
Citric acid 15%	300	149	Nitric acid 20%	250	121
Citric acid, conc.	300	149	Nitric acid 70%	150	66
Copper carbonate	150	66	Nitric acid, anhydrous	150	66
Copper chloride	300	149	Nitrous acid, concentrated	250	121
Copper cyanide	300	149	Oleum	x	x
Copper sulfate	300	149	Perchloric acid 10%	150	66
Cresol	300	149	Perchloric acid 70%	150	66
Cupric chloride 5%	300	149	Phenol	150	66
Cupric chloride 50%	300	149	Phosphoric acid 50–80%	250	121
Cyclohexane	300	149	Picric acid	80	27
Cyclohexanol	300	149	Potassium bromide 30%	300	149
Ethylene glycol	300	149	Salicylic acid	250	121
Ferric chloride	300	149	Sodium carbonate	300	149
Ferric chloride 50% in water	300	149	Sodium chloride	300	149
Ferric nitrate 10–50%	300	149	Sodium hydroxide 10%	300	149
Ferrous chloride	300	149	Sodium hydroxide 50%	250	121
Ferrous nitrate	300	149	Sodium hydroxide, concentrated	150	66
Fluorine gas, dry	x	x	Sodium hypochlorite 20%	300	149
Fluorine gas, moist	80	27	Sodium hypochlorite, concentrated	300	149
Hydrobromic acid, dilute	300	149	Sodium sulfide to 50%	300	149
Hydrobromic acid 20%	300	149	Stannic chloride	300	149
Hydrobromic acid 50%	300	149	Stannous chloride	300	149
Hydrochloric acid 20%	300	149	Sulfuric acid 10%	250	121
Hydrochloric acid 38%	300	149	Sulfuric acid 50%	250	121
Hydrocyanic acid 10%	300	149	Sulfuric acid 70%	250	121
Hydrofluoric acid 30%	250	121	Sulfuric acid 90%	150	66
Hydrofluoric acid 70%	240	116	Sulfuric acid 98%	150	66
Hydrofluoric acid 100%	240	116	Sulfuric acid 100%	80	27
Hypochlorous acid	300	149	Sulfuric acid, fuming	300	149
Iodine solution 10%	250	121	Sulfurous acid	250	121
Lactic acid 25%	150	66	Thionyl chloride	150	66
Lactic acid, concentrated	150	66	Toluene	150	66
Magnesium chloride	300	149	Trichloroacetic acid	150	66
Malic acid	250	121	White liquor	250	121
Methyl chloride	300	149	Zinc chloride	300	149
Methyl ethyl ketone	150	66			
Methyl isobutyl ketone	150	66			

[a]The chemicals listed are in the pure state or in a saturated solution unless otherwise indicated. Compatibility is shown to the maximum allowable temperature for which data are available. Incompatibility is shown by an x. A blank space indicates that the data are unavailable.

Source: Schweitzer, Philip A. *Corrosion Resistance Tables*, 4th ed. Vols 1–3. New York: Marcel Dekker, 1995.

Table E.8 Physical and Mechanical Properties of ECTFE[a] Elastomer

Specific gravity	1.68
Refractive index, nD	1.44
Specific heat, Btu/lb-°F	0.28
Brittle point	–105°F (–76°C)
Insulation resistance, ohms	$>10^{15}$
Thermal conductivity at 203°F (93°C)	
Btu-in./h-ft^2-°F	1.09
Coefficient of linear expansion, °F^{-1} or °C^{-1}	
–22 to 122°F	4.4×10^{-5}
122 to 185°F	5.6×10^{-5}
185 to 257°F	7.5×10^{-5}
257 to 356°F	9.2×10^{-5}
–30 to 50°C	8×10^{-5}
50 to 85°C	10×10^{-5}
85 to 125°C	13.3×10^{-5}
125 to 180°C	16.5×10^{-5}
Dielectric strength, V/mil	
0.0001 in. thick	2000
$^1/_8$ in. thick	490
Dielectric constant	
at 60 Hz	2.6
at 10^3 Hz (1 kHz)	2.5
at 10^6 Hz (1 MHz)	2.5
Dissipation factor at 60 Hz	<0.0009
at 10^3 Hz (1 kHz)	0.005
at 10^6 Hz (1 MHz)	0.003
Arc resistance, s	135
Moisture absorption, %	<0.1
Tensile strength, psi	6000–7000
Elongation at break, % at room temperature	200–300
Hardness, Shore D	75
Impact resistance, ft-lb/in. notch	
at 73°F (23°C)	No break
at –40°F (–40°C)	2–3
Abrasion resistance, Armstrong (ASTM D1242)	
30 lb load, volume loss, cm^3	0.3
Coefficient of friction	
static	0.15
dynamic, 50 cm/s	0.65
Maximum temperature, continuous use	340°F (171°C)
Machining qualities	Excellent
Resistance to sunlight	Excellent
Effect of aging	Good
Resistance to heat	Good

[a]These are representative values since they may be altered by compounding.

ETHYLENE-PROPYLENE RUBBERS (EPDM and EPT)

Ethylene-propylene rubber is a synthetic hydrocarbon based rubber made either from ethylene-propylene diene monomer or from ethylene-propylene terpolymer. These monomers are combined so as to produce an elastomer with a completely saturated backbone and pendant unsaturation for sulfur vulcanization. As a result of this configuration, vulcanizates of EPDM elastomers are extremely resistant to attack by ozone, oxygen, and weather.

Ethylene-propylene rubber possesses many properties superior to those of natural rubber and conventional general-purpose elastomers. In some applications it will perform better than other materials, while in other applications it will last longer or require less maintenance and may even cost less.

EPDM has exceptional heat resistance, being able to operate at temperatures of 300–350°F (148–176°C), while also finding application at temperatures as low as –70°F (–56°C). Experience has shown that EPDM has exceptional resistance to steam. Hoses manufactured from EPDM have had lives several times longer than those of hoses manufactured from other elastomers.

Dynamic properties of EPDM remain constant over a wide temperature range, making this elastomer suitable for a variety of applications. It also has a very high resistance to sunlight, aging, and weather, excellent electrical properties, and good chemical resistance. However, being hydrocarbon based, it is not resistant to petroleum based oils or flame.

This material may be processed and vulcanized by the same techniques and with the same equipment as those used for processing other general-purpose elastomers. As with other elastomers, compounding plays an important part in tailoring the properties of EPDM to meet the needs of a specific application. Each of the properties of the elastomer can be enhanced or reduced by the addition or deletion of chemicals and fillers. Because of this, the properties discussed must be considered in general terms.

Ethylene-propylene terpolymer is a synthetic hydrocarbon based rubber produced from an ethylene-propylene terpolymer. It is very similar in physical and mechanical properties to EPDM.

Physical and Mechanical Properties

It is possible to compound EPDM to provide either higher resilience or higher damping. When compounded to provide high resilience the products are similar to natural rubber in liveliness and minimum hysteresis values. The energy absorbing compounds have low resilience values approaching those of specialty elastomers used for shock and vibration damping. Whether the compound has been compounded for resilience or high damping, its properties remain relatively constant over a wide temperature range. As can be seen from Table E.9, a temperature variation of 200°F (110°C) has little effect on the resilience of the compound. The isolation efficiency (based on the percentage of disturbing force transmitted) over a temperature range of 0–180°F (–18 to 82°C) varies by only 10%. This property is particularly important in vibration isolation applications such as in automotive body mounts.

EPDM exhibits little sensitivity to changes in load. When properly compounded it has excellent resistance to creep under both static and dynamic conditions.

Flexibility at low temperature is another advantage of this elastomer. Standard

Table E.9 Physical and Mechanical Properties of Ethylene-Propylene Rubber (EDPM)[a]

Specific gravity	0.85
Specific heat, cal/g	0.56
Brittle point	–90°F (–68°C)
Resilience, %	
at 212°F (100°C)	78
at 75°F (24°C)	77
at 14°F (–10°C)	63
Dielectric strength, V/mil	800
Insulation resistance, megohms/1000 ft	25,500
Insulation resistance, constant *K*, megohms/1000 ft	76,400
Permeability to air at 86°F (30°C), cm^3/cm^2-cm-s-atm	8.5×10^{-8}
Tensile strength, psi	To 3500
Elongation, % at break	560
Hardness, Shore A	30–90
Abrasion resistance	Good
Maximum temperature, continuous use	300°F (148°C)
Impact resistance	Good
Compression set, %	
at 158°F (70°C)	8–10
at 212°F (100°C)	12–26
Resistance to sunlight	Excellent
Effect of aging	Nil
Resistance to heat	Excellent
Tear resistance	Good

[a]These are representative values since they may be altered by compounding.

compounds have brittle points of –90°F (–68°C) or below. Special compounding can supply material with stiffness values of –90°F (–68°C) and brittle points below –100°F (–73°C).

The electrical properties of EPDM are excellent, particularly for high voltage insulation. The properties are also stable after long periods of immersion in water. Excellent resistance is also provided against cutting caused by high voltage corona discharge.

Ethylene-propylene rubber can be produced in any color including white and pastel shades. The color stability is excellent, with aging characteristics that are available in other elastomers only in black. Since techniques have been developed whereby the material can be painted with permanent waterproof colors, the elastomer can be produced in a black stock providing the maximum physical properties.

Ethylene-propylene rubber has a relatively high resistance to heat. Standard formulations can be used continuously at temperatures of 250–300°F (121–148°C) in air. In the absence of air, such as in a steam hose lining or cable insulation covered with an outer jacket, higher temperatures can be tolerated. It is also possible by special compounding to produce material that can be used in services up to 350°F (176°C). Standard compounds can be used in intermittent service at 350°F (176°C).

Other advantageous properties of EPDM include good resistance to impact, tearing, abrasion, and cut growth over a wide temperature range. These properties make the elastomer suitable for applications that involve continuous flexing or twisting during operation.

Ethylene-propylene rubber exhibits a low degree of permenant deformation. Table E.9 gives examples of these values. The ranges shown are for both standard and special compounds. In addition, compounds can be supplied that will have a permanent deformation of only 26% after compression at 350°F (176°C).

Resistance to Sun, Weather, and Ozone

Ethylene-propylene rubber is particularly resistant to sun, weather, and ozone attack. Excellent weather resistance is obtained whether the material is formulated in color, white, or black. The elastomer remains free of surface crazing and retains a high percentage of its properties after years of exposure. Ozone resistance is inherent in the polymer, and for all practical purposes it can be considered immune to ozone attack. It is not necessary to add any special compounding ingredients to produce this resistance.

Chemical Resistance

Ethylene-propylene rubber resists attack from oxygenated solvents such as acetone, methyl ethyl ketone, ethyl acetate, weak acids and alkalies, detergents, phosphate esters, alcohols, and glycols. It exhibits exceptional resistance to hot water and high pressure steam. The elastomer, being hydrocarbon based, is not resistant to hydrocarbon solvents or oils, chlorinated hydrocarbons, or turpentine. However, by proper compounding, its resistance to oil can be improved to provide adequate service life in many applications where such resistance is required. Ethylene-propylene terpolymer rubbers are in general resistant to most of the same corrodents as EPDM but do not have as broad a resistance to mineral acids and some organics. Table E.10 lists the compatibility of EPDM rubber with selected corrodents.

Applications

Extensive use is made of ethylene-propylene rubber in the automotive industry. Because of its paintability, this elastomer is used as the gap-filling panel between the grills and the bumper, which provides a durable and elastic element. Under-hood components such as radiator hose, ignition wire insulation, overflow tubing, window washing tubing, exhaust emission control tubing, and various other items make use of EPDM because of its resistance to heat, chemicals, and ozone. Other automotive applications include body mounts, spring mounting pads, miscellaneous body seals, floor mats, and pedal pads. Each application takes advantage of one or more specific properties of the elastomer.

Appliance manufacturers, especially washer manufacturers have also found wide use for EPDM. Its heat and chemical resistance combined with its physical properties make it ideal for such applications as door seals and cushions, drain and water circulating hoses, bleach tubing, inlet nozzles, boots, seals, gaskets, diaphragms, vibration isolators, and a variety of grommets. The elastomer is also used in dishwashers, refrigerators, ovens, and a variety of small appliances.

Ethylene-propylene rubber finds application in the electrical industry and in the manufacture of electrical equipment. One of the primary applications is as an insulating material. It is used for medium voltage (up to 35 kV) and secondary network power cable,

Table E.10 Compatiblity of EDPM Rubber with Selected Corrodents[a]

Chemical	Maximum temp. °F	Maximum temp. °C	Chemical	Maximum temp. °F	Maximum temp. °C
Acetaldehyde	200	93	Barium sulfate	200	93
Acetamide	200	93	Barium sulfide	140	60
Acetic acid 10%	140	60	Benzaldehyde	150	66
Acetic acid 50%	140	60	Benzene	x	x
Acetic acid 80%	140	60	Benzene sulfonic acid 10%	x	x
Acetic acid, glacial	140	60	Benzoic acid	x	x
Acetic anhydride	x	x	Benzyl alcohol	x	x
Acetone	200	93	Benzyl chloride	x	x
Acetyl chloride	x	x	Borax	200	93
Acrylonitrile	140	60	Boric acid	190	88
Adipic acid	200	93	Bromine gas, dry	x	x
Allyl alcohol	200	93	Bromine gas, moist	x	x
Allyl chloride	x	x	Bromine liquid	x	x
Alum	200	93	Butadiene	x	x
Aluminum fluoride	190	88	Butyl acetate	140	60
Aluminum hydroxide	200	93	Butyl alcohol	200	93
Aluminum nitrate	200	93	Butyric acid	140	60
Aluminum sulfate	190	88	Calcium bisulfite	x	x
Ammonia gas	200	93	Calcium carbonate	200	93
Ammonium bifluoride	200	93	Calcium chlorate	140	60
Ammonium carbonate	200	93	Calcium chloride	200	93
Ammonium chloride 10%	200	93	Calcium hydroxide 10%	200	93
Ammonium chloride 50%	200	93	Calcium hydroxide, sat.	200	93
Ammonium chloride, sat.	200	93	Calcium hypochlorite	200	93
Ammonium fluoride 10%	200	93	Calcium nitrate	200	93
Ammonium fluoride 25%	200	93	Calcium oxide	200	93
Ammonium hydroxide 25%	100	38	Calcium sulfate	200	93
Ammonium hydroxide, sat.	100	38	Carbon bisulfide	x	x
Ammonium nitrate	200	93	Carbon dioxide, dry	200	93
Ammonium persulfate	200	93	Carbon dioxide, wet	200	93
Ammonium phosphate	200	93	Carbon disulfide	200	93
Ammonium sulfate 10–40%	200	93	Carbon monoxide	x	x
Ammonium sulfide	200	93	Carbon tetrachloride	200	93
Amyl acetate	200	93	Carbonic acid	x	x
Amyl alcohol	200	93	Cellosolve	200	93
Amyl chloride	x	x	Chloracetic acid	160	71
Aniline	140	60	Chlorine gas, dry	x	x
Antimony trichloride	200	93	Chlorine gas, wet	x	x
Aqua regia 3:1	x	x	Chlorine, liquid	x	x
Barium carbonate	200	93	Chlorobenzene	x	x
Barium chloride	200	93	Chloroform	x	x
Barium hydroxide	200	93	Chlorosulfonic acid	x	x

Table E.10 Continued

Chemical	Maximum temp. °F	°C	Chemical	Maximum temp. °F	°C
Chromic acid 50%	x	x	Malic acid	x	x
Citric acid 15%	200	93	Methyl chloride	x	x
Citric acid, concentrated	200	93	Methyl ethyl ketone	80	27
Copper acetate	100	38	Methyl isobutyl ketone	60	16
Copper carbonate	200	93	Nitric acid 5%	60	16
Copper chloride	200	93	Nitric acid 20%	60	16
Copper cyanide	200	93	Nitric acid 70%	x	x
Copper sulfate	200	93	Nitric acid, anhydrous	x	x
Cresol	x	x	Oleum	x	x
Cupric chloride 5%	200	93	Perchloric acid 10%	140	60
Cupric chloride 50%	200	93	Phosphoric acid 50–80%	140	60
Cyclohexane	x	x	Picric acid	200	93
Cyclohexanol	x	x	Potassium bromide 30%	200	93
Dichloroethane			Salicylic acid	200	93
(ethylene dichloride)	x	x	Sodium carbonate	200	93
Ethylene glycol	200	93	Sodium chloride	140	60
Ferric chloride	200	93	Sodium hydroxide 10%	200	93
Ferric chloride 50% in water	200	93	Sodium hydroxide 50%	180	82
Ferric nitrate 10–50%	200	93	Sodium hydroxide,		
Ferrous chloride	200	93	concentrated	180	82
Ferrous nitrate	200	93	Sodium hypochlorite 20%	200	93
Fluorine gas, moist	60	16	Sodium hypochlorite,		
Hydrobromic acid, dilute	90	32	concentrated	200	93
Hydrobromic acid 20%	140	60	Sodium sulfide to 50%	200	93
Hydrobromic acid 50%	140	60	Stannic chloride	200	93
Hydrochloric acid 20%	100	38	Stannous chloride	200	93
Hydrochloric acid 38%	90	32	Sulfuric acid 10%	150	66
Hydrocyanic acid 10%	200	93	Sulfuric acid 50%	150	66
Hydrofluoric acid 30%	60	16	Sulfuric acid 70%	140	60
Hydrofluoric acid 70%	x	x	Sulfuric acid 90%	x	x
Hydrofluoric acid 100%	x	x	Sulfuric acid 98%	x	x
Hypochlorous acid	200	93	Sulfuric acid 100%	x	x
Iodine solution 10%	140	60	Sulfuric acid, fuming	x	x
Ketones, general	x	x	Toluene	x	x
Lactic acid 25%	140	60	Trichloroacetic acid	80	27
Lactic acid, concentrated			White liquor	200	93
Magnesium chloride	200	93	Zinc chloride	200	93

[a]The chemicals listed are in the pure state or in a saturated solution unless otherwise indicated. Compatibility is shown to the maximum allowable temperature for which data are available. Incompatibility is shown by an x. A blank space indicates that the data are unavailable.

Source: Schweitzer, Philip A. *Corrosion Resistance Tables*, 4th ed. Vols 1–3. New York: Marcel Dekker, 1995.

Table E.11 Physical and Mechanical Properties of ETFE

Specific gravity	1.70
Tensile strength, psi	6500
Modulus of elasticity, psi × 10^5	2.17
Elongation, %	300
Flexural modulus, psi × 10^5	1.7
Impact strength, ft-lb/in.	No break
Hardness, Shore D	67
Water absorption, 24 h at 73°F/23°C, %	<0.03
Thermal conductivity, Btu/h/ft^2/°F/in.	1.6
Heat distortion temperatue, °F/°C	
at 66 psi	220/104
at 264 psi	160/71
Limiting oxygen index, %	30
Underwriters lab rating, Sub 94	V-O

coverings for line and multiplex distribution wire, jacketing and insulation for types S and SJ flexible cords, and insulation for automotive ignition cable.

Accessory items such as molded terminal covers, plugs, transformer connectors, line tap and switching devices, splices, and insulating and semiconductor tape are also produced from EPDM.

Medium and high voltage underground power distribution cable insulated with EPDM offers many advantages. It provides excellent resistance to tearing and failure caused by high voltage contaminants and stress. Its excellent electrical properties make it suitable for high voltage cable insulation. It withstands heavy corona discharge without sustaining damage.

Manufacturers of other industrial products take advantage of the heat and chemical resistance, physical durability, ozone resistance, and dynamic properties of EPDM. Typical applications include high pressure steam hose, high temperature conveyor belting, water and chemical hose, hydraulic hose for phosphate type liquids, vibration mounts, industrial tires, tugboat and dock bumpers, tank and pump linings, O-rings, gaskets, and a variety of molded products. Standard formulations of EPDM are also used for such consumer items as garden hose, bicycle tires, sporting goods, and tires for garden equipment.

See Refs. 1 and 3.

ETHYLENE-TETRAFLUOROETHYLENE (ETFE)

This thermoplast is sold under the trade name of Tefzel by DuPont. ETFE is a partially fluorinated copolymer of ethylene and tetrafluoroethylene. It has a maximum service temperature of 300°F (149°C). The physical and mechanical properties are given in Table E.11.

ETFE is inert to strong mineral acids, halogens, inorganic bases, and metal salt solutions. Carboxylic acids, aldehydes, aromatic and aliphatic hydrocarbons, alcohols, aldehydes, ketones, ethers, esters, chlorocarbons, and classic polymer solvents have little effect on Tefzel.

Very strong oxidizing acids such as nitric, and organic bases such as amines and sulfonic acids at high concentrations and near their boiling points will affect ETFE to various degrees. Refer to Table E.12 for the compatibility of ETFE with selected corrodents.

ETHYLENE-TETRAFLUOROETHYLENE (ETFE) ELASTOMER

Ethylene-tetrafluoroethylene (ETFE) is a modified partially fluorinated copolymer of ethylene and polytetrafluoroethylene (PTFE). Since it contains more than 75% TFE by weight, it has better resistance to abrasion and cut-through than TFE while retaining most of the corrosion resistance properties.

Physical and Mechanical Properties

Ethylene-tetrafluorolene has excellent mechanical strength, stiffness, and abrasion resistance, with a service temperature range of –370 to 300°F (–223 to 149°C). It also exhibits good tear resistance and good electrical properties. However, its outstanding property is its resistance to a wide range of corrodents.

The physical and mechanical properties of ETFE are given in Table E.13.

Resistance to Sun, Weather, and Ozone

Ethylene-tetrafluoroethylene has outstanding resistance to sunlight, ozone, and weather. This feature, coupled with its wide range of corrosion resistance, makes the material particularly suitable for outdoor applications subject to atmospheric corrosion.

Chemical Resistance

Ethylene-tetrafluoroethylene is inert to strong mineral acids, inorganic bases, halogens, and metal salt solutions. Even carboxylic acids, anhydrides, aromatic and aliphatic hydrocarbons, alcohols, aldehydes, ketones, ethers, esters, chlorocarbons, and classic polymer solvents have little effect on ETFE.

Very strong oxidizing acids near their boiling points, such as nitric acid at high concentration, will affect ETFE in varying degrees, as will organic bases such as amines and sulfonic acids. Refer to Table E.11 for the compatibility of ETFE with selected corrodents.

Applications

The principal applications for ETFE are found in such products as gaskets, packings, and seals (O-rings, lip, and X-rings) in areas where corrosion is a problem. The material is also

Table E.12 Compatiblity of ETFE with Selected Corrodents[a]

Chemical	Maximum temp. °F	Maximum temp. °C	Chemical	Maximum temp. °F	Maximum temp. °C
Acetaldehyde	200	93	Barium hydroxide	300	149
Acetamide	250	121	Barium sulfate	300	149
Acetic acid 10%	250	121	Barium sulfide	300	149
Acetic acid 50%	250	121	Benzaldehyde	210	99
Acetic acid 80%	230	110	Benzene	210	99
Acetic acid, glacial	230	100	Benzene sulfonic acid 10%	210	99
Acetic anhydride	300	149	Benzoic acid	270	132
Acetone	150	66	Benzyl alcohol	300	149
Acetyl chloride	150	66	Benzyl chloride	300	149
Acrylonitrile	150	66	Borax	300	149
Adipic acid	280	138	Boric acid	300	149
Allyl alcohol	210	99	Bromine gas, dry	150	66
Allyl chloride	190	88	Bromine water 10%	230	110
Alum	300	149	Butadiene	250	121
Aluminum chloride, aqueous	300	149	Butyl acetate	230	110
Aluminum chloride, dry	300	149	Butyl alcohol	300	149
Aluminum fluoride	300	149	*n*-Butylamine	120	49
Aluminum hydroxide	300	149	Butyl phthalate	150	66
Aluminum nitrate	300	149	Butyric acid	250	121
Aluminum oxycloride	300	149	Calcium bisulfide	300	149
Aluminum sulfate	300	149	Calcium carbonate	300	149
Ammonium bifluoride	300	149	Calcium chlorate	300	149
Ammonium carbonate	300	149	Calcium chloride	300	149
Ammonium chloride 10%	300	149	Calcium hydroxide 10%	300	149
Ammonium chloride 50%	290	143	Calcium hydroxide, sat.	300	149
Ammonium chloride, sat.	300	149	Calcium hypochlorite	300	149
Ammonium fluoride 10%	300	149	Calcium nitrate	300	149
Ammonium fluoride 25%	300	149	Calcium oxide	260	127
Ammonium hydroxide 25%	300	149	Calcium sulfate	300	149
Ammonium hydroxide, sat.	300	149	Caprylic acid	210	99
Ammonium nitrate	230	110	Carbon bisulfide	150	66
Ammonium persulfate	300	149	Carbon dioxide, dry	300	149
Ammonium phosphate	300	149	Carbon dioxide, wet	300	149
Ammonium sulfate 10–40%	300	149	Carbon disulfide	150	66
Ammonium sulfide	300	149	Carbon monoxide	300	149
Amyl acetate	250	121	Carbon tetrachloride	270	132
Amyl alcohol	300	149	Carbonic acid	300	149
Amyl chloride	300	149	Cellosolve	300	149
Aniline	230	110	Chloracetic acid, 50% water	230	110
Antimony trichloride	210	99	Chloracetic acid 50 %	230	110
Aqua regia 3:1	210	99	Chlorine gas, dry	210	99
Barium carbonate	300	149	Chlorine gas, wet	250	121
Barium chloride	300	149	Chlorine, water	100	38

Table E.12 Continued

Chemical	Maximum temp. °F	Maximum temp. °C	Chemical	Maximum temp. °F	Maximum temp. °C
Chlorobenzene	210	99	Methyl ethyl ketone	230	110
Chloroform	230	110	Methyl isobutyl ketone	300	149
Chlorosulfonic acid	80	27	Muriatic acid	300	149
Chromic acid 10%	150	66	Nitric acid 5%	150	66
Chromic acid 50%	150	66	Nitric acid 20%	150	66
Chromyl chloride	210	99	Nitric acid 70%	80	27
Citric acid 15%	120	49	Nitric acid, anhydrous	x	x
Copper chloride	300	149	Nitrous acid, concentrated	210	99
Copper cyanide	300	149	Oleum	150	66
Copper sulfate	300	149	Perchloric acid 10%	230	110
Cresol	270	132	Perchloric acid 70%	150	66
Cupric chloride 5%	300	149	Phenol	210	99
Cyclohexane	300	149	Phosphoric acid 50–80%	270	132
Cyclohexanol	250	121	Picric acid	130	54
Dichloroacetic acid	150	66	Potassium bromide 30%	300	149
Ethylene glycol	300	149	Salicylic acid	250	121
Ferric chloride 50% in water	300	149	Sodium carbonate	300	149
Ferric nitrate 10–50%	300	149	Sodium chloride	300	149
Ferrous chloride	300	149	Sodium hydroxide 10%	230	110
Ferrous nitrate	300	149	Sodium hydroxide 50%	230	110
Fluorine gas, dry	100	38	Sodium hypochlorite 20%	300	149
Fluorine gas, moist	100	38	Sodium hypochlorite, concentrated	300	149
Hydrobromic acid, dilute	300	149	Sodium sulfide to 50%	300	149
Hydrobromic acid 20%	300	149	Stannic chloride	300	149
Hydrobromic acid 50%	300	149	Stannous chloride	300	149
Hydrochloric acid 20%	300	149	Sulfuric acid 10%	300	149
Hydrochloric acid 38%	300	149	Sulfuric acid 50%	300	149
Hydrocyanic acid 10%	300	149	Sulfuric acid 70%	300	149
Hydrofluoric acid 30%	270	132	Sulfuric acid 90%	300	149
Hydrofluoric acid 70%	250	121	Sulfuric acid 98%	300	149
Hydrofluoric acid 100%	230	110	Sulfuric acid 100%	300	149
Hypochlorous acid	300	149	Sulfuric acid, fuming	120	49
Lactic acid 25%	250	121	Sulfurous acid	210	99
Lactic acid, concentrated	250	121	Thionyl chloride	210	99
Magnesium chloride	300	149	Toluene	250	121
Malic acid	270	132	Trichloroacetic acid	210	99
Manganese chloride	120	49	Zinc chloride	300	149
Methyl chloride	300	149			

[a]The chemicals listed are in the pure state or in a saturated solution unless otherwise indicated. Compatibility is shown to the maximum allowable temperature for which data are available. Incompatibility is shown by an x. A blank space indicates that the data are unavailable.

Material extracted from: Schweitzer, Philip A. *Corrosion Resistance Tables*, 4th ed. Vols 1–3. New York: Marcel Dekker, 1995.

Table E.13 Physical and Mechanical Properties ETFE Elastomer[a]

Specific gravity	1.7
Hardness range, Rockwell	R-50 to D-75
Tensile strength, psi	6500
Elongation, % at break	100–400
Tear resistance	Good
Maximum temperature, continuous use	300°F (149°C)
Brittle point	–150°F (–101°C)
Water absorption, %/24 h	0.029
Abrasion resistance	Good
Volume resistivity, ohm-cm	$>10^{16}$
Dielectric strength, kV/mm	16 (3 mm)
Dielectric constant (10^{-3} to 10^6 Hz range)	2.6
Dissipation (power) factor	8×10^{-4}
Resistance to sunlight	Excellent
Resistance to heat	Good
Machining qualities	Good

[a]These are representative values since they may be altered by compounding.

used for sleeve, split curled, and thrust bearings, and for bearing pads for pipe and equipment support where expansion and contraction or movement may occur.

EXFOLIATION CORROSION

When intergranular corrosion takes place in a metal with a highly directional grain structure, its propagates internally, parallel to the surface of the metal. The corrosion product formed is about five times as voluminous as the metal consumed, and it is trapped beneath the surface. As a result, an internal stress is produced that splits off the overlying layers of metal. Hence the name "exfoliation."

This is a dangerous form of corrosion, since the splitting off of uncorroded metal rapidly reduces load carrying ability. The splitting action continually exposes film free metal, so the rate of corrosion is not self-limiting.

Exfoliation corrosion is mostly found in certain alloys and tempers of aluminum, particularly in areas of high chloride content such as deicing salts and seacoast atmospheres. In these applications an aluminum with a resistant temper should be used.

REFERENCES

1. PA Schweitzer. Corrosion Resistance of Elastomers. New York: Marcel Dekker, 1990.
2. GT Murray. Introduction to Engineering Materials. New York: Marcel Dekker, 1993.
3. PA Schweitzer. Corrosion Resistance Tables. 4th ed. Vols. 1–3. New York: Marcel Dekker, 1995.

4. D Landolt. Introduction to surface reactions: Electrochemical basis of corrosion. In: P Marcus and J. Oudar, eds. Corrosion Mechanisms in Theory and Practice. New York: Marcel Dekker, 1995, pp.1–8.
5. DM Berger. Fundamentals and prevention of metallic corrosion. In: PA Schweitzer, ed. Corrosion and Corrosion Protection Handbook. 2nd. ed. New York: Marcel Dekker, 1989, 3–11.
6. HH Uhlig. Corrosion and Corrosion Control. New York: John Wiley, 1963.
7. CP Dillon. Corrosion Control in the Chemical Process Industries. 2nd. ed. St. Louis: Materials Technology Institute of the Chemical Process Industries, 1994.
8. CP Dillon. Corrosion Resistance of Stainless Steels. New York: Marcel Dekker, 1995.
9. MR Louthan Jr. The effect of hydrogen on metals. In: F Mansfield, ed. Corrosion Mechanisms. New York: Marcel Dekker, 1987.
10. FP Ford and PL Andersen. Corrosion in nuclear systems: Environmentally assisted cracking in light water reactors. In: Corrosion Mechanisms in Theory and Practice. P Marcus and J Oudar, eds. New York: Marcel Dekker, 1995.
11. JH Mallinson. Corrosion-Resistant Plastic Composites in Chemical Plant Design. New York: Marcel Dekker, 1988.
12. PA Schweitzer. Corrosion Resistant Piping Systems. New York: Marcel Dekker, 1994.
13. PA Schweitzer. In: PA Schweitzer, ed. Corrosion Engineering Handbook. New York: Marcel Dekker, 1996, p. 312.

F

FERRITE

As austenite cools below 1675°F (915°C), ferrite is formed. Ferrite is essentially pure iron. Since ferrite does not contain enough carbon to permit the formation of martensite, it cannot be hardened by heat treatment. Therefore steels composed of only ferrite are not hardenable by heat treatment.

The generic term "ferritic steel" is used to refer to carbon or low alloy steels that contain other phases in addition to ferrite. These steels are usually hardenable by heat treatment.

FERRITIC STAINLESS STEELS

The ferritic stainless steels contain 15–30% chromium with a low carbon content (0.1%). Refer to Table F.1 for the physical and mechanical properties. Corrosion resistance of these alloys is rated good. Mildly corrosive solutions and oxidizing media are handled satisfactorily. These alloys exhibit a high level of resistance to pitting and crevice attack in many chloride-containing environments. They are also essentially immune to stress corrosion cracking in chloride environments. Ferritic stainless steels are especially resistant to organic acids. Their resistance to chloride attack makes these alloys ideal candidates for food processing applications and a wide range of chemical processing uses.

These alloys will not resist reducing media such as hydrochloric acid.

Type 430

This alloy exhibits resistance to chloride stress corrosion cracking and to elevated temperature sulfide attack. It is subject to embrittlement at 885°F (475°C) and loss of ductility at subzero temperatures.

Type 444

As with all ferritic stainless steels this alloy relies on a passive film to resist corrosion but exhibits relatively high corrosion rates when activitated. This explains the abrupt change in corrosion rates that occurs at particular acid concentrations. For example, it is resistant to very dilute solutions of sulfuric acid at boiling.

The corrosion rates of type 444 in strongly concentrated sodium hydroxide solutions are higher than those for austenitic stainless steels.

Table F.1 Physical and Mechanical Properties of Ferritic Stainless Steels

Property	Type of alloy		
	430	444	XM-27
Modulus of elasticity × 10^6	29	29	
Tensile strength × 10^3, psi	60	60	70
Yield strength 0.2% offset × 10^3, psi	35	40	56
Elongation in 2 in., %	20	20	30
Hardness, Brinell	B-165	217	Rock. B-83
Density, lb/in.^{3}	0.278	0.28	0.28
Specific gravity	7.75	7.75	7.66
Specific heat (32–212°F), Btu/lb°F	0.11	0.102	0.102
Thermal conductivity, Btu/lb°F			
at70°F (20°C)	15.1	15.5	
at 1500°F (815°C)	15.2		
Thermal expansion coefficient (32–212°F) × 10^{-6} in./in.°F	6.0	6.1	5.9

Type XM17

This alloy provides excellent resistance to pitting and crevice corrosion in chloride-containing environments, excellent resistance to chloride stress corrosion cracking, resistance to intergranular corrosion, and a resistance to a wide variety of other corrosive environments. This alloy can also be used in direct contact with foods. Refer to Table F.2 for the compatibility of ferritic stainless steels with selected corrodents.

See Refs. 1–3.

Table F.2 Compatibility of Ferritic Stainless Steels with Selected Corrodents[a]

Chemical	430 (°F/°C)	444 (°F/°C)	XM-27 (°F/°C)
Acetaldehyde			
Acetamide			
Acetic acid 10%	70/21	200/93	200/93
Acetic acid 50%	x	200/93	200/93
Acetic acid 80%	70/21	200/93	130/54
Acetic acid, glacial	70/21		140/60
Acetic anhydride 90%	150/66		300/149
Aluminum chloride, aqueous	x		110/43
Aluminum hydroxide	70/21		
Aluminum sulfate	x		
Ammonia gas	212/100		
Ammonium carbonate	70/21		
Ammonium chloride 10%			200/93
Ammonium hydroxide 25%	70/21		
Ammonium hydroxide, sat.	70/21		
Ammonium nitrate	212/100		
Ammonium persulfate 5%	70/21		

Table F.2 Continued

Chemical	430 (°F/°C)	444 (°F/°C)	XM-27 (°F/°C)
Ammonium phosphate	70/21		
Ammonium sulfate 10–40%	x		
Amyl acetate	70/21		
Amyl chloride	x		
Aniline	70/21		
Antimony trichloride	x		
Aqua regia 3:1			x
Barium carbonate	70/21		
Barium chloride	70/21[b]		
Barium sulfate	70/21		
Barium sulfide	70/21		
Benzaldehyde			210/99
Benzene	70/21		
Benzoic acid	70/21		
Borax 5%	200/93		
Boric acid	200/93[a]		
Bromine gas, dry	x		
Bromine gas, moist	x		
Bromine liquid	x		
Butyric acid	200/93		
Calcium carbonate	200/93		
Calcium chloride	x		
Calcium hypochlorite	x		
Calcium sulfate	70/21		
Carbon bisulfide	70/21		
Carbon dioxide, dry	70/21		
Carbon monoxide	1600/871		
Carbon tetrachloride, dry	212/100		
Carbonic acid	x		
Chloracetic acid, 50% water	x		
Chloracetic acid	x		
Chlorine gas, dry	x		
Chlorine gas, wet	x		
Chloroform, dry	70/21		
Chromic acid 10%	70/21		120/49
Chromic acid 50%	x		x
Chromyl chloride			
Citric acid 15%	70/21	200/93	200/93
Citric acid, concentrated	x		
Copper acetate	70/21		
Copper carbonate	70/21		
Copper chloride	x		x
Copper cyanide	212/100		
Copper sulfate	212/100		
Cupric chloride 5%	x		
Cupric chloride 50%	x		
Ethylene glycol	70/21		
Ferric chloride	x		80/27

Table F.2 Continued

Chemical	430 (°F/°C)	444 (°F/°C)	XM-27 (°F/°C)
Ferric chloride 10% in water			75/25
Ferric nitrate 10–50%	70/21		
Ferrous chloride	x		
Ferrous nitrate			
Fluorine gas, dry	x		
Fluorine gas, moist	x		
Hydrobromic acid, dilute	x		
Hydrobromic acid 20%	x		
Hydrobromic acid 50%	x		
Hydrochloric acid 20%	x		
Hydrochloric acid 38%	x		
Hydrocyanic acid 10%	x		
Hydrofluoric acid 30%	x		x
Hydrofluoric acid 70%	x		x
Hydrofluoric acid 100%	x		x
Iodine solution 10%	x		
Lactic acid 20%	x	200/93	200/93
Lactic acid, concentrated	x		
Magnesium chloride			200/93
Malic acid	200/93		
Muriatic acid	x		
Nitric acid 5%	70/21	200/93	320/160
Nitric acid 20%	200/93	200/93	320/160
Nitric acid 70%	70/21	x	210/99
Nitric acid, anhydrous	x	x	
Nitrous acid 5%	70/21		
Phenol	200/93		
Phosphoric acid 50–80%	x	200/93	200/93
Picric acid	x		
Silver bromide 10%	x		
Sodium chloride	70/21[b]		
Sodium hydroxide 10%	70/21	212/100	200/93
Sodium hydroxide 50%		x	180/82
Sodium hydroxide, concentrated		x	
Sodium hypochlorite 30%			90/32
Sodium sulfide to 50%	x		
Stannic chloride	x		
Stannous chloride 10%			90/32
Sulfuric acid 10%	x	x	x
Sulfuric acid 50%	x	x	x
Sulfuric acid 70%		x	x
Sulfuric acid 90%		x	x
Sulfuric acid 98%		x	280/138
Sulfuric acid 100%	70/21	x	
Sulfuric acid, fuming		x	
Sulfurous acid 5%	x		360/182
Thionyl chloride			
Toluene			210/99

Table F.2 Continued

Chemical	430 (°F/°C)	444 (°F/°C)	XM-27 (°F/°C)
Trichloroacetic acid	x		
White liquor			
Zinc chloride 20%	70/21[b]		200/93

[a]The chemicals listed are in the pure state or in a saturated solution unless otherwise indicated. Compatibility is shown to the maximum allowable temperature for which data are available. Incompatibility is shown by an x. A blank space indicates that data are unavailable. When compatible, the corrosion rate is < 20 MPY.

[b]Pitting may occur.

Source: PA Schweitzer. *Corrosion Resistance Tables,* 4th ed. Vols. 1–3. New York: Marcel Dekker, 1995.

FIBERGLASS

Fiberglass was developed during and immediately after World War II. Fiberglass is made by a number of different processes, such as melt spinning or by drawing from a marble. At the present time E and C glass predominate, which are boroalumiosilicate and aluminosilicate glass respectively. The major use is as a reinforcing material for various plastic resins. To improve adhesion of these glasses to resins various so-called binders have been developed, the most common of which are the silanes.

Also refer to "Thermoset Reinforcing Materials."

FIBER REINFORCED PLASTICS (COMPOSITES)

Plastic resins, particularly the thermosets, require some type of reinforcing material to provide strength and stability. Reinforcement is also used with thermoplasts on occasion to provide additional strength.

Many types of fibers are used as reinforcing materials, with glass being the predominant material used. Other materials used include carbon, boron, silicon carbide, polyester fibers, and aramid fibers. It is essential that the polymer and reinforcing fiber both be resistant to the chemical being handled and that they both be suitable for use at the maximum temperature desired.

Additional information can be found by referring to the specific fiber and resin.

FILIFORM CORROSION

Metals with semipermeable coatings or films may undergo a type of corrosion resulting in numerous meandering threadlike filaments of corrosion beneath the coatings or films. The essential conditions for this form of corrosion to develop are generally high humidity (65 to 95% relative humidity at room temperature), sufficient water permeability of the film, stimulation by impurities, and the presence of film defects (mechanical damage, pores, insufficient coverage of localized areas, air bubbles, salt crystals or dust particles).

See Refs. 1 and 4.

FLUOREL

See "Fluoroelastomers."

FLUOROELASTOMERS (FKM)

Fluoroelastomers are fluorine-containing hydrocarbon polymers with a saturated structure obtained by polymerizing fluorinated monomers such as vinylidene fluoride, hexafluoropropene, and tetrafluoroethylene. The result is a high performance synthetic rubber with exceptional resistance to oils and chemicals at elevated temperatures. Initially this material was used to produce O-rings for use in severe conditions. Although this remains a major area of application, these compounds have found wide use in other applications because of their chemical resistance at high temperatures and other desirable properties.

As with other rubbers, fluoroelastomers are capable of being compounded with various additives to enhance specific properties for particular applications. Fluoroelastomers are suitable for all rubber processing applications, including compression molding, injection molding, injection/compression molding, transfer molding, extrusion, calendering, spreading, and dipping.

These compounds possess the rapid recovery from deformation, or resilience, of a true elastomer and exhibit mechanical properties of the same order of magnitude as those of conventional synthetic rubbers.

Fluoroelastomers are manufactured under various trade names by different manufacturers. Three typical materials are listed below.

Trade name	Manufacturer
Viton	DuPont
Technoflon	Ausimont
Fluorel	3M

These elastomers have the ASTM designation of FKM.

Physical and Mechanical Properties

The general physical and mechanical properties of fluoroelastomers are similar to those of other synthetic rubbers. General-purpose compounds have a hardness of 70–75 Shore A. Formulations are produced that have hardnesses ranging from 45 to 95 Shore A. At elevated temperatures, 250–500°F (121–260°C), hardness may decrease by 5–15 points depending upon the polymer and the formulation. These variations must be taken into account when specifying hardness of products to be used at elevated temperatures.

Fluoroelastomer compounds have good tensile strengths, ranging from 188 to 2900 psi. In general the tensile strength of any elastomer tends to decrease at elevated temperatures; however this loss in tensile strength is much less with the fluoroelastomers. Percent elongation at break is an indication of operating life. A high percentage is essential when high resistance to bending stress is required for the application. These elastomers have a range of 100–400%. The ability of fluoroelastomers to recover their original dimension after compression and their exceptional thermal resistance make it possible to fabricate cured items with very low set compression values even under the most severe operating conditions. These values become even more meaningful at elevated temperatures when it is realized that most rubbers have a maximum service temperature of less than 250°F (121°C). Table F.3 lists the physical and mechanical properties of the fluoroelastomers.

Table F.3 Physical and Mechanical Properties of Fluoroelastomers[a]

Specific gravity	1.8
Specific heat	0.395
Brittle point	–25 to –75°F (–32 to –59°C)
Coefficient of linear expansion	88×10^{-6}/°F, 16×10^{-5}/°C
Thermal conductivity	
BTU-in./h-ft^2 °F at 100°F	1.58
kg-cal/cm-cm^2-°C-h at 38°C	1.96
Electrical properties	
Dielectric constant at 1000 Hz	
at 75°F (24°C)	10.5
at 300°F (149°C)	7.1
at 390°F (199°C)	9.1
Dissipation factor at 1000 Hz	
at 75°F (24°C)	0.034
at 300°F (149°C)	0.273
at 390°F (199°C)	0.39–1.19
Permeability, cm^3/cm^2-cm-sec-atm	
at 75°F (24°C)	
to air	0.0099×10^{-7}
to helium	0.892×10^{-7}
to nitrogen	0.0054×10^{-7}
at 86°F (30°C)	
to carbon dioxide	0.59×10^{-7}
to oxygen	0.11×10^{-7}
Tensile strength, psi	1800–2900
Elongation, % at break	400
Hardness, Shore A	45–95
Abrasion resistance	Good
Maximum temperature, continuous use	400°F (205°C)
Compression set, %	
at 70°F (21°C)	21
at 300°F (149°C)	32
at 392°F (200°C)	98
Tear resistance	Good
Resistance to sunlight	Excellent
Effect of aging	Nil
Resistance to heat	Excellent

[a]These are representative values since they may be altered by compounding.

The resilience of the fluoroelastomers makes them suitable for application as vibration isolators at elevated temperatures and as vibration dampers (energy absorbers) at room temperature. In the latter case, because of cost, they would normally be used only in extremely corrosive atmospheres. These rubbers can be applied as coatings to fabrics or adhered to a variety of metals to provide fluid resistance to the substrate.

The temperature resistance of the fluoroelastomers is exceptionally good over a wide temperature range. At high temperatures their mechanical properties are retained better than those of any other elastomer. Compounds remain usefully elastic indefinitely when exposed to aging up to 400°F (204°C). Continuous service limits are generally considered to be as in the table.

3000 h at 450°F (232°C)
1000 h at 500°F (260°C)
240 h at 550°F (288°C)
48 h at 600°F (313°C)

On the low temperature side these rubbers are generally serviceable in dynamic applications down to –10°F (–23°C). Flexibility at low temperature is a function of the material thickness. The thinner the cross-section, the less stiff the material is at every temperature. The brittle point at a thickness of 0.075 in. (1.9 mm) is in the neighborhood of –50°F (–45°C). This temperature can have a range of –25°F to –75°F (–32° to –59°C) depending upon the thickness and hardness of the material. Fluoroelastomers are relatively impermeable to air and gases, ranking about midway between the best and the poorest elastomers in this respect. This permeability can be modified considerably by the way they are compounded. In all cases permeability increases rapidly with increasing temperature. Table F.3 provides some data on the permeability of the fluoroelastomers.

Being halogen-containing polymers, these elastomers are more resistant to burning than are exclusively hydrocarbon rubbers. Normally compounded material will burn when directly exposed to flame but will stop burning when the flame is removed. Natural rubber and synthetic hydrocarbon rubbers under the same conditions will continue to burn when the flame is removed. However, it must be remembered that under an actual fire condition fluoroelastomers will burn. During combustion, fluorinated products such as hydrofluoric acid can be given off. Special compounding can improve the flame resistance. One such formulation has been developed for the space program that will not ignite under conditions of the NASA test, which specifies 100% oxygen at 6.2 psi absolute.

The fluoroelastomers will increase in stiffness and hardness when exposed to gamma radiation from a cobalt-60 source. For dynamic applications radiation exposure should not exceed 1×10^7 roentgens. Higher dosages are permissible for static applications. There is no evidence of radiation induced stress cracking. There are other elastomers that exhibit superior radiation resistance. However, high temperatures are frequently encountered along with exposure to radiation, and in many cases these elevated temperatures will rule out the more radiation-resistant elastomers.

Fluoroelastomers are particularly recommended when resistance to ozone, high temperatures, or highly corrosive fluids is required in addition to radiation resistance.

The dielectric properties of the fluoroelastomers permit them to be used as insulating materials at low tension and frequency in high temperature applications and in the presence of higher concentrations of ozone and highly aggressive chemicals. The values of the individual properties can be greatly influenced by formulation but are generally in the following ranges:

Direct current resistivity	2×10^{13} ohm-cm
Dielectric constant	10–15
Dissipation factor	0.01–0.05
Dielectric strength	500 V/mil (2000 V/mm)

Fluoroelastomers have been approved by the U.S. Food and Drug Administration for use in repeated contact with food products. More details are available in the Federal Register Vol. 33 No. 5, Tuesday January 9, 1968, Part 121—Food Additives, Subpart F—Food additives Resulting from Contact with Containers or Equipment and Food Additives Otherwise Affecting Food—Rubber Articles Intended for Repeated Use.

The biological resistance of fluoroelastomers is excellent. A typical compound tested against specification MIL E-5272C showed no fungus growth for 30 days. This specification covers four common fungus groups.

Resistance to Sun, Weather, and Ozone

Because of their chemically saturated structure, the fluoroelastomers exhibit excellent weathering resistance to sunlight and especially to ozone. After 13 years of exposure in Florida in direct sunlight, samples showed little or no change in properties or appearance. Similar results were experienced with samples exposed to various tropical conditions in Panama for a period of 10 years. Products made of this elastomer are unaffected by ozone concentrations as high as 100 ppm. No cracking occurred in a bent loop test after 1 year exposure to 100 ppm of ozone in air at 100°F (38°C) or in a sample held at 356°F (180°C) for several hundred hours. This property is particularly important considering that standard tests, for example in the autumotive industry, require resistance to only 0.5 ppm ozone.

Chemical Resistance

The fluoroelastomers provide excellent resistance to oils, fuels, lubricants, most mineral acids, many aliphatic and aromatic hydrocarbons (carbon tetrachloride, benzene, toluene, xylene) that act as solvents for other rubbers, gasoline, naphtha, chlorinated solvents, and pesticides. Special formulations can be produced to obtain resistance to hot mineral acids, steam, and hot water.

These elastomers are not suitable for use with low molecular weight esters and ethers, ketones, certain amines, or hot anhydrous hydrofluoric or chlorosulfonic acids. Their solubility in low molecular weight ketones is an advantage in producing solution coatings of fluoroelastomers. Table F.4 provides the compatibility of fluoroelastomers with selected corrodents.

Table F.4 Compatibility of Fluoroelastomers with Selected Corrodents[a]

Chemical	Maximum temp. °F	Maximum temp. °C	Chemical	Maximum temp. °F	Maximum temp. °C
Acetaldehyde	x	x	Barium hydroxide	400	204
Acetamide	210	199	Barium sulfate	400	204
Acetic acid 10%	190	88	Barium sulfide	400	204
Acetic acid 50%	180	82	Benzaldehyde	x	x
Acetic acid 80%	180	82	Benzene	400	204
Acetic acid, glacial	x	x	Benzene sulfonic acid 10%	190	88
Acetic anhydride	x	x	Benzoic acid	400	204
Acetone	x	x	Benzyl alcohol	400	204
Acetyl chloride	400	204	Benzyl chloride	400	204
Acrylic acid	x	x	Borax	190	88
Acrylonitrile	x	x	Boric acid	400	204
Adipic acid	190	88	Bromine gas, dry, 25%	180	82
Allyl alcohol	190	88	Bromine gas, moist, 25%	180	82
Allyl chloride	100	38	Bromine liquid	350	177
Alum	190	88	Butadiene	400	204
Aluminum acetate	180	82	Butyl acetate	x	x
Aluminum chloride, aqueous	400	204	Butyl alcohol	400	204
Aluminum fluoride	400	204	*n*-Butylamine	x	x
Aluminum hydroxide	190	88	Butyl phthalate	80	27
Aluminum nitrate	400	204	Butyric acid	120	49
Aluminum oxychloride	x	x	Calcium bisulfide	400	204
Aluminum sulfate	390	199	Calcium bisulfite	400	204
Ammonia gas	x	x	Calcium carbonate	190	88
Ammonium bifluoride	140	60	Calcium chlorate	190	88
Ammonium carbonate	190	88	Calcium chloride	300	149
Ammonium chloride 10%	400	204	Calcium hydroxide 10%	300	149
Ammonium chloride 50%	300	149	Calcium hydroxide, sat.	400	204
Ammonium chloride, sat.	300	149	Calcium hypochlorite	400	204
Ammonium fluoride 10%	140	60	Calcium nitrate	400	204
Ammonium fluoride 25%	140	60	Calcium sulfate	200	93
Ammonium hydroxide 25%	190	88	Carbon bisulfide	400	204
Ammonium hydroxide, sat.	190	88	Carbon dioxide, dry	80	27
Ammonium nitrate	x	x	Carbon dioxide, wet	x	x
Ammonium persulfate	140	60	Carbon disulfide	400	204
Ammonium phosphate	180	82	Carbon monoxide	400	204
Ammonium sulfate 10–40%	180	82	Carbon tetrachloride	350	177
Ammonium sulfide	x	x	Carbonic acid	400	204
Amyl acetate	x	x	Cellosolve	x	x
Amyl alcohol	200	93	Chloracetic acid, 50% water	x	x
Amyl chloride	190	88	Chloracetic acid	x	x
Aniline	230	110	Chlorine gas, dry	190	88
Antimony trichloride	190	88	Chlorine gas, wet	190	88
Aqua regia 3:1	190	88	Chlorine, liquid	190	88
Barium carbonate	250	121	Chlorobenzene	400	204
Barium chloride	400	204	Chloroform	400	204

Table F.4 Continued

Chemical	Maximum temp. °F	Maximum temp. °C	Chemical	Maximum temp. °F	Maximum temp. °C
Chlorosulfonic acid	x	x	Methyl chloride	190	88
Chromic acid 10%	350	177	Methyl ethyl ketone	x	x
Chromic acid 50%	350	177	Methyl isobutyl ketone	x	x
Citric acid 15%	300	149	Muriatic acid	350	149
Citric acid, concentrated	400	204	Nitric acid 5%	400	204
Copper acetate	x	x	Nitric acid 20%	400	204
Copper carbonate	190	88	Nitric acid 70%	190	88
Copper chloride	400	204	Nitric acid, anhydrous	190	88
Copper cyanide	400	204	Nitrous acid, concentrated	90	32
Copper sulfate	400	204	Oleum	190	88
Cresol	x	x	Perchloric acid 10%	400	204
Cupric chloride 5%	180	82	Perchloric acid 70%	400	204
Cupric chloride 50%	180	82	Phenol	210	99
Cyclohexane	400	204	Phosphoric acid 50–80%	300	149
Cyclohexanol	400	204	Picric acid	400	204
Dichloroethane (ethylene dichloride)	190	88	Potassium bromide 30%	190	88
Ethylene glycol	400	204	Salicylic acid	300	149
Ferric chloride	400	204	Sodium carbonate	190	88
Ferric chloride 50% in water	400	204	Sodium chloride	400	204
Ferric nitrate 10–50%	400	204	Sodium hydroxide 10%	x	x
Ferrous chloride	180	82	Sodium hydroxide 50%	x	x
Ferrous nitrate	210	99	Sodium hydroxide, concentrated	x	x
Fluorine gas, dry	x	x	Sodium hypochlorite 20%	400	204
Fluorine gas, moist	x	x	Sodium hypochlorite, concentrated	400	204
Hydrobromic acid, dilute	400	204	Sodium sulfide to 50%	190	88
Hydrobromic acid 20%	400	204	Stannic chloride	400	204
Hydrobromic acid 50%	400	204	Stannous chloride	400	204
Hydrochloric acid 20%	350	177	Sulfuric acid 10%	350	149
Hydrochloric acid 38%	350	177	Sulfuric acid 50%	350	149
Hydrocyanic acid 10%	400	204	Sulfuric acid 70%	350	149
Hydrofluoric acid 30%	210	99	Sulfuric acid 90%	350	149
Hydrofluoric acid 70%	350	177	Sulfuric acid 98%	350	149
Hydrofluoric acid 100%	x	x	Sulfuric acid 100%	180	88
Hypochlorous acid	400	204	Sulfuric acid, fuming	200	93
Iodine solution 10%	190	88	Sulfurous acid	400	204
Ketones, general	x	x	Thionyl chloride	x	x
Lactic acid 25%	300	149	Toluene	400	204
Lactic acid, concentrated	400	204	Trichloroacetic acid	190	88
Magnesium chloride	390	199	White liquor	190	88
Malic acid	390	199	Zinc chloride	400	204
Manganese chloride	180	82			

[a]The chemicals listed are in the pure state or in a saturated solution unless otherwise indicated. Compatibility is shown to the maximum allowable temperature for which data are available. Incompatibility is shown by an x. A blank space indicates that data are unavailable.

Source: PA Schweitzer. *Corrosion Resistance Tables*, 4th ed. Vols. 1–3. New York: Marcel Dekker, 1995.

Applications

The main applications for the fluoroelastomers are in those products requiring resistance to high operating temperatures together with high chemical resistance to aggressive fluids and to those characterized by severe operating conditions that no other elastomer can withstand. By proper formulation, cured items can be produced that will meet the rigid specifications of the industrial, aerospace, and military communities.

Recent changes in the automotive industry that have required reduction in environmental pollution, reduced costs, energy saving, and improved reliability have resulted in higher operating temperatures, which in turn require a higher performance elastomer. The main innovations resulting from these requirements are

Turbocharging
More compact, more efficient and faster engines
Catalytic exhausts
C_X reduction
Soundproofing

In addition, the use of lead free fuels, alternative fuels, sour gasoline lubricants, and antifreeze fluids have caused automotive fluids to be more corrosive to elastomers. At the present time fluoroelastomers are being applied as shaft seals, valve stem seals, O-rings (water-cooled cylinders and injection pumps), engine head gaskets, filter casing gaskets, diaphragms for fuel pumps, water pump gaskets, turbocharge lubricating circuit bellows, carburetor accelerating pump diaphragms, carburetor needle-valve tips, fuel hoses, and seals for exhaust gas pollution control equipment.

In the field of aerospace applications, the reliability of materials under extreme exposure conditions is of prime importance. The high and low temperature properties of the fluoroelastomers have permitted them to give reliable performance in a number of aircraft and missle components, specifically manifold gaskets, coated manifold gaskets, coated fabrics, firewall seals, heat shrinkable tubing and fittings for wire and cable, mastic adhesive sealants, protective coatings, and numerous types of O-ring seals.

The ability of the fluoroelastomers to seal under extreme vacuum conditions in the range of 10^{-9} mm Hg is an additional feature that makes these materials useful for components used in space.

The exploitation of oil fields in difficult areas such as desert or offshore sites has increased the problems of high temperatures and pressures, high viscosities, and high alkalinity. These extreme operating conditions require elastomers that have a high chemical resistance, thermal stability, and overall reliability to reduce maintenance. The same problems exist in the chemical industry. The fluoroplastics provide a solution to these problems and are used for O-rings, V-rings, U-rings, gaskets, valve seats, diaphragms for metering pumps, hoses, expansion joints, safety clothing and gloves, linings for valves, and maintenance coatings.

An important application for these elastomers is in the production of coatings and linings. Their chemical stability solves the problem of chemical corrosion by making it possible to use them for such purposes as

A protective lining for power station stacks operated with high-sulfur fuels
A coating on rolls for the textile industry to permit scouring of fabrics

Tank linings for the chemical industry

See Refs. 5 and 6.

FLUORINATED ETHYLENE PROPYLENE (FEP)

FEP is a fully fluorinated thermoplast with some branching but consists mainly of linear chains having the following formula:

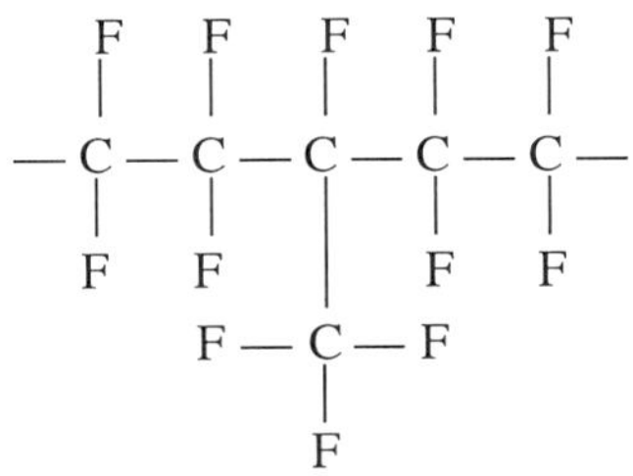

FEP has a maximum operating temperature of 375°F (190°C). After prolonged exposure at 400°F (204°C) it exhibits changes in physical strength. It is a relatively soft plastic with lower tensile strength, wear resistance, and creep resistance than other plastics. It is insensitive to notched impact forces and has excellent permeation resistance except to some chlorinated hydrocarbons. Table F.5 lists the physical and mechanical properties. FEP may be subject to permeation by specific materials. Refer to "Permeation" for details.

FEP basically exhibits the same corrosion resistance as PTFE, with a few exceptions, but at lower operating temperatures. It is resistant to practically all chemicals, the exceptions being extremely potent oxidizers, such as chlorine trifluoride and related compounds. Some chemicals will attack FEP when present in high concentrations at or near the service temperature limit. Refer to Table F.6 for the compatibility of FEP with selected corrodents. Reference 6 lists the compatibility of FEP with a wide range of selected corrodents.

Table F.5 Physical and Mechanical Properties of FEP

Specific gravity	2.15
Water absorption 24 h at 73°F (23°C), %	< 0.01
Tensile strength at 73°F (23°C), psi	2700–3100
Modulus of elasticity in tension at 73°F (23°C) $\times 10^5$ psi	0.9
Compressive strength, psi	16,000
Flexural strength, psi	3000
Izod impact strength, notched at 73°F (23°C)	no break
Coefficient of thermal expansion, in./in. °F $\times 10^{-5}$	8.3–10.5
Thermal conductivity, Btu/h/ft^2/°F/in.	0.11
Heat distortion temperature, at 66 psi °F/°C	158/70
Resistance to heat at continuous drainage, °F/°C	400/204
Limiting oxygen index, %	95
Flame spread	Nonflammable

Table F.6 Compatibility of FEP with Selected Corrodents[a]

Chemical	Maximum temp. °F	Maximum temp. °C
Acetaldehyde	200	93
Acetamide	400	204
Acetic acid 10%	400	204
Acetic acid 50%	400	204
Acetic acid 80%	400	204
Acetic acid, glacial	400	204
Acetic anhydride	400	204
Acetone[b]	400	204
Acetyl chloride	400	204
Acrylic acid	200	93
Acrylonitrile	400	204
Adipic acid	400	204
Allyl alcohol	400	204
Allyl chloride	400	204
Alum	400	204
Aluminum acetate	400	204
Aluminum chloride, aqueous	400	204
Aluminum chloride, dry	300	149
Aluminum fluoride[c]	400	204
Aluminum hydroxide	400	204
Aluminum nitrate	400	204
Aluminum oxychloride	400	204
Aluminum sulfate	400	204
Ammonia gas[c]	400	204
Ammonium bifluoride[c]	400	204
Ammonium carbonate	400	204
Ammonium chloride 10%	400	204
Ammonium chloride 50%	400	204
Ammonium chloride, sat.	400	204
Ammonium fluoride 10%[c]	400	204
Ammonium fluoride 25%[c]	400	204
Ammonium hydroxide 25%	400	204
Ammonium hydroxide, sat.	400	204
Ammonium nitrate	400	204
Ammonium persulfate	400	204
Ammonium phosphate	400	204
Ammonium sulfate 10–40%	400	204
Ammonium sulfide	400	204
Ammonium sulfite	400	204
Amyl acetate	400	204
Amyl alcohol	400	204
Amyl chloride	400	204
Aniline[b]	400	204
Antimony trichloride	250	121
Aqua regia 3:1	400	204
Barium carbonate	400	204
Barium chloride	400	204
Barium hydroxide	400	204
Barium sulfate	400	204
Barium sulfide	400	204
Benzaldehyde[b]	400	204
Benzene[b,c]	400	204
Benzene sulfonic acid 10%	400	204
Benzoic acid	400	204
Benzyl alcohol	400	204
Benzyl chloride	400	204
Borax	400	204
Boric acid	400	204
Bromine gas, dry[c]	200	93
Bromine gas, moist[c]	200	93
Bromine liquid[b,c]	400	204
Butadiene[c]	400	204
Butyl acetate	400	204
Butyl alcohol	400	204
n-Butylamine[b]	400	204
Butyl phthalate	400	204
Butyric acid	400	204
Calcium bisulfide	400	204
Calcium bisulfite	400	204
Calcium carbonate	400	204
Calcium chlorate	400	204
Calcium chloride	400	204
Calcium hydroxide 10%	400	204
Calcium hydroxide, sat.	400	204
Calcium hypochlorite	400	204
Calcium nitrate	400	204
Calcium oxide	400	204
Calcium sulfate	400	204
Caprylic acid	400	204
Carbon bisulfide[c]	400	204
Carbon dioxide, dry	400	204
Carbon dioxide, wet	400	204
Carbon disulfide	400	204
Carbon monoxide	400	204
Carbon tetrachloride[b,c,d]	400	204
Carbonic acid	400	204
Cellosolve	400	204
Chloracetic acid, 50% water	400	204
Chloracetic acid	400	204
Chlorine gas, dry	x	x
Chlorine gas, wet[c]	400	204
Chlorine, liquid[b]	400	204
Chlorobenzene[c]	400	204
Chloroform[c]	400	204

Table F.6 Continued

Chemical	Maximum temp. °F	°C	Chemical	Maximum temp. °F	°C
Chlorosulfonic acid[b]	400	204	Manganese chloride	300	149
Chromic acid 10%	400	204	Methyl chloride[c]	400	204
Chromic acid 50%[b]	400	204	Methyl ethyl ketone[c]	400	204
Chromyl chloride	400	204	Methyl isobutyl ketone[c]	400	204
Citric acid 15%	400	204	Muriatic acid[c]	400	204
Citric acid, concentrated	400	204	Nitric acid 5%[c]	400	204
Copper acetate	400	204	Nitric acid 20%[c]	400	204
Copper carbonate	400	204	Nitric acid 70%[c]	400	204
Copper chloride	400	204	Nitric acid, anhydrous[c]	400	204
Copper cyanide	400	204	Nitrous acid, concentrated	400	204
Copper sulfate	400	204	Oleum	400	204
Cresol	400	204	Perchloric acid 10%	400	204
Cupric chloride 5%	400	204	Perchloric acid 70%	400	204
Cupric chloride 50%	400	204	Phenol[c]	400	204
Cyclohexane	400	204	Phosphoric acid 50–80%	400	204
Cyclohexanol	400	204	Picric acid	400	204
Dichloroacetic acid	400	204	Potassium bromide 30%	400	204
Dichloroethane (ethylene dichloride)[c]	400	204	Salicylic acid	400	204
Ethylene glycol	400	204	Silver bromide 10%	400	204
Ferric chloride	400	204	Sodium carbonate	400	204
Ferric chloride 50% in water[b]	260	127	Sodium chloride	400	204
Ferric nitrate 10–50%	260	127	Sodium hydroxide 10%[b]	400	204
Ferrous chloride	400	204	Sodium hydroxide 50%	400	204
Ferrous nitrate	400	204	Sodium hydroxide, concentrated	400	204
Fluorine gas, dry	200	93	Sodium hypochlorite 20%	400	204
Fluorine gas, moist	x	x	Sodium hypochlorite, concentrated	400	204
Hydrobromic acid, dilute	400	204	Sodium sulfide to 50%	400	204
Hydrobromic acid 20%[c,d]	400	204	Stannic chloride	400	204
Hydrobromic acid 50%[c,d]	400	204	Stannous chloride	400	204
Hydrochloric acid 20%[c,d]	400	204	Sulfuric acid 10%	400	204
Hydrochloric acid 38%[c,d]	400	204	Sulfuric acid 50%	400	204
Hydrocyanic acid 10%	400	204	Sulfuric acid 70%	400	204
Hydrofluoric acid 30%[c]	400	204	Sulfuric acid 90%	400	204
Hydrofluoric acid 70%[c]	400	204	Sulfuric acid 98%	400	204
Hydrofluoric acid 100%[c]	400	204	Sulfuric acid 100%	400	204
Hypochlorous acid	400	204	Sulfuric acid, fuming[c]	400	204
Iodine solution 10%[c]	400	204	Sulfurous acid	400	204
Ketones, general	400	204	Thionyl chloride[c]	400	204
Lactic acid 25%	400	204	Toluene[c]	400	204
Lactic acid, concentrated	400	204	Trichloroacetic acid	400	204
Magnesium chloride	400	204	White liquor	400	204
Malic acid	400	204	Zinc chloride[d]	400	204

[a]The chemicals listed are in the pure state or in a saturated solution unless otherwise indicated. Compatibility is shown to the maximum allowable temperature for which data are available. Incompatibility is shown by an x. A blank space indicates that data are unavailable.
[b]Material will be absorbed.
[c]Material will permeate.
[d]Material can cause stress cracking.
Source: PA Schweitzer. *Corrosion Resistance Tables,* 4th ed. Vols. 1–3. New York: Marcel Dekker, 1995.

FLUOROCARBON RESINS

Fluorocarbon resins are organic compounds of which the hydrogen atoms have been replaced by fluorine. They are fully fluorinated, while fluoropolymer resins are only partially fluorinated. Included in this group of resins are polytetrafluoroethylene (PTFE), fluorinated ethylene propylene (FEP) and perfluoralkoxy (PFA). They are characterized by the following properties:

1. Nonpolarity: the carbon backbone of the linear polymer is completely sheathed by the tightly held electron cloud of fluorine atoms, with electro-negatives balanced.
2. High C—F and C—C bond strengths.
3. Low interchain forces: interactive forces between the two adjacent polymer chains are significantly lower than the bond forces within one chain.
4. Crystallinity.
5. High degree of polymerization.

These properties provide the following advantages to these materials:

1. High melting point
2. High thermal stability
3. High upper service temperatures
4. Inertness to chemical attack by almost all chemicals
5. Low coefficient of friction
6. Low water absorbability
7. Weatherability
8. Flame resistance
9. Toughness

Listed in the table are the properties of the fluorocarbon resins.

Property	ASTM Standard	PTFE	FEP	PFA
Specific gravity	D792	2.13–2.22	2.15	2.15
Tensile strength, psi	D638	2500–4000	3400	3600
Elongation, %	D638	200–400	325	300
Flexural modulus, psi	D790	27000	90000	90000
Impact strength, ft-lb/in.	D256	3.5	no break	no break
Hardness, Shore D	D2240	50–65	56	60
Coefficient of friction	D1894	0.1	0.2	0.2
Upper service temp., °F/°C	UL746B	500/260	400/204	500/260
Flame rating	UL94	VO	VO	VO
Limiting oxygen index, %	D2863	95	95	95
Chemical/solvent resistance	D543		outstanding	
Water absorption, 24 h	D570	0.01	0.01	0.03

For more information on the fluorocarbon resins refer to the specific resin and thermoplasts.

FLUOROPOLYMER RESINS

The fluoropolymers are resistant to a broader range of chemicals at higher temperatures than chlorinated or hydrogenated polymers, polyesters, and polyamides. However, their properties are significantly different from those of fully fluorinated resins (fluorocarbons). Included in this category are ethylene tetrafluoroethylene (ETFE) sold under the trade name of Tefzel by DuPont, polyvinylidene fluoride (PVDF) sold under the trade names of Kynar by Elf Atochem, Solef by Solvay, Hylar by Ausimont USA, and Super Pro and Iso by Asahi/America, and ethylene chlorotrifluoroethylene (ECTFE) sold under the trade name of Halar by Ausimont USA.

The polarity of these resins is increased as the result of substituting hydrogen or chlorine, which have different electronegatives relative to fluorine. The length of their bonds to the carbon backbone also differs from those with fluorine. The centers of electronegativity and electropositivity are not held as tightly as with carbon–fluorine bonds. As a result, differential separation of charge can be induced chemically between atoms in adjacent chains to permit electrostatic interaction between chains. Higher mechanical properties are produced because of the increased interpolymer chain attraction along with the interlocking of differently sized atoms. In addition, the increased polarity/interpolymer attraction reduces the permeation of penetrants through the resin. Because of the substitution of hydrogen or of hydrogen and chlorine for fluorine, chemical and thermal stability are sacrificed.

The chemical stability is affected by the arrangement of the substituting elements along the polymer chain. Solubility can be a leading indicator. For example ETFE has no known solvent under ordinary circumstances, while PVDF is soluble in common industrial ketones. (e.g. methyl ethyl ketone) and ECTFE is soluble in some fluorinated solvents. While the fully fluorinated polymers are resistant to strong acids and alkalies, the substituted polymers are adversely affected. However, these resins do possess advantageous properties both mechanical and chemically resistant. Typical properties of the fluoropolymer resins are shown in the table.

Property	ASTM Std	ETFE	PVDF	ECTFE
Specific gravity	D792	1.70	1.78	1.68
Tensile strength, psi	D638	6300	4500	7000
Elongation, %	D638	300	50	210
Flexural modulus, psi × 10^5	D790	1.7	2.5	2.4
Impact strength, ft-lb/in.	D256	no break	2	no break
Hardness, Shore D	D2240	67	78	75
Coefficient of friction	D1894	0.4		
Upper service temperature, °F/°C	UL746	300/150	300/150	300/150
Flame rating	UL94	VO	VO	VO
Limiting oxygen index, %	D2863	30	30	30
Chemical/solvent resistance	D543	excellent	fair	good
Water absorption, 24 h	D570	< 0.03	< 0.03	< 0.1

For additional information see the specific fluoropolymer and thermoplast.

FLUOROSILICONE RUBBER

See "Silicone Rubber."

FORMS OF CORROSION

The several forms of corrosion to which a metal may be subjected are

1. Electrochemical corrosion
2. Uniform corrosion
3. Intergranular corrosion
4. Galvanic corrosion
5. Crevice corrosion
6. Pitting
7. Erosion corrosion
8. Stress corrosion cracking (SCC)
9. Biological corrosion
10. Dezincification (dealloying)
11. Concentration cell
12. Embrittlement
13. Filiform corrosion
14. Corrosion fatigue
15. Fretting corrosion
16. Graphitization

Not all of these forms of corrosion are present in all applications, but it is possible to have more than one form present. In addition, not all metals are subject to all of these forms of corrosion. Understanding when each of these forms of corrosion could be present will permit the designer to take steps to eliminate the condition or to keep the corrosion within acceptable limits. Refer to the specific form of corrosion for details.

FRETTING CORROSION

Fretting corrosion occurs because of mechanical stresses, and in the extreme may lead to failure by fatigue or corrosion fatigue. It is defined as damage occuring at the interface of two surfaces in contact, one or both of which are metals, subject to slight relative slip.

Damage by fretting corrosion is characterized by discoloration of the metal surface and in the case of oscillating motion by formation of pits. It is at such pits that fatigue cracks eventually start.

See Ref. 1

FUEL ASH CORROSION

Also see "High Temperature Corrosion." The combustion of heavier petroleum fuels, both liquid and solid, containing alkaline and sulfur impurities can result in the dissolving of the protective oxide film, subjecting the metal to corrosion.

FURAN RESINS

Also see "Polymers" and "Thermoset Polymers." Since there are different formulations of the furan resins, the supplier should be checked as to the compatibility of a particular resin with the corrodents to be encountered. Corrosion charts will indicate the compatibility of at least one formulation.

The strong point of the furan resins is their excellent resistance to solvents in combination with acids and alkalies. They are compatible with the following corrodents:

Solvents	
Acetone	Methyl ethyl ketone
Benzene	Perchlorethylene
Carbon disulfide	Styrene
Chlorobenzene	Toluene
Ethanol	Trichloroethylene
Ethyl acetate	Xylene
Methanol	
Acids	
Acetic	Phosphoric
Hydrochloric	60% Sulfuric
5% Nitric	
Bases	
Dimethylamine	Sodium sulfide
Sodium carbonate	50% Sodium hydroxide

The furans are not resistant to bleaches, such as peroxides and hypochlorites, concentrated sulfuric acid, phenol and free chlorine, or to higher concentrations of chromic or nitric acids.

Refer to Table F.7 for the compatibility of furan resins with selected corrodents.

See Refs. 6–8, 9.

Table F.7 Compatibility of Furan Resins with Selected Corrodents[a]

	Maximum temp.			Maximum temp.	
Chemical	°F	°C	Chemical	°F	°C
Acetaldehyde	x	x	Acetic acid, glacial	80	27
Acetic acid 10%	212	100	Acetic anhydride	80	27
Acetic acid 50%	160	71	Acetone	80	27
Acetic acid 80%	80	27	Acetyl chloride	200	93

Table F.7 Continued

Chemical	Maximum temp. °F	°C	Chemical	Maximum temp. °F	°C
Acrylic acid	80	27	*n*-Butylamine	x	x
Acrylonitrile	80	27	Butyric acid	260	127
Adipic acid 25%	280	138	Calcium bisulfite	260	127
Allyl alcohol	300	149	Calcium chloride	160	71
Allyl chloride	300	149	Calcium hydroxide, sat.	260	127
Alum 5%	140	60	Calcium hypochlorite	x	x
Aluminum chloride, aqueous	300	149	Calcium nitrate	260	127
Aluminum chloride, dry	300	149	Calcium oxide		
Aluminum fluoride	280	138	Calcium sulfate	260	127
Aluminum hydroxide	260	127	Caprylic acid	250	121
Aluminum sulfate	160	71	Carbon bisulfide	160	71
Ammonium carbonate	240	116	Carbon dioxide, dry	90	32
Ammonium hydroxide 25%	250	121	Carbon dioxide, wet	80	27
Ammonium hydroxide, sat.	200	93	Carbon disulfide	260	127
Ammonium nitrate	250	121	Carbon tetrachloride	212	100
Ammonium persulfate	260	127	Cellosolve	240	116
Ammonium phosphate	260	127	Chloracetic acid, 50% water	100	38
Ammonium sulfate 10–40%	260	127	Chloracetic acid	240	116
Ammonium sulfide	260	127	Chlorine gas, dry	260	127
Ammonium sulfite	240	116	Chlorine gas, wet	260	127
Amyl acetate	260	127	Chlorine, liquid	x	x
Amyl alcohol	278	137	Chlorobenzene	260	127
Amyl chloride	x	x	Chloroform	x	x
Aniline	80	27	Chlorosulfonic acid	260	127
Antimony trichloride	250	121	Chromic acid 10%	x	x
Aqua regia 3:1	x	x	Chromic acid 50%	x	x
Barium carbonate	240	116	Chromyl chloride	250	121
Barium chloride	260	127	Citric acid 15%	250	121
Barium hydroxide	260	127	Citric acid, concentrated	250	121
Barium sulfide	260	127	Copper acetate	260	127
Benzaldehyde	80	27	Copper carbonate		
Benzene	160	71	Copper chloride	260	127
Benzene sulfonic acid 10%	160	71	Copper cyanide	240	116
Benzoic acid	260	127	Copper sulfate	300	149
Benzyl alcohol	80	27	Cresol	260	127
Benzyl chloride	140	60	Cupric chloride 5%	300	149
Borax	140	60	Cupric chloride 50%	300	149
Boric acid	300	149	Cyclohexane	140	60
Bromine gas, dry	x	x	Cyclohexanol		
Bromine gas, moist	x	x	Dichloroacetic acid	x	x
Bromine liquid 3% max.	300	149	Dichloroethane (ethylene dichloride)	250	121
Butadiene			Ethylene glycol	160	71
Butyl acetate	260	127	Ferric chloride	260	127
Butyl alcohol	212	100	Ferric chloride 50% in water	160	71

Table F.7 Continued

Chemical	Maximum temp. °F	°C	Chemical	Maximum temp. °F	°C
Ferric nitrate 10–50%	160	71	Perchloric acid 10%	x	x
Ferrous chloride	160	71	Perchloric acid 70%	260	127
Ferrous nitrate			Phenol	x	x
Fluorine gas, dry	x	x	Phosphoric acid 50	212	100
Fluorine gas, moist	x	x	Picric acid		
Hydrobromic acid, dilute	212	100	Potassium bromide 30%	260	127
Hydrobromic acid 20%	212	100	Salicylic acid	260	127
Hydrobromic acid 50%	212	100	Silver bromide 10%		
Hydrochloric acid 20%	212	100	Sodium carbonate	212	100
Hydrochloric acid 38%	80	27	Sodium chloride	260	127
Hydrocyanic acid 10%	160	71	Sodium hydroxide 10%	x	x
Hydrofluoric acid 30%	230	110	Sodium hydroxide 50%	x	x
Hydrofluoric acid 70%	140	60	Sodium hydroxide, concentrated	x	x
Hydrofluoric acid 100%	140	60	Sodium hypochlorite 15%	x	x
Hypochlorous acid	x	x	Sodium hypochlorite, concentrated	x	x
Iodine solution 10%	x	x	Sodium sulfide to 10%	260	127
Ketones, general	100	38	Stannic chloride	260	127
Lactic acid 25%	212	100	Stannous chloride	250	121
Lactic acid, concentrated	160	71	Sulfuric acid 10%	160	71
Magnesium chloride	260	127	Sulfuric acid 50%	80	27
Malic acid 10%	260	127	Sulfuric acid 70%	80	27
Manganese chloride	200	93	Sulfuric acid 90%	x	x
Methyl chloride	120	49	Sulfuric acid 98%	x	x
Methyl ethyl ketone	80	27	Sulfuric acid 100%	x	x
Methyl isobutyl ketone	160	71	Sulfuric acid, fuming	x	x
Muriatic acid	80	27	Sulfurous acid	160	71
Nitric acid 5%	x	x	Thionyl chloride	x	x
Nitric acid 20%	x	x	Toluene	212	100
Nitric acid 70%	x	x	Trichloroacetic acid 30%	80	27
Nitric acid, anhydrous	x	x	White liquor	140	60
Nitrous acid, concentrated	x	x	Zinc chloride	160	71
Oleum	190	88			

[a]The chemicals listed are in the pure state or in a saturated solution unless otherwise indicated. Compatibility is shown to the maximum allowable temperature for which data are available. Incompatibility is shown by an x. A blank space indicates that data are unavailable.

Source: PA Schweitzer. *Corrosion Resistance Tables,* 4th ed. Vols. 1–3. New York: Marcel Dekker, 1995.

REFERENCES

1. HH Ulhig. Corrosion and Corrosion Control. New York: John Wiley, 1963.
2. CP Dillon. Corrosion Resistance of Stainless Steels. New York: Marcel Dekker, 1995.
3. PA Schweitzer. Stainless steel. In: PA Schweitzer, ed. Corrosion and Corrosion Protection Handbook. 2nd. ed. New York: Marcel Dekker, 1989, pp. 69–71.

4. JHW deWit. Inorganic and organic coatings. In: P Marcus and J Oudar, eds Corrosion Mechanisms in Theory and Practice. New York: Marcel Dekker, 1995, pp. 602–609.
5. PA Schweitzer. Corrosion Resistance of Elastomers. New York: Marcel Dekker, 1990.
6. PA Schweitzer. Corrosion Resistance Tables. 4th ed. Vol. 1–3. New York: Marcel Dekker, 1995.
7. GT Murray. Introduction to Engineering Materials. New York: Marcel Dekker, 1993.
8. JH Mallinson. Corrosion Resistant Plastics in Chemical Plant Design. New York: Marcel Dekker, 1988.
9. PA Schweitzer. Corrosion Resistant Piping Systems. New York: Marcel Dekker, 1994.

G

GALVANIC CORROSION

This form of corrosion is sometimes referred to as dissimilar metal corrosion and is found in the most unusual places, often causing the most painful professional headaches. The Galvanic Series of metals provides details of how the galvanic current will flow between two metals, and which metal will corrode when they are in contact or near each other and an electrolyte is present. Table G.1 lists the Galvanic Series.

You will note that several materials are shown in two places in the Galvanic Series, being indicated as either active or passive. This is a result of the tendency of some metals and alloys to form surface films, especially in oxidizing environments. These films shift the measured potential in the noble direction. In this state the material is said to be passive.

The particular way in which materials will react can be predicted from the relative positions of the materials in the Galvanic Series. When it is necessary to use dissimilar metals, two materials should be selected which are relatively close in the Galvanic Series. The further apart the metals are in the Galvanic Series, the greater the rate of corrosion.

The rate of corrosion is also affected by the relative areas between the anode and the cathode. Since the flow of current is from the anode to the cathode, the combination of a large cathode area and a small anode area is undesirable. Corrosion of the anode can be 100–1000 times greater than if the two areas were equal. Ideally the anode area should be larger than the cathode area.

The passivity of stainless steel is the result of the presence of a corrosion resistant oxide film on the surface. In most natural environments, it will remain in the passive state and tend to be cathodic to ordinary iron or steel. When chloride concentrations are high, such as in seawater or in reducing solutions, a change to the active state will usually take place. Oxygen starvation also causes a change to the active state. This occurs where there is no free access to oxygen, such as in crevices and beneath contamination of partially fouled surfaces.

Differences in soil conditions, such as moisture content and resistivity, can be responsible for creating anodic and cathodic areas. Where there is a difference in concentration of oxygen in the water or in moist soils in contact with metal at different areas, cathodes will develop at relatively high oxygen concentrations, and anodes at points of low concentrations. Strained portions of metals tend to be anodic and unstrained portions cathodic.

When joining two dissimilar metals together, galvanic corrosion can be prevented by insulating the two metals from each other. For example, when bolting flanges of dissimilar metals together, plastic washers can be used to separate the two metals.

See Refs. 1–5.

Table G.1 Galvanic Series of Metals and Alloys

Corroded end (anodic)	
Magnesium	Muntz metal
Magnesium alloys	Naval bronze
Zinc	Nickel (active)
Galvanized steel	Inconel (active)
Aluminum 6053	Hastelloy C (active)
Aluminum 3003	Yellow brass
Aluminum 2024	Admiralty brass
Aluminum	Aluminum bronze
Alclad	Red brass
Cadmium	Copper
Mild steel	Silicon bronze
Wrought iron	70–30 Cupro-nickel
Cast iron	Nickel (passive)
Ni-resist	Inconel (passive)
13% Chromium stainless steel (active)	Monel
50-50 Lead tin solder	18-8 Stainless steel type 304 (passive)
Ferretic stainless steel 400 series	18-8-3 Stainless steel type 316 (passive)
18-8 Stainless steel type 304 (active)	Silver
18-8-3 Stainless steel type (316 (active)	Graphite
Lead	Gold
Tin	Platinum
	Protected end (cathodic)

GALVANIC PROTECTION

See "Cathodic Protection."

GALVANIZED IRON

See "Galvanized Steel."

GALVANIZED STEEL

See "Zinc" also.

Galvanized steel is steel that has been coated with zinc. The galvanizing process is widely used to protect steel from atmospheric corrosion. Structures, sheet steel, wire, and piping are all forms that are protected by galvanizing. The protection afforded in rural atmospheres is greater than that in urban or industrial atmospheres. In the latter areas there is a greater concentration of industrial pollutants. The air in these areas is contaminated with various sulfur compounds, which together with the mositure in the air convert the normally impervious corrosion resistant zinc carbonate and zinc oxide layer into zinc sulfate and zinc sulfite. These water-soluble compounds have poor adhesion to the zinc surface and therefore are washed away relatively easily by rain. This exposes the underlying surface to attack by the oxygen in the air.

Galvanized steel is widely used in contact with many chemical specialty products such as detergents, agricultural chemicals, and similar materials. In most cases galvanized steel comes into contact with these chemicals during the handling, packaging, and storage of the finished products. Table G-2 shows the compatibility of galvanized steel with selected corrodents.

See Refs. 4 and 5.

Table G.2 Compatibility of Galvanized Steel with Selected Corrodents

Acetic acid	U	Chloric acid 20%	U
Acetone	G	Chlorine, dry	G
Acetonitrile	G	Chlorine water	U
Acrylonitrile	G	Chromium chloride	U
Acrylic latex	U	Chromium sulfate solution	U
Aluminum chloride 26%	U	Copper chloride solution	U
Aluminum hydroxide	U	Decyl acrylate	G
Aluminum nitrate	U	Diamylamine	G
Ammonia, dry vapor	U	Dibutylamine	G
Ammonium acetate solution	U	Dibutyl cellosolve	G
Ammonium bisulfate	U	Dibutyl phthalate	G
Ammonium bromide	U	Dichloroethyl ether	G
Ammonium carbonate	U	Diethylene glycol	G
Ammonium chloride 10%	U	Dipropylene glycol	G
Ammonium dichloride	U	Ethanol	G
Ammonium hydroxide		Ethyl acetate	G
Vapor	U	Ethyl acrylate	G
Reagent	U	Ethyl amine 69%	G
Ammonium molybdate	G	*N*-Ethyl butylamine	G
Ammonium nitrate	U	2-Ethyl butyric acid	G
Argon	G	Ethyl ether	G
Barium hydroxide		Ethyl hexanol	G
Barium nitrate solution	S	Fluorine, dry, pure	G
Barium sulfate solution	S	Formaldehyde	G
Beeswax	U	Fruit juices	S
Borax	S	Hexanol	G
Bromine, moist	U	Hexylamine	G
2-Butanol	G	Hexylene glycol	G
Butyl acetate	G	Hydrochloric acid	U
Butyl chloride	G	Hydrogen peroxide	S
Butyl ether	G	Iodine, gas	U
Butylphenol	G	Isohexanol	G
Cadmium chloride solution	U	Isooctanol	G
Cadmium nitrate solution	U	Isopropyl ether	G
Cadmium sulfate solution	U	Lead sulfate	U
Calcium hydroxide		Lead sulfite	S
sat. solution	U	Magnesium carbonate	S
20% solution	S	Magnesium chloride 42.5%	U
Calcium sulfate, sat. solution	U	Magnesium fluoride	G
Cellosolve acetate	G	Magnesium hydroxide sat.	S

Table G.2 Continued

Magnesium sulfate		Potassium nitrate	
2% solution	S	5–10% solution	S
10% solution	U	Potassium peroxide	U
Methyl amyl alcohol	G	Potassium persulfate 10%	U
Methyl ethyl ketone	G	Propyl acetate	G
Methyl propyl ketone	G	Propylene glycol	G
Methyl isobutyl ketone	G	Propionaldehyde	G
Nickel ammonium sulfate	U	Propionic acid	U
Nickel chloride	U	Silver bromide	U
Nickel sulfate	S	Silver chloride	
Nitric acid	U	pure, dry	S
Nitrogen, dry, pure	G	moist, wet	U
Nonylphenol	G	Silver nitrate solution	U
Oxygen		Sodium acetate	S
dry, pure	G	Sodium aluminum sulfate	U
moist	U	Sodium bicarbonate solution	U
Paraldehyde	G	Sodium bisulfate	U
Perchloric acid solution	S	Sodium carbonate solution	U
Permanganate solution	S	Sodium chloride solution	U
Peroxide		Sodium hydroxide solution	U
pure, dry	S	Sodium nitrate solution	U
moist	U	Sodium sulfate solution	U
Phosphoric acid 0.3–3%	G	Sodium sulfide	U
Polyvinyl acetate latex	U	Sodium sulfite	U
Potassium carbonate		Styrene, monomeric	G
10% solution	U	Styrene oxide	G
50% solution	U	Tetraethylene glycol	G
Potassium chloride solution	U	1,1,2 Trichloroethane	G
Potassium bichromate		1,2,3 Trichloropropane	G
14.7%	G	Vinyl acetate	G
20%	S	Vinyl ethyl ether	G
Potassium disulfate	S	Vinyl butyl ether	G
Potassium fluoride 5–20%	G	Water	
Potassium hydroxide	U	potable, hard	G

G = Suitable application; S = Borderline application; U = Not suitable

GASEOUS CORROSION

Gaseous corrosion is a corrosive reaction with only a gas present as the corrodent. There is no aqueous phase present. It is also known as dry corrosion.

GASEOUS PHASE CORROSION

See "Reducing Atmosphere Corrosion."

GENERAL CORROSION

See "Uniform Corrosion."

GLASS COATINGS

See "Glass Linings."

GLASS FIBER REINFORCEMENT

See "Thermoset reinforcing material."

GLASS LININGS

The bonding of glass to metal is not a recent development. Relics of glass bonded to gold jewelry (enameling) have been found which date back to 400 B.C. Until the early 1800s, enameling continued as an art form. The first application of a glass coating for any purpose other than an art form took place when cast iron sanitary ware was first coated.

The brewing industry was responsible for the development of the first large scale glassed steel equipment. It was the result of the need to improve the consistency of the quality of the beer. This took place in the early 1880s. Between that time and the start of the Second World War there were no notable developments in glassed steel composite.

With the advent of war the need developed for critical chemicals that were often corrosive and sticky. This sparked development programs to find materials to meet these needs. It was during this time that the importance of characteristics other than corrosion resistance of the composite were recognized, specifically thermal loading and mechanical stressing. Since that time, continued research efforts have culminated in a glassed steel product that is one of the major materials of construction used by the chemical processing industry.

There are a variety of glass linings, each of which has been developed for specific needs. The metal substrate provides the required strength and base thermal expansion for the glass lining while the majority of the end use requirements must be met by making adjustments to the glass composition. These end use requirements include resistance to corrosion, adhesion of glass to metal, thermal, mechanical, electrical type stressing influence, and reduced product adherence.

There are five basic types of glass lining formulations that are available. The most commonly specified is the standard lining, which is widely used by the chemical process industry. This lining represents a balance between chemical and physical property serviceability. The maximum thermal shock (cold-to-hot or hot-to-cold) is in the range of 260°F (127°C). Glass linings are available in blue or white. The white coloration is useful in ensuring the complete clean-out of a dark colored product.

When low temperature service is encountered, the substrate is of stainless steel construction, in conjunction with a special glass lining. This combination permits operation down to –200°F (–129°C). In modifying the glass composition to allow the low temperature operation, some corrosion resistance is lost.

Glass compositions can also be modified to allow high temperature operation. These systems extend the operation to 650°F (343°C) with the thermal shock differential increased to 360°F (182°C). There is no reduction in corrosion resistance.

Some applications require low product adherence to the lining. To meet this requirement a special surface layer free of many microscopic surface imperfections is required. Other properties remain unchanged. If a corrosive etch develops, the surface layer is no longer effective.

Glass is extremely weak in tension but strong in compression. By incorporating crystals in the glass lining, any applied tension type stresses will be effectively transferred to the stronger crystals. Quartz (SiO_2) crystals or other relatively glass-insoluble particles or fibers are used. The addition of these crystals will also improve impact resistance, abrasion resistance, and heat transfer.

In addition to improving the mechanical and physical properties mentioned above they also have the outstanding ability to inhibit crack propagation. Once a crack has formed in a pure glass system the crack will continue to propagate, especially under the influence of motion stressing, temperature change, and water. Crystals result in limiting the damage to a relatively small area and permit an economical repair to be made.

Glass Structure

A glass system used to produce lining can be considered as a three-dimensionsl network type structure, consisting of one or more oxide groups. The network formers are acidic type oxides that form the backbone of the glass structure. Silicon dioxide (SiO_2) is the primary network former and is obtained from relatively inexpensive beach sand. It is usually present in glasses in amounts exceeding 50 weight percent. The network modifiers are base type oxides but are not part of the network forming structure. They cannot form glasses by themselves. As the name implies, these oxides modify the properties of the network formers. The intermediates are amphoteric in nature and can act as either network formers or modifiers, depending upon the concentration, nature, and amounts of the other constituents. Aluminum oxide and titanium dioxide improve general corrosion resistance while zirconium dioxide improves alkali resistance.

The cover coat systems for glass linings are complex mixtures of up to 15 oxides taken from the above three groups and built around the framework of the silica network.

Corrosion of Glass

The corrosion of glass linings takes place by means of two basic mechanisms. One relates to the removal of the modifiers and/or the intermediates, while the other is the removal of the network formers.

Acids (fluorine and phosphorus compounds excepted), small ions, and the first stage of water attack invoke removal of the intermediates and modifiers, by a diffusion-controlled ion exhange mechanism in which the small ion, e.g., hydrogen from the acid, exchanges for the larger modifier/intermediate ion. This results in stress relief cracking in the network former that eventually leads to the dulling of the glass.

Corrosive attack by alkalies, fluorine, and phosphorus-containing compounds and the second stage of water attack removes compounds from the network former through a regenerated dissolution type reaction.

Dulling of the glass surface is usually the first sign of corrosion. This may be caused by either the ion exchange or the dissolution type of corrosion reaction and is usually uniform in appearance. The corrosive upset of the glass surface leads to localized stress differences, which in turn, lead to varying degrees of stress relief and eventually glass chipping.

One of the most serious types of glass corrosion damage is pitting. This type of attack is both corrosive and glass-composition sensitive, and can be caused by fluorine based chemistries and alkalies. Older glass systems and glass/crystal composites are more susceptible to this type of attack.

Corrosion Resistance of Glass

When considering the application of glassed steel equipment to handle a specific corrosive, thought must also be given to the effects of the corrodent on any tantalum repair plugs that may be present in the vessel. For example while sulfuric acid can be handled in glassed steel equipment the presence of small amounts of sulfur trioxide would attack any tantalum present. The same would apply to chlorosulfonic acid and oleum.

In general the following corrodents may be handled safely in glassed steel equipment:

Hydrochloric acid up to 300°F (149°C)
Chlorides in general
Bromides
Solids
Sulfuric acid up to 450°F (232°C)
Chlorosulfonic acid
Acetic acid
Organic compounds

Corrosives that will attack glassed steel are

Fluorides.
Alkaline compounds.
Salts with small cations, e.g., lithium, magnesium, aluminum in aqueous media should be used with caution above 150°F (66°C).
Phosphorus compounds frequently contain fluorides, and some phosphorus compounds possess a mutual solubility for glass.

See Refs. 6 and 12.

GLASSED STEEL

See "Glass Linings."

GRAPHITE FIBERS

Graphite fibers are used as reinforcing in FRP laminates and to provide conductivity. See "Thermoset Reinforcing Materials."

GRAPHITIZATION (GRAPHITIC CORROSION)

Cast iron is composed of a mixture of ferrite phase (almost pure iron) and graphite flakes. When it corrodes it forms corrosion products, which in some soils or waters cement together the residual graphite flakes. The resulting structure (e.g., water pipe), although corroded completely, may have sufficient remaining strength, even with low ductility, to continue to operate under the required pressures and stresses.

This form of corrosion occurs only with gray cast iron or ductile cast iron containing spheroidal graphite and not with white iron, which contains cementite and ferrite.

See Refs. 1, 7, and 8.

GRAY IRON

Gray iron is the least expensive of all the cast metals and consequently is the most widely used cast material on a weight basis. In general the corrosion resistance properties of gray iron are similar to those of carbon steel. In the presence of seawater, water, and various atmospheric environments its corrosion resistance is superior to that of carbon steel. Gray iron exhibits good resistance to alkaline solutions, such as sodium hydroxide and molten caustic soda. It is also resistant to alkaline salt solutions, such as cyanides, carbonates, sulfides, and silicates. Acid and oxidizing salts rapidly attack gray iron.

Gray iron is subject to a form of corrosion known as graphitization, which is a selective leaching of the iron matrix, leaving only a graphite network. No apparent dimensional changes may be present but there can be sufficient loss of material cross-section and strength that failure may occur.

Gray iron is a brittle alloy with poor shock resistance. This is the major drawback to a more widespread use.

See Refs. 8 and 9.

GREEN PLAGUE

Green corrosion deposits formed in copper hot water piping and on water faucets are known as green plague. It is thought to be caused by electrical grounding on copper dissolution.

GREEN ROT

Green rot is the formation of green chromium oxide on chromium-bearing alloys resulting from high temperature corrosion. The other alloy constituents remain unaffected.

GREEN RUST

Green rust is a greenish corrosion product formed on ferrous metals. It contains iron in two oxidation states and has a variable anion content and is related structurally to the pyroaurite group of naturally occurring minerals.

GROOVING CORROSION

Grooving corrosion is a form of localized corrosion appearing as grooves. When electric resistance welded carbon steel pipe is exposed to aggressive waters, grooves caused by the redistribution of sulfide inclusions along the weld line during the welding process are formed in the weld.

GROUT

There are two types of grout, each of which is used for a specific purpose. The first is a thin soupy mortar used for filling joints between previously laid tile or brick. These joints are usually approximately 1/4 in. (6 mm) wide. The grout is applied by squeegeeing it into the open joints using a flat rectangular rubber faced trowel. This is usually associated with tilesetting.

The second form of grout is used for setting machinery. These grouts are similar to the tilesetting grouts but utilize larger aggregate than the tile grouts. Resin viscosities can also vary from those of the tile grouts.

Both forms of grout are available in the same corrosion resistant formulations. For additional information regarding the various formulations and their corrosion resistant properties see "Mortars," since grouts and mortars are of the same chemical formulations.

See Refs. 10 and 11.

REFERENCES

1. HH Uhlig. Corrosion and Corrosion Control. New York: John Wiley, 1963.
2. DM Berger. Fundamentals and prevention of metallic corrosion. In: PA Schweitzer, ed. Corrosion and Corrosion Protection Handbook. 2nd ed. New York: Marcel Dekker, 1989, pp. 5–9.
3. CP Dillon. Corrosion Resistance of Stainless Steels. New York: Marcel Dekker, 1995.
4. FC Porter. Corrosion Resistance of Zinc and Zinc Alloys. New York: Marcel Dekker, 1994.
5. I Suzuki. Corrosion Resistant Coatings Technology. New York: Marcel Dekker, 1989.
6. DH De Clerk. Glass linings. In: PA Schweiter, ed. Corrosion Engineering Handbook. New York: Marcel Dekker, 1996, pp. 489–544.
7. CP Dillon. Corrosion Control in the Chemical Process Industries. 2nd. ed. St. Louis: Materials Technology Institute of the Chemical Process Industries, 1994.
8. GW George and PG Breig. Cast alloys. In: PA Schweitzer, ed. Corrosion and Corrosion Protection Handbook. 2nd. ed. New York: Marcel Dekker, 1989, pp. 285–289.
9. JL Gossett. Corrosion resistance of cast alloys. In: PA Schweitzer, ed. Corrosion Engineering Handbook New York: Marcel Dekker, 1996, pp. 258–259.
10. AA Boova. Chemical-resistant mortars, grouts, and monolithic surfacings. In: PA Schweitzer, ed. Corrosion Engineering Handbook. New York: Marcel Dekker, 1996, pp. 459–487.
11. WL Sheppard, Jr. Chemically Resistant Masonry 2nd ed., New York: Marcel Dekker, 1982.
12. PA Schweitzer, Corrosion Resistant Piping Systems. New York: Marcel Dekker, 1994.

H

HALAR

See "Ethylene Chlorotrifluoroethylene."

HALIDE

A halide is a compound of a halogen with another element or group.

HALOGEN

Halogen refers to any element in the family that consists of fluorine, chlorine, bromine, iodine, and astame (group V11A in the periodic chart of the elements).

HALOGENATED POLYESTER RESINS

Also see "Polymers" and "Thermoset Polymers." These consist of either chlorinated or brominated polyesters. Excellent resistance is exhibited in contact with oxidizing acids and solutions, such as 35% nitric acid at elevated temperatures and 70% nitric acid at room temperature, 40% chromic acid, chlorine water, wet chlorine, and 15% hypochlorites. They are also resistant to neutral and acid salts, nonoxidizing acids, organic acids, mercaptans, ketones, alcohols, glycols, organic esters, and fats and oils.

These polyesters are not resistant to highly alkaline solutions of sodium hydroxide, concentrated sulfuric acid, alkaline solutions with pH greater than 10, aliphatic, primary and aromatic amines, amides and other alkaline organics, phenol, and acid halides. Refer to Table H.1 for the compatibility of halogenated polyesters with selected corrodents. Reference 1 provides the compatibility of halogenated resins with a wider range of corrodents.

See also Refs. 2–4.

HASTELLOY

Hastelloy is the trademark of Haynes International Inc. and is prefixed to a series of high nickel alloys designed for corrosion resistance. These alloys include Hastelloy C-276, Hastelloy B and B-2, Hastelloy G, G-3, and G-30, Hastelloy alloy X, Hastelloy C-2000, and Hastelloy alloy No. 230.

Table H.1 Compatibility of Halogenated Polyester with Selected Corrodents[a]

Chemical	Maximum temp. °F	Maximum temp. °C	Chemical	Maximum temp. °F	Maximum temp. °C
Acetaldehyde	x	x	Benzene	90	32
Acetic acid 10%	140	60	Benzene sulfonic acid 10%	120	49
Acetic acid 50%	90	32	Benzoic acid	250	121
Acetic acid, glacial	110	43	Benzyl alcohol	x	x
Acetic anhydride	100	38	Benzyl chloride	x	x
Acetone	x	x	Borax	190	88
Acetyl chloride	x	x	Boric acid	180	82
Acrylic acid	x	x	Bromine gas, dry	100	38
Acrylonitrile	x	x	Bromine gas, moist	100	38
Adipic acid	220	104	Bromine liquid	x	x
Allyl alcohol	x	x	Butyl acetate	80	27
Allyl chloride	x	x	Butyl alcohol	100	38
Alum 10%	200	93	*n*-Butylamine	x	x
Aluminum chloride, aqueous	120	49	Butyl phthalate	100	38
Aluminum fluoride 10%	90	32	Butyric acid 20%	200	93
Aluminum hydroxide	170	77	Calcium bisulfide	x	x
Aluminum nitrate	160	71	Calcium bisulfite	150	66
Aluminum oxychloride			Calcium carbonate	210	99
Aluminum sulfate	250	121	Calcium chlorate	250	121
Ammonia gas	150	66	Calcium chloride	250	121
Ammonium carbonate	140	60	Calcium hydroxide, sat.	x	x
Ammonium chloride 10%	200	93	Calcium hypochlorite 20%	80	27
Ammjonium chloride 50%	200	93	Calcium nitrate	220	104
Ammonium chloride, sat.	200	93	Calcium oxide	150	66
Ammonium fluoride 10%	140	60	Calcium sulfate	250	121
Ammonium fluoride 25%	140	60	Caprylic acid	140	60
Ammonium hydroxide 25%	90	32	Carbon bisulfide	x	x
Ammonium hydroxide, sat.	90	32	Carbon dioxide, dry	250	121
Ammonium nitrate	200	93	Carbon dioxide, wet	250	121
Ammonium persulfate	140	60	Carbon disulfide	x	x
Ammonium phosphate	150	66	Carbon monoxide	170	77
Ammonium sulfate 10–40%	200	93	Carbon tetrachloride	120	49
Ammonium sulfide	120	49	Carbonic acid	160	71
Ammonium sulfite	100	38	Cellosolve	80	27
Amyl acetate	190	85	Chloracetic acid, 50% water	100	38
Amyl alcohol	200	93	Chloracetic acid 25%	90	32
Amyl chloride	x	x	Chlorine gas, dry	200	93
Aniline	120	49	Chlorine gas, wet	220	104
Antimony trichloride 50%	200	93	Chlorine, liquid	x	x
Aqua regia 3:1	x	x	Chlorobenzene	x	x
Barium carbonate	250	121	Chloroform	x	x
Barium chloride	250	121	Chlorosulfonic acid	x	x
Barium hydroxide	x	x	Chromic acid 10%	180	82
Barium sulfate	180	82	Chromic acid 50%	140	60
Barium sulfide	x	x	Chromyl chloride	210	99
Benzaldehyde	x	x	Citric acid 15%	250	121

Table H.1 Continued

Chemical	Maximum temp. °F	Maximum temp. °C	Chemical	Maximum temp. °F	Maximum temp. °C
Citric acid, concentrated	250	121	Nitric acid 70%	80	27
Copper acetate	210	99	Nitrous acid, concentrated	90	32
Copper chloride	250	121	Oleum	x	x
Copper cyanide	250	121	Perchloric acid 10%	90	32
Copper sulfate	250	121	Perchloric acid 70%	90	32
Cresol	x	x	Phenol 5%	90	32
Cyclohexane	140	60	Phosphoric acid 50–80%	250	121
Dichloroacetic acid	100	38	Picric acid	100	38
Dichloroethane (ethylene dichloride)	x	x	Potassium bromide 30%	230	110
Ethylene glycol	250	121	Salicylic acid	130	54
Ferric chloride	250	121	Sodium carbonate 10%	190	88
Ferric chloride 50% in water	250	121	Sodium chloride	250	121
Ferric nitrate 10–50%	250	121	Sodium hydroxide 10%	110	43
Ferrous chloride	250	121	Sodium hydroxide 50%	x	x
Ferrous nitrate	160	71	Sodium hydroxide, concentrated	x	x
Hydrobromic acid, dilute	200	93	Sodium hypochlorite 20%	x	x
Hydrobromic acid 20%	160	71	Sodium hypochlorite, concentrated	x	x
Hydrobromic acid 50%	200	93	Sodium sulfide to 50%	x	x
Hydrochloric acid 20%	230	110	Stannic chloride	80	27
Hydrochloric acid 38%	180	82	Stannous chloride	250	121
Hydrocyanic acid 10%	150	66	Sulfuric acid 10%	260	127
Hydrofluoric acid 30%	120	49	Sulfuric acid 50%	200	93
Hypochlorous acid 10%	100	38	Sulfuric acid 70%	190	88
Lactic acid 25%	200	93	Sulfuric acid 90%	x	x
Lactic acid, concentrated	200	93	Sulfuric acid 98%	x	x
Magnesium chloride	250	121	Sulfuric acid 100%	x	x
Malic acid 10%	90	32	Sulfuric acid, fuming	x	x
Methyl chloride	80	27	Sulfurous acid 10%	80	27
Methyl ethyl ketone	x	x	Thionyl chloride	x	x
Methyl isobutyl ketone	80	27	Toluene	110	43
Muriatic acid	190	88	Trichloroacetic acid 50%	200	93
Nitric acid 5%	210	99	White liquor	x	x
Nitric acid 20%	80	27	Zinc chloride	200	93

[a]The chemicals listed are in the pure state or in a saturated solution unless otherwise indicated. Compatibility is shown to the maximum allowable temperature for which data are available. Incompatibility is shown by an x. A blank space indicates that data are unavailable.
Source: PA Schweitzer. *Corrosion Resistance Tables,* 4th ed. Vols. 1–3. New York: Marcel Dekker, 1995.

Refer to "High Nickel Alloys" or to the specific alloy for detailed information.

HASTELLOY ALLOY C-2000

Hastelloy C-2999 is a nickel–chromium–molybdenum–copper alloy containing 23% chromium, 16% molybdenum, 1.6% copper, 0.01% carbon (max.), 0.08% silicon (max.), and the remainder nickel. It has the unique property of possessing outstanding resistance to oxidizing media with superior resistance to reducing environments. In addition the alloy also exhibits pitting resistance and crevice corrosion resistance superior to C-276 alloy.

In contact with a 0–60 weight percent of hot sulfuric acid, alloy C-2000 has better corrosion resistance than alloy C-276. When exposed to boiling hydrochloric acid (weight percent of 1 to 15), alloy C-276 exceeds a corrosion rate of 20 mpy, while alloy C-2000 provides good resistance up to a concentration of 3 weight percent. Excellent resistance is also displayed in oxidizing media such as nitric acid and solutions containing ferric ions, cupric ions or dissolved oxygen.

Hastelloy and alloy C-2000 are trademarks of Haynes International Inc.

HASTELLOY ALLOY G-30

This alloy is a modification of alloy G with a chromium content of 29.5%. This higher chromium content provides alloy G-30 with a greater resistance to oxidizing environments than other alloys in this class. It also contains 5.5% molybdenum and 2.5% tungsten, which improves the alloy's resistance to moderate amounts of chlorides and fluorides.

In acid mixtures such as sulfuric plus nitric and nitric plus hydrofluoric alloy G-30 shows the highest resistance of this class of alloys as illustrated in the table.

Alloy	Corrosion rate mm/yr	(mpy)
Hastelloy alloy G-30	4.43	(177)
Hastelloy alloy G	13.5	(540)
Haynes alloy 625	60.0	(2400)

Alloy G-30 finds application in the evaporators of commercial wet-process phosphate manufacturing systems. In this application the alloy is subjected to complex mixtures of phosphoric, sulfuric, and hydrofluoric acids that also contain various oxides. The resistance of alloy G-30 as compared to other alloys in this application is shown in the table.

Alloy	Corrosion rate mm/yr	(mpy)
Hastelloy alloy G-30	0.15	(6)
Hastelloy alloy G	0.40	(16)
Haynes alloy 625	0.40	(16)

See Ref. 5.

HAYNES ALLOY NO. 214

Alloy 214 develops a tenacious aluminum oxide layer at the surface that provides superior resistance to carburizing environments containing chorine and oxygen. This alloy also possesses the highest oxidation resistance of any of the nickel based alloys as a result of this film.

As typical of many high temperature alloys, this alloy does not possess good resistance to aqueous chloride solutions, so dew point conditions must be avoided.

Typical applications include mesh belts for supporting chinaware while being heated in a kiln, strand annealing tubes for making medical grade wire, and honeycomb seals in turbine engines.

See Ref. 5.

HAYNES ALLOY NO. 230

This alloy was originally developed for its high temperature strength, thermal stability, and oxidation resistance. However, its nitridation resistance, along with its high creep strength, has enabled this alloy to be used as a catalyst support grid in the manufacture of nitric acid.

This alloy also possesses good resistance to carburization, but it does not have adequate resistance to sulfadizing environments.

See Ref. 5.

HAYNES ALLOY No. 556

The presence of 18% cobalt in this alloy provides greater resistance to sulfidation than many nickel based alloys such as alloy X or alloy 800H. In pure oxidation alloy 556 shows good resistance, but it is superseded in performance by other alloys such as alloys X and 214.

In chloride bearing oxidizing environments alloy 556 shows better resistance than alloys 800H and alloy X but not as good as alloy 214.

In carburizing environments alloy 556 is better than 310 stainless steel and some nickel based alloys such as alloy X and 617 too, but not as good as the aluminum-containing alloys such as alloy 214.

Typical applications include internals of municipal waste incinerators and refractory anchors in a refinery train-gas-burning unit.

See Ref. 5.

HASTELLOY ALLOY X

Hastelloy alloy X possesses a combination of high strength and excellent oxidation resistance. Its oxidation resistance is due to the formation of a complex chromium oxide spinel that provides good resistance up to temperatures of 2150°F (1177°C). The high temperature strength and resistance to warpage and distortion provide outstanding performance as distributor plates and catalyst grid supports.

Typical applications in the process industries are as distributor plates in the production of magnesium chloride and grid supports in the manufacture of nitric acid. Alloy X is a major material of construction in the hot section of gas turbine engines for such parts as burner cans and transition ducts.

Other high temperature applications include flare nozzles, thermowell protection tubes, expansion bellows, furnace internals retorts, muffles, and trays.

See Refs. 5.

HEAT AFFECTED ZONE (HAZ)

A heat affected zone is an area of a section of metal in which the mechanical properties and/or the microstructure has been changed by the heat of welding or thermal cutting. For most welds in carbon and low-alloy steels the heat affected zone is a band, usually approximately 1/8 in. (3 mm) wide, adjacent to the fusion line of the weld. In austenitic stainless steels there may be generated a secondary HAZ some distance from the fusion line as a result of welding-induced sensitization.

HIGH ALLOY CAST IRONS

There are three types of high alloy cast irons, namely austenitic gray cast irons, austenitic ductile cast irons, and high silicon cast irons. The austenitic gray cast irons are often referred to as Ni-Resist, a trademark of Inco Alloys International. The austenitic ductile cast irons are commonly called Ductile Ni-Resist, also a trademark of Inco Alloys International. The high silicon cast irons are commonly referred to as Duriron and Durichlor 51, both trademarks of Duriron Company, Inc.

HIGH NICKEL ALLOYS

This class of alloys has the widest range of applications of any class of alloys. The electrochemical properties of nickel and its crystallographic characteristics enable it to accommodate large amounts of alloying elements. The alloys used for corrosion resistant purposes are primarily solid-solution strengthened alloys supplied in the fully annealed condition.

The element nickel is nobler than iron but more active than copper in the electromotive series. In reducing environments, such as dilute sulfuric acid, nickel is more corrosion resistant than iron but not as resistant as copper or nickel copper alloys. The nickel molybdenum alloys are more corrosion resistant in reducing acids than nickel or nickel copper alloys.

Nickel can form a protective passive film in some environments. However, this passive film is not particularly stable, and therefore nickel cannot generally be used in oxidizing environments such as nitric acid. If it is alloyed with chromium, a much more stable film is formed, and corrosion resistance is exhibited to a variety of oxidizing environments. However, these alloys can corrode in environments containing appreciable amounts of chlorides or other halides, especially if oxidizing species are present. Alloying with molybdenum or tungsten is necessary to improve the resistance to corrosion by oxidizing chloride solutions.

Monel Alloy 400

Also see "Monel." The alloying of 30–33% copper with nickel, producing Monel 400, provides an alloy with many of the characteristics of chemically pure nickel, with improvements. A major area of application is in water handling, including brackish water and seawater. As with nickel 200, the alloy can pit in stagnant seawater, but the rate of attack is significantly reduced. The absence of chloride stress corrosion cracking is also a factor in the selection of this alloy.

The general corrosion resistance of Monel 400 in the nonoxidizing acids such as sulfuric, hydrochloric, and phosphoric is improved over that of pure nickel. The alloy is not resistant to oxidizing media such as nitric acid, ferric chloride, chromic acid, wet chlorine, sulfur dioxide, or ammonia.

Monel 400 exhibits excellent resistance to hydrofluoric acid solutions at various concentrations and temperatures. It is subject to stress corrosion cracking in moist, aerated hydrofluoric or hydrofluorosilicic acids and vapor. The corrosion of Monel is negligible in all types of atmospheres. Indoor exposure produces a very light tarnish that is easily removed by occasional wiping. Outdoor surfaces that are exposed to rain produce a thin gray-green patina. In sulfurous atmospheres a smooth brown adherent film forms.

Because of its high nickel content Monel 400 is practically as resistant as nickel 200 to caustic soda throughout most of the concentration range. It is also resistant to anhydrous ammonia and to ammonium hydroxide solutions of up to 3% concentration. The alloy is subject to stress corrosion cracking at high temperatures, in concentrated caustic, and in mercury.

Inconel Alloy 600

Also see "Inconel." This is a nickel–chromium–iron alloy for use in environments requiring resistance to heat and corrosion. Alloy 600 is practically free from corrosion by fresh waters, including the most corrosive of natural waters containing free carbon dioxide, iron compounds, and dissolved air. It remains free from stress corrosion cracking even in boiling magnesium chloride.

Alloy 600 has excellent resistance to dry halogens at elevated temperatures and has been used successfully for chlorination equipment at temperatures up to 1000°F (538°C). Alloy 600 is not subject to stress corrosion cracking in any of the chloride salts and has excellent resistance to all of the nonoxidizing halides.

Resistance to stress-corrosion cracking is imparted to alloy 600 by virtue of its nickel base. Therefore it finds application in handling water environments where stainless steel fails by cracking.

Because of its chromium content, alloy 600 exhibits greater resistance to sulfuric acid under oxidizing conditions than either nickel 200 or Monel 400. The addition of oxidizing salts to sulfuric acid tends to passivate alloy 600, which makes it suitable for use with acid mine waters or brass pickling solutions, applications for which Monel 400 cannot be used.

Inconel Alloy 625

Also see "Inconel." This is a nickel-chromium-molybdenum-columbium alloy. It is resistant to oxidation, general corrosion, and pitting and crevice corrosion, and it is virtually immune to chloride ion stress corrosion cracking.

In fresh and distilled water, corrosion rates are essentially zero; in seawater, under flowing stagnant conditions, weight losses are extremely low.

Resistance to aqueous solutions is good in a variety of applications including organic acids, sulfuric acid, and hydrochloric acid at temperatures below 150°F (60°C). The alloy also exhibits satisfactory resistance to hydrofluoric acid. Although nickel based alloys are not normally used in nitric acid service because of cost, alloy 625 is resistant to mixtures of nitric and hydrofluoric, where stainless steel loses its resistance.

Alloy 625 also has excellent resistance to phosphoric acid solutions, including commercial grades of acids that contain fluorides, sulfates, and chlorides in the production of superphosphoric acid (72% P_2O_8).

In general, alloy 625 can be used wherever alloy 600 is suitable, and in some applications where alloy 600 does not have a satisfactory resistance.

Incoloy Alloy 800

Also see "Incoloy." This alloy contains approximately 20% chromium, 32% nickel, and 46% iron as balance. It is used primarily for its oxidation resistance at elevated temperatures.

At moderate temperatures the general corrosion resistance of alloy 800 is similar to that of the other austenitic nickel–iron–chromium alloys. However, as the temperature increases, alloy 800 continues to exhibit good corrosion resistance, while other austenitic alloys are unsatisfactory for the service.

Alloy 800 has excellent resistance to nitric acid at concentrations up to about 70%. It resists a variety of oxidizing salts, but not halide salts. It also has good resistance to organic acids, such as formic, acetic, and propionic. Alloy 800 is particularly suited for the handling of hot corrosive gases, such as hydrogen sulfide.

In aqueous service, alloy 800 has a general resistance to corrosion that falls between that of types 304 and 316 stainless steels. Therefore the alloy is not widely used for aqueous service. The stress corrosion cracking resistance of alloy 800 is, though it is not immune to it, better than that of the 300 series of stainless steels, and, it may be substituted for it on this basis.

Incoloy Alloy 825

Also see "Incoloy Alloy 825." Alloy 825 is very similar to alloy 800, but the composition has been modified to provide for improved aqueous corrosion resistance.

The higher nickel content of alloy 825, compared to alloy 800, makes it resistant to chloride ion stress corrosion cracking. Additions of molybdenum and copper give resistance to pitting and to corrosion in reducing acid environments such as sulfuric or phosphoric acid solutions. Alloy 825 is resistant to pure sulfuric acid solution up to 40% by weight at boiling temperatures and at all concentrations at a maximum temperature of 150°F (60°C). In dilute solutions the presence of oxidizing salts such as cupric or ferric actually reduces the corrosion rates. It has limited use in hydrochloric or hydrofluoric acids.

The chromium content of alloy 825 gives it resistance to various oxidizing environments such as nitrates, nitric acid solutions, and oxidizing salts. The alloy is not fully resistant to stress corrosion cracking when tested in boiling magnesium chloride, but it has good resistance in neutral chloride environments.

If localized corrosion is a problem with the 300 series stainless steels, alloy 825 may be substituted. Alloy 825 also provides excellent resistance to corrosion by seawater.

Alloy 800H

This alloy is a controlled carbon version of alloy 800. The carbon content is maintained between 0.05 and 0.1% to provide the alloy with better elevated temperature creep and stress rupture properties. It is used in a variety of high temperature applications in the refining and heat treatment industries.

Alloy B-2

Also see "Alloy B-2." Within the nickel–molybdenum series one alloy, B-2, is a low carbon and low silicon version of alloy B. These alloys are uniquely different from other corrosion resistant alloys because they do not contain chromium. Molybdenum is the primary alloying ingredient and provides significant corrosion resistance to reducing environments.

Alloy B-2 is recommended for the handling of all concentrations of hydrochloric acid in the temperature range of 158–212°F (70–100°C) and for the handling of wet hydrogen chloride gas.

Alloy B-2 has excellent resistance to pure sulfuric acid at all concentrations and temperatures below 60% acid and good resistance to 212°F (100°C) above 60% acid. It is also resistant to nonoxidizing environments such as hydrofluoric and phosphoric acids, and several organic acids such as acetic, formic, and crysilic acids. It is also resistant to many nonoxidizing chloride-bearing salts such as aluminum chloride, magnesium chloride, and antimony chloride.

Because alloy B-2 is nickel rich (approximately 70%), it is resistant to chloride-induced stress corrosion cracking. By virtue of the high molybdenum content it is highly resistant to pitting attack in acid chloride environments.

The major disadvantage of alloy B-2 is its lack of corrosion resistance in oxidizing environments. It has virtually no corrosion resistance to oxidizing acids such as nitric and chromic or to oxidizing salts such as ferric chloride or cupric chloride. The presence of oxidizing salts such as ferric chloride, ferric sulfate, or cupric chloride, even when present in the part per million range, can accelerate the attack in hydrochloric or sulfuric acids. Even dissolved oxygen has sufficient oxidizing power to affect the corrosion rates in hydrochloric acid.

Alloy C-276

Also see "Alloy C-276." This is a nickel–molybdenum–chromium–tungsten alloy with outstanding corrosion resistance that is maintained even in the welded condition.

Alloy C-276 possesses good resistance in both oxidizing and reducing media, including conditions with halogen ion contamination. The pitting and crevice corrosion resistance of this alloy makes it an excellent choice when dealing with acid chloride salts.

Alloy C-276 has exceptional resistance to highly oxidizing neutral and acid chlorides, solvents, chlorine, formic and acetic acids, and acetic anhydride. It also resists highly corrosive agents such as wet chlorine gas, hypochlorites, and chlorine solutions.

Alloy C-276 has excellent resistance to phosphoric acid. At all temperatures below the boiling point of phosphoric acid, when concentrations are less than 65% by weight, tests have shown corrosion rates of less than 5 mpy. At acid concentrations above 65% by weight and up to 85%, alloy C-276 displays similar corrosive behavior, except at temperatures between 240°F (116°C) and the boiling point, where corrosion rates may be erratic and may reach 25 mpy.

Nickel–Chromium–Iron–Molybdenum Alloys

The addition of molybdenum to the Ni–Cr–Fe alloys provides for an increase in the corrosion resistance to enviroumnents containing chlorides and to moderately corrosive reducing environments such as dilute sulfuric and phosphoric acids. The amount of molybdenum that can be alloyed is limited because of the presence of iron, which leads to embrittlement resulting from precipitation of intermetallic phases. The iron present also tends to reduce the cost of these alloys.

Alloy G/Hastelloy Alloy G-3

Also see "Alloy G" and "Alloy G-3." Alloy G-3 is a low carbon version of alloy G. The corrosion resistance of alloy G and alloy G-3 are approximately the same, but the thermal stability of alloy G-3 is greater. These alloys are highly resistant to pitting and stress corrosion cracking in both acid and alkaline environments, including hot sulfuric and phosphoric acids, hydrofluoric and contaminated nitric acids, mixed acids, and sulfate compounds.

Alloy G has found many applications in pollution control equipment. Excellent performance has been demonstrated in municipal garbage incinerator systems, including fans, ducts, and scrubber equipment. The SO_2 scrubber systems for power plants that use water or alkaline quench have been incorporating alloy G as a major material of construction because of its resistance to sulfuric acid and conditions during which the chloride ion can concentrate.

Hastelloy Alloy X

Also see "Hastelloy alloy X." Alloy X provides high strength and excellent oxidation resistance at elevated temperatures. A typical application in the process industries is the nitric acid catalyst grid supports operating at 1650°F (900°C). The high temperature strength and resistance to warpage and distortion provide outstanding performance.

Alloy X is also used for distributor plates in the manufacture of magnesium chloride. Other high temperature applications include flare nozzles, thermowell protection tubes, expansion bellows, furnace internal retorts, muffles, and trays.

Hastelloy Alloy G-30

Also see "Hastelloy Alloy G-30." This alloy is a modification of alloy G-3 with a higher chromium content, which provides it with a higher resistance to oxidizing environments than other alloys in this series.

In acid mixtures such as nitric plus hydrofluoric and sulfuric plus nitric, alloy G-30 shows the highest resistance of this class of alloys.

This alloy is used in the evaporators of commercial wet-process phosphoric manufacturing systems. In these environments, which are complex mixtures of phosphoric, sulfuric, and hydrofluoric acids and contain various oxides, alloy G-30 has shown the lowest corrosion rate of a number of alloys tested. The corrosion rate for alloy G-30 was 6 mpy as compared to 16 mpy for alloy G-3 and alloy 625.

Alloys for High Temperature Corrosion

Alloys that are designed to resist high temperature corrosion are basically oxidation resistant materials, since all forms of attack at elevated temperatures are considered to be oxidation. As with aqueous corrosion, a protective oxide film is formed. The rate at which the metal will oxidize will depend upon the stability of the film. If the film is stable and remains in place, the rate will be logarithmic, diminishing with time.

Cycling temperatures will tend to spall off the surface film, leading to a stepwise oxidation of the alloy. Changes in the environment can also have the same effect.

Although all high temperature corrosion is considered to be oxidation, there are other terms that are also encountered, such as oxidation/reduction, sulfidation, fuel ash corrosion, carburization, and nitridation, to mention a few.

While many of the high nickel alloys previously discussed can be utilized at elevated temperatures, there are some instances where these materials are not satisfactory. Consequently other alloys have been developed to overcome these shortcomings.

Haynes Alloy No. 556

The presence of 18% cobalt in this alloy provides greater resistance to sulfidation than many nickel based alloys such as alloy X or alloy 800H. In pure oxidation, alloy 556 shows good resistance, but it is superseded in performance by other alloys, such as alloys X and 214.

In chloride bearing oxidizing environments, the alloy shows better resistance than alloys 800H and X, but not as good as alloy 214.

In carburizing environments, the alloy is better than 310 stainless steel and some nickel based alloys such as alloy X and 617, but not as good as the aluminum containing alloys such as alloy 214.

Typical applications include internals of municipal waste incinerators and refractory anchors in a refinery train-gas-burning unit.

Haynes Alloy No. 214

This alloy possesses the highest oxidation resistance of any of the nickel based alloys to both static and dynamic environments. Alloy 214 develops a tenacious aluminum oxide layer at the surface. The aluminum film also provides superior resistance to carburizing environments containing chlorine and oxygen.

As is typical of many high temperature alloys, this alloy does not have good resistance to aqueous chloride solutions, so dew point conditions must be avoided.

Typical applications of this alloy include mesh belts for supporting chinaware while being heated in a kiln, strand annealing tubes for making medical grade stainless wire, and honeycomb seals in turbine engines.

Hastelloy Alloy No. 230

The outstanding feature of this alloy is its superior nitridation resistance. This property, with its high creep strength, has enabled use of the alloy as a catalyst support grid in the manufacture of nitric acid. It also exhibits good resistance to carburization. However, the alloy does not possess adequate resistance to sulfidizing environments.

Reference 1 provides an extensive listing of the compatibility of high nickel alloys with various corrodents.

HIGH SILICON IRON

There are two high silicon iron alloys that are of value in the corrosion field, both of which are manufactured by the Duriron Company. The first alloy, known as Duriron, contains nominally 14.5% silicon, 1% carbon, with the balance iron. When 4% chromium is added, the product is known as Durichlor. These are cast alloys.

The high silicon content improves corrosion resistance, but it also lowers some of the mechanical properties as compared with gray iron. Silicon irons are hard and brittle and therefore do not stand up well under shock and impact. Because of their hardness, high silicon irons are good for combined corrosion–erosion service. These alloys cannot withstand any substantial stressing or impact, and they cannot be subjected to sudden fluctuations in temperature. Refer to Table H.2 for the mechanical and physical properties.

High silicon irons have excellent corrosion resistance to a wide range of chemicals. One of the major applications is in handling sulfuric acid. It is resistant to all concentrations of sulfuric acid, up to and including the normal boiling point. This alloy will handle nitric acid above 30% to the boiling point. Below 30% the temperature is limited to about 180°F (27°C). Refer to Table H.3 for the compatibility of high silicon iron with selected corrodents.

Refer to Ref. 1 for a broader range of the compatibility of high silicon iron with selected corrodents.

See also Ref. 7.

Table H.2 Mechanical and Physical Properties of High Silicon Iron

Property	Duriron	Durichlor
Modulus of elasticity × 10^6, psi	23	23
Tensile strength × 10^3, psi	16	16
Elongation in 2 in., %	nil	nil
Hardness, Brinell	520	520
Density, lb/in.^{3}	0.255	0.255
Specific gravity	7.0	7.0
Specific heat at 32–212°F, Btu/lb °F	0.13	0.13
Coefficient of thermal expansion × 10^{-6}, Btu/ft^2 h/°F/in.		
at 32–212°F		7.2
at 68–392°F	7.4	

Table H.3 Compatibility of High Silicon Iron[a] with Selected Corrodents[b]

Chemical	Maximum temp. °F	°C	Chemical	Maximum temp. °F	°C
Acetaldehyde	90	32	Boric acid	80	27
Acetic acid 10%	200	93	Bromine gas, dry	x	x
Acetic acid 50%	200	93	Bromine gas, moist	80	27
Acetic acid 80%	260	127	Butyl alcohol	80	27
Acetic acid, glacial	230	110	Butyl phthalate	80	27
Acetic anhydride	120	49	Butyric acid	80	27
Acetone	80	27	Calcium bisulfite	x	x
Acetyl chloride	80	27	Calcium carbonate	90	32
Acrylonitrile	80	27	Calcium chlorate	80	27
Adipic acid	80	27	Calcium chloride	210	99
Allyl alcohol	80	27	Calcium hydroxide, sat.	200	93
Allyl chloride	90	32	Calcium hypochlorite	80	27
Alum	240	116	Calcium sulfate	80	27
Aluminum acetate	200	93	Caprylic acid	90	32
Aluminum fluoride	x	x	Carbon bisulfide	210	99
Aluminum hydroxide	80	27	Carbon dioxide, dry	570	299
Aluminum nitrate	80	27	Carbon dioxide, wet	80	27
Aluminum sulfate	80	27	Carbon tetrachloride	210	99
Ammonium bifluoride	x	x	Carbonic acid	80	27
Ammonium carbonate	200	93	Cellosolve	90	32
Ammonium chloride 10%			Chloracetic acid, 50% water	80	27
Ammonium chloride 50%	200	93	Chloracetic acid	90	32
Ammonium fluoride 10%	x	x	Chlorobenzene	80	27
Ammonium fluoride 25%	x	x	Chloroform	90	32
Ammonium hydroxide 25%	210	99	Chromic acid 10%	200	93
Ammonium nitrate	90	32	Chromic acid 50%	200	93
Ammonium persulfate	80	27	Chromyl chloride	210	99
Ammonium phosphate	90	32	Citric acid, concentrated	200	93
Ammonium sulfate 10–40%	80	27	Copper chloride	x	x
Amyl acetate	90	32	Copper cyanide	80	27
Amyl alcohol	90	32	Copper sulfate	100	38
Amyl chloride	90	32	Cyclohexane	80	27
Aniline	250	121	Cyclohexanol	80	27
Antimony trichloride	80	27	Dichloroethane (ethylene dichloride)	80	27
Aqua regia 3:1	x	x	Ethylene glycol	210	99
Barium carbonate	80	27	Ferric chloride	x	x
Barium chloride	80	27	Ferric nitrate 10–50%	90	32
Barium sulfate	80	27	Ferrous chloride	100	38
Barium sulfide	80	27	Fluorine gas, dry	x	x
Benzaldehyde	120	49	Hydrobromic acid, dilute	x	x
Benzene	210	99	Hydrobromic acid 50%	x	x
Benzene sulfonic acid 10%	90	32	Hydrochloric acid 20%[c]	80	27
Benzoic acid	90	32	Hydrocyanic acid 10%	x	x
Benzyl alcohol	80	27	Hydrofluoric acid 30%	x	x
Benzyl chloride	90	32	Hydrofluoric acid 70%	x	x
Borax	90	32	Hydrofluoric acid 100%	x	x

Table H.3 Continued

Chemical	Maximum temp. °F	°C	Chemical	Maximum temp. °F	°C
Ketones, general	90	32	Salicylic acid	80	27
Lactic acid 25%	90	32	Sodium chloride to 30%	150	66
Lactic acid, concentrated	90	32	Sodium hydroxide 10%	170	77
Magnesium chloride 30%	250	121	Sodium hydroxide 50%	x	x
Malic acid	90	32	Sodium hydroxide, concentrated	x	x
Methyl ethyl ketone	80	27	Sodium hypochlorite 20%	60	16
Methyl isobutyl ketone	80	27	Sodium hypochlorite, concentrated		
Nitric acid 5%	180	82	Sodium sulfide to 50%	90	32
Nitric acid 20%	180	82	Stannic chloride	x	x
Nitric acid 70%	186	86	Stannous chloride	x	x
Nitric acid, anhydrous	150	66	Sulfuric acid 10%	212	100
Nitrous acid, concentrated	80	27	Sulfuric acid 50%	295	146
Oleum	x	x	Sulfuric acid 70%	386	197
Perchloric acid 10%	80	27	Sulfuric acid 90%	485	252
Perchloric acid 70%	80	27	Sulfuric acid 98%	538	281
Phenol	100	38	Sulfuric acid 100%	644	340
Phosphoric acid 50–80%	210	99	Sulfurous acid	x	x
Picric acid	80	27	Trichloroacetic acid	80	27
Potassium bromide 30%	100	38			

[a]Resistance applies to Duriron unless otherwise noted.
[b]The chemicals listed are in the pure state or in a saturated solution unless otherwise indicated. Compatibility is shown to the maximum allowable temperature for which data are available. Incompatibility is shown by an x. A blank space indicates that data are unavailable when compatible corrosion rate is < 20 mpy.
[c]Resistance applies only to Durichlor.
Source: PA Schweitzer. *Corrosion Resistance Tables,* 4th ed. Vols. 1–3. New York: Marcel Dekker, 1995.

HIGH TEMPERATURE CORROSION

High temperature environments may be oxidizing or reducing, in a manner analogous to aqueous corrosion. There is an even electron transfer involved, and an "electrolyte" (i.e., the semiconductive layer of corrosion products). There is a migration of ions and electrons between the metal, corrosion product layer, and environment. The overall nature of the corrosion is dependent upon the ratio of the specific gases, vapors, or molten materials present. The common materials encountered in gaseous media are as in the table.

Oxidizing	Reducing
Oxygen	Hydrogen
Steam	Hydrogen sulfide
Sulfurous oxides (SO_2, SO_3)	Carbon disulfide
Sulfur	Carbon monoxide
Carbon dioxide	Carbon
Chlorine	Hydrocarbons
Oxides of nitrogen	Hydrogen chloride
	Ammonia

Molten metals can cause chemical reactions, whereas molten salts can be either oxidizing or reducing.

As with aqueous corrosion, a protective oxide film is formed. The rate at which the metal oxidizes will depend on the stability of the film.

If the film is stable and remains in place, the rate will be logarithmic, diminishing with time. Cycling temperatures will tend to spall off the surface film, leading to a stepwise oxidation of the alloy.

Although most high temperature corrosion is considered to be oxidation, there are other forms that are also encountered such as oxidation–reduction, sulfidation, fuel ash corrosion, carburization, and nitridation.

Oxidation–Reduction

An environment of hot air, oxygen, steam, carbon dioxide, etc. will tend to oxidize a metal, while an environment of hydrogen, hydrogen rich gases, or carbon monoxide is reducing and tends to convert the oxides back to the metallic state. In mixtures, the ratio of carbon monoxide to carbon dioxide determines the carburization or decarburization conditions. The same analogy holds for other combinations of oxidizing and reducing species such as hydrogen and water vapor, nitric oxides and ammonia, and sulfurous oxides with hydrogen sulfide. Further complicating the situation is the fact that one atmosphere may be reducing to one component such as nickel but oxidizing to another such as chromium or silicon.

Sulfidation

Sulfidation is analogeus to oxidation insofar as a sulfide film is formed on the surface of the metal. However sulfide films are less protective than the corresponding oxide films.

Carburization

Carburizing is not a specific type of corrosion, but it does reduce the efficiency of a prior oxide film by the formation of chromium carbides, which deplete the matrix of chromium. The most common corrosion phenomenon associated with carburization is general absorption. Another form of attack is metal dusting where under alternating oxidizing and reducing conditions localized high carbon areas are burned out during the oxidation period.

Nitriding

Nitriding takes place when active nitrogen reacts with hot surfaces. Because elements like chromium, aluminum, and titanium readily form nitrides, the integrity of the protective oxide film is at risk.

Whenever the oxide film is damaged or removed, the metal becomes subject to corrosion.

Fuel Ash Corrosion

Fuel ash corrosion is the result of the dissolving of the protective oxide film by the alkaline and sulfur constituents present in some heavier liquid petroleum and solid fuels.

While many of the austenitic stainless steels and high nickel alloys can be utilized at elevated temperatures, there are some instances where these materials are not suitable.

Consequently other alloys have been developed. Refer to "Haynes Alloy No. 230," "Haynes Alloy No. 214," "Haynes Alloy No. 556," and "High Nickel Alloys."

See Refs. 6 and 8.

HOMOPOLYMER

A homopolymer is a polymer formed from a single monomer.

HYDROGEN ATTACK

The carbon content of steel determines its hardness and strength. The tensile strength of steel is reduced when it loses carbon (decarburization).

High temperature hydrogen attack refers to a reaction between hydrogen and a component of the alloy. In steel at high temperature, hydrogen will react with iron carbide to form methane gas. Being unable to diffuse out of the steel, the methane gas accumulates and causes blistering and fissuring. This reduces the alloy's strength and ductility. When chromium and molybdenum are used as alloying elements, the carbides formed are more stable than iron carbides and resist hydrogen attack. At high temperatures, water vapor and carbon dioxide will also decarburize steel.

Carbon and low alloy steels are also subject to low temperature hydrogen attack, causing hydrogen blistering and hydrogen embrittlement that can lead to hydrogen assisted cracking.

See Refs. 9–11.

HYDROGEN BLISTERING

In some acid services, hydrogen atoms may penetrate the crystal lattice and collect in fissures or cavities in steel. Acid pickling is one such service. These atoms then combine to form hydrogen molecules, eventually reaching pressures in the range of several hundred thousand atmospheres and forming blisters on the surface of the steel. In closed systems such as pickling operations, inhibitors can be added to the acid to reduce the hydrogen penetration.

See Refs. 9 and 11.

HYDROGEN CRACKING

Hydrogen cracking is also referred to as hydrogen stress cracking and is the result of atomic or nascent hydrogen entering the metal lattice and building up sufficient internal stresses to initiate cracking. For hydrogen cracking to occur it is not necessary for the steel to be under internal or applied stress, but if the steel is under stress, this will enhance the movement of the hydrogen atoms. The cracks tend to be mostly transgranular. In martensite they follow the former austenite grain boundaries. Ferrous metals are especially susceptible to hydrogen cracking when heat treated to form martensite. Ferritic or body centered cubic structures are less susceptible to hydrogen cracking. Under most conditions of exposure, austenitic steels, in which hydrogen is more soluble than in ferrite, and in which the diffusion rate is lower, are immune to hydrogen cracking.

Hydrogen cracking can be caused by many sources including welding, contact with hydrogen gas, surface cleaning acids, electroplating, hydrogen sulfide, and water (particularly salt water). Hydrogen penetration will be retarded by oxides on the steel or other surface protection. Another preventative measure is to determine the critical stress level and ensure that this stress is not exceeded. Refer to "Hydrogen Embrittlement."

See Refs, 10, 12, 9, and 11.

HYDROGEN EMBRITTLEMENT

Molecular hydrogen cannot diffuse through steel, but atomic or nascent hydrogen can diffuse through steels. When this occurs the hydrogen atoms enter the metal lattice causing hydrogen embrittlement (loss of ductility). In some ferrous alloys high internal stresses may be induced sufficient to cause spontaneous cracking (hydrogen cracking). Low alloy steels are most susceptible in their highest strength levels. Alloys containing nickel or molybdenum are less susceptible. Under normal circumstances hydrogen embrittlement is not usually a problem in steels with yield strengths below approximately 150 ksi (1000 MPa), but if hydrofluoric acid or hydrogen sulfide is present the yield strength must be below 80 ksi (550 MPa) for good resistance. To minimize hydrogen embrittlement, welding conditions should be dry and low hydrogen filler metal should be used.

Common sources of atomic hydrogen include corrosion (including acid pickling of steel), high temperature moist atmospheres, electroplating, misapplied cathodic protection, and welding.

See Refs. 9, 12, 10, and 11.

HYDROGEN PROBES

Hydrogen probes are used to detect the penetration of atomic or nascent hydrogen into metal pipes and vessels. There are three basic types of hydrogen probes. For more details refer to "Monitoring Corrosion."

HYDROLYSIS

Hydrolysis is a reaction of a salt with water to form an acid and a base, also the chemical reaction of any compound with water. It also refers to a decomposition process in the presence of water, particularly of coatings or paints.

HYLAR

See "Polyvinylidene Fluoride."

HYPALON

See "Chlorosulfonated Polyethylene."

REFERENCES

1. PA Schweitzer. Corrosion Resistance Tables. 4th ed. Vols. 1–3. New York: Marcel Dekker, 1995.
2. FT Murray. Introduction to Engineering Materials. New York: Marcel Dekker, 1993.
3. JH Mallinson. Corrosion Resistant Plastic Composites in Chemical Plant Design. New York: Marcel Dekker, 1988.
4. PA Schweitzer. Corrosion Resistant Piping Systems. New York: Marcel Dekker, 1994.
5. M Sridhar and G Hodge. Nickel and high nickel alloys. In: PA Schweitzer, ed. Corrosion and Corrosion Protection Handbook. 2nd ed. New York: Marcel Dekker, 1989, pp. 96–124.
6. CP Dillon. Corrosion Resistance of Stainless Steels. New York: Marcel Dekker, 1995.
7. RB Norden. Materials of construction. In: Perry and Chilten, eds. Chemical Engineers Handbook. 5th ed. New York: McGraw-Hill, 1973, Sec. 22, p.15.
8. CP Dillon. Corrosion Control in the Chemical Process Industry. 2nd ed. St. Louis: Materials Technology Institute of the Chemical Process Industries, 1994.
9. GM Kirby. The corrosion of carbon and low allow steels. In: PA Schweitzer, ed. Corrosion Engineering Handbook. New York: Marcel Dekker, 1996, pp. 32–52.
10. GT Murray. Introduction to Engineering Materials. New York: Marcel Dekker, 1993.
11. MR Louthan Jr. The effect of hydrogen on metals. In: F Mansfield, ed. Corrosion Mechanisms. New York: Marcel Dekker, 1989, pp. 329–365.
12. HH Uhlig. Corrosion and Corrosion Control. New York: John Wiley 1963.

I

IMMERSION COATINGS

Immersion coatings are coatings for marine service. The most common formulations are coal tar epoxies and straight epoxies. These coatings resist moisture absorption, moisture transfer, and electroendosmosis (electrochemically induced diffusion of moisture through the coating).

Cathodic protection is usually used in conjunction with these coatings to supplement the protection supplied by the immersion coating.

IMPERVIOUS GRAPHITE

Graphite is a crystalline form of carbon produced from carbon particles bonded with materials that carbonize when produced at processing temperature in excess of 3600°F (1980°C). The normal fine grain graphite is porous. By impregnating with organic resins such as phenolic or furan prior to the final heat treatment the graphite is made impervious.

Impervious graphite has a wide range of chemical resistance. Its stability depends upon the impregnating resin. A phenolic resin provides chemical resistance to most acids, salt solutions, and organic compounds. A furan resin imparts resistance against alkaline and oxidizing media. The normal maximum operating temperature is 338°F (170°C), depending upon the corrosive media. Higher temperatures can be achieved by impregnating with PTFE.

Table I.1 shows the compatibility of impervious graphite with selected corrodents. Additional data are available in Refs. 1 and 2.

Houston Tx, 1986 pgs 703-770.

IMPINGEMENT CORROSION ATTACK

Impingement corrosion is a localized pitting type of erosion corrosion resulting from the impinging or turbulent flow of liquids.

INCOLOY ALLOY 800

Alloy 800 contains about 20% chromium, 32% nickel, and 46% iron as balance. The alloy is used primarily for its oxidation resistance and strength at elevated temperatures. It is particularly useful for high temperature applications because the alloy does not form the embrittling sigma phase after long exposures at 1200–1600°F (649–871°). High creep and

Table I.1 Compatibility of Impervious Graphite with Selected Corrodents[a]

Chemical	Resin	Max. Temp. °F	Max. Temp. °C
Acetaldehyde	phenolic	460	238
Acetamide	phenolic	460	238
Acetic acid 10%	furan	400	204
Acetic acid 50%	furan	400	204
Acetic acid 80%	furan	400	204
Acetic acid, glacial	furan	400	204
Acetic anhydride	furan	400	204
Acetone	furan	400	204
Acetyl chloride	furan	460	238
Acrylic acid			
Acrylonitrile	furan	400	204
Adipic acid	furan	460	238
Allyl alcohol		x	x
Aluminum chloride, aqueous	phenolic	120	49
Aluminum chloride, dry	furan	460	238
Aluminum fluoride	furan	460	238
Aluminum hydroxide	furan	250	121
Aluminum nitrate	furan	460	238
Aluminum sulfate	phenolic	460	238
Ammonium bifluoride	phenolic	390	199
Ammonium carbonate	phenolic	460	238
Ammonium chloride 10%	phenolic	400	204
Ammonium hydroxide 25%	phenolic	400	204
Ammonium hydroxide, sat.	phenolic	400	204
Ammonium nitrate	furan	460	238
Ammonium persulfate	furan	250	121
Ammonium phosphate	furan	210	99
Ammonium sulfate 10–40%	phenolic	400	204
Amyl acetate	furan	460	238
Amyl alcohol	furan	400	204
Amyl chloride	phenolic	210	99
Aniline	furan	400	204
Aqua regia 3:1		x	x
Barium chloride	furan	250	121
Barium hydroxide	phenolic	250	121
Barium sulfate	phenolic	250	121
Barium sulfide	phenolic	250	121
Benzaldehyde	furan	460	238
Benzene	furan	400	204
Benzene sulfonic acid 10%	phenolic	460	238
Borax	furan	460	238
Boric acid	phenolic	460	238
Bromine gas, dry		x	x
Bromine gas, moist		x	x
Butadiene	furan	460	238
Butyl acetate	furan	460	238
Butyl alcohol	furan	400	204
n-Butylamine	phenolic	210	99

Table I.1 Continued

Chemical	Resin	Max. Temp. °F	Max. Temp. °C
Butyric acid	furan	460	238
Calcium bisulfite	furan	460	238
Calcium carbonate	phenolic	460	238
Calcium chloride	furan	460	238
Calcium hydroxide 10%	furan	250	121
Calcium hydroxide, sat.	furan	250	121
Calcium hypochlorite	furan	170	77
Calcium nitrate	furan	460	238
Calcium oxide			
Calcium sulfate	furan	460	238
Carbon bisulfide	furan	400	204
Carbon dioxide, dry	phenolic	460	238
Carbon dioxide, wet	phenolic	460	238
Carbon disulfide	furan	400	204
Carbon monoxide	phenolic	460	238
Carbon tetrachloride	furan	400	204
Carbonic acid	phenolic	400	204
Cellosolve	furan	460	238
Chloracetic acid, 50% water	furan	400	204
Chloracetic acid	furan	400	204
Chlorine gas, dry	furan	400	204
Chlorine, liquid	phenolic	130	54
Chlorobenzene	phenolic	400	204
Chloroform	furan	400	204
Chlorosulfonic acid		x	x
Chromic acid 10%		x	x
Chromic acid 50%		x	x
Citric acid 15%	furan	400	204
Citric acid, conc.	furan	400	204
Copper chloride	furan	400	204
Copper cyanide	furan	460	238
Copper sulfate	phenolic	400	204
Cresol	furan	400	204
Cupric chloride 5%	furan	400	204
Cupric chloride 50%	furan	400	204
Cyclohexane	furan	460	238
Ethylene glycol	furan	330	166
Ferric chloride 60%	phenolic	210	99
Ferric chloride 50% in water	furan	260	127
Ferric nitrate 10–50%	furan	210	99
Ferrous chloride	furan	400	204
Fluorine gas, dry	phenolic	300	149
Fluorine gas, moist		x	x
Hydrobromic acid, dilute	furan	120	49
Hydrobromic acid 20%	furan	250	121
Hydrobromic acid 50%	furan	120	49
Hydrochloric acid 20%	phenolic	400	204
Hydrochloric acid 38%	phenolic	400	204
Hydrocyanic acid 10%	furan	460	238

Table I.1 Continued

Chemical	Resin	Max. Temp. °F	Max. Temp. °C
Hydrofluoric acid 30%	phenolic	460	238
Hydrofluoric acid 70%		x	x
Hydrofluoric acid 100%		x	x
Iodine solution 10%	phenolic	120	49
Ketones, general	furan	400	204
Lactic acid 25%	furan	400	204
Lactic acid, concentrated	furan	400	204
Magnesium chloride	furan	170	77
Manganese chloride	furan	460	238
Methyl chloride	phenolic	460	238
Methyl ethyl ketone	furan	460	238
Methyl isobutyl ketone	furan	460	238
Muriatic acid	phenolic	400	204
Nitric acid 5%	phenolic	220	104
Nitric acid 20%	phenolic	220	104
Nitric acid 70%		x	x
Nitric acid, anhydrous		x	x
Nitrous acid, concentrated		x	x
Oleum		x	x
Phenol	furan	400	204
Phosphoric acid 50–80%	phenolic	400	204
Potassium bromide 30%	furan	460	238
Salicylic acid	furan	340	171
Sodium carbonate	furan	400	204
Sodium chloride	phenolic	400	204
Sodium hydroxide 10%	furan	400	204
Sodium hydroxide 50%	furan	400	204
Sodium hydroxide, concentrated		x	x
Sodium hypochlorite 20%		x	x
Sodium hypochlorite, concentrated		x	x
Stannic chloride	furan	400	204
Sulfuric acid 10%	phenolic	400	204
Sulfuric acid 50%	phenolic	400	204
Sulfuric acid 70%	phenolic	400	204
Sulfuric acid 90%	phenolic	400	204
Sulfuric acid 98%		x	x
Sulfuric acid 100%		x	x
Sulfuric acid, fuming		x	x
Sulfurous acid	phenolic	400	204
Thionyl chloride	phenolic	320	160
Toluene	furan	400	204
Trichloroacetic acid	furan	340	171
Zinc chloride	phenolic	400	204

[a]The chemicals listed are in the pure state or in a saturated solution unless otherwise indicated. Compatibility is shown to the maximum allowable temperature for which data are available. Incompatibility is shown by an x. A blank space indicates that data are unavailable.

Material extracted from PA Schweitzer. *Corrosion Resistance Tables*, 4th ed. Vols. 1–3. New York: Marcel Dekker, 1995.

rupture strengths are other factors that contribute to its performance in many other applications. Refer to Table I.2 for the mechanical and physical properties of alloy 800.

At moderate temperatures the general corrosion resistance of alloy 800 is similar to that of the other austenitic nickel–iron–chromium alloys. However, as the temperature increases, alloy 800 continues to exhibit good corrosion resistance, while other austenitic alloys are unsatisfactory for the service.

Alloy 800 has excellent resistance to nitric acid at concentrations up to about 70%. It resists a variety of oxidizing salts, but not halide salts. It also has good resistance to organic acids, such as formic, acetic, and propionic. Alloy 800 is particularly suited for the handling of hot corrosive gases such as hydrogen sulfide.

In aqueous corrosion service alloy 800 has general resistance that falls between 304 and 316 stainless steels. Thus the alloy is not widely used for aqueous service. The stress corrosion cracking resistance of alloy 800 is, while not immune, better than that of the 300 series of stainless steels and may be substituted on this basis. Table I.3 provides the compatibility of alloy 800 with selected corrodents.

See Refs. 1 and 3.

INCOLOY ALLOY 825

Alloy 825 is very similar to alloy 800, but the composition has been modified to provide for improved aqueous corrosion resistance. Table I.4 gives the mechanical and physical properties of the alloy.

Table I.2 Mechanical and Physical Properties of Alloy 800

Modulus of elasticity $\times 10^6$, psi	28.5
Tensile strength $\times 10^3$, psi	75
Yield strength 0.2% offset $\times 10^3$, psi	26.6
Elongation in 2 in., %	30
Hardness, Brinell	152
Density, lb/in.^{3}	0.287
Specific gravity	7.94
Specific heat, Btu/lb°F	0.11
Thermal conductivity, Btu-in/h/ft^2/°F/in.	
at 0°F	75
at 70°F	80
at 100°F	83
at 200°F	89
at 400°F	103
at 600°F	115
at 1200°F	152
at 1600°F	181
Coefficient of thermal expansion, in./in. °F $\times 10^{-6}$	
at 70–200°F	7.9
at 70–400°F	8.8
at 70–600°F	9.0
at 70–800°F	9.2
at 70–1000°F	9.4
at 70–1200°F	9.6
at 70–1400°F	9.9
at 70–1600°F	10.2

Table I.3 Compatibility of Alloy 800 and Alloy 825 with Selected Corrodents[a]

Chemical	Maximum temp. °F	Maximum temp. °C	Chemical	Maximum temp. °F	Maximum temp. °C
Acetic acid 10%[b]	200	93	Chloroform	90	32
Acetic acid 50%[b]	220	104	Chlorosulfonic acid	x	x
Acetic acid 80%[b]	210	99	Chromic acid 10%[b]	210	99
Acetic acid, glacial[b]	220	104	Chromic acid 50%	x	x
Acetic anhydride	230	110	Citric acid 15%	210	99
Acetone	210	99	Citric acid, concentrated[b]	210	99
Acetyl chloride	210	99	Copper acetate	90	32
Aluminum acetate	60	16	Copper carbonate	90	32
Aluminum chloride, aqueous	60	16	Copper chloride 5%[b]	80	27
Aluminum fluoride 5%	80	27	Copper cyanide	210	99
Aluminum hydroxide	80	27	Copper sulfate	210	99
Aluminum sulfate	210	99	Cupric chloride 5%	x	x
Ammonium carbonate	190	88	Ferric chloride	x	x
Ammonium chloride 10%[b]	230	110	Ferric chloride 50% in water	x	x
Ammonium chloride, sat.	200	93	Ferric nitrate 10–50%	90	32
Ammonium hydroxide, sat.	110	43	Ferrous chloride[b,c]	90	32
Ammonium nitrate	90	32	Fluorine gas, dry	x	x
Ammonium persulfate	90	32	Fluorine gas, moist	x	x
Ammonium sulfate 10–40%	210	99	Hydrobromic acid 20%	x	x
Ammonium sulfite	210	99	Hydrobromic acid 50%	x	x
Amyl acetate[b]	200	93	Hydrochloric acid 20%[b]	90	32
Amyl chloride	90	32	Hydrochloric acid 38%	x	x
Aniline	90	32	Hydrocyanic acid 10%	60	16
Antimony trichloride	90	32	Hydrofluoric acid 30%	x	x
Barium carbonate	90	32	Hydrofluoric acid 70%	x	x
Barium sulfate	90	32	Hydrofluoric acid 100%	x	x
Benzene	190	88	Magnesium chloride 1–5%	170	77
Benzoic acid 5%	90	32	Malic acid	170	77
Borax	190	88	Manganese chloride 10–50%	210	99
Boric acid 5%	210	99	Muriatic acid[b]	90	32
Bromine gas, dry[b]	90	32	Nitric acid 5%	90	32
Butyl acetate[b]	90	32	Nitric acid 20%	60	16
Butyric acid 5%	90	32	Nitric acid, anhydrous	210	99
Calcium carbonate	90	32	Phenol	90	32
Calcium chlorate	80	27	Picric acid	90	32
Calcium chloride[b,c]	60	16	Potassium bromide 5%	90	32
Calcium hydroxide 10%	200	93	Salicylic acid	90	32
Calcium hypochlorite	x	x	Silver bromide 10%[b]	90	32
Calcium sulfate	90	32	Sodium carbonate	90	32
Carbon monoxide	570	299	Sodium chloride[c]	200	93
Carbon tetrachloride	90	32	Sodium hydroxide 10%	90	32
Carbonic acid	90	32	Sodium hydroxide, concentrated	90	32
Chloracetic acid	x	x	Sodium sulfide to 50%	90	32
Chlorine gas, dry[b]	90	32	Stannic chloride	x	x
Chlorine gas, wet	x	x	Stannous chloride 5%	90	32
Chlorobenzene	90	32	Sulfuric acid 10%[b]	230	110

Table I.3 Continued

Chemical	Maximum temp. °F	Maximum temp. °C	Chemical	Maximum temp. °F	Maximum temp. °C
Sulfuric acid 50%[b]	210	99	Sulfuric acid 100%[b]	230	110
Sulfuric acid 70%[b]	150	66	Sulfuric acid, fuming	x	x
Sulfuric acid 90%[b]	180	82	Sulfurous acid[a]	370	188
Sulfuric acid 98%[b]	220	104	Zinc chloride 5%	140	60

[a]The chemicals listed are in the pure state or in a saturated solution unless otherwise indicated. Compatibility is shown to the maximum allowable temperature for which data are available. Incompatibility is shown by an x. A blank space indicates that data are unavailable. When compatible, corrosion rate is < 20 mpy.
[b]Applicable to alloy 825 only.
[c]Material subject to pitting.
Source: Schweitzer, Philip A. *Corrosion Resistance Tables,* 4th ed. Vols. 1–3. New York: Marcel Dekker, 1995.

The higher nickel content of alloy 825, as compared to alloy 800, makes it resistant to chloride ion stress corrosion cracking. Additions of molybdenum and copper give resistance to pitting and to corrosion in reducing acid environments such as sulfuric or phosphoric acid solutions. Alloy 825 is resistant to pure sulfuric acid solutions up to 40% by weight at boiling temperatures and at all concentrations at a maximum temperature of 150°F (60°C). In dilute solutions, the presence of oxidizing salts, such as cupric or ferric, actually lowers the corrosion rates. It has limited use in hydrochloric or hydrofluoric acids.

The alloy's chromium content gives it resistance to various oxidizing environments such as nitrates, nitric acid solutions, and oxidizing salts. Alloy 825 is not fully resistant to stress corrosion cracking when tested in boiling magnesium chloride, but it has good resistance in neutral chloride environments. If localized corrosion is a problem with 300 series stainless steel, alloy 825 may be substituted. In addition alloy 825 offers excellent

Table I.4 Mechanical and Physical Properties of Alloy 825

Modulus of elasticity × 10^6, psi	28.3
Tensile strength × 10^3, psi	85
Yield strength 0.2% offset × 10^3, psi	35
Elongation in 2 in., %	30
Hardness, Brinell	150
Density, lb/in.^{3}	0.294
Specific gravity	8.14
Coefficient of thermal expansion, in./in. °F × 10^{-6}	
at 80–200°F	7.8
at 80–400°F	8.3
at 80–600°F	8.5
at 80–800°F	8.7
at 80–1000°F	8.8
at 80–1200°F	9.1
at 80–1400°F	9.5
at 80–1600°F	9.7

resistance to corrosion by seawater. Refer to Table I.3 for the compatibility of alloy 825 with selected corrodents under the heading "Incoloy 800 Alloy."

See Refs. 1 and 3.

INCONEL ALLOY 600

Alloy 600 is a nickel–chromium–iron alloy for use in environments requiring resistance to heat and corrosion. The high nickel content of this alloy makes it resistant to corrosion by a number of organic and inorganic compounds and gives it excellent resistance to chloride ion stress corrosion cracking. Its chromium content gives the alloy resistance to sulfur compounds and various oxidizing environments.

In addition, alloy 600 has excellent mechanical properties and a combination of high strength and good workability. It performs well in applications with temperatures from cryogenic to 1200°F (649°C). Refer to Table I.5 for mechanical and physical properties.

Alloy 600 is practically free from corrosion by freshwaters, including the most corrosive of natural waters containing free carbon dioxide, iron compounds, and dissolved air. It remains free from stress corrosion cracking even in boiling magnesium chloride.

Because of its chromium content, alloy 600 exhibits greater resistance to sulfuric acid under oxidizing conditions than either nickel 200 or Monel alloy 400. The addition of oxidizing salts to sulfuric acid tends to passivate alloy 600, which makes it suitable for

Table I.5 Mechanical and Physical Properties of Alloy 600

Modulus of elasticity $\times 10^6$, psi	30–31
Tensile strength $\times 10^3$, psi	80
Yield strength 0.2% offset $\times 10^3$, psi	30–35
Elongation in 2 in., %	40
Hardness, Rockwell	B-120-170
Density, lb/in.^{3}	0.306
Specific gravity	8.42
Specific heat, Btu/lb°F	0.106
Thermal conductivity, Btu/h/ft^2/°F/in.	
at 70°F	103
at 200°F	109
at 400°F	121
at 600°F	133
at 800°F	145
at 1000°F	158
at 1200°F	172
Thermal expansion coefficient, in./in./°F $\times 10^{-6}$	
at 70–200°F	7.4
at 70–400°F	7.7
at 70–600°F	7.9
at 70–800°F	8.1
at 70–1000°F	8.4
at 70–1200°F	8.6

use with acid mine waters or brass pickling solutions, applications where Monel alloy 400 cannot be used.

Alloy 600 is not subject to stress corrosion cracking in any of the chloride salts and has excellent resistance to all of the nonoxidizing halides. Refer to Table I.6 for the compatibility of alloy 600 with selected corrodents.

See Refs. 1 and 3.

Table I.6 Compatibility of Alloy 600 and Alloy 625 with Selected Corrodents[a]

Chemical	Maximum temp. °F	Maximum temp. °C	Chemical	Maximum temp. °F	Maximum temp. °C
Acetaldehyde	140	60	Aqua regia 3:1	x	x
Acetic acid 10%	80	27	Barium carbonate	80	27
Acetic acid 50%	x	x	Barium chloride	570	299
Acetic acid 80%	x	x	Barium hydroxide	90	32
Acetic acid, glacial	220	104	Barium sulfate	210	99
Acetic anhydride	200	93	Benzaldehyde	210	99
Acetone	190	88	Benzene	210	99
Acetyl chloride	80	27	Benzoic acid 10%	90	32
Acrylonitrile	210	99	Benzyl alcohol	210	99
Adipic acid	210	99	Benzyl chloride	210	99
Allyl alcohol	200	93	Borax	90	32
Allyl chloride	150	66	Boric acid	80	27
Alum	200	93	Bromine gas, dry	60	16
Aluminum acetate	80	27	Bromine gas, moist	x	x
Aluminum chloride, aqueous	x	x	Butadiene	80	27
Aluminum chloride, dry	x	x	Butyl acetate	80	27
Aluminum fluoride	80	27	Butyl alcohol	80	27
Aluminum hydroxide	80	27	*n*-Butylamine		
Aluminum sulfate	x	x	Butyl phthalate	210	99
Ammonium carbonate	190	88	Butyric acid	x	x
Ammonium chloride 10%[b]	230	110	Calcium bisulfite	x	x
Ammonium chloride 50%	170	77	Calcium carbonate	90	32
Ammonium chloride, sat.	200	93	Calcium chlorate	80	27
Ammonium fluoride 10%	90	32	Calcium chloride	80	27
Ammonium fluoride 25%	90	32	Calcium hydroxide 10%	210	99
Ammonium hydroxide 25%	80	27	Calcium hydroxide, sat.	90	32
Ammonium hydroxide, sat.	90	32	Calcium hypochlorite	x	x
Ammonium nitrate	x	x	Calcium sulfate[c]	210	99
Ammonium persulfate	80	27	Caprylic acid	230	110
Ammonium phosphate 10%	210	99	Carbon bisulfide	80	27
Ammonium sulfate 10–40%[c]	210	99	Carbon dioxide, dry	210	99
Ammonium sulfide			Carbon dioxide, wet	200	93
Ammonium sulfite	90	32	Carbon disulfide	80	27
Amyl acetate	300	149	Carbon monoxide	570	299
Amyl chloride	x	x	Carbon tetrachloride	210	99
Aniline	210	99	Carbonic acid	210	99
Antimony trichloride	90	32	Cellosolve	210	99

Table I.6 Continued

Chemical	Maximum temp. °F	Maximum temp. °C	Chemical	Maximum temp. °F	Maximum temp. °C
Chloracetic acid	x	x	Malic acid	210	99
Chlorine gas, dry	90	32	Manganese chloride 37%	x	x
Chlorine gas, wet	x	x	Methyl chloride	210	99
Chlorobenzene	210	99	Methyl ethyl ketone	210	99
Chloroform	210	99	Methyl isobutyl ketone	200	93
Chromic acid 10%	130	54	Muriatic acid	x	x
Chromic acid 50%	90	32	Nitric acid 5%	90	32
Chromyl chloride	210	99	Nitric acid 20%	80	27
Citric acid 15%	210	99	Nitric acid 70%	x	x
Citric acid, concentrated	210	99	Nitric acid, anhydrous	x	x
Copper acetate	100	38	Nitrous acid, concentrated	x	x
Copper carbonate	80	27	Oleum	x	x
Copper chloride	x	x	Phenol	570	299
Copper cyanide	80	27	Phosphoric acid 50–80%	190	88
Copper sulfate	80	27	Picric acid	x	x
Cresol	100	38	Potassium bromide 30%	210	99
Cupric chloride 5%	x	x	Salicylic acid	80	27
Cupric chloride 50%	x	x	Sodium carbonate to 30%	210	99
Cyclohexanol	80	27	Sodium chloride to 30%	210	99
Dichloroethane (ethylene dichloride)	200	93	Sodium hydroxide 10%	300	149
Ethylene glycol	210	99	Sodium hydroxide 50%[b]	300	149
Ferric chloride	x	x	Sodium hydroxide, concentrated	80	27
Ferric chloride 50% in water	x	x	Sodium hypochlorite 20%	x	x
Ferric nitrate 10–50%	x	x	Sodium hypochlorite, concentrated	x	x
Ferrous chloride	x	x	Sodium sulfide to 50%	210	99
Fluorine gas, dry	570	299	Stannic chloride	x	x
Fluorine gas, moist	60	16	Stannous chloride, dry	570	299
Hydrobromic acid, dilute	90	32	Sulfuric acid 10%	x	x
Hydrobromic acid 20%	80	27	Sulfuric acid 50%	x	x
Hydrobromic acid 50%	x	x	Sulfuric acid 70%	x	x
Hydrochloric acid 20%	80	27	Sulfuric acid 90%	x	x
Hydrochloric acid 38%	x	x	Sulfuric acid 98%	x	x
Hydrofluoric acid 30%	x	x	Sulfuric acid 100%	x	x
Hydrofluoric acid 70%	x	x	Sulfuric acid, fuming	x	x
Hydrofluoric acid 100%	120	49	Sulfurous acid	90	32
Lactic acid 25%	210	99	Toluene	210	99
Lactic acid, concentrated	90	32	Trichloroacetic acid	80	27
Magnesium chloride 50%	130	54	Zinc chloride, dry	80	27

[a]The chemicals listed are in the pure state or in a saturated solution unless otherwise indicated. Compatibility is shown to the maximum allowable temperature for which data are available. Incompatibility is shown by an x. A blank space indicates that data are unavailable when compatible corrosion rate is < 20 mpy.
[b]Material is subject to stress cracking.
[c]Material subject to pitting.
Source: PA Schweitzer. *Corrosion Resistance Tables,* 4th ed. Vols. 1–3. New York: Marcel Dekker, 1995.

INCONEL ALLOY 625

Alloy 625 is a nickel–chromium–molybdenum–columbium alloy well suited for applications where strength and corrosion resistance are required. The alloy exhibits exceptional fatigue strength and superior strength and toughness at temperatures ranging from cryogenic to 2000°F (1093°). Mechanical and physical properties are shown in Table I.7.

Alloy 625 is resistant to oxidation, general corrosion, pitting, and crevice corrosion and is virtually immune to chloride ion stress corrosion cracking. These properties are derived from additions of molybdenum and columbium to the alloy's basic nickel–chromium composition.

Resistance to aqueous solutions is good in a variety of applications including organic acids, sulfuric acid, and hydrofluoric acid. Although nickel based alloys are not normally used in nitric acid service because of cost, alloy 625 is resistant to mixtures of nitric–hydrofluoric, where stainless steel loses its resistance.

Alloy 625 has excellent resistance to phosphoric acid solutions, including commercial grades of acids that contain fluorides, sulfates, and chlorides in the production of superphosphoric acid (72% P_2O_5). In general, alloy 625 can be used in applications that alloy 600 is not suitable for. Refer to Table I.6 under the heading "Inconel 600" for the compatibility of alloy 625 with selected corrodents.

See Refs. 1 and 3.

Table I.7 Mechanical and Physical Properties of Alloy 625

Modulus of elasticity $\times 10^6$, psi	30.1
Tensile strength $\times 10^3$, psi	100–120
Yield strength 0.2% offset $\times 10^3$, psi	60
Elongation in 2 in., %	30
Hardness, Brinell	192
Density, lb/in.^{3}	0.305
Specific gravity	8.44
Specific heat, Btu/lb°F	0.098
Thermal conductivity, Btu-in/ft^2 h °F	
at –250°F	50
at –100°F	58
at 0°F	64
at 70°F	68
at 100°F	70
at 200°F	75
at 400°F	87
at 600°F	98
at 1000°F	121
at 1400°F	144
Thermal expansion coefficient, in./in. °F $\times 10^{-6}$	
at 70–200°F	7.1
at 70–400°F	7.3
at 70–600°F	7.4
at 70–800°F	7.6
at 70–1000°F	7.8
at 70–1200°F	8.2
at 70–1400°F	8.5
at 70–1600°F	8.8

INCONEL* ALLOY 686

Inconel alloy 686 is an austenitic, nickel–chromium–molybdenum–tungsten alloy. It is resistant to both reducing and oxidizing media. The low carbon content along with other alloying control ingredients help to minimize grain boundary precipitation to maintain corrosion resistance in heat affected zones of welded joints.

Alloy 686 is resistant to general corrosion, stress corrosion cracking, pitting, and crevice corrosion in a wide variety of aggressive corrodents. It has excellent resistance to mixed acids as well as reducing acids and oxidizing acids and to mixed acids containing high concentrations of halides. Good resistance has been shown to mixed acid media having pH levels of 1 or less and chloride levels over 100,000 ppm.

The ability of alloy 686 to resist pitting can be seen from its Pitting Resistance Equivalent, which is 51. This compares to the PRE numbers as follows for other pitting resistant alloys (the higher the PRE the better the pitting resistance).

Alloy 654	63.09
25-6MO	47.45
Inconel alloy 622	47
UNS N-06059	47
A1-6XN	46.96
UNS N-06022	46
Alloy C-276	46
Inconel alloy 625	40

Another method used to compare the resistance of alloys to localized attack are their respective Critical Pitting and Critical Crevice Temperatures. These are the temperatures at which pitting and/or crevice attack is initiated. Critical temperatures for several alloys are given in the table.

	Critical Temperature °F/°C	
Alloy	Pitting attack	Crevice attack
Inconel alloy 686	> 185/> 85	> 185/> 85
UNS N-06059	> 185/> 85	> 185/> 85
Inconel alloy 622	≥ 185/≥ 85	≥ 185/≥ 85
UNS N-06022	> 185/> 85	> 136/> 58
Alloy C-276	> 185/> 85	113/45
Alloy 625	> 185/> 85	95/35
Alloy 25-6MO	158/70	86/30
Alloy 825	86/30	41/5
UNS N-31600	68/20	< 32/< 0

*Inconel is the Trademark of Inco Alloys International.

In addition to the wide range of corrosion resistance Inconel alloy 686 also exhibits good mechanical and physical properties.

INHIBITORS

See "Corrosion Inhibitors."

INORGANIC COATINGS

See "Coatings" also.

Inorganic coatings are coatings (paint) employing inorganic binders or vehicles such as silicates or phosphates, which are usually pigmented with metallic zinc. Unlike organic coatings, in which the pigment is secondary to the resistance of the organic binders, the role of the zinc pigment in the zinc rich coating predominates, and the high amount of zinc dust metal in the dried film determines the coating's basic property, galvanic protection.

Also cold-applied ceramic coatings, but not as dense or glossy as a fused ceramic coating.

INTERGRANULAR CORROSION

This is a localized form of attack taking place at the grain boundaries of a metal with little or no attack on the grain boundaries themselves. This results in loss of strength and ductility. The attack is often rapid, penetrating deeply into the metal and causing failure.

In the case of austenitic stainless steels the attack is the result of carbide precipitation during welding operations. Carbide precipitation can be prevented by using alloys containing less than 0.03% carbon, or by using alloys that have been stabilized with columbium or titanium or by specifying solution heat treatment followed by a rapid quench that will keep the carbides in solution. The most practical approach is to use either a low carbon content or stabilized austenitic stainless steel.

Nickel based alloys can also be subjected to carbide precipitation and precipitation of intermetallic phases, when exposed to temperatures lower than their annealing temperatures. As with the austenitic stainless steels, low carbon content alloys are recommended to delay precipitation of carbides. In some alloys, such as alloy 625, niobium, tantalum, or titanium is added to stabilize the alloy against precipitation of chromium, or molybdenum carbides. These elements combine with carbon instead of chromium or molybdenum.

See Refs. 4–7.

ISO

See "Polyvinylidene Fluoride."

ISOCORROSION DIAGRAM

An isocorrosion diagram is a chart or graph showing constant corrosion behavior with changing conditions such as temperature, solution concentration, or other environment composition. The isocorrosion diagram for Monel 400 in hydrofluoric acid is shown in Fig. I.1.

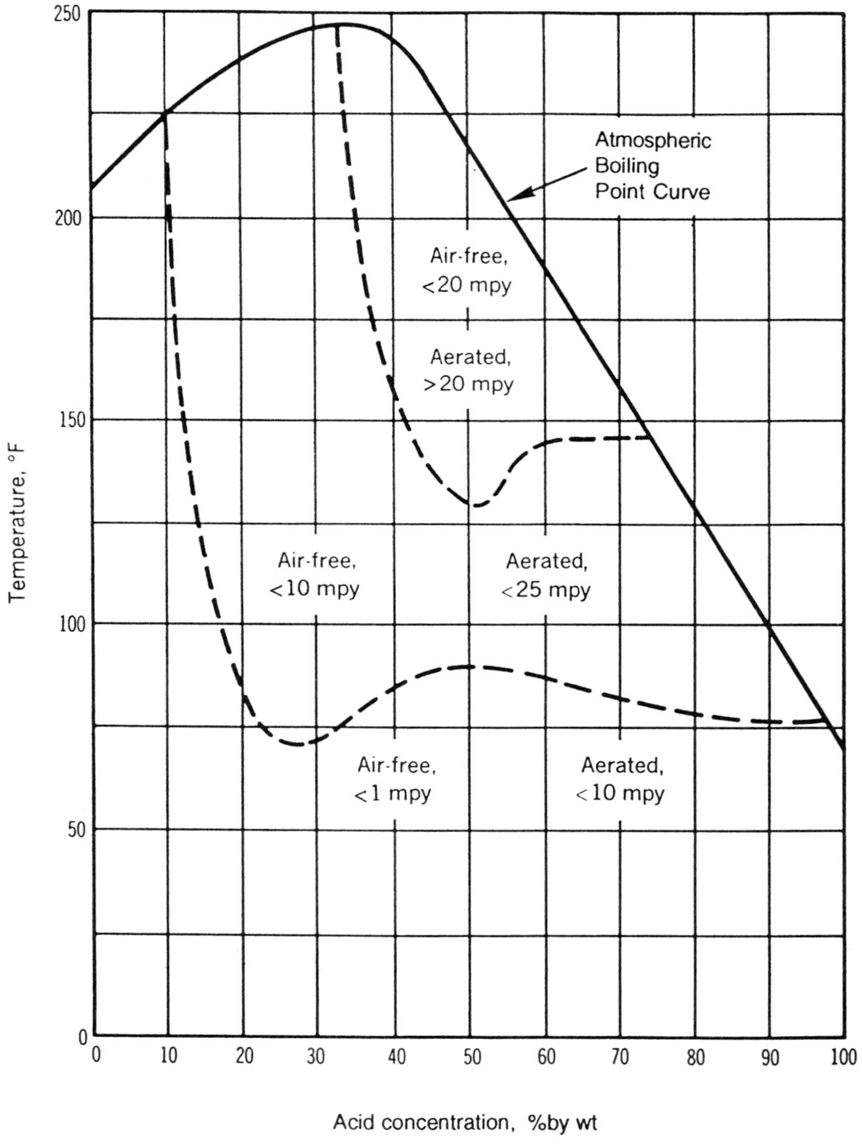

Figure I.1 Isocorrosion diagram for Monel 400.

ISOPHTHALIC ESTERS

Also see "Polymers and Thermoset Polymers."

The isophthalic esters have a relatively wide range of corrosion resistance. They are satisfactory for use up to 125°F (52°C) in such acids as 10% acetic, benzoic, boric, citric, oleic, 25% phosphoric, tartaric, 10–25% sulfuric, and fatty acids. Most inorganic salts are also compatible with the isophthalic esters. Solvents such as amyl alcohols, ethylene glycol, formaldehyde, gasoline, kerosene, and naphtha are also compatible.

The isophthalic resins are not resistant to acetone, amyl acetate, benzene, carbon disulfides, solutions of alkaline salts of potassium and sodium, hot distilled water, or higher concentrations of oxidizing acids. Refer to Table I.8 for the compatibility of the isophthalic resins with selected corrodents. Refer to Ref. 1 for a wider range of compatibility of the isophthalic esters with selected corrodents.

Also See Refs. 8–10.

Table I.8 Compatibility of Isophthalic Polyester with Selected Corrodents[a]

	Maximum temp.			Maximum temp.	
Chemical	°F	°C	Chemical	°F	°C
Acetaldehyde	x	x	Ammonium nitrate	160	71
Acetic acid 10%	180	82	Ammonium persulfate	160	71
Acetic acid 50%	110	43	Ammonium phosphate	160	71
Acetic acid 80%	x	x	Ammonium sulfate 10%	180	82
Acetic acid, glacial	x	x	Ammonium sulfide	x	x
Acetic anhydride	x	x	Ammonium sulfite	x	x
Acetone	x	x	Amyl acetate	x	x
Acetyl chloride	x	x	Amyl alcohol	160	71
Acrylic acid	x	x	Amyl chloride	x	x
Acrylonitrile	x	x	Aniline	x	x
Adipic acid	220	104	Antimony trichloride	160	71
Allyl alcohol	x	x	Aqua regia 3:1	x	x
Allyl chloride	x	x	Barium carbonate	190	88
Alum	250	121	Barium chloride	140	60
Aluminum chloride, aqueous	180	82	Barium hydroxide	x	x
Aluminum chloride, dry	170	77	Barium sulfate	160	71
Aluminum fluoride 10%	140	60	Barium sulfide	90	32
Aluminum hydroxide	160	71	Benzaldehyde	x	x
Aluminum nitrate	160	71	Benzene	x	x
Aluminum sulfate	180	82	Benzene sulfonic acid 10%	180	82
Ammonia gas	90	32	Benzoic acid	180	82
Ammonium carbonate	x	x	Benzyl alcohol	x	x
Ammonium chloride 10%	160	71	Benzyl chloride	x	x
Ammonium chloride 50%	160	71	Borax	140	60
Ammonium chloride, sat.	180	82	Boric acid	180	82
Ammonium fluoride 10%	90	32	Bromine gas, dry	x	x
Ammonium fluoride 25%	90	32	Bromine gas, moist	x	x
Ammonium hydroxide 25%	x	x	Bromine liquid	x	x
Ammonium hydroxide, sat.	x	x	Butyl acetate	x	x

Table I.8 Continued

Chemical	Maximum temp. °F	Maximum temp. °C	Chemical	Maximum temp. °F	Maximum temp. °C
Butyl alcohol	80	27	Ethylene glycol	120	49
n-Butylamine	x	x	Ferric chloride	180	82
Butyric acid 25%	120	49	Ferric chloride 50% in water	160	71
Calcium bisulfide	160	71	Ferric nitrate 10–50%	180	82
Calcium bisulfite	150	66	Ferrous chloride	180	82
Calcium carbonate	160	71	Ferrous nitrate	160	71
Calcium chlorate	160	71	Fluorine gas, dry	x	x
Calcium chloride	180	82	Fluorine gas, moist	x	x
Calcium hydroxide 10%	160	71	Hydrobromic acid, dilute	120	49
Calcium hydroxide, sat.	160	71	Hydrobromic acid 20%	140	60
Calcium hypochlorite 10%	120	49	Hydrobromic acid 50%	140	60
Calcium nitrate	140	60	Hydrochloric acid 20%	160	71
Calcium oxide	160	71	Hydrochloric acid 38%	160	71
Calcium sulfate	160	71	Hydrocyanic acid 10%	90	32
Caprylic acid	160	71	Hydrofluoric acid 30%	x	x
Carbon bisulfide	x	x	Hydrofluoric acid 70%	x	x
Carbon dioxide, dry	160	71	Hydrofluoric acid 100%	x	x
Carbon dioxide, wet	160	71	Hypochlorous acid	90	32
Carbon disulfide	x	x	Ketones, general	x	x
Carbon monoxide	160	71	Lactic acid 25%	160	71
Carbon tetrachloride	x	x	Lactic acid, concentrated	160	71
Carbonic acid	160	71	Magnesium chloride	180	82
Cellosolve	x	x	Malic acid	90	32
Chloracetic acid, 50% water	x	x	Methyl ethyl ketone	x	x
Chloracetic acid to 25%	150	66	Methyl isobutyl ketone	x	x
Chlorine gas, dry	160	71	Muriatic acid	160	71
Chlorine gas, wet	160	71	Nitric acid 5%	120	49
Chlorine, liquid	x	x	Nitric acid 20%	x	x
Chlorobenzene	x	x	Nitric acid 70%	x	x
Chloroform	x	x	Nitric acid, anhydrous	x	x
Chlorosulfonic acid	x	x	Nitrous acid, concentrated	120	49
Chromic acid 10%	x	x	Oleum	x	x
Chromic acid 50%	x	x	Perchloric acid 10%	x	x
Chromyl chloride	140	60	Perchloric acid 70%	x	x
Citric acid 15%	160	71	Phenol	x	x
Citric acid, concentrated	200	93	Phosphoric acid 50–80%	180	82
Copper acetate	160	71	Picric acid	x	x
Copper chloride	180	82	Potassium bromide 30%	160	71
Copper cyanide	160	71	Salicylic acid	100	38
Copper sulfate	200	93	Sodium carbonate 20%	90	32
Cresol	x	x	Sodium chloride	200	93
Cupric chloride 5%	170	77	Sodium hydroxide 10%	x	x
Cupric chloride 50%	170	77	Sodium hydroxide 50%	x	x
Cyclohexane	80	27	Sodium hydroxide, concentrated	x	x
Dichloroacetic acid	x	x	Sodium hypochlorite 20%	x	x
Dichloroethane (ethylene dichloride)	x	x	Sodium hypochlorite, concentrated	x	x

Table I.8 Continued

Chemical	Maximum temp. °F	Maximum temp. °C	Chemical	Maximum temp. °F	Maximum temp. °C
Sodium sulfide to 50%	x	x	Sulfuric acid 100%	x	x
Stannic chloride	180	82	Sulfuric acid, fuming	x	x
Stannous chloride	180	82	Sulfurous acid	x	x
Sulfuric acid 10%	160	71	Thionyl chloride	x	x
Sulfuric acid 50%	150	66	Toluene	110	43
Sulfuric acid 70%	x	x	Trichloroacetic acid 50%	170	77
Sulfuric acid 90%	x	x	White liquor	x	x
Sulfuric acid 98%	x	x	Zinc chloride	180	82

[a]The chemicals listed are in the pure state or in a saturated solution unless otherwise indicated. Compatibility is shown to the maximum allowable temperature for which data are available. Incompatibility is shown by an x. A blank space indicates that data are unavailable.

ISOPRENE RUBBER (IR)

Natural rubber chemically is natural *cis*-polyisoprene. The synthetic form of natural rubber, synthetic *cis*-polyisoprene is called isoprene rubber. The physical and mechanical properties of isoprene rubber are similar to the physical and mechanical properties of natural rubber, the one major difference being that isoprene does not have an odor. This feature permits the use of isoprene rubber in certain food-handling applications.

Isoprene rubber can be compounded, processed, and used in the same manner as natural rubber. Other than the lack of odor, isoprene rubber has no advantages over natural rubber.

See Refs. 11 and 1.

REFERENCES

1. PA Schweitzer. Corrosion Resistance Tables. 4th ed. Vol. 1–3. New York: Marcel Dekker, 1995.
2. PF Lafyatis. Carbon and graphite. In: BJ Monig and WI Pollock, eds. Process Industries Corrosion—Theory and Practice: Houston: NACE International, 1986, pp. 703–770.
3. N Sridhar and G Hodge. Nickel and high nickel alloys. In: PA Schweitzer, ed. Corrosion and Corrosion Protection Handbook. 2nd ed. New York: Marcel Dekker, 1989, pp. 96–124.
4 HH Uhlig. Corrosion and Corrosion Control. New York: John Wiley, 1963.
5. FG Porter. Corrosion Resistance of Zinc and Zinc Alloys. New York: Marcel Dekker, 1994.
6. CP Dillon. Corrosion Control in the Chemical Process Industries. 2nd ed. St. Louis: Materials Technology Institute of the Chemical Process Industries, 1994.
7. CP Dillon. Corrosion Resistance of Stainless Steels. New York: Marcel Dekker, 1995.
8. JH Mallinson. Corrosion-Resistant Plastic Composites in Chemical Plant Design. New York: Marcel Dekker, 1988.
9. GT Murray. Introduction to Engineering Materials. New York: Marcel Dekker, 1993.
10. PA Schweitzer. Corrosion Resistant Piping Systems. New York: Marcel Dekker, 1994.
11. PA Schweitzer. Corrosion Resistance of Elastomers. New York: Marcel Dekker, 1990.

K

KEVLAR

Kevlar is an aramid fiber (trademark of E. I. DuPont de Nemours) used as reinforcing in laminates. See "Thermoset Reinforcing Materials."

KILLED CARBON STEEL

Raw liquid steel contains oxygen as iron oxides or as dissolved gas. Carbon is also contained in the liquid steel. The oxygen and carbon can react to form carbon monoxide. This reaction can cause violent boiling during pouring and the solidification process. The excess oxygen can be removed as slag by adding an oxygen scavenger such as silicon to the molten steel prior to pouring. The resulting material does not boil during pouring and cooling, thereby producing a more homogenous "killed steel." These steels are cleaner and contain fewer defects than "unkilled" or "wild" steels.

Steels can also be killed with a combination of silicon and aluminum or with aluminum alone. Silicon when used alone to deoxidize the steel tends to produce a coarse grain structure. These steels have a relatively high brittle–ductile transition temperature. This precludes their use for applications requiring low temperature toughness. However, the coarse grained steels are more resistant to creep, graphitization, and some forms of corrosion.

When the steel is deoxidized with a combination of silicon and aluminum or with aluminum alone, a fine austenitic grain size is produced. These steels are used for applications requiring low temperature toughness.

Using an oxygen scavenger is the primary method of deoxidation. A less common method is vacuum degassing. Vacuum degassing not only controls oxidizing gasses such as oxygen and carbon dioxide but also will help to limit nonoxidizing gasses such as nitrogen and hydrogen. This treatment substantially decreases the number of nonmetallic inclusions in the steel and is used for bearings and other high quality steel.

When steel cools in a mold, shrinkage of the steel on solidifying causes "piping." The cavity of a "pipe" is usually found in the upper portion of the ingot. To minimize this condition, a large-end-up mold is used together with a refractory "hot top" that supplies molten steel to the main body of the ingot while solidification takes place. This minimizes the quantity of metal that has to be discarded because of pipe.

To reduce the large percentage of metal that must be discarded when making steel for structural purposes, the steel is not fully oxidized. This results in blowholes in the steel on solidification. These blowholes minimize the piping by distributing small voids throughout the ingot instead of having one large one in the upper center of the ingot. As long as the blowholes are not exposed on the surface they will weld together during rolling. Steel deoxidized in this manner is called semikilled steel.

If the molten steel is deoxidized still less, a reaction takes place during solidification in which the oxygen and carbon react to form carbon monoxide, which is freely evolved from the ingot. This evolution affects the structure of the ingot. When the reaction is allowed to go to completion the product is called "rimmed steel." If the reaction is stopped in a mechanical manner after a short period of time preventing further evolution of the gas from the top of the ingot, the product is called "capped steel."

Rimmed steel ingots have an outer skin that is clean and very low in carbon. In capped steel the skin is thinner and there is less segregation or concentration of impurities than in rimmed steel.

KNIFE LINE ATTACK

Knife line attack is a highly localized form of intergranular corrosion. Titanium or columbium is added to the type 300 stainless steels to prevent the precipitation of chromium carbide by permitting the alloy to precipitate titanium or columbium carbide, instead of chromium carbide.

There are two problems with this approach. Titanium will lose much of its effectiveness in multipass welding or cross-welding (intercepting horizontal and vertical welds). Columbium does not react in this manner, but the columbium carbides, as well as the titanium carbides, can be redissolved by the heat of welding, particularly with alloys of higher nickel content. Because of this, cross-welding or multiple pass welding can first redissolve titanium or columbium carbides and then allow chromium carbide precipitation in the "fusion zone" and not the heat-affected zone. This results in knife line attack at the "fusion zone." The alloys most subject to this form of attack are type 347 stainless steel and alloy 825.

KYNAR

See "Polyvinylidene Fluoride."

L

LAMELLAR CORROSION

See "Exfoliation Corrosion."

LAYER CORROSION

Layer corrosion is a localized attack of a metal surface. The attack is in thin parallel layers oriented to the direction of metal processing and leads to unattached metal layers being released like pages in a book.

LEAD

Lead being amphoteric is susceptible to attack by both acids and alkalies under certain conditions. Since it is a weak metal, being unable to support its own weight, alloys have been developed to improve its physical and mechanical properties.

Chemical lead is lead with traces of copper and silver left in from the original ore. It is not economical to recover the copper and silver, and the copper content is believed to improve the general corrosion resistance.

Antimonial lead (also called hard lead) is an alloy containing from 2 to 6% antimony to improve the mechanical properties. Hard lead can be used in services up to 200°F (93°C). Above this temperature strength and corrosion resistance are reduced.

Tellurium lead is a lead alloy containing a fraction of a percent of tellurium. This alloy has better resistance to fatigue failure caused by vibration because of its ability to work harden under strain.

There are also a number of proprietary alloys to which copper and other elements have been added to improve corrosion resistance and creep resistance.

Corrosion Resistance

The corrosion resistance of lead is due primarily to the protective film formed by the insolubility of some of its corrosion products. Lead will be attacked by hydrochloric and nitric acids as well as organic acids if they are dilute or if they contain oxidizing agents. It is resistant to sulfuric, sulfurous, chromic, and phosphoric acid and cold hydrofluoric acid.

In caustics, lead is limited to concentrations of 10% maximum up to 195°F (90°C). It will resist cold strong amines but is attacked by dilute aqueous amine solutions.

Lead will be attacked by soft aggressive waters but is resistant to most natural waters. Because of the toxicity of lead salts, lead should not be used to handle potable (drinking) water.

Lead is initially anodic to more highly alloyed materials, but due to a film of insoluble corrosion products on its surface it may become cathodic in time. There have been instances where alloy 20Cb3 valves have undergone accelerated attack in lead piping systems handling sulfuric acid services. In such applications it is necessary electrically to isolate the valves from the piping to prevent galvanic action.

Because of the toxicity problems associated with lead burning (joining process) applications of lead have been greatly reduced in modern practice.

LININGS, SHEET

See "Sheet Linings."

LIQUID APPLIED LININGS

Of the various coating applications the most critical is that of a tank lining. The coating must be resistant to the corrodent and be free of pinholes through which the corrosive could penetrate and reach the substrate. The severe attack that many corrosives have on the bare tank emphasizes the importance of using the correct procedure in lining a tank to obtain a perfect coating. It is also essential that the tank be designed and constructed in the proper manner to permit a perfect lining to be applied.

Liquid applied linings or coatings may be trowled on or spray applied. In a lined tank there are usually four areas of contact with the stored material. Each area has the potential of developing a different form of corrosive attack. These areas are the bottom of the tank (where moisture and other contaminants of greater density may settle), the liquid phase (the area constantly immersed), the interphase (the area where liquid phase meets the vapor phase), and the vapor phase (the area above the liquid). Each of these areas can be more severely attacked than the rest at one time or another. The type of material contained, the nature of impurities that may be present, and the amount of water and oxygen present are all factors affecting the attack. In view of this it is necessary to understand the corrosion resistance of the lining material under each condition and not only the immersed condition.

Other factors that have an effect on the performance of the lining material are vessel design, vessel preparation prior to lining, application techniques of the coating, curing of the coating, inspection, operating instructions, and temperature limitations.

Vessel Design

All vessels to be lined should be of welded construction. Riveted tanks will expand or contract, thus damaging the liner and causing leakage. Butt welding is preferred, but lap welding can be used, providing a fillet weld is used and all sharp edges are ground smooth. Butt welds need not be ground flush but they must be ground to a smooth rounded contour. All weld splatter must be removed. Any sharp prominence may result in a spot where the film thickness will be inadequate and noncontinuous, thus causing premature failure. Other design considerations are as follows:

1. Do not use construction that will result in the creation of pockets or crevices that will not drain or that cannot be properly sandblasted and lined.
2. All joints must be continuous and solid welded. All welds must be smooth with no porosity, holes, high spots, lumps or pockets.
3. All sharp edges must be ground to a minimum of $1\frac{1}{8}$ in. radius.
4. Outlets must be flanged or pad type rather than threaded.
5. Stiffening members should be on the outside of the vessel.
6. Tanks larger than 25 feet in diameter may require three manways for working entrances. Usually two are located at the bottom (180° apart) and one at the top. The minimum opening diameter should be 20 in., but 30 in. openings are preferable.

Concrete tanks should be located above the water table. Unless absolutely necessary, expansion joints should be avoided. Small tanks do not normally require expansion joints. Larger tanks can make use of a chemical resistant joint such as polyvinyl chloride (PVC). Any concrete curing compound used must be compatible with the lining material or removed before coating. Form joints must be made as smooth as possible. Adequate steel reinforcement must be used in a strong, dense, concrete mix to reduce movement and cracking. The coating manufacturer should be consulted for special instructions.

Vessel Preparation

In order for the lining material to obtain maximum adhesion to the substrate surface it is essential that the surface be absolutely clean. All steel surfaces to be coated must be abrasive blasted to a white metal in accordance with SSPC Specifications SP5-63 or NACE Specification 1. A white metal blast is defined as removing all rust, scales, paints, and so on, to a clean white metal that has a uniform gray white appearance. No streaks or stains of rust or any other contaminants are allowed. At times a near white, blast cleaned surface equal to SSPC SP-10 may be used. If this is permitted by the manufacturer of the lining material it should be used as it is less expensive.

All dust and spent abrasive must be removed from the surface by vacuum cleaning or brushing. After blasting, all workers coming into contact with the clean surface should wear clean, protective gloves and clothing to prevent contamination of the cleaned surface. Any contamination may cause premature failure by osmotic blistering or adhesion loss. The first coat must be applied before the surface starts to rust.

Concrete surfaces must be clean, dry, and properly cured before applying the lining. All protrusions and form joints must be removed. All surfaces must be toughened by sand blasting to remove all loose, weak or powdery concrete to open all voids and provide the necessary profile for mechanical adhesion of the coating. All dust must be removed by brushing or vacuuming. The coating manufacturer should be contacted for special priming and caulking methods.

Lining Selection

In order properly to specify a lining material it is necessary to know specifically what is being handled and under what conditions. The following information must be known about the material being handled:

1. What are the primary chemicals to be handled and at what concentrations?

2. Are there any secondary chemicals, and if so at what concentrations?
3. Are there any trace impurities or chemicals?
4. Are there any solids present, and if so what are the particle sizes and concentrations?
5. Will there be any agitation?
6. What are the fluid purity requirements?

The answers to these questions will narrow the selection to those coatings that are compatible. However, the answers to the next set of questions will narrow the selection down to those materials that are compatible as well as to those coatings that have the required mechanical and/or physical properties.

1. What is the normal operating temperature and temperature range?
2. What peak temperatures can be reached during shutdown, startup, process upset, etc.?
3. Will any mixing areas exist where exothermic or heat of mixing temperatures may develop?
4. What is the normal operating pressure?
5. What vacuum conditions and range are possible during operation, startup, shutdown, and upset conditions?
6. Is grounding necessary?

Other factors must also be considered before the final decision can be made as to which coating to use. After the previous questions have been answered, there will still be several potential materials from which to choose.

Service life expectation must be considered. Different protective coating options afford different degrees of protection for different periods of time at a variety of costs. Such factors as maintenance cycle, operating cycles, and reliability of the coating must all be considered. Can the facility tolerate any downtime for inspection and maintenance? If so, how often and for how long?

When these questions have all been answered an appropriate decision can be made as to which coating material will be used.

Lining Application

In lining a vessel the primary concern is to deposit a void free film of the specified thickness on the surface. Any area that is considerably less than the specified thickness may have a noncontinuous film. In addition, pinholes in the coating may cause premature failure.

If the film is too thick there is always the danger of solvents being entrapped, which can lead to bad adhesion, excessive brittleness, improper cure, and subsequent poor performance. Dry spraying of the coating should be avoided because it causes the coating to be porous. Poor film formation may be caused if thinners other than those recommended by the coating manufacturer are used. Do not permit application to take place below the temperature recommended by the manufacturer.

During application the film thickness should be checked. This can be accomplished by use of an Elcometer or Nordsen wet film thickness gauge. If the wet film thickness meets specifications, in all probability the dry film will also be within specification limits.

All gauges used to measure dry film thickness must be calibrated before use, following the manufacturer's recommended procedure. Readings should be taken at random locations on a frequent basis. Special attention should be given to hard-to-coat areas.

Inspection

Proper inspection requires that the inspector be involved with the job from the beginning. An understanding of the design criteria of the vessel and the reasons for the specific design configuration is helpful. The inspector should participate in the prework meeting, prejob inspection, and coating application inspection. Daily inspection reports should be prepared along with a final acceptance report.

Before the coating is applied the inspector should verify that the vessel has been properly prepared for lining. Welds must be ground smooth with a rounded contour. It is not necessary that they be ground flush. Sharp protrusions should be rounded and weld crevices opened up manually so that the coating can penetrate. If this is not possible the projection should be removed by grinding. Back-to-back angles, tape, or stitch welding and so forth cannot be properly cleaned and coated. They should be sealed with caulking to prevent crevice corrosion.

Once the vessel has been sand blasted the inspector should work quickly so that the application of the coating to the surface is not delayed.

The inspector should examine coatings during and after application. It is important to check for porosities. The first visual inspection is mandatory to detect pinholding and provide recoat instructions. Visual inspections are performed either with the unaided eye or by the use of a magnifying glass.

After repairs of visible defects, inspection should be done using low voltage (75 V or less) holiday detectors that ring, buzz, or light up to show electrical contact through a porosity within the coating to the metal surface. This check should be performed after the primer or second coat, so that these areas can be touched up and made free of porosities before the final top coat. These visual techniques permit the inspector to identify areas that have been missed, damaged areas, or thin areas. In some instances white primers have been used to spot areas of low film thickness or inadequate coverage of the substrate. The final test employing instruments will provide the inspector with an accurate appraisal as to whether the proper film thickness has been met.

The inspector should also be involved in the selection of the applicator. Since a tank lining requires nearly perfect application a knowledgeable and conscientious contractor is required. The lowest bidder is not necessarily the best choice. Evaluate the applicator before awarding the contract to assure that the tank lining contractor is experienced in applying the specified lining. Before placing a contractor on the bidder's list, review his qualifications. Inquire as to what jobs he has done using the selected lining material and follow up with those particular applications. If possible, visit his facilities and inspect his workmanship. An experienced inspector is helpful during this evaluation. Taking these steps will help to ensure installation of a tank lining that will provide the desired performance.

Curing

Proper curing is essential if the lining is to provide the corrosion protection for which it was selected. Each coat must be cured using proper air circulation techniques. Fresh air over 50°F (10°C) and having a relative humidity of less than 89% should be supplied to an opening at the top of the vessel with an exhaust at the bottom. The air flow should be by forced air fans and should be downward because the solvents used in coatings are usually heavier than air. This is why proper exhaustion can only be obtained with downward flow.

To prevent solvent entrapment between coats and to ensure a proper final cure, the curing time and temperature must be in accordance with the manufacturer's instructions for the specific coating material. A warm forced air cure between coats and as a final cure will provide a dense film and tighter cross-linking, which provides superior resistance to solvents and moisture permeability. Before placing the vessel in service, the lining should be washed down with water to remove any loose overspray.

Linings must be allowed sufficient time to obtain a full cure before being placed in service. This usually requires 3–7 days. Do not skimp on this time.

When the tank is placed in service, operating instructions should be prepared and should include the maximum temperature to be used. The outside of the tank should be labeled: "Do not exceed x°F (x°C). This tank has been lined with ————. It is to be used only for ———— service."

Properties of Lining Materials

A wide variety of lining materials are available. Information as to the corrosion resistance of specific lining materials is available from the manufacturer. If data on the resistance of a lining material to a specific corrodent are not available, then tests should be conducted by evaluating sample panels of several coating systems for at least 90 days. A six-month test would be preferable. The tests should simulate as closely as possible the actual operating conditions to which the coating will be subjected. This should include maximum operating temperatures, any temperature cycling, washing cycles, and whatever other conditions the coating will be exposed to.

Table L.1 lists the more common lining materials and their general area of application. Because most of these coatings are formulated products their performance will be based on their formulation. Therefore the resistance of these generic coatings may vary between manufacturers. If a generic coating is selected, resistance should be verified with the manufacturer.

By the same token, any testing done should be with material supplied by several manufacturers.

See Refs. 1–6.

Table L.1 Corrosion Resistance of Lining Materials

Type of coating	Use
Baked phenolic	Most widely used lining material. Has dry heat resistance of 400°F (204°C). Excellent resistance to acids, solvents, food products, beverages, water. Compared with other coatings its flexibility is poor
Air-dried phenolic	Can be formulated for excellent resistance to alkalies, solvents, fresh water, salt water, deionized water; mild acid resistance. Excellent for dry products. Has dry heat resistance of 150°F (65°C)
Vinyl ester	Excellent resistance to strong acids. Resistant up to 400°F (204°C) depending upon thickness

Table L.1 Continued

Type of coating	Use
Vinyl polyester	Excellent resistance to acids, alkalies, solvents and water. Resistant up to 400°F (204°C) for short duration. Continuous maximum temperature of 220°F (104°C)
Polyester unsaturated	Excellent resistance to strong mineral and organic acids and oxidizing materials. Very poor aromatic solvent and alkali resistance
Epoxy (amine catalyst)	Good alkali resistance, fair to good resistance to mild acids, solvents and dry food products. Widely used for covered hopper-car linings and nuclear containment facilities. Maximum temperature 275°F (135°C)
Baked epoxy	Excellent resistance to acids, alkalies, solvents, inorganic salts, and water. Maximum continuous temperature 325°F (163°C)
Epoxy polyamide	Poor acid resistance, fair alkali resistance, good resistance to water and brines. Used in storage tanks and nuclear containment facilities
Epoxy polyester	Poor solvent resistance, good abrasion resistance. Used for covered hopper-car linings
Coal tar	Excellent water resistance. Used for water tanks
Coal tar epoxy	Excellent resistance to salt water, freshwater, mild acids, mild alkalies. Poor solvent resistance. Used for crude oil storage tanks, sewage disposal plants, and water works
Asphalts	Good water and acid resistance
Modified polyvinyl chloride (polyvinyl chloracetates) air-cured	Excellent resistance to strong mineral acids and water. Poor solvent resistance. Used in water immersion service, potable and marine. Used extensively for water storage tanks (beverage processing)
Polyvinyl chloride (PVC) plastisols	Popular acid-resistant lining. Must be heat-cured
Chlorinated rubber	Excellent water resistance. Poor solvent resistance
Hypalon	Chemical salts
Neoprene	Good acid and flame resistance. Chemical processing
Polysulfide	Good solvent resistance and water resistance. Used for lining jet fuel tanks
Butyl rubber	Good resistance to strong acids, alkalies, aggressive salts, and oxidizing agents. Flexible coating absorbs vibration, resists wear/abrasion
ECTFE	Excellent resistance to mineral and organic acids at moderate temperatures
PFA	Recommended where severe corrosion is a problem. Can withstand full vacuum at 300°F (149°C). Ideal for high-purity applications
Styrene-butadiene polymers	Food and beverage processing, concrete tanks
Urethanes	Excellent resistance to strong mineral acids and alkalies. Fair solvent resistance. Used to line dishwashers and washing machines
Rubber latex	Excellent alkali resistance. Used to line 50% and 73% caustic tanks 180–250°F (82–121°C)
Inorganic zinc	Jet fuel storage tanks, petroleum products
Inorganic zinc solvent based self-curing	Excellent resistance to most organic solvents (i.e., aromatic, ketones, and hydrocarbons), excellent water resistance, difficult to clean. Often sensitive to decomposition products in tanks
Alkyds, epoxy esters, oleoresinous primers	Water immersion service. Primers for other top coats

LIQUID IMPINGEMENT CORROSION

See "Impingement Corrosion."

LIQUID METAL ATTACK

The attack by liquid metals on solid metals may result in one or more of the following:

1. High temperature corrosion of the solid metal.
2. When the solid metal is under applied or residual stresses and is contacted by a liquid metal instantaneous failure may result.
3. When the static stress level of the solid metal is below the tensile stress of the solid metal, contact with a liquid metal may result in a delayed failure.
4. Stress-independent grain boundary penetration of the solid metal by the liquid metal may occur.

Conditions 2 and 3 above are referred to as liquid metal embrittlement (LME) or liquid metal induced embrittlement (LMIE).

When solid metals are in contact with liquid metals for extended periods of time, corrosion becomes a concern. This is particularly true when liquid metals, because of their excellent heat transfer properties, are used in heat pipes. Such applications make use of liquid metals such as lithium, sodium, or sodium potassium.

The four processes by which liquid metal corrosion can take place are

1. Direct dissolution
2. Corrosion product formation
3. Elemental transfer
4. Alloying

The release of atoms from the solid metal into the melt is known as direct dissolution. In an isothermal system as the adjacent melt becomes saturated with the dissolving metal, the dissolution rate decreases or may stop completely. However, in a non isothermal liquid metal system such a condition may never be reached because of convection from the hotter to the colder regions. As the dissolved metal moves from the hot area to the cold area the dissolved metal is deposited. Plugging of the coolant pipes may result.

The dissolution may be uniform or selective. For example, it has been reported that molten sodium will preferentially leach silicon from stainless steels, and that molten lithium sodium or bismuth will preferentially leach nickel from type 316 stainless steel. The selective leaching may proceed to such an extent that voids are left in the solid metal.

At times the corrosion products form protective films on the solid metal surface, reducing further attack. The addition of aluminum or silicon to steel helps to form corrosion products that will protect the surface. The addition of zirconium to liquid bismuth or mercury will inhibit the corrosion of steel in contact with these liquid metals.

Elemental transfer is the transfer of impurities to or from a liquid metal. Under these conditions the atoms of the liquid metal do not react with the atoms of the solid metal. Liquid sodium systems contaminated with carbon have resulted in the carburization of austenitic stainless steels and refractory materials. Another example of elemental transfer is the decarburization of iron–chromium–molybdenum steels, particularly lower chromium steels in liquid sodium or lithium.

Intermetallic compounds may be formed when two solid metals are in contact with a liquid metal as a result of elemental transfer. When aluminum and molybdenum are exposed to molten bismuth, intermetallic compounds Al_3Mo and Al_5Mo may be formed.

In some instances alloys or stable intermetallic compounds may be formed such as nickel in contact with molten aluminum. It is recommended that these systems be avoided.

Failures resulting from liquid metal embrittlement may be immediate or may occur after a period of time during which the stressed metal has been in contact with the liquid metal. The former is considered the typical liquid metal embrittlement and the latter is referred to as delayed failure or static fatigue. In either case stress must be present in the solid metal. This stress may be shear, tortional, or tensile but not compressive. LME is similar to stress corrosion cracking except that the propagation of the fracture take place much faster.

Even when stress is not present, a solid metal may become brittle if sufficient time is allowed as a result of intergranular penetration of the liquid metal.

LIQUID METAL EMBRITTLEMENT

Liquid metal embrittlement, also known as liquid metal cracking, is a form of environmental cracking resulting in brittle failure of a normally ductile metal as a result of the ductile metal contacting a liquid or molten metal. Typical examples are the action of mercury on cold-worked brass or aluminum and molten zinc on austenitic stainless steel.

See Refs. 6–9.

LOCAL CORROSION CELL

Corrosion in metals is caused by the flow of electricity from one metal to another metal or a recipient of some kind; or from one part of the surface of one piece of metal to another part of the same metal, when conditions permit the flow of electricity. A moist conductor or electrolyte must be present.

A local corrosion cell is formed when one area or region on a metal surface has a negative charge relative to a second area which has a positive charge in opposition, and an electrolyte is present. These cells can be formed by surface imperfections, grain orientation, lack of homogeneity of the metal, variation in the environment, localized shear and torque during manufacture, mill scale, and existing red iron oxide rust. The result is usually a pitting type of corrosion.

LOCALIZED CORROSION

Localized corrosion is any form of corrosion that takes place at discrete sites. This includes pitting, crevice corrosion, intergranular attack, corrosion fatigue, and stress corrosion cracking.

LOW ALLOY STEELS

There are two basic types of low alloy steels; weathering steels and hardenable steels. The weathering steels contain small additions of copper, chromium, and nickel to form a more adherent oxide during atmospheric exposure. A typical example is U.S. Steel's Cor-Ten steel. Hardenable steels contain additions of chromium or molybdenum and possibly nickel. These steels offer higher strength and hardness after proper heat treatment. Typical examples are 4130 and 4340 steels.

See Ref. 10.

REFERENCES

1. National Association of Corrosion Engineers. Laboratory Methods for Evaluation of Protective Coatings Used as Lining Materials in Immersion Service. Materials Performance. Houston: NACE, 1978, TM-01-74.
2. National Association of Corrosion Engineers. Coatings and Linings for Immersion Service. Houston: NACE TCP2, 1972.
3. National Association of Corrosion Engineers. Recommended Practice, Design, Fabrication, and Surface Finish of Metal Tanks and Vessels to be Lined for Chemical Immersion Service. Materials Performance. Houston: NACE, 1978, RP-01-78.
4. DM Berger and SE Mroz. Instruments for Inspection of Coatings. 1 Test Eval 4:29–39, 1976.
5. RA Mixer and SI Oechsle Jr. Materials of construction. 1. Protective lining systems. Chem Eng 181–182, 1956, November.
6. PA Schweitzer. Corrosion Resistance Tables. 4th ed. New York: Marcel Dekker, 1995.
7. CP Dillon. Corrosion Control in the Chemical Process Industries. 2nd ed. St. Louis: Materials Technology Institute of the Chemical Process Industries, 1994.
8. RD Kane. Corrosion testing. In: PA Schweitzer, ed. Corrosion Engineering Handbook. New York: Marcel Dekker, 1996, pp. 607-621.
9. CP Dillon. Corrosion Resistance of Stainless Steels. New York: Marcel Dekker, 1995.
10. GN Kirby. The corrosion of carbon and low-alloy steels. In: PA Schweitzer, ed. Corrosion Engineering Handbook. New York: Marcel Dekker, 1996, pp. 35-52.

M

MALLEABLE IRON

Malleable iron is relatively expensive to produce in comparison to ductile iron whose properties are similar. For this reason it is declining in popularity. In general the corrosion resistance of malleable iron and gray iron are the same, as are the properties of ductile iron.

See Ref. 1.

MARINE COATINGS

Marine atmospheres include salt particles that attack steel substrate, saltwater spray and splashes, and immersion. Paint coatings used to protect against salt spray include chlorinated rubber, epoxy, and vinyl resin. For protection against seawater zinc coatings are also effective. Each mil (0.03 mm) of zinc will provide protection against rusting for approximately one year. Epoxy coal tar will also provide protection. See "Coatings."

MARINE ENVIRONMENT

Marine environment is an atmospheric exposure that is frequently wetted by salt mist but which is not in direct contact with salt spray or splashing waves.

MARTENSITE

When austenite is cooled rapidly, preventing the formation of ferrite, martensite is formed. Since martensite is a brittle material, it is normally tempered. The tempering is done to permit some carbon to diffuse from the martensite. Tempered martensite is considerably stronger and tougher than the parent ferritic alloy. A tempered material should never be stress relieved or post weld heat treated at a temperature in excess of the final tempering temperature.

The specific alloy will determine what procedure is required to effect cooling to produce martensite from austenite. Some heat treatable alloys require quenching in water or some other liquid such as oil or a molten salt in order to obtain the cooling rate necessary. Some steels have sufficient alloying additions that quenching is not necessary to produce martensite. The martensitic microstructure can be produced in type 410 stainless steel by air cooling.

Heat treated martensitic steels become brittle at low temperatures even though they have superior fracture toughness. Most martensitic steels are very sensitive to hydrogen embrittlement.

See "Austenite."

MARTENSITIC STAINLESS STEELS

Martensitic stainless steels contain relatively lower levels of chromium with sufficient carbon to permit martensite formation with rapid cooling. Martensite is a body centered tetragonal structure that provides increased strength and hardness versus the annealed stainless with other lattice structures. Other elements, such as nickel and molybdenum, may be added for improved corrosion resistance and mechanical properties.

Precipitation-hardening stainless steels contain chromium and nickel and are strengthened by aging due to the presence of elements such as copper, columbium, titanium, or aluminum. Molybdenum may be added for corrosion resistance, and columbium may be used to stabilize carbon.

The corrosion resistance of the martensitic stainless steels is inferior to that of the austenitic stainless steels. These alloys are generally used in mildly corrosive services such as atmospheric, fresh water, and organic exposures. Their greatest weakness is their susceptibility to absorption of atomic hydrogen, resulting in hydrogen-assisted cracking, particularly in sulfide environments.

The primary areas of application are for cutlery, turbine blades, and high temperature parts.

Alloy 350

This is an austenitic/martensitic alloy. It may be subject to intergranular attack unless cooled to subzero temperature before aging. The corrosion resistance of alloy 350 is similar to that of type 304 stainless steel. This alloy is used where high strength and corrosion resistance at room temperature are required. The mechanical and physical properties will be found in Table M.1. To prevent overaging, operating temperatures should be kept below 1000°F (540°C).

Table M.1 Mechanical and Physical Properties of Alloy 350 Stainless Steel

Modulus of elasticity $\times 10^6$, psi	(aged) 29.4
Tensile strength $\times 10^3$, psi	(aged) 200
Yield strength 0.2% offset $\times 10^3$, psi	(aged) 585
Elongation in 2 in., %	(aged) 12
Hardness, Rockwell	(aged) C-30
Density, lb/in.^{3}	0.286
Thermal conductivity, Btu/ft h °F	
at 70°F (20°C)	8.4
at 1500°F (815°C)	12.2

410 Stainless Steel

This alloy contains 11.5–13.5% chromium with 1% manganese, 1% silicon, and a maximum of 0.03% sulfur. Mechanical and physical properties are shown in Table M.2.

In general, corrosion resistance is somewhat less than that for type 304 stainless steel. Refer to Table M.3 for the compatibility of type 410 stainless steel with selected corrodents.

Custom 450* Stainless Steel

This is a martensitic age-hardenable stainless steel with good corrosion resistance and moderate strength. Its corrosion resistance is similar to that of type 304 stainless steel. This alloy finds applications where type 304 stainless steel is not strong enough or where type 410 is insufficiently corrosion resistant. Refer to Table M.4 for the mechanical and physical properties of custom 450 stainless steel.

Custom 455* Stainless Steel

This is a martensitic age-hardenable stainless steel having high strength with corrosion resistance better than type 410 stainless steel. It may be subject to hydrogen embrittlement under some conditions. This alloy finds application when high strength and corrosion resistance is required. It resists chloride cracking as well as alkalies and nitric acid. Refer to Table M.5 for the mechanical and physical properties of custom alloy 455.

Table M.2 Mechanical and Physical Properties of Alloy 410 Stainless Steel

Modulus of elasticity $\times 10^6$, psi	29
Tensile strength $\times 10^3$, psi	
annealed	75
heat-treated	150
Yield strength 0.2% offset $\times 10^3$, psi	
annealed	40
heat-treated	115
Elongation in 2 in., %	
annealed	30
heat-treated	15
Hardness, Brinell	
annealed	150
heat-treated	410
Density, lb/in.^{3}	0.28
Specific gravity	7.75
Specific heat, (32–212°F) Btu/lb °F	0.11
Thermal expansion coefficient (32–212°F), in./in. °F $\times 10^{-6}$	173

*Custom 450 and custom 455 are registered trademarks of Carpenter Technology Corp.

Table M.3 Compatibility of Type 410 Stainless Steel with Selected Corrodents[a]

Chemical	Maximum temp. °F	Maximum temp. °C	Chemical	Maximum temp. °F	Maximum temp. °C
Acetaldehyde	60	16	Bromine gas, moist	x	x
Acetamide	60	16	Bromine liquid	x	x
Acetic acid 10%	70	21	Butadiene	60	16
Acetic acid 50%	70	21	Butyl acetate	90	32
Acetic acid 80%	70	21	Butyl alcohol	60	16
Acetic acid, glacial	x	x	Butyric acid	150	66
Acetic anhydride	x	x	Calcium bisulfite	x	x
Acetone	210	99	Calcium carbonate	210	99
Acrylonitrile	110	43	Calcium chloride[b]	150	66
Allyl alcohol	90	27	Calcium hydroxide 10%	210	99
Alum	x	x	Calcium hypochlorite	x	x
Aluminum chloride, aqueous	x	x	Calcium sulfate	210	99
Aluminum chloride, dry	150	66	Carbon bisulfide	60	16
Aluminum fluoride	x	x	Carbon dioxide, dry	570	299
Aluminum hydroxide	60	16	Carbon dioxide, wet	570	299
Aluminum nitrate	210	99	Carbon disulfide	60	16
Aluminum oxychloride	x	x	Carbon monoxide	570	299
Aluminum sulfate	x	x	Carbon tetrachloride[b]	210	99
Ammonium bifluoride	x	x	Carbonic acid	60	16
Ammonium carbonate	210	99	Chloracetic acid	x	x
Ammonium chloride 10%[b]	230	110	Chlorine gas, dry	x	x
Ammonium chloride 50%	x	x	Chlorine gas, wet	x	x
Ammonium chloride, sat.	x	x	Chlorine, liquid	x	x
Ammonium hydroxide, sat.	70	21	Chlorobenzene, dry	60	16
Ammonium nitrate	210	99	Chloroform	150	66
Ammonium persulfate 5%	60	16	Chlorosulfonic acid	x	x
Ammonium phosphate 5%	90	32	Chromic acid 10%	x	x
Ammonium sulfate 10–40%	60	16	Chromic acid 50%	x	x
Ammonium sulfite	x	x	Citric acid 15%	210	99
Amyl acetate[b]	60	16	Citric acid 50%	140	60
Amyl alcohol	110	43	Copper acetate	90	32
Amyl chloride	x	x	Copper carbonate	80	27
Aniline	210	99	Copper chloride	x	x
Antimony trichloride	x	x	Copper cyanide	210	99
Barium carbonate 10%	210	99	Copper sulfate	210	99
Barium chloride[b]	60	16	Cupric chloride 5%	x	x
Barium hydroxide	230	110	Cupric chloride 50%	x	x
Barium sulfate	210	99	Cyclohexane	80	27
Barium sulfide	70	21	Cyclohexanol	90	32
Benzaldehyde			Ethylene glycol	210	99
Benzene	230	110	Ferric chloride	x	x
Benzoic acid	210	99	Ferric chloride 50% in water	x	x
Benzyl alcohol	130	54	Ferric nitrate 10–50%	60	16
Borax	150	66	Ferrous chloride	x	x
Boric acid	130	54	Fluorine gas, dry	570	299
Bromine gas, dry	x	x	Fluorine gas, moist	x	x

Table M.3 Continued

Chemical	Maximum temp. °F	Maximum temp. °C	Chemical	Maximum temp. °F	Maximum temp. °C
Hydrobromic acid, dilute	x	x	Phosphoric acid 50–80%	x	x
Hydrobromic acid 20%	x	x	Picric acid	60	16
Hydrobromic acid 50%	x	x	Potassium bromide 30%	210	99
Hydrochloric acid 20%	x	x	Salicylic acid	210	99
Hydrochloric acid 38%	x	x	Silver bromide 10%	x	x
Hydrocyanic acid 10%	210	99	Sodium carbonate 10–30%	210	99
Hydrofluoric acid 30%	x	x	Sodium chloride[b]	210	99
Hydrofluoric acid 70%	x	x	Sodium hydroxide 10%	210	99
Hydrofluoric acid 100%	x	x	Sodium hydroxide 50%	60	16
Ketones, general	60	16	Sodium hypochlorite 20%	x	x
Lactic acid 25%	60	16	Sodium hypochlorite, concentrated	x	x
Lactic acid, concentrated	60	16	Sodium sulfide to 50%	x	x
Magnesium chloride 50%	210	99	Stannic chloride	x	x
Malic acid	210	99	Stannous chloride	x	x
Methyl chloride, dry	210	99	Sulfuric acid 10%	x	x
Methyl ethyl ketone	60	16	Sulfuric acid 50%	x	x
Muriatic acid	x	x	Sulfuric acid 70%	x	x
Nitric acid 5%	90	32	Sulfuric acid 90%	x	x
Nitric acid 20%	160	71	Sulfuric acid 98%	x	x
Nitric acid 70%	60	16	Sulfuric acid 100%	x	x
Nitric acid, anhydrous	x	x	Sulfurous acid	x	x
Nitrous acid, concentrated	60	16	Toluene	210	99
Perchloric acid 10%	x	x	Trichloroacetic acid	x	x
Perchloric acid 70%	x	x	Zinc chloride	x	x
Phenol[b]	210	99			

[a]The chemicals listed are in the pure state or in a saturated solution unless otherwise indicated. Compatibility is shown to the maximum allowable temperature for which data are available. Incompatibility is shown by an x. A blank space indicates that data are unavailable. When compatible the corrosion rate is < 20 mpy.
[b]Material is subject to pitting.
Source: PA Schweitzer. *Corrosion Resistance Tables,* 4th ed. Vols. 1–3. New York: Marcel Dekker, 1995.

Alloy 718

This alloy is an age-hardenable martensitic alloy having high strength and good tensile, fatigue, creep, and rupture strength. It has an operating temperature range of –423 to 1300°F (–253 to 704°C). Alloy 718 possesses good corrosion resistance, being compatible with alkalies, organic acids, and sulfuric acid. Excellent oxidation resistance is demonstrated throughout its operating temperature range. Refer to Table M.6 for the mechanical and physical properties

Alloy 17-7PH

This is a semiaustenitic stainless steel. In the annealed condition it is austenitic, but in the aged condition it is martensitic.

In rural and mild industrial atmospheres it has excellent resistance to general

Table M.4 Mechanical and Physical Properties of Custom 450[a] Stainless Steel

Modulus of elasticity × 10^3, psi	
annealed	28
aged	29
Tensile strength × 10^3, psi	
annealed	142
aged	196
Yield strength 0.2% offset × 10^3, psi	
annealed	118
aged	188
Elongation in 2 in., %	
annealed	13
aged	14
Hardness, Rockwell	
annealed	C-28
aged	C-42.5
Density, lb/in.^{3}	0.28

[a]Custom 450 is the registered trademark of Carpenter Technology Corp.

Table M.5 Mechanical and Physical Properties of Custom 455[a] Stainless Steel

Modulus of elasticity × 10^6, psi	
aged	29
Tensile strength × 10^3, psi	
annealed	140
aged	230
Yield strength 0.2% offset × 10^3, psi	
annealed	115
aged	220
Elongation in 2 in., %	
annealed	12
aged	10
Hardness, Rockwell	
annealed	C-31
aged	C-48
Density, lb/in.^{3}	0.28
Thermal conductivity, Btu/ft h °F	
at 70°F (20°C)	10.4
at 1500°F (815°C)	14.3

[a]Custom 455 is the registered trademark of Carpenter Technology Corp.

Table M.6 Mechanical and Physical Properties of Alloy 718 Stainless Steel

Modulus of elasticity $\times 10^6$, psi	
aged	29
Tensile strength $\times 10^3$, psi	
annealed	140
aged	180
Yield strength 0.2% offset $\times 10^3$, psi	
annealed	80
aged	150
Elongation in 2 in., %	
annealed	30
aged	15
Hardness, Rockwell	
annealed	C-20
aged	C-36
Density, lb/in.^{3}	0.296
Thermal conductivity, Btu/ft h °F	
at 70°F (20°C)	6.6
at 1500°F (815°C)	13.9

corrosion, being equivalent to type 304 stainless steel. In seacoast applications alloy 17-7PH will generally develop a light rusting and pitting in all heat treated conditions. It is almost equal to type 304 in these applications.

Alloy 17-7PH is quite susceptible to stress corrosion cracking when in condition H900. When hardened at temperatures of 1025°F/552°C and higher, the alloy is highly resistant to stress corrosion cracking.

Refer to Table M.7 for the mechanical and physical properties of alloy 17.7PH.

See Refs. 2–4.

MEASURING CORROSION

See "Monitoring Corrosion" and "Corrosion Testing."

MEMBRANE

A membrane is any sheet or layer that acts as a barrier to prevent the passage of corrodents, specifically a sheet installed between the brick or masonry lining of a vessel and the vessel substrate. Membranes can be classified as either rigid or non-rigid. Included under the classification of rigid are baked epoxy, phenolic, and furan resin coatings and glass linings. Coatings are not really true membranes. Also included in the rigid classification are unplasticized PVC, fiberglass reinforced built-up resin linings, and glass flake filled sprayed polyester resin coatings.

Under the classification of nonrigid membranes are materials such as natural rubber sheet, neoprene, plasticized PVC, butyl rubber, polyisobutylene sheet, and polyurethane.

Natural rubber and neoprene are the most commonly used with a steel substrate. If

Table M.7 Mechanical and Physical Properties of 17-7PH Stainless Steel

Modulus of elasticity $\times 10^6$, psi	
annealed	30.5
aged	32.5
Tensile strength $\times 10^3$, psi	
annealed	133
aged	210
Yield strength 0.2% offset $\times 10^3$, psi	
annealed	42
aged	190
Elongation in 2 in., %	
annealed	19
Hardness, Rockwell	
annealed	B-85
aged	C-48
Density, lb/in.^{3}	0.282
Thermal conductivity, Btu/ft h °F	
at 70°F (20°C)	9.75
at 1500°F (815°C)	12.2

oxidizing agents are present, plasticized PVC is the better choice. For higher temperature applications butyl rubber can be used, but it will be damaged if petroleum products are present. For strong oxidizing conditions, polyisobutylene sheet is recommended.

At times, rubber, neoprene, and polyurethane have been employed in multicoat built-up membrane applications.

MERCURY CORROSION

See "Liquid Metal Embrittlement."

METAL DUSTING

Also see "High Temperature Corrosion." Metal dusting is a form of high temperature corrosion in which, under alternating oxidizing and reducing conditions, localized high carbon areas are burned out during the oxidation period.

METALLIC COATINGS

The development of new and improved coatings and linings has been increasing over the past several years. New technologies have evolved that have expanded the usage of these materials. By incorporating these coatings and linings with a substrate having the required physical and mechanical properties, it is possible to obtain the desired strength and the optimum corrosion resistance at an economical cost.

Some of the materials can be used as either coatings or linings. The terminology of coating or lining is usually associated with the specific material and is not necessarily indicative of its actual usage.

There are several methods by which metalic coatings may be applied

1. Brief immersion in a molten bath of metal, called hot dipping
2. Electroplating from an aqueous electrolyte
3. Spraying in which a gun is used that simultaneously melts and propels small droplets of metal onto the surface to be coated, similar to spray painting
4. Cementation, in which the material to be coated is tumbled in a mixture of metal powder and an appropriate flux at elevated temperatures, which allows the metal to diffuse into the base metal
5. Gas phase reaction
6. Chemical reduction of metal–salt solutions, the precipitated metal forming an overlay on the base metal (nickel coatings of this type are referred to as "electroless" nickel plate)

Coatings from a corrosion viewpoint are classified as either noble or sacrificial. All metal coatings contain some degree of porosity. Coating performance is therefore determined by the degree of Galvanic action that takes place at the base of a pore, scratch, or other imperfection in the coating.

Noble coatings, consisting of nickel, silver, copper, lead, or chromium on steel, are noble in the Galvanic Series with respect to steel, resulting in Galvanic current attack at the base of the pores of the base metal and eventually undermining the coating (see Fig. M.1).

In order to reduce this rate of attack, it is important that this type of coating be prepared with a minimum number of pores and that any pores present be as small as possible. This can be accomplished by increased coating thickness.

In sacrificial coatings consisting of zinc, cadmium, and in certain environments aluminum and tin on steel, the base metal is noble in the Galvanic Series to the coating material, resulting in cathodic protection to the base metal and attack on the coating material (see Fig M.2).

As long as sufficient current flows and the coating remains in electrical contact, the base metal will be protected from corrosion. Contrary to noble coatings, the degree of porosity of sacrificial coatings is relatively insignificant. However, the thicker the coating the longer cathodic protection will be provided to the base metal.

For information on other metallic coatings refer to the specific metal (e.g., "Tin Coatings," "Nickel Coatings," etc.)

See refs. 5 and 6.

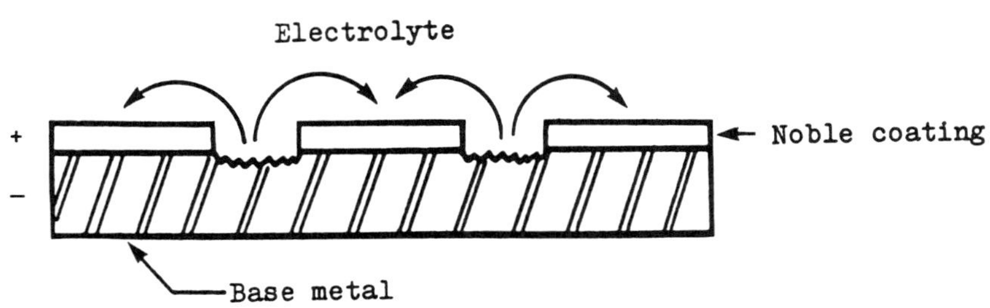

Figure M.1 Galvanic action with a noble coating.

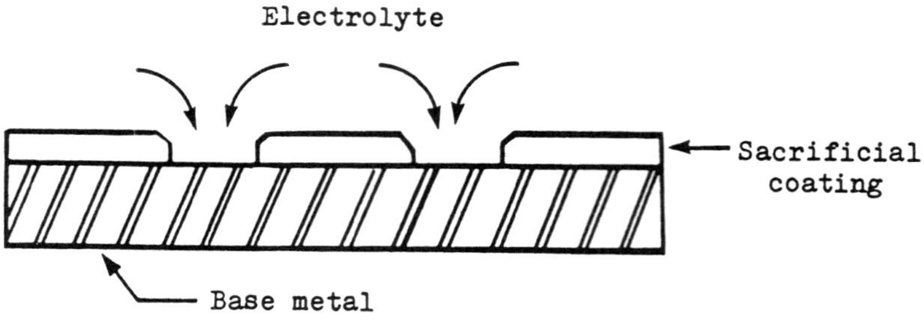

Figure M.2 Galvanic action with a sacrificial coating.

MICROALLOYED STEELS

Microalloyed steels are killed steels that have been alloyed with small amounts of vanadium, titanium, or niobium. The total of such alloying ingredients is usually less than 0.1 weight percent. The purpose of these elements is to modify the microstructure and refine the grain size, which results in a relatively small and uniform grain size.

The addition of these elements improves toughness and strength. Applications for these steels exist where section thickness or gross weight is a concern as in large diameter, long pipelines, where pipe tonnage is a major cost factor. At times these steels are selected for use in applications where improved toughness is a requirement. Most such applications are for plate steels used for improved piping and vessel toughness.

Microalloyed steels require some care in selecting weld joint geometrics and welding procedures since they have a tendency to produce an excessively hard heat-affected zone. These heat-affected zones make the steel more susceptible to various forms of hydrogen stress cracking. The hard heat-affected zone is of no concern if the intended service does not involve the threat of hydrogen stress cracking. Double-sided welds are more likely to produce a hard heat-affected zone than single-side welds. In multiple-pass single-side welds, the previously deposited bead weld is tempered by the following bead(s). Because of this, pipe welds are less likely to retain hard heat-affected zones.

These steels are sometimes referred to as high strength, low alloy steels or HSLA steels.

MICROBIAL CORROSION

The term microorganism covers a wide variety of life forms including bacteria, blue-green cyanobacteria, algae, lichens, fungi, and protozoa. All microorganisms may be involved in the biodeterioration of materials. Pure cultures never occur under natural conditions, rather mixed cultures prevail. Of the mixed culture only a few may actually be actively involved in the process of corrosion. The other organisms support the active ones by adjusting the environmental conditions in such a manner as to support their growth. For example, in the case of metal corrosion caused by sulfate reducing bacteria (SRB), the accompanying organisms remove toxic oxygen and produce simple carbon compounds like acetic and/or lactic acid as nutrients for the SBR.

Bacteria

Bacteria are the smallest living organisms on this planet. Some can live with and others without oxygen. Some can adapt to changing conditions and live either aerobically or anaerobically. There is a wide diversity with regard to their metabolism. They are classified as to their source of metabolic energy as in the table.

Energy Source	Classification
Light	Phototrophs
Chemical reactions	Chemotrophs
Inorganic hydrogen donators	Lithotrophs
Organic hydrogen donators	Organotrophs
Carbon dioxide (cell source)	Autotrophs
Organic molecules (cell source)	Heterotrophs

These six terms may be combined to describe easily the nutritional requirements of a bacterium. For example, if energy is derived from inorganic hydrogen donators and biomass is derived from organic molecules, they are called mirotrophs (chemolitho-organotrophs).

An important feature of microbial life is the ability to degrade any naturally occurring compound. Exceptions to this rule are a few manmade materials like highly polymerized and halogenated compounds.

In addition to energy and carbon sources, nitrogen, phosphorous, and trace elements are needed by microorganisms. Nitrogen compounds may be inorganic ammonium nitrate as well as organically bound nitrogen (e.g., amino acids, nucleotides). With the help of an enzyme called nitrogenase, bacteria are able to fix nitrogen from atmospheric nitrogen, producing ammonia, which is incorporated in cell constituents.

Phosphorus is taken up as inorganic phosphate or as organically bound phosphoxylated compounds such as phosphorus-containing sugars and lipids. Phosphorus in the form of adenosine triphosphate (ATP) is the main energy storage compound.

For many of the metabolic purposes, trace elements are needed. Cobalt aids in the transfer of methyl groups from/to organic or inorganic molecules (vitamin B_{12}, cobalamine, is involved in the methylation of heavy metals such as mercury); iron as Fe^{2+} or Fe^{3+} is required for the electron transport system where it acts as an oxidizable/reducible central atom in cytochromes or in nonheme iron sulfur proteins: Magnesium acts in a similar manner in the chlorophyll molecule; copper is an essential part of a cytochrome, which at the terminal end of the electron transport system is responsible for the reduction of oxygen to water.

Since life cannot exist without water, water is an essential requirement for microbial life and growth. Different microorganisms have different requirements as to the amount of water needed. A solid material is surrounded by three types of water, hygroscopic, pellicular, and gravitational. Only pellicular and gravitational water are biologically available and can be used by microorganisms. The biologically available water is usually measured as the water activity a_w of a sample

$$a_w = \frac{\text{vapor pressure of a solution}}{\text{vapor pressure of pure water}} = \frac{V_s}{P_w}$$

at the same temperature. Most bacteria require an a_w value in excess of 0.90.

Hydrogen ion concentration is another important factor affecting growth. Microorganisms are classified as to their ability to grow under acidic, neutral, or alkaline conditions, being given such titles as acidophiles, neutrophiles, or alkalophiles. Most microorganisms thrive in a neutral pH range of 6–8.

Microbial growth is also affected by redox potential. Under standard conditions, hydrogen is assumed to have a redox potential of –421 mV, and oxygen has a redox potential of +820 mV. Metabolism can take place within this range.

Available oxygen is another factor that influences microbial growth. Microbial growth is possible under well aerated as well as under totally oxygen free conditions. Those organisms living with the amount of oxygen contained in the air are called aerobes, while those that perform their metabolism without any free oxygen are called anaerobes. These latter are able to use only bound oxygen (sulfate, carbon dioxide) or to ferment organic compounds.

Temperature is another important factor affecting microbial growth. Microbial life is possible within the temperature range of –5°C to +110°C. Microorganisms are also classified as to the temperature range in which they thrive as in the table.

Microorganism	Temperature/Range
Psychrophiles	–5°C to +20°C
Psychotrophes	5°C to 30°C
Mesophiles	20°C to 45°C
Moderate thermophiles	40°C to 55°C
Thermophiles	55°C to 85°C
Extreme thermophiles	up to 110°C

Most of the organisms live in the mesophilic range of 20°C to 45°C, which corresponds to the usual temperature on the surface of the earth.

Corrosion of Specific Materials

Microbially induced corrosion (MIC) may occur for metallic materials in many industrial applications. MIC has been reported in the following industrial applications in the table.

Industry	Location of MIC
Chemical processing industry	Pipelines, stainless steel tanks, flanged joints, welded areas, after hydrotesting with natural river or well waters.
Nuclear power generation	Copper–nickel, stainless steel, brass, and aluminum–bronze cooling water pipes, carbon and stainless steel piping and tanks
Underground pipeline industry	Water-saturated clay type soils of near neutral pH with decaying organic matter and a source of sulfur reducing bacteria
Metalworking industry	Increased wear from breakdown of machinery oils and emulsions

Industry	Location of MIC
Onshore and offshore oil and gas industries	Mothballed and water flooded systems, oil and gas handling systems, particularly in environments soured by sulfate-reducing bacteria produced sulfides
Water treatment industry	Heat exchanges and piping
Sewage handling and treatment industry	Concrete and reinforced concrete structures
Highway maintenance industry	Culvert piping
Aviation industry	Aluminum integral wing tanks and fuel storage tanks

MIC of metallic materials is not a new form of corrosion. The methods by which microorganisms increase the rate of corrosion of metals and/or the susceptibility to localized corrosion in an aqueous environment are

1. "Production of corrosive metabolites." Bacteria may produce inorganic acids, organic acids, sulfides, and ammonia, all of which may be corrosive to metallic materials.
2. "Destruction of protective layers." Organic coatings may be attacked by various microorganisms leading to the corrosion of the underlying metal.
3. "Hydrogen embrittlement." By acting as a source of hydrogen and/or through the production of hydrogen sulfide, microorganisms may influence hydrogen embrittlement of metals.
4. "Formation of concentration cells at the metal surface and in particular oxygen concentration cells." A concentration cell may be formed when a biofilm or bacterial growth develops heterogeneously on the metal surface. Some bacteria tend to trap heavy metals such as copper and cadmium within their extracellular polymeric substance, causing the formation of ionic concentration cells. These lead to localized corrosion.
5. "Modification of corrosion inhibitors." Certain bacteria may convert nitrite corrosion inhibitors used to protect mild steel to nitrate, while other bacteria may convert nitrate inhibitors used to protect aluminum and aluminum alloys to nitrite and ammonia.
6. "Stimulation of electrochemical reactions." An example of this type is the evolution of cathodic hydrogen from microbially produced hydrogen sulfide.

MIC can result from

1. Production of sulfuric acid by bacteria of the genus thiobacillus through the oxidation of various inorganic sulfur compounds. The concentration of the sulfuric acid may be as high as 10–12%.
2. Production of hydrogen sulfide by sulfate-reducing bacteria.
3. Production of organic acids.
4. Production of nitric acid.
5. Production of hydrogen sulfide.
6. Production of ammonia.

Prevention

There are a large number of approaches that may be used to prevent or minimize MIC. Among the choices are

1. Change or modify the material
2. Modify the environment or process parameters
3. Use of organic coatings
4. Cathodic protection
5. Use of biocides
6. Microbiological methods
7. Physical methods

Which approach to follow will depend upon the type of bacteria present. A technique that has gained importance in addition to the preventative methods is that of "simulation of biogenic attack." By simulation of the biogenic attack, a quick-motion effect can be produced that will allow materials to be tested for their compatibility for a specific application. In order to conduct the simulation properly it is necessary that a thorough knowledge of all the processes and participating microorganisms be known. The situation may be modeled under conditions that will be optimum for the microorganisms resulting in a reduced time span for the corrosion to become detectable.

See Refs. 7–10.

MILS PER YEAR (MPY)

This is an expression for the uniform corrosion rate of a metal. See "Uniform Corrosion" for more details.

MINERAL ACID

Mineral acids are acids derived from inorganic materials. Typical mineral acids include hydrochloric, sulfuric, nitric, and phosphoric.

MONEL

The first nickel alloy, invented in 1905, was approximately two-thirds nickel and one-third copper. The present equivalent of that alloy, Monel alloy 400, remains one of the widely used nickel alloys.

Nickel copper alloys offer somewhat higher strength than unalloyed nickel, with no sacrifice of ductility. The thermal conductivity of Monel, although lower than that of nickel, is significantly higher than that of nickel alloys containing substantial amounts of chromium or iron. Table M.8 provides the mechanical and physical properties of Monel 400.

The alloying of 30–33% copper with nickel, producing Monel 400, provides an alloy with many of the characteristics of chemically pure nickel but improves on others. Water handling, including brackish and seawaters, is a major area of application. It gives excellent service under high velocity conditions. The alloy can pit in stagnant seawater, as

Table M.8 Mechanical and Physical Properties of Monel 400

Modulus of elasticity × 10^6, psi	26
Tensile strength × 10^3, psi	70
Yield strength 0.2% offset × 10^3, psi	25–28
Elongation in 2 in., %	48
Hardness, Brinell	130
Density, lb/in.^{3}	0.318
Specific gravity	8.84
Specific heat, at 32–212°F Btu/lb °F	0.102
Thermal conductivity, Btu/h/ft^2/in.°F	
at 70°F	151
at 200°F	167
at 500°F	204
Thermal expansion coefficient, in/in °F × 10^{-6}	
at 70–200°F	7.7
at 70–400°F	8.6
at 70–500°F	8.7
at 70–1000°F	9.1

does nickel 200, but the rates of attack are significantly diminished. The absence of chloride stress corrosion cracking is also a factor in the selection of this alloy.

The general corrosion resistance of Monel 400 in the nonoxidizing acids such as sulfuric, hydrochloric, and phosphoric is improved over that of pure nickel. The alloy is not resistant to oxidizing media, such as nitric acid, ferric chloride, chromic acid, wet chlorine, sulfur dioxide, or ammonia.

Monel 400 does have excellent resistance to hydrofluoric acid solutions at various concentrations and temperatures. It is subject to stress corrosion cracking in moist, aerated hydrofluoric or hydrofluosilicic acid vapor. Cracking is unlikely when completely immersed in the acid. The corrosion of Monel 400 is negligible in all types of atmospheres. Indoor exposure produces a very light tarnish that is easily removed by occasional wiping. Outdoor surfaces that are exposed to rain produce a thin gray-green patina. In sulfurous atmospheres a smooth, brown, adherent film forms.

Because of its high nickel content, Monel 400 is practically as resistant as nickel 200 to caustic soda throughout most of the concentration range. It is also resistant to anhydrous ammonia and to ammonium hydroxide solutions of up to 3% concentration. Refer to Table M.9 for the compatibility of Monel 400 with selected corrodents.

See Refs. 11 and 12.

MONITORING CORROSION

There are many operating and environmental parameters that can affect the corrosion of a metal, changes in any of which can greatly affect the corrosion rate. When construction materials are initially selected, some compromise may be involved between hard-to-obtain alloys highly resistant to corrosion by the process under any conceivable operating conditions and less expensive, more readily obtainable materials that offer corrosion

Table M.9 Compatibility of Monel 400 with Selected Corrodents[a]

Chemical	Maximum temp. °F	Maximum temp. °C	Chemical	Maximum temp. °F	Maximum temp. °C
Acetaldehyde	170	77	Benzaldehyde	210	99
Acetamide	340	171	Benzene	210	99
Acetic acid 10%	80	27	Benzene sulfonic acid 10%	210	99
Acetic acid 50%	200	93	Benzoic acid	210	99
Acetic acid 80%	200	93	Benzyl alcohol	400	204
Acetic acid, glacial	290	143	Benzyl chloride	210	99
Acetic anhydride	190	88	Borax	90	32
Acetone	190	88	Boric acid	210	99
Acetyl chloride	400	204	Bromine gas, dry	120	49
Acrylonitrile	210	99	Bromine gas, moist	x	x
Adipic acid	210	99	Butadiene	180	82
Allyl alcohol	400	204	Butyl acetate	380	193
Allyl chloride	200	93	Butyl alcohol	200	93
Alum	100	38	Butyl phthalate	210	99
Aluminum acetate	80	27	Butyric acid	210	99
Aluminum chloride, aqueous	x	x	Calcium bisulfide	60	16
Aluminum chloride, dry	150	66	Calcium bisulfite	x	x
Aluminum fluoride	90	32	Calcium carbonate	200	93
Aluminum hydroxide	80	27	Calcium chlorate	140	60
Aluminum sulfate	210	99	Calcium chloride	350	177
Ammonia gas	x	x	Calcium hydroxide 10%	210	99
Ammonium bifluoride	400	204	Calcium hydroxide, sat.	200	93
Ammonium carbonate	190	88	Calcium hypochlorite	x	x
Ammonium chloride 10%	230	110	Calcium oxide	90	32
Ammonium chloride 50%	170	77	Calcium sulfate	80	27
Ammonium chloride, sat.	570	299	Caprylic acid[b]	210	99
Ammonium fluoride 10%	400	204	Carbon bisulfide	x	x
Ammonium fluoride 25%	400	204	Carbon dioxide, dry	570	299
Ammonium hydroxide 25%	x	x	Carbon dioxide, wet[b]	400	204
Ammonium hydroxide, sat.	x	x	Carbon disulfide	x	x
Ammonium nitrate	x	x	Carbon monoxide	570	299
Ammonium persulfate	x	x	Carbon tetrachloride	400	204
Ammonium phosphate 30%	210	99	Carbonic acid	x	x
Ammonium sulfate 10–40%	400	204	Cellosolve	210	99
Ammonium sulfite	90	32	Chloracetic acid, 50% water	180	82
Amyl acetate	300	149	Chloracetic acid	x	x
Amyl alcohol	180	82	Chlorine gas, dry	570	299
Amyl chloride	400	204	Chlorine gas, wet	x	x
Aniline	210	99	Chlorine, liquid	150	66
Antimony trichloride	350	177	Chlorobenzene, dry	400	204
Aqua regia 3:1	x	x	Chloroform	210	99
Barium carbonate	210	99	Chlorosulfonic acid	80	27
Barium chloride	210	99	Chromic acid 10%	130	54
Barium hydroxide	80	27	Chromic acid 50%	x	x
Barium sulfate	210	99	Chromyl chloride	210	99
Barium sulfide	x	x	Citric acid 15%	210	99

Table M.9 Continued

Chemical	Maximum temp. °F	°C	Chemical	Maximum temp. °F	°C
Citric acid, concentrated	80	27	Methyl isobutyl ketone	200	93
Copper acetate	x	x	Muriatic acid	x	x
Copper carbonate	x	x	Nitric acid 5%	x	x
Copper chloride	x	x	Nitric acid 20%	x	x
Copper cyanide	x	x	Nitric acid 70%	x	x
Copper sulfate	x	x	Nitric acid, anhydrous	x	x
Cresol	100	38	Nitrous acid, concentrated	x	x
Cupric chloride 5%	x	x	Oleum	x	x
Cupric chloride 50%	x	x	Perchloric acid 10%	x	x
Cyclohexane	180	82	Perchloric acid 70%	x	x
Cyclohexanol	80	27	Phenol	570	299
Dichloroethane (ethylene dichloride)	200	93	Phosphoric acid 50–80%	x	x
Ethylene glycol	210	99	Picric acid	x	x
Ferric chloride	x	x	Potassium bromide 30%, air free	210	99
Ferric chloride 50% in water	x	x	Salicylic acid	210	99
Ferric nitrate 10–50%	x	x	Silver bromide 10%	80	27
Ferrous chloride	x	x	Sodium carbonate	210	99
Ferrous nitrate			Sodium chloride to 30%	210	99
Fluorine gas, dry	570	299	Sodium hydroxide 10%[c]	350	177
Fluorine gas, moist	x	x	Sodium hydroxide 50%[c]	300	149
Hydrobromic acid, dilute	x	x	Sodium hydroxide, concentrated	350	177
Hydrobromic acid 20%	x	x	Sodium hypochlorite 20%	x	x
Hydrobromic acid 50%	x	x	Sodium hypochlorite, concentrated	x	x
Hydrochloric acid 20%	80	27	Sodium sulfide to 50%	210	99
Hydrochloric acid 38%	x	x	Stannic chloride	x	x
Hydrocyanic acid 10%	80	27	Stannous chloride, dry	570	299
Hydrofluoric acid 30%[c]	400	204	Sulfuric acid 10%	x	x
Hydrofluoric acid 70%[c]	400	204	Sulfuric acid 50%	80	27
Hydrofluoric acid 100%[c]	210	99	Sulfuric acid 70%	80	27
Hypochlorous acid	x	x	Sulfuric acid 90%	x	x
Iodine solution 10%	x	x	Sulfuric acid 98%	x	x
Ketones, general	100	38	Sulfuric acid 100%	x	x
Lactic acid 25%	x	x	Sulfuric acid, fuming	x	x
Lactic acid, concentrated	x	x	Sulfurous acid	x	x
Magnesium chloride 50%	350	177	Thionyl chloride	300	149
Malic acid	210	99	Toluene	210	99
Manganese chloride 40%	100	38	Trichloroacetic acid	170	77
Methyl chloride	210	99	White liquor	x	x
Methyl ethyl ketone	200	93	Zinc chloride to 80%	200	93

[a]The chemicals listed are in the pure state or in a saturated solution unless otherwise indicated. Compatibility is shown to the maximum allowable temperature for which data are available. Incompatibility is shown by an x. A blank space indicates that data are unavailable. When compatible the corrosion rate is < 20 mpy.
[b]Not for use with carbonated beverages.
[c]Material is subject to stress cracking.
Source: PA Schweitzer. *Corrosion Resistance Tables,* 4th ed. Vols. 1–3. New York: Marcel Dekker, 1995.

resistance under normal operating conditions but would have a high corrosion rate during process upsets. Under these conditions direct on-line monitoring systems are essential to detect and measure the effect of these changes. Without these data, costly and potentially hazardous damage can occur before corrections are made. Accurate and timely corrosion measurement is an essential part of almost all corrosion control programs.

To monitor the condition of process equipment, both onstream and offstream nondestructive testing must be relied upon. Nondestructive testing is testing to detect internal, external, and concealed flaws in materials by use of techniques that do not damage or destroy the items being tested. Nondestructive testing can be utilized as an early warning system to indicate when process equipment is approaching the end of its safe serviceability, or when changed process conditions have increased the corrosion rates.

Measurement of corrosion refers to any technique that can be used to determine the effects of corrosion. Monitoring refers to those measurement techniques that are suitable for use while the equipment is in operation. Included in nondestructive measurement techniques are the following:

1. Radiography (x-ray). Radiography is not normally used in continuous on-line monitoring since it cannot detect small changes in residual wall thickness due to accuracy limitations. It is suitable for detecting major flaws or severe corrosive attack.

2. Ultrasonic Measurement. There are several types of ultrasonic equipment, primarily A-scan, B-scan, and C-scan. A-scan generally measures wall thickness but can be fooled by mid-wall flaws. B-scan is more powerful and produces cross-sectional images similar to x-rays, while C-scan produces a three-dimensional view of a surface using complex and expensive equipment. These units are used almost exclusively for measurement rather than monitoring.

3. Visual Inspection. This can only be performed when the equipment is out of service during a plant shutdown. When possible an internal inspection should be made to verify the results of any on-line monitoring program. Visual inspection will also permit detection of forms of corrosion not detectable by on-line monitoring equipment such as pitting or crevice corrosion.

4. Coupons. Coupons are the simplest device used for the monitoring of corrosion. They are small pieces of metal which are inserted into the process stream and removed after a period of time (at least 30 days) for study. Coupons are used to determine the average corrosion rate over the period of exposure. An advantage of coupons is their ability to indicate the forms of corrosion present. They can be examined for evidence of pitting and other forms of localized attack. Refer also to "Corrosion Testing."

5. Hydrogen Probes. Hydrogen probes are used to detect the penetration of elemental hydrogen into metal such as pipe or vessel walls. There are three types of hydrogen probes. The most common hydrogen probe consists of a thin walled carbon steel tube inserted into the flow stream with a solid rod inside the tube forming a small annular space. Hydrogen atoms small enough to permeate the carbon steel collect in the annular space and combine to form molecular hydrogen gas, which is too large to pass back into the process. Pressure in the annular space builds up as the gas collects and registers on an external pressure gauge.

Patch probes operate the same way, except that the patch is sealed to the outside of a pipe or vessel and collects the hydrogen atoms that penetrate the wall. The third type is the palladium foil type, which produces an electrical output proportional to the hydrogen evolution rate. These probes are used when hydrogen-induced corrosion is a concern, such as in cathodic reactions in acid solutions, particuarly when hydrogen sulfide is present.

6. Polarization Studies. Polarization studies are electrochemical techniques used to study corrosion phenomena, especially pitting. In the past these were primarily used in the laboratory, but with advances in computer technology some of these systems are now being used in the field. These studies can be conducted using a variety of methods and equipment such as in the table.

Potentiostatic	Potential held constant
Galvanostatic	Current held constant
Potentiodynamic	a. Potential changed constantly at a specified rate
	b. Potential changed in steps and held constant at each step
Galvanodynamic	a. Current changed constantly at a specified rate
	b. Current changed in steps and held constant at each step.

Because of the high polarization potentials required, the results are less accurate than those gotten using linear polarization resistance.

7. Electrical Impedance Spectroscopy. Electrical impedance spectroscopy is a laboratory technique, but with the development of more rugged computers, some investigation of this work is being made in the field.

One of the requirements of making impedance measurements is to have a common cell of known geometry, a reference electrode, and instrumentation capable of measuring and recording electrical response of the test corrosion cell over a wide range of AC excitation frequencies.

8. Electrochemical Noise. Electrochemical noise is a monitoring technique of current and electrochemical potential disturbances as a result of ongoing corrosion activity. It is not as quantitative as linear polarization resistance for corrosion rate calculations. Laboratory and field interpretation is still in the development stage.

9. Electrical Resistance. Electrical resistance systems work by measuring the resistance of a thin metal probe. The probe is exposed to the gas or liquid stream. As the measuring element corrodes, the cross-section reduces and the electrical resistance increases. The thickness of the measuring element is directly proportional to a corrosion dial reading. With an automated system, continuous readings are made, and through the use of sophisticated data analysis techniques, detection of significant changes in corrosion rates can be made in as little as two hours.

The actual corrosion rate in mils per year (mpy) can be determined by

$$\text{mpy} = \frac{\text{dial reading}}{\text{time, in days}} \times 0.365 \times \text{probe multiplier}$$

10. Linear Polarization Resistance. Linear polarization resistance is an electrochemical technique that measures the DC current that flows between one or two electrodes of the material under study by application of a small electrical potential. The current is measured on a microammeter that has been converted to read the corrosion rate directly (in mils per year) of the test electrode. Measurements cannot be made in nonconductive fluids or fluids that contain compounds that coat the electrodes (e.g., crude oil).

See Refs. 13–15.

MONOLITHIC SURFACINGS

See "Polymer Concrete."

MONOMER

A monomer is a single molecule or a substance consisting of single molecules. A relatively low mass molecular structure that undergoes a polymerization reaction to form a polymer. Organic molecule or compound capable of polymerizing or linking together with itself or with other monomers to form a dimer, trimer, or polymer.

MORTARS

Mortar is used to bond brick or tile. It must have a heavy enough consistency to support the weight of the tile or brick without being squeezed from the joints while the joint is curing. Application is made by buttering each unit. Joints are usually 1/8 in. (3 mm) wide.

Chemically resistant mortars and grouts are formulated using an inorganic binder or a liquid resin system, fillers such as silica, carbon or combinations thereof, and a hardener or catalyst system.

Carbon is the most inert of the fillers, having a wide resistance to most chemicals; consequently it is the filler most often used. It is resistant to strong alkalies, hydrochloric acid, and other fluorine chemicals. The general resistance of these fillers to acids, alkalies and salts is shown in the table.

Corrodent at 20% concentration	Filler		
	Carbon	Silica	Carbon-silica
Hydrochloric acid	R	R	R
Hydrofluoric acid	R	N	N
Sulfuric acid	R	R	R
Potassium hydroxide	R	N	N
Sodium hydroxide	R	N	N
Neutral salts	R	R	R
Solvents, conc.	R	R	N

R = recommended. N = not recommended.

Inorganic Mortars

The inorganic materials are the original mortars. These mortars are commonly referred to as acid-proof mortars, primarily because they are limited in application to a maximum pH of 7. They cannot be used in alkaline or alternate alkaline–acid service. There are two general types of inorganic mortars, the hot-pour sulfur and the ambiently mixed and applied silicate mortars.

Sulfur Mortars

Sulfur mortars are hot-melt compounds, available in flake, powder, and ingot forms. They must be heated to a temperature of 250°F (120°C) and poured into the joints while hot. Sulfur mortars are particularly useful against oxidizing acids. When they are carbon filled they are suitable for use against combinations of oxidizing acids and hydrofluoric acid. Chemical resistance to strong alkaline solutions and certain organic solvents is poor.

The sulfur mortars possess certain advantages over some of the resin mortars, primarily their resistance to oxidizing, nonoxidizing, and mixed acids, ease of use, resistance to thermal shock, high early strength, unlimited shelf life, and economy.

Sodium Silicate Mortars

Sodium silicate mortars are available as either two component systems, which consist of the liquid sodium solution and the filler powder containing settling agents and selected aggregates, or they may be a one part system in powder form to be mixed with water when used. There are some differences in the chemical resistance between the two types. The differences in the chemical resistance of the two types of mortars are shown in the table.

	Type of mortar	
Corrodent at room temp.	Normal	Water-resistant
Acetic acid, glacial	P	P
Chlorine dioxide, water sol.	N	N
Hydrogen peroxide	N	R
Nitric acid 5%	C	R
Nitric acid 20%	C	R
Nitric acid, over 20%	R	R
Sodium bicarbonate	N	N
Sodium sulfite	R	R
Sulfates, aluminum	R	R
Sulfates, copper	P	P
Sulfates, iron	P	P
Sulfates, magnesium	P	P
Sulfates, nickel	P	P
Sulfates, zinc	P	P
Sulfuric acid, to 93%	P	P
Sulfuric acid, over 93%	P	P

R = recommended. N = not recommended. P = potential failure.
C = conditional.

Sodium silicate mortars are useful in the pH range of 0–6, except where sulfuric acid exposures exist in vapor phase, wet–dry exposures, or in concentrations above 93%.

Potassium Silicate Mortars

Potassium silicate mortars are preferable to sodium silicate mortars. They have better workability because of their smoothness and lack of tackiness. They do not run or flow from the brickwork and they do not stick to the trowel. The potassium silicate mortars have a greater resistance to strong acid solutions as well as to sulfation. These mortars are available with halogen-free hardening systems which eliminate the remote possibility of catalyst poisoning in certain chemical operations. The general corrosion resistance of the two types of mortars are shown in the table.

	Type of mortar	
Corrodent at room temp.	Normal	Halogen-free
Acetic acid, glacial	R	R
Chlorine dioxide, water sol.	R	R
Hydrogen peroxide	N	N
Nitric acid, 5%	R	R
Nitric acid, 20%	R	R
Nitric acid, over 20%	R	R
Sodium bicarbonate	N	N
Sodium sulfite	N	N
Sulfates, aluminum	R	R
Sulfates, copper	R	R
Sulfates, iron	R	R
Sulfates, magnesium	R	R
Sulfates, nickel	R	R
Sulfates, zinc	R	R
Sulfuric acid, to 93%	R	R
Sulfuric acid, over 93%	R	R

R = recommended. N = not recommended.

Silica Mortars

The silica type mortars consist of a colloidal silica binder with quartz fillers. The main difference compared to the other mortars is total freedom from metal ions that could contribute to sulfation hydration within the mortar joints in high concentrations of sulfuric acid. This is a unique system. It can be used up to 2000°F (1093°C). The silica type mortars used in the pH range of 0–7 are resistant to all materials except hydrofluoric acid and acid fluorides.

Organic Mortars

Epoxy

The three main epoxy resins used in the formulation of corrosion resistant mortars are based on Bisphenol A, Bisphenol F, and epoxy phenol. The corrosion resistance as well as the physical and mechanical properties will be determined by the type of hardener used.

The three main types of hardeners used with the Bisphenol A resin are aliphatic amines, modified aliphatic amines, and aromatic amines. Silica is the filler most often used with epoxy mortars. This prohibits the use of the epoxy resins with hydrofluoric acid, other fluorine chemicals, and strong, hot alkalies. Carbon fillers can be substituted but with some sacrifice to the working properties.

The Bisphenol F series of mortars are similar to the Bisphenol A series. They both use alkaline hardeners and the same fillers. The main advantage of the Bisphenol F resins is their improved resistance to aliphatic and aromatic solvents, and higher concentrations of oxidizing and nonoxidizing scids.

The general corrosion resistance is shown in the table.

	Hardeners			
			Aromatic amines	
			Bisphenol	
Corrodent at room temp.	Aliphatic amines	Modified aliphatic amines	A	F
Acetic acid 5–10%	C	U	R	
Acetone	U	U	U	U
Benzene	U	U	R	R
Butyl acetate	U	U	U	R
Butyl alcohol	R	R	R	R
Chromic acid 5%	U	U	R	R
Chromic acid 10%	U	U	U	R
Formaldehyde 35%	R	R	R	R
Gasoline	R	R	R	R
Hydrochloric acid to 36%	U	U	R	R
Nitric acid 30%	U	U	U	U
Phosphoric acid 50%	U	U	R	R
Sulfuric acid 25%	R	U	R	R
Sulfuric acid 50%	U	U	R	R
Sulfuric acid 75%	U	U	U	U
Trichloroethylene	U	U	U	R

R = recommended. U = unsatisfactory.

More extensive compatibility data will be found in Ref. 11.

Phenolic Mortars

The phenolic mortars provide resistance to high concentrations of acids and to sulfuric acid at elevated temperatures. Fillers for the phenolic resins are 100% carbon, 100% silica, or part carbon and part silica. For high concentrations of sulfuric acid, silica is the filler of choice. Carbon fillers are used where resistance to high concentrations of hydrofluoric acid is required. Listed below are some typical compatibilities of phenolic mortars. A more comprehensive listing will be found in Ref. 11.

	Filler	
Corrodent at room temp.	Carbon	Silica
Amyl alcohol	R	R
Chromic acid 10%	U	U
Gasoline	R	R
Hydrofluoric acid to 50%	R	U
Hydrofluoric acid 93%	R	U
Methyl ethyl ketone	R	R
Nitric acid 10%	U	U
Sodium hydroxide to 5%	U	U
Sodium hydroxide 30%	U	U
Sodium hypochlorite 5%	U	U
Sulfuric acid to 50%	R	R
Sulfuric acid 93%	R	R
Xylene	R	R

R = recommended. U = unsatisfactory.

Furan Mortars

The furan mortars are resistant to most nonoxidizing organic and inorganic acids, alkalies, salts, oils, greases, and solvents to temperatures of 360°F (182°C). Fillers are either 100% carbon, 100% silica, or part carbon and part silica. The 100% carbon filled resins provide the widest range of corrosion resistance. Typical corrosion resistance compatibilities are given in the table.

Corrodent at room temp.	100% carbon filler	Part carbon filler part silica
Acetic acid, glacial	R	R
Benzene	R	R
Cadmium salts	R	R
Chlorine dioxide	U	U
Chromic acid	U	U
Copper salts	R	R
Ethyl acetate	R	R
Ethyl alcohol	R	R
Formaldehyde	R	R
Fatty acids	R	R
Gasoline	R	R
Hydrochloric acid	R	R
Hydrofluoric acid	R	U
Iron salts	R	R
Lactic acid	R	R
Methyl ethyl ketone	R	R
Nitric acid	U	U
Phosphoric acid	R	R
Sodium chloride	R	R

Corrodent at room temp.	100% carbon filler	Part carbon filler part silica
Sodium hydroxide to 20%	R	U
Sodium hydroxide 40%	R	U
Sulfuric acid 50%	R	R
Sulfuric acid 80%	U	U
Trichloroethylene	R	R

R = recommended. N = unsatisfactory.

A more complete listing will be found in Ref. 11.

Polyester Mortars

The polyester mortars were originally developed to resist chlorine dioxide. There are a number of types of polyester resins available. The ones most commonly used are the isophthalic, chlorendic, and Bisphenol A fumurate. Depending on the application the polyester mortars can be formulated to incorporate carbon and silica fillers. 100% carbon fillers are used to resist hydrofluoric acid, fluorine chemicals, and strong alkalies such as sodium and potassium hydroxide. The chlorendic and Bisphenol A fumarate resins have improved chemical resistance and higher thermal capabilities than the isophthalic resins. The Bisphenol A fumarate resins exhibit greatly improved resistance to strong alkalies. A comparison of the corrosion resistance between the chlorendic and Bisphenol A fumarate resins is shown below. A more comprehensive compatibility chart will be found in Ref. 11.

	Polyester	
Corrodent at room temp.	Chlorendic	Bisphenol A fumarate
Acetic acid, glacial	U	U
Benzene	U	U
Chlorine dioxide	R	R
Ethyl alcohol	R	R
Hydrochloric acid 36%	R	R
Hydrogen peroxide	R	U
Methanol	R	R
Methyl ethyl ketone	U	U
Motor oil and gasoline	R	R
Nitric acid 40%	R	U
Phenol 5%	R	R
Sodium hydroxide 50%	U	R
Sulfuric acid 75%	R	U
Toluene	U	U
Triethanolamine	U	R
Vinyl toluene	U	U

R = recommended. U = unsatisfactory.

Vinyl Ester and Vinyl Ester Novolac Mortars

These resins have many of the same properties as the epoxy, acrylic, and Bisphenol A fumarate resins. The vinyl ester resins have replaced the polyester resins in mortars for bleach towers in the pulp and paper industry. The major advantages of these resin systems are their resistance to most oxidizing mediums and high concentrations of sulfuric acid, sodium hydroxide, and many solvents. Comparative resistance of the two types of vinyl ester resin systems are shown in the table.

	Vinyl ester	
Corrodent	Vinyl ester	Novolac
Acetic acid, glacial	U	R
Benzene	R	R
Chlorine dioxide	R	R
Ethyl alcohol	R	R
Hydrochloric acid 36%	R	R
Hydrogen peroxide	R	R
Methanol	U	R
Methyl ethyl ketone	U	U
Motor oil and gasoline	R	R
Nitric acid 40%	U	R
Phenol 5%	R	R
Sodium hydroxide 50%	R	R
Sulfuric acid, 75%	R	R
Toluene	U	R
Triethanolamine	R	R
Vinyl toluene	U	R
Max. temp. °F (°C)	220 (104)	230 (110)

R = recommended. U = unsatisfactory.

Table M.10 provides the compatibility of various mortars with selected corrodents. See Refs. 11, 16, and 17.

Table M.10 Compatibility of Various Mortars with Selected Corrodents[a]

Mortars	Acetic acid 10%																				
Silicate	U																				
Sodium silicate	R	—	—	—	—	—	—	—	—	—	—	—	—	—	—	—	—	—	—	—	—
Potassium silicate	R	—	—	—	—	—	—	—	—	—	—	—	—	—	—	—	—	—	—	—	—
Silica	R	—	—	—	—	—	—	—	—	—	—	—	—	—	—	—	—	—	—	—	—
Sulfur	R	—	—	—	—	—	—	—													
Furan resin	R	—	—	—	—	—	—	—	—	—	—	—	—	—	—	—					
Polyester	R	—	—	—	—	—	—														
Epoxy	R	—	—	—	—																
°F	60	80	100	120	140	160	180	200	220	240	260	280	300	320	340	360	380	400	420	440	460
°C	15	26	38	49	60	71	82	93	104	116	127	138	149	160	171	182	193	204	216	227	238

Mortars	Acetic acid 50%																				
Silicate	U																				
Sodium silicate	R	—	—	—	—	—	—	—	—	—	—	—	—	—	—	—	—	—	—	—	—
Potassium silicate	R	—	—	—	—	—	—	—	—	—	—	—	—	—	—	—	—	—	—	—	—
Silica	R	—	—	—	—	—	—	—	—	—	—	—	—	—	—	—	—	—	—	—	—
Sulfur	U																				
Furan resin	R	—	—	—	—	—	—	—	—	—	—	—	—	—	—	—					
Polyester	U																				
Epoxy	U																				
°F	60	80	100	120	140	160	180	200	220	240	260	280	300	320	340	360	380	400	420	440	460
°C	15	26	38	49	60	71	82	93	104	116	127	138	149	160	171	182	193	204	216	227	238

Mortars	Acetic acid 80%																				
Silicate	U																				
Sodium silicate	R	—	—	—	—	—	—	—	—	—	—	—	—	—	—	—	—	—	—	—	—
Potassium silicate	R	—	—	—	—	—	—	—	—	—	—	—	—	—	—	—	—	—	—	—	—
Silica	R	—	—	—	—	—	—	—	—	—	—	—	—	—	—	—	—	—	—	—	—
Sulfur	U																				
Furan resin	R	—	—	—	—	—	—	—	—	—	—	—	—	—	—	—					
Polyester	U																				
Epoxy	U																				
°F	60	80	100	120	140	160	180	200	220	240	260	280	300	320	340	360	380	400	420	440	460
°C	15	26	38	49	60	71	82	93	104	116	127	138	149	160	171	182	193	204	216	227	238

Table M.10 Continued

Mortars	Acetic acid glacial
Silicate	U
Sodium silicate	R (to 460°F / 238°C)
Potassium silicate	R (to 460°F / 238°C)
Silica	R (to 460°F / 238°C)
Sulfur	U
Furan resin	R (to 360°F / 182°C)
Polyester	U
Epoxy	U

Mortars	Acetic anhydride
Silicate	
Sodium silicate	R (to 460°F / 238°C)
Potassium silicate	R (to 460°F / 238°C)
Silica	R (to 460°F / 238°C)
Sulfur	U
Furan resin	U
Polyester	U
Epoxy	U

Mortars	Aluminum chloride, aqueous
Silicate	R (to 460°F / 238°C)
Sodium silicate	R (to 460°F / 238°C)
Potassium silicate	R (to 460°F / 238°C)
Silica	R (to 460°F / 238°C)
Sulfur	R (to 200°F / 93°C)
Furan resin	R (to 360°F / 182°C)
Polyester	R (to 240°F / 116°C)
Epoxy	R (to 100°F / 38°C)

Temperature scale (each table): °F 60, 80, 100, 120, 140, 160, 180, 200, 220, 240, 260, 280, 300, 320, 340, 360, 380, 400, 420, 440, 460; °C 15, 26, 38, 49, 60, 71, 82, 93, 104, 116, 127, 138, 149, 160, 171, 182, 193, 204, 216, 227, 238

Table M.10 Continued

Aluminum fluoride

Mortars / °F	60	80	100	120	140	160	180	200	220	240	260	280	300	320	340	360	380	400	420	440	460
°C	15	26	38	49	60	71	82	93	104	116	127	138	149	160	171	182	193	204	216	227	238
Silicate	U																				
Sodium silicate	U																				
Potassium silicate	U																				
Silica	U																				
Sulfur	R	—	—	—	—	—	—	—													
Furan resin	R	—	—	—	—	—	—	—	—	—	—	—	—	—	—	—					
Polyester	R	—	—	—	—	—	—	—	—	—											
Epoxy	R	—	—	—	—	—	—	—	—	—											

Ammonium chloride 10%

Mortars / °F	60	80	100	120	140	160	180	200	220	240	260	280	300	320	340	360	380	400	420	440	460
°C	15	26	38	49	60	71	82	93	104	116	127	138	149	160	171	182	193	204	216	227	238
Silicate	R	—	—	—	—	—	—	—	—	—	—	—	—	—	—	—	—	—	—	—	—
Sodium silicate	R	—	—																		
Potassium silicate	R	—	—	—	—	—	—	—	—	—	—	—	—	—	—	—	—	—	—	—	—
Silica	R	—	—	—	—	—	—	—	—	—	—	—	—	—	—	—	—	—	—	—	—
Sulfur	R	—	—	—	—	—	—	—													
Furan resin	R	—	—	—	—	—	—	—	—	—	—	—	—	—	—	—					
Polyester	R	—	—	—	—	—	—	—	—	—											
Epoxy	R	—	—	—	—	—	—	—	—	—											

Ammonium chloride 50%

Mortars / °F	60	80	100	120	140	160	180	200	220	240	260	280	300	320	340	360	380	400	420	440	460
°C	15	26	38	49	60	71	82	93	104	116	127	138	149	160	171	182	193	204	216	227	238
Silicate	R	—	—	—	—	—	—	—	—	—	—	—	—	—	—	—	—	—	—	—	—
Sodium silicate	R	—	—																		
Potassium silicate	R	—	—	—	—	—	—	—	—	—	—	—	—	—	—	—	—	—	—	—	—
Silica	R	—	—	—	—	—	—	—	—	—	—	—	—	—	—	—	—	—	—	—	—
Sulfur	R	—	—	—	—	—	—	—													
Furan resin	R	—	—	—	—	—	—	—	—	—	—	—	—	—	—	—					
Polyester	R	—	—	—	—	—	—	—	—	—											
Epoxy	R	—	—	—	—	—	—	—	—	—											

Table M.10 Continued

Mortars	Ammonium chloride, saturated
Silicate	R————————————————————
Sodium silicate	R——
Potassium silicate	R————————————————————
Silica	R————————————————————
Sulfur	R———————
Furan resin	R———————————————
Polyester	R——————————
Epoxy	R——————————

°F	60	80	100	120	140	160	180	200	220	240	260	280	300	320	340	360	380	400	420	440	460
°C	15	26	38	49	60	71	82	93	104	116	127	138	149	160	171	182	193	204	216	227	238

Mortars	Ammonium fluoride, 10%
Silicate	U
Sodium silicate	U
Potassium silicate	U
Silica	U
Sulfur	U
Furan resin	R———————————————
Polyester	R————
Epoxy	R———

°F	60	80	100	120	140	160	180	200	220	240	260	280	300	320	340	360	380	400	420	440	460
°C	15	26	38	49	60	71	82	93	104	116	127	138	149	160	171	182	193	204	216	227	238

Mortars	Ammonium fluoride 25%
Silicate	U
Sodium silicate	U
Potassium silicate	U
Silica	U
Sulfur	U
Furan resin	R———————————————
Polyester	R—————
Epoxy	R——

°F	60	80	100	120	140	160	180	200	220	240	260	280	300	320	340	360	380	400	420	440	460
°C	15	26	38	49	60	71	82	93	104	116	127	138	149	160	171	182	193	204	216	227	238

Table M.10 Continued

Mortars	Ammonium hydroxide 25%																				
Silicate	U																				
Sodium silicate	U																				
Potassium silicate	U																				
Silica	U																				
Sulfur	U																				
Furan resin	R	—	—	—	—	—	—	—	—	—	—	—	—	—	—	—	—				
Polyester	R	—	—	—	—	—	—	—	—	—											
Epoxy	R	—	—	—	—	—	—	—	—	—											
°F	60	80	100	120	140	160	180	200	220	240	260	280	300	320	340	360	380	400	420	440	460
°C	15	26	38	49	60	71	82	93	104	116	127	138	149	160	171	182	193	204	216	227	238

Mortars	Ammonium hydroxide, saturated																				
Silicate	U																				
Sodium silicate																					
Potassium silicate	U																				
Silica	U																				
Sulfur	U																				
Furan resin	R	—	—	—	—	—	—	—	—	—	—	—	—	—							
Polyester	R	—	—																		
Epoxy	R	—	—																		
°F	60	80	100	120	140	160	180	200	220	240	260	280	300	320	340	360	380	400	420	440	460
°C	15	26	38	49	60	71	82	93	104	116	127	138	149	160	171	182	193	204	216	227	238

Mortars	Aqua regia 3:1																				
Silicate	R	—	—	—	—	—	—	—	—	—	—	—	—	—	—	—	—	—	—	—	—
Sodium silicate	R	—	—	—	—	—	—	—	—	—	—	—	—	—	—	—	—	—	—	—	—
Potassium silicate	R	—	—	—	—	—	—	—	—	—	—	—	—	—	—	—	—	—	—	—	—
Silica	R	—	—	—	—	—	—	—	—	—	—	—	—	—	—	—	—	—	—	—	—
Sulfur	U																				
Furan resin	U																				
Polyester	U																				
Epoxy	U																				
°F	60	80	100	120	140	160	180	200	220	240	260	280	300	320	340	360	380	400	420	440	460
°C	15	26	38	49	60	71	82	93	104	116	127	138	149	160	171	182	193	204	216	227	238

Table M.10 Continued

Mortars	Bromine gas, dry
Silicate	
Sodium silicate	R (line to 460°F/238°C)
Potassium silicate	
Silica	
Sulfur	U
Furan resin	U
Polyester	U
Epoxy	U

°F	60	80	100	120	140	160	180	200	220	240	260	280	300	320	340	360	380	400	420	440	460
°C	15	26	38	49	60	71	82	93	104	116	127	138	149	160	171	182	193	204	216	227	238

Mortars	Bromine gas, moist
Silicate	
Sodium silicate	R (line to 460°F/238°C)
Potassium silicate	
Silica	
Sulfur	U
Furan resin	U
Polyester	U
Epoxy	U

°F	60	80	100	120	140	160	180	200	220	240	260	280	300	320	340	360	380	400	420	440	460
°C	15	26	38	49	60	71	82	93	104	116	127	138	149	160	171	182	193	204	216	227	238

Mortars	Bromine liquid
Silicate	R (line to 460°F/238°C)
Sodium silicate	
Potassium silicate	R (line to 460°F/238°C)
Silica	R (line to 460°F/238°C)
Sulfur	R (line to 200°F/93°C)
Furan resin	U
Polyester	U
Epoxy	U

°F	60	80	100	120	140	160	180	200	220	240	260	280	300	320	340	360	380	400	420	440	460
°C	15	26	38	49	60	71	82	93	104	116	127	138	149	160	171	182	193	204	216	227	238

Table M.10 Continued

Mortars	Calcium hypochlorite																				
Silicate																					
Sodium silicate	U																				
Potassium silicate	R	—	—	—	—	—	—	—	—	—	—	—	—	—	—	—	—	—	—	—	—
Silica	R	—	—	—	—	—	—	—	—	—	—	—	—	—	—	—	—	—	—	—	—
Sulfur	U																				
Furan resin	U																				
Polyester																					
Epoxy	U																				
°F	60	80	100	120	140	160	180	200	220	240	260	280	300	320	340	360	380	400	420	440	460
°C	15	26	38	49	60	71	82	93	104	116	127	138	149	160	171	182	193	204	216	227	238

Mortars	Carbon tetrachloride																				
Silicate	R	—	—	—	—	—	—	—	—	—	—	—	—	—	—	—	—	—	—	—	—
Sodium silicate	R	—	—	—	—	—	—	—	—	—	—	—	—	—	—	—	—	—	—	—	—
Potassium silicate	R	—	—	—	—	—	—	—	—	—	—	—	—	—	—	—	—	—	—	—	—
Silica	R	—	—	—	—	—	—	—	—	—	—	—	—	—	—	—	—	—	—	—	—
Sulfur	U																				
Furan resin	R	—	—	—	—	—	—	—	—	—	—	—	—	—	—	—					
Polyester	R	—	—	—	—																
Epoxy	R	—	—	—	—																
°F	60	80	100	120	140	160	180	200	220	240	260	280	300	320	340	360	380	400	420	440	460
°C	15	26	38	49	60	71	82	93	104	116	127	138	149	160	171	182	193	204	216	227	238

Mortars	Chlorine gas, dry																				
Silicate	R	—	—	—	—	—	—	—	—	—	—	—	—	—	—	—	—	—	—	—	—
Sodium silicate	R	—	—	—	—	—	—	—	—	—	—	—	—	—	—	—	—	—	—	—	—
Potassium silicate	R	—	—	—	—	—	—	—	—	—	—	—	—	—	—	—	—	—	—	—	—
Silica	R	—	—	—	—	—	—	—	—	—	—	—	—	—	—	—	—	—	—	—	—
Sulfur	U																				
Furan resin	U																				
Polyester	U																				
Epoxy	U																				
°F	60	80	100	120	140	160	180	200	220	240	260	280	300	320	340	360	380	400	420	440	460
°C	15	26	38	49	60	71	82	93	104	116	127	138	149	160	171	182	193	204	216	227	238

Table M.10 Continued

Mortars	Chlorine gas, wet																				
Silicate	R	—	—	—	—	—	—	—	—	—	—	—	—	—	—	—	—	—	—	—	—
Sodium silicate	R	—	—	—	—	—	—	—	—	—	—	—	—	—	—	—	—	—	—	—	—
Potassium silicate	R	—	—	—	—	—	—	—	—	—	—	—	—	—	—	—	—	—	—	—	—
Silica	R	—	—	—	—	—	—	—	—	—	—	—	—	—	—	—	—	—	—	—	—
Sulfur	U																				
Furan resin	U																				
Polyester	U																				
Epoxy	U																				
°F	60	80	100	120	140	160	180	200	220	240	260	280	300	320	340	360	380	400	420	440	460
°C	15	26	38	49	60	71	82	93	104	116	127	138	149	160	171	182	193	204	216	227	238

Mortars	Chlorine liquid																				
Silicate	R	—	—	—	—	—	—	—	—	—	—	—	—	—	—	—	—	—	—	—	—
Sodium silicate																					
Potassium silicate	R	—	—	—	—	—	—	—	—	—	—	—	—	—	—	—	—	—	—	—	—
Silica	R	—	—	—	—	—	—	—	—	—	—	—	—	—	—	—	—	—	—	—	—
Sulfur	U																				
Furan resin	U																				
Polyester	U																				
Epoxy	U																				
°F	60	80	100	120	140	160	180	200	220	240	260	280	300	320	340	360	380	400	420	440	460
°C	15	26	38	49	60	71	82	93	104	116	127	138	149	160	171	182	193	204	216	227	238

Mortars	Chromic acid 10%																				
Silicate	U																				
Sodium silicate	R	—	—	—	—	—	—	—	—	—	—	—	—	—	—	—	—	—	—	—	—
Potassium silicate	R	—	—	—	—	—	—	—	—	—	—	—	—	—	—	—	—	—	—	—	—
Silica	R	—	—	—	—	—	—	—	—	—	—	—	—	—	—	—	—	—	—	—	—
Sulfur	U																				
Furan resin	U																				
Polyester	R	—	—																		
Epoxy	U																				
°F	60	80	100	120	140	160	180	200	220	240	260	280	300	320	340	360	380	400	420	440	460
°C	15	26	38	49	60	71	82	93	104	116	127	138	149	160	171	182	193	204	216	227	238

Table M.10 Continued

Mortars		Chromic acid 50%
Silicate		R—
Sodium silicate		
Potassium silicate		
Silica		R—
Sulfur		U
Furan resin		U
Polyester	30%	R—
Epoxy		U
	°F	60 80 100 120 140 160 180 200 220 240 260 280 300 320 340 360 380 400 420 440 460
	°C	15 26 38 49 60 71 82 93 104 116 127 138 149 160 171 182 193 204 216 227 238

Mortars		Ferric chloride
Silicate		R—
Sodium silicate		U
Potassium silicate		R—
Silica		R—
Sulfur		R—
Furan resin		R—
Polyester		R—
Epoxy		R—
	°F	60 80 100 120 140 160 180 200 220 240 260 280 300 320 340 360 380 400 420 440 460
	°C	15 26 38 49 60 71 82 93 104 116 127 138 149 160 171 182 193 204 216 227 238

Mortars		Ferric chloride, 50% in water
Silicate		R—
Sodium silicate		U
Potassium silicate		R—
Silica		R—
Sulfur		R—
Furan resin		R—
Polyester		R—
Epoxy		R—
	°F	60 80 100 120 140 160 180 200 220 240 260 280 300 320 340 360 380 400 420 440 460
	°C	15 26 38 49 60 71 82 93 104 116 127 138 149 160 171 182 193 204 216 227 238

Table M.10 Continued

Mortars	Hydrobromic acid, 20%
Silicate	R————————————————————
Sodium silicate	R————————————————————
Potassium silicate	R————————————————————
Silica	R————————————————————
Sulfur	R————————
Furan resin	R——
Polyester	R——
Epoxy	U

°F	60	80	100	120	140	160	180	200	220	240	260	280	300	320	340	360	380	400	420	440	460
°C	15	26	38	49	60	71	82	93	104	116	127	138	149	160	171	182	193	204	216	227	238

Mortars	Hydrobromic acid, 50%
Silicate	R————————————————————
Sodium silicate	R————————————————————
Potassium silicate	R————————————————————
Silica	R————————————————————
Sulfur	R———————
Furan resin	R——
Polyester	R————
Epoxy	U

°F	60	80	100	120	140	160	180	200	220	240	260	280	300	320	340	360	380	400	420	440	460
°C	15	26	38	49	60	71	82	93	104	116	127	138	149	160	171	182	193	204	216	227	238

Mortars	Hydrochloric acid, 20%
Silicate	R————————————————————
Sodium silicate	R————————————————————
Potassium silicate	R————————————————————
Silica	R————————————————————
Sulfur	R———————
Furan resin	R———————————————
Polyester	R————
Epoxy	U

°F	60	80	100	120	140	160	180	200	220	240	260	280	300	320	340	360	380	400	420	440	460
°C	15	26	38	49	60	71	82	93	104	116	127	138	149	160	171	182	193	204	216	227	238

Table M.10 Continued

Mortars	Hydrochloric acid, 38%
Silicate	R————————————————————————————————————
Sodium silicate	R————————————————————————————————————
Potassium silicate	R————————————————————————————————————
Silica	R————————————————————————————————————
Sulfur	R————————————
Furan resin	R——————————————————————————
Polyester	R————
Epoxy	U

°F	60	80	100	120	140	160	180	200	220	240	260	280	300	320	340	360	380	400	420	440	460
°C	15	26	38	49	60	71	82	93	104	116	127	138	149	160	171	182	193	204	216	227	238

Mortars	Hydrofluoric acid 30%
Silicate	U
Sodium silicate	U
Potassium silicate	U
Silica	U
Sulfur	R——————
Furan resin	R——————————————————————————
Polyester	R————
Epoxy	U

°F	60	80	100	120	140	160	180	200	220	240	260	280	300	320	340	360	380	400	420	440	460
°C	15	26	38	49	60	71	82	93	104	116	127	138	149	160	171	182	193	204	216	227	238

Mortars	Hydrofluoric acid 70%
Silicate	U
Sodium silicate	U
Potassium silicate	U
Silica	U
Sulfur	U
Furan resin	U
Polyester	
Epoxy	U

°F	60	80	100	120	140	160	180	200	220	240	260	280	300	320	340	360	380	400	420	440	460
°C	15	26	38	49	60	71	82	93	104	116	127	138	149	160	171	182	193	204	216	227	238

Table M.10 Continued

Mortars	Hydrofluoric acid 100%																				
Silicate	U																				
Sodium silicate	U																				
Potassium silicate	U																				
Silica	U																				
Sulfur	U																				
Furan resin	U																				
Polyester																					
Epoxy	U																				
°F	60	80	100	120	140	160	180	200	220	240	260	280	300	320	340	360	380	400	420	440	460
°C	15	26	38	49	60	71	82	93	104	116	127	138	149	160	171	182	193	204	216	227	238

Mortars	Magnesium chloride																				
Silicate	R	—	—	—	—	—	—	—	—	—	—	—	—	—	—	—	—	—	—	—	—
Sodium silicate																					
Potassium silicate	R	—	—	—	—	—	—	—	—	—	—	—	—	—	—	—	—	—	—	—	—
Silica	R	—	—	—	—	—	—	—	—	—	—	—	—	—	—	—	—	—	—	—	—
Sulfur	R	—	—	—	—	—	—	—	—												
Furan resin	R	—	—	—	—	—	—	—	—	—	—	—	—	—	—	—					
Polyester	R	—	—	—	—	—	—	—	—												
Epoxy	R	—	—	—	—	—	—	—													
°F	60	80	100	120	140	160	180	200	220	240	260	280	300	320	340	360	380	400	420	440	460
°C	15	26	38	49	60	71	82	93	104	116	127	138	149	160	171	182	193	204	216	227	238

Mortars	Nitric acid 5%																				
Silicate	R	—	—	—	—	—	—	—	—	—	—	—	—	—	—	—	—	—	—	—	—
Sodium silicate	R	—	—	—	—	—	—	—	—	—	—	—	—	—	—	—	—	—	—	—	—
Potassium silicate	R	—	—	—	—	—	—	—	—	—	—	—	—	—	—	—	—	—	—	—	—
Silica	R	—	—	—	—	—	—	—	—	—	—	—	—	—	—	—	—	—	—	—	—
Sulfur	R	—	—	—	—																
Furan resin	U																				
Polyester	R	—	—	—	—																
Epoxy	U																				
°F	60	80	100	120	140	160	180	200	220	240	260	280	300	320	340	360	380	400	420	440	460
°C	15	26	38	49	60	71	82	93	104	116	127	138	149	160	171	182	193	204	216	227	238

Table M.10 Continued

Mortars (Nitric acid 20%)	60°F (15°C)	80°F (26°C)	100°F (38°C)	120°F (49°C)	140°F (60°C)	160°F (71°C)	180°F (82°C)	200°F (93°C)	220°F (104°C)	240°F (116°C)	260°F (127°C)	280°F (138°C)	300°F (149°C)	320°F (160°C)	340°F (171°C)	360°F (182°C)	380°F (193°C)	400°F (204°C)	420°F (216°C)	440°F (227°C)	460°F (238°C)
Silicate	R	R	R	R	R	R	R	R	R	R	R	R	R	R	R	R	R	R	R	R	R
Sodium silicate	R	R	R	R	R	R	R	R	R	R	R	R	R	R	R	R	R	R	R	R	R
Potassium silicate																					
Silica																					
Sulfur	R	R	R																		
Furan resin	U																				
Polyester	R	R	R																		
Epoxy	U																				

Mortars (Nitric acid 70%)	60°F (15°C)	80°F (26°C)	100°F (38°C)	120°F (49°C)	140°F (60°C)	160°F (71°C)	180°F (82°C)	200°F (93°C)	220°F (104°C)	240°F (116°C)	260°F (127°C)	280°F (138°C)	300°F (149°C)	320°F (160°C)	340°F (171°C)	360°F (182°C)	380°F (193°C)	400°F (204°C)	420°F (216°C)	440°F (227°C)	460°F (238°C)
Silicate	R	R	R	R	R	R	R	R	R	R	R	R	R	R	R	R	R	R	R	R	R
Sodium silicate	R	R	R	R	R	R	R	R	R	R	R	R	R	R	R	R	R	R	R	R	R
Potassium silicate																					
Silica																					
Sulfur	U																				
Furan resin	U																				
Polyester	U																				
Epoxy	U																				

Mortars (Nitric acid, anhydrous)	60°F (15°C)	80°F (26°C)	100°F (38°C)	120°F (49°C)	140°F (60°C)	160°F (71°C)	180°F (82°C)	200°F (93°C)	220°F (104°C)	240°F (116°C)	260°F (127°C)	280°F (138°C)	300°F (149°C)	320°F (160°C)	340°F (171°C)	360°F (182°C)	380°F (193°C)	400°F (204°C)	420°F (216°C)	440°F (227°C)	460°F (238°C)
Silicate	R	R	R	R	R	R	R	R	R	R	R	R	R	R	R	R	R	R	R	R	R
Sodium silicate	R	R	R	R	R	R	R	R	R	R	R	R	R	R	R	R	R	R	R	R	R
Potassium silicate																					
Silica																					
Sulfur	U																				
Furan resin	U																				
Polyester	U																				
Epoxy	U																				

Table M.10 Continued

Mortars	Oleum																				
°F	60	80	100	120	140	160	180	200	220	240	260	280	300	320	340	360	380	400	420	440	460
°C	15	26	38	49	60	71	82	93	104	116	127	138	149	160	171	182	193	204	216	227	238
Silicate																					
Sodium silicate	R	—	—	—																	
Potassium silicate																					
Silica																					
Sulfur	U																				
Furan resin	U																				
Polyester	U																				
Epoxy	U																				

Mortars	Phosphoric acid, 50–80%																				
°F	60	80	100	120	140	160	180	200	220	240	260	280	300	320	340	360	380	400	420	440	460
°C	15	26	38	49	60	71	82	93	104	116	127	138	149	160	171	182	193	204	216	227	238
Silicate	R	—	—	—	—	—	—	—	—	—	—	—	—	—	—	—	—	—	—	—	—
Sodium silicate	U																				
Potassium silicate	R	—	—	—	—	—	—	—	—	—	—	—	—	—	—	—	—	—	—	—	—
Silica	R	—	—	—	—	—	—	—	—	—	—	—	—	—	—	—	—	—	—	—	—
Sulfur	R	—	—	—	—	—	—	—													
Furan resin	R	—	—	—	—	—	—	—	—	—	—	—	—	—	—	—					
Polyester	R	—	—	—	—	—	—	—	—	—											
Epoxy	U																				

Mortars	Sodium chloride																				
°F	60	80	100	120	140	160	180	200	220	240	260	280	300	320	340	360	380	400	420	440	460
°C	15	26	38	49	60	71	82	93	104	116	127	138	149	160	171	182	193	204	216	227	238
Silicate	U																				
Sodium silicate	R	—	—	—	—	—	—	—	—	—	—	—	—	—	—	—	—	—	—	—	—
Potassium silicate	R	—	—	—	—	—	—	—	—	—	—	—	—	—	—	—	—	—	—	—	—
Silica	R	—	—	—	—	—	—	—	—	—	—	—	—	—	—	—	—	—	—	—	—
Sulfur	R	—	—	—	—	—	—	—													
Furan resin	R	—	—	—	—	—	—	—	—	—	—	—	—	—	—	—					
Polyester	R	—	—	—	—	—	—	—	—	—											
Epoxy	R	—	—	—	—	—	—	—	—	—											

Table M.10 Continued

Mortars	Sodium hydroxide, 10%																				
Silicate	U																				
Sodium silicate	U																				
Potassium silicate	U																				
Silica	U																				
Sulfur	U																				
Furan resin	R	—	—	—	—	—	—	—	—	—	—	—	—	—	—	—					
Polyester	R	—	—																		
Epoxy	R	—	—	—	—	—	—	—	—	—											
°F	60	80	100	120	140	160	180	200	220	240	260	280	300	320	340	360	380	400	420	440	460
°C	15	26	38	49	60	71	82	93	104	116	127	138	149	160	171	182	193	204	216	227	238

Mortars	Sodium hydroxide, 50%																				
Silicate	U																				
Sodium silicate	U																				
Potassium silicate	U																				
Silica	U																				
Sulfur	U																				
Furan resin	R	—	—	—	—	—	—	—	—	—	—	—	—	—	—	—					
Polyester	R	—	—																		
Epoxy	R	—	—	—	—	—	—	—	—	—											
°F	60	80	100	120	140	160	180	200	220	240	260	280	300	320	340	360	380	400	420	440	460
°C	15	26	38	49	60	71	82	93	104	116	127	138	149	160	171	182	193	204	216	227	238

Mortars	Sodium hydroxide, concentrated																				
Silicate	U																				
Sodium silicate	U																				
Potassium silicate	U																				
Silica	U																				
Sulfur	U																				
Furan resin																					
Polyester																					
Epoxy	R	—																			
°F	60	80	100	120	140	160	180	200	220	240	260	280	300	320	340	360	380	400	420	440	460
°C	15	26	38	49	60	71	82	93	104	116	127	138	149	160	171	182	193	204	216	227	238

Table M.10 Continued

Mortars	Sodium hypochlorite, 20%																				
Silicate	U																				
Sodium silicate	U																				
Potassium silicate	U																				
Silica	U																				
Sulfur	U																				
Furan resin	U																				
Polyester	U																				
Epoxy	U																				
°F	60	80	100	120	140	160	180	200	220	240	260	280	300	320	340	360	380	400	420	440	460
°C	15	26	38	49	60	71	82	93	104	116	127	138	149	160	171	182	193	204	216	227	238

Mortars	Sodium hypochlorite, concentrated																				
Silicate	U																				
Sodium silicate	U																				
Potassium silicate	U																				
Silica	U																				
Sulfur	U																				
Furan resin	U																				
Polyester	U																				
Epoxy	U																				
°F	60	80	100	120	140	160	180	200	220	240	260	280	300	320	340	360	380	400	420	440	460
°C	15	26	38	49	60	71	82	93	104	116	127	138	149	160	171	182	193	204	216	227	238

Mortars	Sulfuric acid, 10%																				
Silicate	U																				
Sodium silicate	R	—	—	—	—	—	—	—	—	—	—	—	—	—	—	—	—	—	—	—	—
Potassium silicate																					
Silica																					
Sulfur	R	—	—	—	—	—	—	—													
Furan resin	R	—	—	—	—	—	—	—	—	—	—	—	—	—	—	—					
Polyester	R	—	—	—	—	—	—	—	—	—											
Epoxy	R	—	—																		
°F	60	80	100	120	140	160	180	200	220	240	260	280	300	320	340	360	380	400	420	440	460
°C	15	26	38	49	60	71	82	93	104	116	127	138	149	160	171	182	193	204	216	227	238

Table M.10 Continued

Mortars	Sulfuric acid, 50%
Silicate	U
Sodium silicate	R—— (to 460°F / 238°C)
Potassium silicate	
Silica	
Sulfur	R—— (to 140°F / 60°C)
Furan resin	R—— (to 360°F / 182°C)
Polyester	R—— (to 260°F / 127°C)
Epoxy	U

°F	60	80	100	120	140	160	180	200	220	240	260	280	300	320	340	360	380	400	420	440	460
°C	15	26	38	49	60	71	82	93	104	116	127	138	149	160	171	182	193	204	216	227	238

Mortars	Sulfuric acid, 70%
Silicate	U
Sodium silicate	R—— (to 140°F / 60°C)
Potassium silicate	R—— (to 200°F / 93°C)
Silica	R—— (to 200°F / 93°C)
Sulfur	R—— (to 100°F / 38°C)
Furan resin	R—— (to 100°F / 38°C)
Polyester	R—— (to 100°F / 38°C)
Epoxy	U

°F	60	80	100	120	140	160	180	200	220	240	260	280	300	320	340	360	380	400	420	440	460
°C	15	26	38	49	60	71	82	93	104	116	127	138	149	160	171	182	193	204	216	227	238

Mortars	Sulfuric acid, 90%
Silicate	U
Sodium silicate	R—— (to 100°F / 38°C)
Potassium silicate	
Silica	
Sulfur	U
Furan resin	U
Polyester	U
Epoxy	U

°F	60	80	100	120	140	160	180	200	220	240	260	280	300	320	340	360	380	400	420	440	460
°C	15	26	38	49	60	71	82	93	104	116	127	138	149	160	171	182	193	204	216	227	238

Table M.10 Continued

Mortars	Sulfuric acid, 98%																				
Silicate	U																				
Sodium silicate	R——																				
Potassium silicate																					
Silica																					
Sulfur	U																				
Furan resin	U																				
Polyester	U																				
Epoxy	U																				
°F	60	80	100	120	140	160	180	200	220	240	260	280	300	320	340	360	380	400	420	440	460
°C	15	26	38	49	60	71	82	93	104	116	127	138	149	160	171	182	193	204	216	227	238

[a]The table is arranged alphabetically according to corrodent. Unless otherwise noted, the corrodent is considered pure, in the case of liquids, and a saturated aqueous solution in the case of solids. All percentages shown are weight percents.

Corrosion is a function of temperature. When using the tables note that the vertical lines refer to temperatures midway between the temperatures cited. An entry of R indicates that the material is resistant to the maximum temperature shown. An entry of U indicates that the material is unsatisfactory. A blank indicates that no data are available.

Source: PA Schweitzer. *Corrosion Resistance Tables,* 4th ed. Vols. 1–3. New York: Marcel Dekker, 1995.

REFERENCES

1. JL Gossett. Corrosion resistance of cast alloys. In: PA Schweitzer, ed. Corrosion Engineering Handbook. New York: Marcel Dekker, 1996, p. 260.
2. HH Uhlig. Corrosion and Corrosion Control. New York: John Wiley, 1963.
3. CP Dillon. Corrosion Resistance of Stainless Steels. New York: Marcel Dekker, 1995.
4. PA Schweitzer. Stainless steel. In: PA Schweitzer, ed. Corrosion and Corrosion Protection Handbook. New York: Marcel Dekker, 1989, pp. 76–79.
5. I Suzuki. Corrosion Resistant Coatings Technology. New York: Marcel Dekker, 1989.
6. H Leidheiser Jr. Coatings. In: F Mansfield, ed. Corrosion Mechanisms. New York: Marcel Dekker, 1987, pp. 165–209.
7. D Thierry and W Sand. Microbially influenced corrosion. In: P Marcus and J Oudar, eds. Corrosion Mechanisms in Theory and Practice. New York: Marcel Dekker, 1995, pp. 457–499.
8. G Cragnolino and OH Tuovinen. The role of reducing and sulphur oxidizing bacteria sulphate in the localized corrosion of iron-base alloys. Int Biodeterioration 20:9–26, 1984.
9. WA Hamilton. Sulphate reducing bacteria and anaerobic corrosion. Annu Rev Microbiol 39:195–217, 1985.
10. WP Iverson. Anaerobic corrosion mechanisms. Corrosion 83, NACE, paper 243, 1983.
11. PA Schweitzer. Corrosion Resistance Tables. 4th ed. Vols. 1–3. New York: Marcel Dekker, 1995.
12. M Sridhar and G Hodge. Nickel and high nickel alloys. In: PA Schweitzer, ed. Corrosion and Corrosion Protection Handbook. 2nd ed. New York: Marcel Dekker, 1989, pp. 96–124.
13. CT Arnold and PA Schweitzer. Corrosion testing techniques. In: PA Schweitzer, ed. Corrosion and Corrosion Protection Handbook. New York: Marcel Dekker, 1989, pp. 587–618.

14. A Perkins. Corrosion monitoring. In: PA Schweitzer, ed. Corrosion Engineering Handbook. New York: Marcel Dekker, 1996, pp. 623–652.
15. GF Rak and PA Schweitzer. Corrosion monitoring. In: PA Schweitzer, ed. Corrosion and Corrosion Protection Handbook. 2nd ed. New York: Marcel Dekker, 1989, pp. 547–586.
16. AA Boova. Chemical-resistant mortars, grouts, and monolithic surfacings. In: PA Schweitzer, ed. Corrosion Engineering Handbook. New York: Marcel Dekker, 1996, pp. 459–487.
17. WL Sheppard Jr. Chemical Resistant Masonry. 2nd ed. New York: Marcel Dekker, 1982.

N

NATURAL RUBBER (NR)

Natural rubber of the best quality is prepared by coagulating the latex of the Heve brasiliensis tree, which is cultivated primarily in the Far East.

Chemically, natural rubber is a polymer of methylbutadiene (isoprene):

$$\begin{array}{l} \qquad\quad CH_3 \\ \qquad\quad | \\ CH_2 = C - CH = CH_2 \end{array}$$

When polymerized, the units link together, forming long chains that each contain over 1000 units. Simple butadiene does not yield a good grade of rubber, apparently because the chains are too smooth and do not form a strong enough interlock. Synthetic rubbers are produced by introducing side groups into the chain either by modifying butadiene or by making a copolymer of butadiene and some other compound.

Purified raw rubber becomes sticky in hot weather and brittle in cold weather. Its valuable properties become apparent after vulcanization.

Depending upon the degree of curing, natural rubber is classified as soft, semihard, or hard rubber. Only soft rubber meets the ASTM definition of an elastomer.

Most rubber is made to combine with sulfur or sulfur-bearing organic compounds or with other cross-linking chemical agents in a process known as vulcanization, which was invented by Charles Goodyear in 1839 and forms the basis of all later developments in the rubber industry.

When properly carried out, vulcanization improves mechanical properties, eliminates tackiness, renders the rubber less susceptible to temperature changes, and makes it insoluble in all known solvents. Other materials are added for various purposes as follows:

Carbon blacks, precipitated pigments, and organic vulcanization accelerators are added to increase tensile strength and resistance to abrasion.

Whiting, barite, talc, silica, silicates, clays, and fibrous materials are added to cheapen and stiffen.

Bituminous substances, coal tar and its products, vegetable and mineral oils, paraffin, petrolatum, petroleum, oils and asphalt are added to soften (for purposes of processing or for final properties).

Condensation amines and waxes are added as protective agents against natural aging, sunlight, heat, and flexing.

Pigments are added to provide coloration.

Physical and Mechanical Properties

The physical and mechanical properties of natural rubber are shown in Table N.1. It is these properties that are responsible for the many varied applications of natural rubber. Many of these properties are modified somewhat through the process of vulcanization. Freshly cut or torn rubber has the power of self-adhesion. This property is for all intents and purposes absent in vulcanized rubber.

Dry heat up to 120°F (49°C) has little deteriorating effect on natural rubber. At temperatures of 300–400°F (148–205°C) rubber begins to melt and becomes sticky; at higher temperatures it becomes entirely carbonized. Natural rubber has good electrical insulation properties but poor flame resistance.

Vulcanization has the greatest effect on the mechanical properties of natural rubber. Vulcanized rubber can be stretched to approximately ten times its length and at this point will bear a load of 10 tons/in^2. It can be compressed to one-third of its thickness thousands of times without injury. When most types of vulcanized rubbers are stretched, their resistance increases in greater proportion than their extension. Even when stretched just short of their rupture they recover almost all of their original dimensions on being released and then gradually recover a portion of the residual distortion. The outstanding property

Table N.1 Physical and Mechanical Properties of Natural Rubber[a]

Specific gravity	0.92
Refractive index	1.52
Specific heat, cal/g	0.452
Swelling, % by volume	
in kerosene at 77°F (25°C)	200
in benzene at 77°F (25°C)	200
in acetone at 77°F (25°C)	25
in mineral oil at 100°F (70°C)	120
Brittle point	–68°F (–56°C)
Relative permeability to hydrogen	50
Relative permeability to air	11
Insulation resistance, ohms/cm	10
Resilience, %	90
Tear resistance, psi	1640
Coefficient of linear expansion at 32–140°F, in./in.-°F	0.000036
Coefficient of heat conduction K, BTU/ft^2-in.-°F	1.07
Tensile strength, psi	3000–4500
Elongation, % at break	775–780
Hardness, Shore A	40–100
Abrasion resistance	Excellent
Maximum temperature, continuous use	175°F (80°C)
Impact resistance	Excellent
Compression set	Good
Machining qualities	Can be ground
Effect of sunlight	Deteriorates
Effect of aging	Moderately resistant
Effect of heat	Softens

[a]These are representative values since they may be altered by compounding.

of natural rubber in comparison to synthetic rubbers is its resilence. It has excellent rebound properties, either hot or cold.

Resistance to Sun, Weather, and Ozone

Cold water preserves natural rubber, but if it is exposed to the air, particularly in sunlight, rubber tends to become hard and brittle. It has only fair resistance to ozone. Unlike the synthetic elastomers, natural rubber softens and reverts with aging to sunlight. In general it has relatively poor weathering and aging properties.

Chemical Resistance

Natural rubber offers excellent resistance to most inorganic salt solutions, alkalies, and nonoxidizing acids. Hydrochloric acid will react with soft rubber to form rubber hydrochloride, and therefore it is not recommended that natural rubber be used for items that will come into contact with that acid. Strong oxidizing media such as nitric acid, concentrated sulfuric acid, permanganates, dichromates, chlorine dioxide, and sodium hypochlorite will severely attack rubber. Mineral and vegetable oils, gasoline, benzene, toluene, and chlorinated hydrocarbons also affect rubber. Cold water tends to preserve natural rubber. Natural rubber offers good resistance to radiation and alcohols.

Unvulcanized rubber is soluble in gasoline, naphtha, carbon bisulfide, benzene, petroleum ether, turpentine, and other liquids.

Refer to Tables N.2, N.3, and N.4 for the compatibility of natural rubber with selected corrodents.

Applications

Natural rubber finds its major use in the manufacture of pneumatic tires and tubes, power transmission belts, conveyor belts, gaskets, mountings, hose, chemical tank linings, printing press platens, sound and/or shock absorbers, and seals against air, moisture, sound, and dirt.

Rubber has been used for many years as a lining material for steel tanks, particularly for protection against corrosion of inorganic salt solutions, especially brine, alkalies, and nonoxidizing acids. These linings have the advantage of being readily repaired in place. Natural rubber is also used for lining pipelines used to convey these types of materials. Some of these applications have been replaced by synthetic rubbers that have been developed over the years.

See Refs. 1 and 2.

NEOPRENE (CR)

Neoprene is one of the oldest and most versatile of the synthetic rubbers. Chemically it is polychloroprene. Its basic unit is a chlorinated butadiene whose formula is

$$\begin{array}{c} \quad\;\; Cl \\ \quad\;\; | \\ CH_2 - C - CH = CH_2 \end{array}$$

Table N.2 Compatibility of Soft Natural Rubber with Selected Corrodents[a]

Chemical	Maximum temp. °F	Maximum temp. °C	Chemical	Maximum temp. °F	Maximum temp. °C
Acetaldehyde	x	x	Barium hydroxide	140	60
Acetamide	x	x	Barium sulfate	140	60
Acetic acid 10%	150	66	Barium sulfide	140	60
Acetic acid 50%	x	x	Benzaldehyde	x	x
Acetic acid 80%	x	x	Benzene	x	x
Acetic acid, glacial	x	x	Benzene sulfonic acid 10%	x	x
Acetic anhydride	x	x	Benzoic acid	140	60
Acetone	140	60	Benzyl alcohol	x	x
Acetyl chloride	x	x	Benzyl chloride	x	x
Acrylic acid			Borax	140	60
Acrylonitrile			Boric acid	140	60
Adipic acid			Bromine gas, dry		
Allyl alcohol			Bromine gas, moist		
Allyl chloride			Bromine liquid		
Alum	140	60	Butadiene		
Aluminum acetate			Butyl acetate	x	x
Aluminum chloride, aqueous	140	60	Butyl alcohol	140	60
Aluminum chloride, dry	160	71	*n*-Butylamine		
Aluminum fluoride	x	x	Butyl phthalate		
Aluminum hydroxide			Butyric acid	x	x
Aluminum nitrate	x	x	Calcium bisulfide		
Aluminum oxychloride			Calcium bisulfite	140	60
Aluminum sulfate	140	60	Calcium carbonate	140	60
Ammonia gas			Calcium chlorate	140	60
Ammonium bifluoride			Calcium chloride	140	60
Ammonium carbonate	140	60	Calcium hydroxide 10%	140	60
Ammonium chloride 10%	140	60	Calcium hydroxide, sat.	140	60
Ammonium chloride 50%	140	60	Calcium hypochlorite	x	x
Ammonium chloride, sat.	140	60	Calcium nitrate	x	x
Ammonium fluoride 10%	x	x	Calcium oxide	140	60
Ammonium fluoride 25%	x	x	Calcium sulfate	140	60
Ammonium hydroxide 25%	140	60	Caprylic acid		
Ammonium hydroxide, sat.	140	60	Carbon bisulfide	x	x
Ammonium nitrate	140	60	Carbon dioxide, dry		
Ammonium persulfate			Carbon dioxide, wet		
Ammonium phosphate	140	60	Carbon disulfide	x	x
Ammonium sulfate 10–40%	140	60	Carbon monoxide	x	x
Ammonium sulfide	140	60	Carbon tetrachloride	x	x
Ammonium sulfite			Carbonic acid	140	60
Amyl acetate	x	x	Cellosolve	x	x
Amyl alcohol	140	60	Chloracetic acid, 50% water	x	x
Amyl chloride	x	x	Chloracetic acid	x	x
Aniline	x	x	Chlorine gas, dry	x	x
Antimony trichloride			Chlorine gas, wet	x	x
Aqua regia 3:1	x	x	Chlorine, liquid	x	x
Barium carbonate	140	60	Chlorobenzene	x	x
Barium chloride	140	60	Chloroform	x	x

Table N.2 Continued

Chemical	Maximum temp. °F	Maximum temp. °C	Chemical	Maximum temp. °F	Maximum temp. °C
Chlorosulfonic acid	x	x	Manganese chloride		
Chromic acid 10%	x	x	Methyl chloride	x	x
Chromic acid 50%	x	x	Methyl ethyl ketone	x	x
Chromyl chloride			Methyl isobutyl ketone	x	x
Citric acid 15%	140	60	Muriatic acid	140	60
Citric acid, concentrated	x	x	Nitric acid 5%	x	x
Copper acetate			Nitric acid 20%	x	x
Copper carbonate	x	x	Nitric acid 70%	x	x
Copper chloride	x	x	Nitric acid, anhydrous	x	x
Copper cyanide	140	60	Nitrous acid, concentrated	x	x
Copper sulfate	140	60	Oleum		
Cresol	x	x	Perchloric acid 10%		
Cupric chloride 5%	x	x	Perchloric acid 70%		
Cupric chloride 50%	x	x	Phenol	x	x
Cyclohexane	x	x	Phosphoric acid 50–80%	140	60
Cyclohexanol			Picric acid		
Dichloroacetic acid			Potassium bromide 30%	140	60
Dichloroethane (ethylene dichloride)	x	x	Salicylic acid		
Ethylene glycol	140	60	Silver bromide 10%		
Ferric chloride	140	60	Sodium carbonate	140	60
Ferric chloride 50% in water	140	60	Sodium chloride	140	60
Ferric nitrate 10–50%	x	x	Sodium hydroxide 10%	140	60
Ferrous chloride	140	60	Sodium hydroxide 50%	x	x
Ferrous nitrate	x	x	Sodium hydroxide, concentrated	x	x
Fluorine gas, dry	x	x	Sodium hypochlorite 20%	x	x
Fluorine gas, moist			Sodium hypochlorite, concentrated	x	x
Hydrobromic acid, dilute	140	60	Sodium sulfide to 50%	140	60
Hydrobromic acid 20%	140	60	Stannic chloride	140	60
Hydrobromic acid 50%	140	60	Stannous chloride	140	60
Hydrochloric acid 20%	x	x	Sulfuric acid 10%	140	60
Hydrochloric acid 38%	140	60	Sulfuric acid 50%	x	x
Hydrocyanic acid 10%			Sulfuric acid 70%	x	x
Hydrofluoric acid 30%	x	x	Sulfuric acid 90%	x	x
Hydrofluoric acid 70%	x	x	Sulfuric acid 98%	x	x
Hydrofluoric acid 100%	x	x	Sulfuric acid 100%	x	x
Hypochlorous acid			Sulfuric acid, fuming	x	x
Iodine solution 10%			Sulfurous acid	x	x
Ketones, general			Thionyl chloride		
Lactic acid 25%	x	x	Toluene		
Lactic acid, concentrated	x	x	Trichloroacetic acid		
Magnesium chloride	140	60	White liquor		
Malic acid	x	x	Zinc chloride	140	60

[a]The chemicals listed are in the pure state or in a saturated solution unless otherwise indicated. Compatibility is shown to the maximum allowable temperature for which data are available. Incompatibility is shown by an x. A blank space indicates that data are unavailable.

Source: PA Schweitzer. *Corrosion Resistance Tables,* 4th ed. Vols. 1–3. New York: Marcel Dekker, 1995.

Table N.3 Compatibility of Semi-Hard Natural Rubber with Selected Corrodents[a]

Chemical	Maximum temp. °F	Maximum temp. °C	Chemical	Maximum temp. °F	Maximum temp. °C
Acetaldehyde			Barium hydroxide	180	82
Acetamide	x	x	Barium sulfate	180	82
Acetic acid 10%			Barium sulfide		
Acetic acid 50%			Benzaldehyde		
Acetic acid 80%			Benzene	x	x
Acetic acid, glacial	x	x	Benzene sulfonic acid 10%		
Acetic anhydride	x	x	Benzoic acid	180	82
Acetone	x	x	Benzyl alcohol		
Acetyl chloride			Benzyl chloride		
Acrylic acid			Borax	180	82
Acrylonitrile			Boric acid	180	82
Adipic acid			Bromine gas, dry		
Allyl alcohol			Bromine gas, moist		
Allyl chloride			Bromine liquid		
Alum	180	82	Butadiene		
Aluminum acetate			Butyl acetate	100	38
Aluminum chloride, aqueous	180	82	Butyl alcohol	180	82
Aluminum chloride, dry			*n*-Butylamine		
Aluminum fluoride	x	x	Butyl phthalate		
Aluminum hydroxide			Butyric acid	100	38
Aluminum nitrate	100	38	Calcium bisulfide		
Aluminum oxychloride			Calcium bisulfite	180	82
Aluminum sulfate	180	82	Calcium carbonate	180	82
Ammonia gas			Calcium chlorate		
Ammonium bifluoride			Calcium chloride	180	82
Ammonium carbonate	180	82	Calcium hydroxide 10%	180	82
Ammonium chloride 10%	180	82	Calcium hydroxide, sat.	180	82
Ammonium chloride 50%	180	82	Calcium hypochlorite	x	x
Ammonium chloride, sat.	180	82	Calcium nitrate		
Ammonium fluoride 10%	x	x	Calcium oxide		
Ammonium fluoride 25%	x	x	Calcium sulfate		
Ammonium hydroxide 25%			Caprylic acid		
Ammonium hydroxide, sat.			Carbon bisulfide		
Ammonium nitrate	180	82	Carbon dioxide, dry		
Ammonium persulfate			Carbon dioxide, wet		
Ammonium phosphate	180	82	Carbon disulfide		
Ammonium sulfate 10–40%	180	82	Carbon monoxide		
Ammonium sulfide	180	82	Carbon tetrachloride		
Ammonium sulfite			Carbonic acid		
Amyl acetate			Cellosolve		
Amyl alcohol	180	82	Chloracetic acid, 50% water	100	38
Amyl chloride			Chloracetic acid	100	38
Aniline	x	x	Chlorine gas, dry		
Antimony trichloride			Chlorine gas, wet		
Aqua regia 3:1			Chlorine, liquid		
Barium carbonate	180	82	Chlorobenzene		
Barium chloride	180	82	Chloroform		

Table N.3 Continued

Chemical	Maximum temp. °F	Maximum temp. °C	Chemical	Maximum temp. °F	Maximum temp. °C
Chlorosulfonic acid			Manganese chloride		
Chromic acid 10%	x	x	Methyl chloride	100	38
Chromic acid 50%	x	x	Methyl ethyl ketone		
Chromyl chloride			Methyl isobutyl ketone		
Citric acid 15%	100	38	Muriatic acid		
Citric acid, concentrated	100	38	Nitric acid 5%	100	38
Copper acetate			Nitric acid 20%	x	x
Copper carbonate	180	82	Nitric acid 70%	x	x
Copper chloride	x	x	Nitric acid, anhydrous	x	x
Copper cyanide	180	82	Nitrous acid, concentrated	100	38
Copper sulfate	180	82	Oleum		
Cresol			Perchloric acid 10%		
Cupric chloride 5%	x	x	Perchloric acid 70%		
Cupric chloride 50%	x	x	Phenol	x	x
Cyclohexane			Phosphoric acid 50–80%	180	82
Cyclohexanol			Picric acid		
Dichloroacetic acid			Potassium bromide 30%	180	82
Dichloroethane (ethylene dichloride)			Salicylic acid		
Ethylene glycol	180	82	Silver bromide 10%		
Ferric chloride	180	82	Sodium carbonate	180	82
Ferric chloride 50% in water	180	82	Sodium chloride	180	82
Ferric nitrate 10–50%	100	38	Sodium hydroxide 10%	180	82
Ferrous chloride	180	82	Sodium hydroxide 50%	100	38
Ferrous nitrate	100	38	Sodium hydroxide, concentrated	100	38
Fluorine gas, dry			Sodium hypochlorite 20%	x	x
Fluorine gas, moist			Sodium hypochlorite, concentrated	x	x
Hydrobromic acid, dilute	180	82	Sodium sulfide to 50%	180	82
Hydrobromic acid 20%	180	82	Stannic chloride	180	82
Hydrobromic acid 50%	180	82	Stannous chloride	180	82
Hydrochloric acid 20%	180	82	Sulfuric acid 10%	180	82
Hydrochloric acid 38%	180	82	Sulfuric acid 50%	100	38
Hydrocyanic acid 10%			Sulfuric acid 70%	x	x
Hydrofluoric acid 30%			Sulfuric acid 90%	x	x
Hydrofluoric acid 70%	x	x	Sulfuric acid 98%	x	x
Hydrofluoric acid 100%	x	x	Sulfuric acid 100%	x	x
Hypochlorous acid			Sulfuric acid, fuming	x	x
Iodine solution 10%			Sulfurous acid	150	66
Ketones, general			Thionyl chloride		
Lactic acid 25%	100	38	Toluene		
Lactic acid, concentrated	100	38	Trichloroacetic acid		
Magnesium chloride	180	82	White liquor		
Malic acid	100	38	Zinc chloride	180	82

[a]The chemicals listed are in the pure state or in a saturated solution unless otherwise indicated. Compatibility is shown to the maximum allowable temperature for which data are available. Incompatibility is shown by an x. A blank space indicates that data are unavailable.

Source: PA Schweitzer. *Corrosion Resistance Tables,* 4th ed. Vols. 1–3. New York: Marcel Dekker, 1995.

Table N.4 Compatibility of Hard Natural Rubber with Selected Corrodents[a]

Chemical	Maximum temp. °F	Maximum temp. °C	Chemical	Maximum temp. °F	Maximum temp. °C
Acetaldehyde			Barium hydroxide	x	x
Acetamide	x	x	Barium sulfate	200	93
Acetic acid 10%	200	93	Barium sulfide	200	93
Acetic acid 50%	200	93	Benzaldehyde	x	x
Acetic acid 80%	150	66	Benzene	x	x
Acetic acid, glacial	100	38	Benzene sulfonic acid 10%		
Acetic anhydride	100	38	Benzoic acid	200	93
Acetone	x	x	Benzyl alcohol		
Acetyl chloride			Benzyl chloride		
Acrylic acid			Borax	200	93
Acrylonitrile	90	32	Boric acid	200	93
Adipic acid	80	27	Bromine gas, dry		
Allyl alcohol	x	x	Bromine gas, moist		
Allyl chloride			Bromine liquid		
Alum	200	93	Butadiene		
Aluminum acetate			Butyl acetate	x	x
Aluminum chloride, aqueous	200	93	Butyl alcohol	160	71
Aluminum chloride, dry			*n*-Butylamine		
Aluminum fluoride	x	x	Butyl phthalate		
Aluminum hydroxide	200	93	Butyric acid	150	66
Aluminum nitrate	190	88	Calcium bisulfide		
Aluminum oxychloride			Calcium bisulfite	200	93
Aluminum sulfate	200	93	Calcium carbonate	200	93
Ammonia gas			Calcium chlorate		
Ammonium bifluoride	200	93	Calcium chloride	200	93
Ammonium carbonate	200	93	Calcium hydroxide 10%	200	93
Ammonium chloride 10%	200	93	Calcium hydroxide, sat.	200	93
Ammonium chloride 50%	200	93	Calcium hypochlorite	x	x
Ammonium chloride, sat.	200	93	Calcium nitrate	200	93
Ammonium fluoride 10%	x	x	Calcium oxide	200	93
Ammonium fluoride 25%	x	x	Calcium sulfate	200	93
Ammonium hydroxide 25%	x	x	Caprylic acid		
Ammonium hydroxide, sat.	x	x	Carbon bisulfide	x	x
Ammonium nitrate	150	66	Carbon dioxide, dry		
Ammonium persulfate	200	93	Carbon dioxide, wet		
Ammonium phosphate	200	93	Carbon disulfide	x	x
Ammonium sulfate 10–40%	200	93	Carbon monoxide	x	x
Ammonium sulfide	200	93	Carbon tetrachloride	x	x
Ammonium sulfite			Carbonic acid	200	93
Amyl acetate			Cellosolve		
Amyl alcohol	200	93	Chloracetic acid, 50% water	120	49
Amyl chloride			Chloracetic acid	120	49
Aniline	x	x	Chlorine gas, dry	x	x
Antimony trichloride			Chlorine gas, wet	190	88
Aqua regia 3:1	x	x	Chlorine, liquid	x	x
Barium carbonate	200	93	Chlorobenzene	x	x
Barium chloride	200	93	Chloroform	x	x

Table N.4 Continued

Chemical	Maximum temp. °F	Maximum temp. °C	Chemical	Maximum temp. °F	Maximum temp. °C
Chlorosulfonic acid	x	x	Manganese chloride		
Chromic acid 10%	x	x	Methyl chloride		
Chromic acid 50%	x	x	Methyl ethyl ketone	x	x
Chromyl chloride			Methyl isobutyl ketone		
Citric acid 15%	150	66	Muriatic acid	200	93
Citric acid, concentrated	150	66	Nitric acid 5%	150	66
Copper acetate			Nitric acid 20%	x	x
Copper carbonate	200	93	Nitric acid 70%	x	x
Copper chloride	100	38	Nitric acid, anhydrous	x	x
Copper cyanide	200	93	Nitrous acid, concentrated	150	66
Copper sulfate	200	93	Oleum		
Cresol			Perchloric acid 10%		
Cupric chloride 5%			Perchloric acid 70%		
Cupric chloride 50%			Phenol	x	x
Cyclohexane			Phosphoric acid 50–80%	200	93
Cyclohexanol			Picric acid		
Dichloroacetic acid			Potassium bromide 30%	200	93
Dichloroethane (ethylene dichloride)			Salicylic acid		
Ethylene glycol	200	93	Silver bromide 10%		
Ferric chloride	200	93	Sodium carbonate	200	93
Ferric chloride 50% in water	200	93	Sodium chloride	200	93
Ferric nitrate 10–50%	150	66	Sodium hydroxide 10%	200	93
Ferrous chloride	200	93	Sodium hydroxide 50%	150	66
Ferrous nitrate	150	66	Sodium hydroxide, concentrated	150	66
Fluorine gas, dry			Sodium hypochlorite 20%	x	x
Fluorine gas, moist			Sodium hypochlorite, concentrated	x	x
Hydrobromic acid, dilute	200	93	Sodium sulfide to 50%	200	93
Hydrobromic acid 20%	200	93	Stannic chloride	200	93
Hydrobromic acid 50%	200	93	Stannous chloride	200	93
Hydrochloric acid 20%	200	93	Sulfuric acid 10%	200	93
Hydrochloric acid 38%	200	93	Sulfuric acid 50%		
Hydrocyanic acid 10%	200	93	Sulfuric acid 70%	x	x
Hydrofluoric acid 30%	x	x	Sulfuric acid 90%		
Hydrofluoric acid 70%	x	x	Sulfuric acid 98%		
Hydrofluoric acid 100%	x	x	Sulfuric acid 100%		
Hypochlorous acid	150	66	Sulfuric acid, fuming		
Iodine solution 10%			Sulfurous acid	200	93
Ketones, general			Thionyl chloride		
Lactic acid 25%	150	66	Toluene		
Lactic acid, concentrated	150	66	Trichloroacetic acid		
Magnesium chloride	200	93	White liquor		
Malic acid	150	66	Zinc chloride	200	93

[a]The chemicals listed are in the pure state or in a saturated solution unless otherwise indicated. Compatibility is shown to the maximum allowable temperature for which data are available. Incompatibility is shown by an x. A blank space indicates that data are unavailable.

Source: PA Schweitzer. *Corrosion Resistance Tables,* 4th ed. Vols. 1–3. New York: Marcel Dekker, 1995.

The raw material is acetylene, which makes this product more expensive than some of the other elastomeric materials.

Neoprene was introduced commercially by DuPont in 1932 as an oil resistant substitute for natural rubber. Its dynamic properties are very similar to those of natural rubber, but its range of chemical resistance overcomes many of the shortcomings of natural rubber.

As with other elastomeric materials, neoprene is available in a variety of formulations. Depending on the compounding procedure, material can be produced to impart specific properties to meet specific application needs.

Neoprene is also available in a variety of forms. In addition to a neoprene latex that is similar to natural rubber latex, neoprene is produced in a "fluid" form as either a compounded latex dispersion or a solvent solution. Once these materials have solidified or cured they have the same physical and chemical properties as the solid or cellular forms.

Physical and Mechanical Properties

The properties discussed here are attainable with neoprene but may not necessarily be incorporated into every neoprene product. Nor will every neoprene product perform the same in all environments. The reason for this variation is compounding. By selective addition and/or deletion of specific ingredients during compounding, specific properties can be enhanced or reduced to provide the neoprene formulation best suited for the application. A neoprene compound can be produced that will provide whichever of the properties discussed are desired. When the hardness of neoprene is above 55 Shore A its resilience exceeds that of natural rubber by approximately 5%. At hardnesses below 50 Shore A its resilience is not as good as that of natural rubber even though its resilience is measured at 75%, which is a high value. Because of its high resilience, neoprene products have low hysteresis and a minimum heat buildup during dynamic operations.

Solid neoprene products can be ignited by an open flame but will stop burning when the flame is removed. Because of its chlorine content, neoprene is more resistant to burning than exclusively hydrocarbon elastomers. Natural rubber and many of the other synthetic elastomers will continue to burn once ignited even if the flame is removed. In an actual fire situation, neoprene will burn. Although compounding can improve the flame resistance of neoprene it cannot make it immune to burning.

Compared to natural rubber, neoprene is relatively impermeable to gases. Table N.5 lists typical permeability constants. Because of this impermeability, neoprene can be used to seal against freon blowing agents, propane, butane, and other gases.

Neoprene is used in many electrical applications, although its dielectric characteristics limit its use as an insulation to low voltage (600 V) and low frequency (60 HZ). Because of its high degree of resistance to indoor and outdoor aging and its resistance to weathering, neoprene is often used as a protective outer jacket to insulation at all voltages. It is also immune to high voltage corona discharge effects that cause severe surface cutting in many types of elastomers.

At the maximum operating temperature of 200°F (93°C), neoprene continues to maintain good physical properties and has excellent resistance to long-term heat degradation. Unlike other elastomers, neoprene does not soften or melt when heated, regardless of the degree of heat. Heat failure results from the hardening of the product and lack of resilience. Neoprene products display little change in performance characteristics down to approximately 0°F (–18°C). As temperatures decrease further, the material stiffens until the

Table N.5 Physical and Mechanical Properties of Neoprene (CR)[a]

Specific gravity	1.4
Refractive index	1.56
Specific heat, cal/g	0.40
Volumetric coefficient of thermal expansions	
at 77°F	403×10^{-6}/°F
at 25°F	725×10^{-6}/°C
Thermal conductivity	
Btu/h-ft^2-in. °F	1.45
g-cal/h-cm^{2}cm °C	1.80
Brittle point	–40°F (–40°C)
DC resistivity, ohm-cm	2×10^{13}
Dielectric strength, V/mil	600
Permeability (cm^3/cm^2-cm-sec-atm) at 77°F (25°C)	
to nitrogen	1×10^{-8}
to methane	2×10^{-8}
to oxygen	3×10^{-8}
to helium	10×10^{-8}
to carbon dioxide	19×10^{-8}
Tensile strength, psi	1000–2500
Elongation, % at break	200–600
Hardness, Shore A	40–95
Abrasion resistance	Excellent
Maximum temperature, continuous use	180–200°F (82–93°C)
Impact resistance	Excellent
Compression set, %	15–35
Machining qualities	Can be ground
Resistance to sunlight	Excellent
Effect of aging	Little effect
Resistance to heat	Good

[a]These are representative values since they may be altered by compounding.

brittle point is reached. Although the brittle point for standard neoprene products is –40°F (–40°C), special compounding can produce materials that can be used at temperatures as low as –67°F (–55°C).

Neoprene will form an extremely strong mechanical bond with cotton fabric. If suitable treatments or additives are provided, it can also be made to adhere to such man-made fibers as glass, nylon, rayon, acrylic, and polyester. It can also be molded in contact with metals, particularly carbon and alloy steels, stainless steels, aluminum and aluminum alloys, brass, and copper, using any one of the commercially available bonding agents.

Neoprene provides no nourishment for microorganisms, but it will not deter them from consuming other ingredients in the compound. Consequently, products containing metabolizable compounding ingredients require the inclusion of a fungicide, bactercide, or pesticide in the formulation to provide protection.

Pigmentation of neoprene products is a simple matter since the elastomer readily accepts color additives. However, the lighter shades, such as tones of yellow, red, blue, and other bright colors, will eventually discolor with prolonged exposure to sunlight and ultraviolet light. Because of this, products intended for prolonged outdoor service are usually produced in shades of grey, maroon, brown, or black. The lighter shades are used for products having limited exposure to sunlight, such as rainwear and appliance parts.

Resistance to Sun, Weather, and Ozone

Neoprene displays excellent resistance to sun, weather, and ozone. Because of its low rate of oxidation, products made of neoprene have a high resistance to both outdoor and indoor aging. Over prolonged periods of time in an outdoor environment, the physical properties of neoprene display insignificant change. If neoprene is properly compounded, ozone in atmospheric concentrations has little effect on the product. When severe ozone exposure is expected, as for example around electrical equipment, compositions of neoprene can be provided to resist thousands of parts per million of ozone for hours without surface cracking. Natural rubber will crack within minutes when subjected to ozone concentrations of only 50 ppm.

Chemical Resistance

Neoprene's resistance to attack from solvents, waxes, fats, oils, greases, and many other petroleum based products is one of its outstanding properties. Excellent service is also experienced when it is in contact with aliphatic compounds (methyl and ethyl alcohols, ethylene glycols, etc.), aliphatic hydrocarbons, and most freon refrigerants. A minimum amount of swelling and relatively little loss of strength occur when neoprene is in contact with these fluids.

When exposed to dilute mineral acids, inorganic salt solutions, or alkalies, neoprene products show little if any change in appearance or change in properties.

Chlorinated and aromatic hydrocarbons, organic esters, aromatic hydroxy compounds, and certain ketones have an adverse effect on neoprene, and consequently only limited serviceability can be expected with them. Highly oxidizing acid and salt solutions also cause surface deterioration and loss of strength. Included in this category are nitric acid and concentrated sulfuric acid.

Neoprene formulations can be produced that provide products with outstanding resistance to water absorption. These products can be used in continuous or periodic immersion in either fresh water or salt water without loss of properties.

Properly compounded neoprene can be buried underground successfully, since moisture, bacteria, and soil chemicals usually found in the earth have little effect on its properties. It is unaffected by soils saturated with water, sea water, chemicals, oils, gasolines, wastes, and other industrial byproducts. Refer to Table N.6 for the compatibility of neoprene with selected corrodents.

Applications

Neoprene products are available in three basic forms:

1. Conventional solid rubber parts
2. Highly compressed cellular materials
3. Free-flowing fluids

Each form has certain specific properties that can be of advantage to final products.

Table N.6 Compatibility of Neoprene with Selected Corrodents[a]

Chemical	Maximum temp. °F	Maximum temp. °C	Chemical	Maximum temp. °F	Maximum temp. °C
Acetaldehyde	200	93	Barium hydroxide	230	110
Acetamide	200	93	Barium sulfate	200	93
Acetic acid 10%	160	71	Barium sulfide	200	93
Acetic acid 50%	160	71	Benzaldehyde	x	x
Acetic acid 80%	160	71	Benzene	x	x
Acetic acid, glacial	x	x	Benzene sulfonic acid 10%	100	38
Acetic anhydride	x	x	Benzoic acid	150	66
Acetone	x	x	Benzyl alcohol	x	x
Acetyl chloride	x	x	Benzyl chloride	x	x
Acrylic acid	x	x	Borax	200	93
Acrylonitrile	140	60	Boric acid	150	66
Adipic acid	160	71	Bromine gas, dry	x	x
Allyl alcohol	120	49	Bromine gas, moist	x	x
Allyl chloride	x	x	Bromine liquid	x	x
Alum	200	93	Butadiene	140	60
Aluminum acetate			Butyl acetate	60	16
Aluminum chloride, aqueous	150	66	Butyl alcohol	200	93
Aluminum chloride, dry			*n*-Butylamine		
Aluminum fluoride	200	93	Butyl phthalate		
Aluminum hydroxide	180	82	Butyric acid	x	x
Aluminum nitrate	200	93	Calcium bisulfide		
Aluminum oxychloride			Calcium bisulfite	x	x
Aluminum sulfate	200	93	Calcium carbonate	200	93
Ammonia gas	140	60	Calcium chlorate	200	93
Ammonium bifluoride	x	x	Calcium chloride	150	66
Ammonium carbonate	200	93	Calcium hydroxide 10%	230	110
Ammonium chloride 10%	150	66	Calcium hydroxide, sat.	230	110
Ammonium chloride 50%	150	66	Calcium hypochlorite	x	x
Ammonium chloride, sat.	150	66	Calcium nitrate	150	66
Ammonium fluoride 10%	200	93	Calcium oxide	200	93
Ammonium fluoride 25%	200	93	Calcium sulfate	150	66
Ammonium hydroxide 25%	200	93	Caprylic acid		
Ammonium hydroxide, sat.	200	93	Carbon bisulfide	x	x
Ammonium nitrate	200	93	Carbon dioxide, dry	200	93
Ammonium persulfate	200	93	Carbon dioxide, wet	200	93
Ammonium phosphate	150	66	Carbon disulfide	x	x
Ammonium sulfate 10–40%	150	66	Carbon monoxide	x	x
Ammonium sulfide	160	71	Carbon tetrachloride	x	x
Ammonium sulfite			Carbonic acid	150	66
Amyl acetate	x	x	Cellosolve	x	x
Amyl alcohol	200	93	Chloracetic acid, 50% water	x	x
Amyl chloride	x	x	Chloracetic acid	x	x
Aniline	x	x	Chlorine gas, dry	x	x
Antimony trichloride	140	60	Chlorine gas, wet	x	x
Aqua regia 3:1	x	x	Chlorine, liquid	x	x
Barium carbonate	150	66	Chlorobenzene	x	x
Barium chloride	150	66	Chloroform	x	x

Table N.6 Continued

Chemical	Maximum temp. °F	°C	Chemical	Maximum temp. °F	°C
Chlorosulfonic acid	x	x	Manganese chloride	200	93
Chromic acid 10%	140	60	Methyl chloride	x	x
Chromic acid 50%	100	38	Methyl ethyl ketone	x	x
Chromyl chloride			Methyl isobutyl ketone	x	x
Citric acid 15%	150	66	Muriatic acid	x	x
Citric acid, concentrated	150	66	Nitric acid 5%	x	x
Copper acetate	160	71	Nitric acid 20%	x	x
Copper carbonate			Nitric acid 70%	x	x
Copper chloride	200	93	Nitric acid, anhydrous	x	x
Copper cyanide	160	71	Nitrous acid, concentrated	x	x
Copper sulfate	200	93	Oleum	x	x
Cresol	x	x	Perchloric acid 10%		
Cupric chloride 5%	200	93	Perchloric acid 70%	x	x
Cupric chloride 50%	160	71	Phenol	x	x
Cyclohexane	x	x	Phosphoric acid 50–80%	150	66
Cyclohexanol	x	x	Picric acid	200	93
Dichloroacetic acid	x	x	Potassium bromide 30%	160	71
Dichloroethane (ethylene dichloride)	x	x	Salicylic acid		
Ethylene glycol	100	38	Silver bromide 10%		
Ferric chloride	160	71	Sodium carbonate	200	93
Ferric chloride 50% in water	160	71	Sodium chloride	200	93
Ferric nitrate 10–50%	200	93	Sodium hydroxide 10%	230	110
Ferrous chloride	90	32	Sodium hydroxide 50%	230	110
Ferrous nitrate	200	93	Sodium hydroxide, concentrated	230	110
Fluorine gas, dry	x	x	Sodium hypochlorite 20%	x	x
Fluorine gas, moist	x	x	Sodium hypochlorite, concentrated	x	x
Hydrobromic acid, dilute	x	x	Sodium sulfide to 50%	200	93
Hydrobromic acid 20%	x	x	Stannic chloride	200	93
Hydrobromic acid 50%	x	x	Stannous chloride	x	x
Hydrochloric acid 20%	x	x	Sulfuric acid 10%	150	66
Hydrochloric acid 38%	x	x	Sulfuric acid 50%	100	38
Hydrocyanic acid 10%	x	x	Sulfuric acid 70%	x	x
Hydrofluoric acid 30%	x	x	Sulfuric acid 90%	x	x
Hydrofluoric acid 70%	x	x	Sulfuric acid 98%	x	x
Hydrofluoric acid 100%	x	x	Sulfuric acid 100%	x	x
Hypochlorous acid	x	x	Sulfuric acid, fuming	x	x
Iodine solution 10%	80	27	Sulfurous acid	100	38
Ketones, general	x	x	Thionyl chloride	x	x
Lactic acid 25%	140	60	Toluene	x	x
Lactic acid, concentrated	90	32	Trichloroacetic acid	x	x
Magnesium chloride	200	93	White liquor	140	60
Malic acid			Zinc chloride	160	71

[a]The chemicals listed are in the pure state or in a saturated solution unless otherwise indicated. Compatibility is shown to the maximum allowable temperature for which data are available. Incompatibility is shown by an x. A blank space indicates that data are unavailable.

Source: PA Schweitzer. *Corrosion Resistance Tables,* 4th ed. Vols. 1–3. New York: Marcel Dekker, 1995.

Solid products can be produced by molding, extruding, or calendering. Molding can be accomplished by compression, transfer, injection, blow, vacuum, or wrapped-mandrel methods. Typical products produced by these methods are instrument seals, shoe soles and heels, auto sparkplug boots, radiator hose, boating accessories, appliance parts, O-rings, and other miscellaneous components.

Extrusion processes provide means of economically and uniformly mass producing products quickly. Neoprene products manufactured by these processes include tubing, sealing strips, wire jacketing, filaments, rods, and many types of hose.

Calendered products include sheet stock, belting, and friction and coated fabrics. A large proportion of sheet stock is later die-cut into finished products such as pads, gaskets, and diaphragms.

Cellular forms of neoprene are used primarily for gasketing, insulation, cushioning, and sound and vibration damping. This material provides compressibility not found in solid rubber but still retains the advantageous properties of neoprene. It is available as an open-cell sponge, a closed-cell neoprene, and a foam neoprene.

Open-cell neoprene is a compressible, absorbent material whose cells are uniform and connected. This is particularly useful for gasketing and dust-proofing applications where exposure to fluids is not expected.

Closed-cell neoprene is a resilient complex of individual nonconnecting cells that impart an added advantage of nonabsorbency. This property makes closed-cell neoprene especially suitable for sealing applications where fluid contact is expected, for products such as wet suits for divers, shoe soles, automotive deck lid seals, and for other applications where a compressible nonabsorbent weather resistent material is required.

Foam neoprene is similar to open-cell neoprene in that it is a compressible material with connecting cells. Its main area of application is for cushioning, for example, in mattresses and seating and carpet underlay. Because of the good heat and oil resistance of neoprene, it has also found application as a railroad car lubricator. This absorbent open-cell structure provides a wicking action to deliver oil to the journal bearings.

Fluid forms of neoprene are important in the manufacture of many products because of their versatility. "Fluid" neoprene is the primary component in such products as adhesives, coatings and paints, sealants, caulks, and fiber binders. It is available in two forms, as a neoprene latex or as a solvent solution. Neoprene latex is an elastomer–water dispersion. It is used primarily in the manufacture of dipped products, such as household and industrial gloves, meteorological ballons, sealed fractional horsepower motors, and a variety of rubber covered metal parts. Other applications include use as a binder for curled animal hair in resilient furniture cushioning, transportation seating, acoustical filtering, and packaging. It is also used extensively in latex based gloves, foams, protective coatings, and knife-coated fabrics, as a binder for cellulose and asbestos, and as an elasticizing additive for concrete, mortar, and asphalt. These products produced from neoprene latex possess the same properties as those associated with solid neoprene, including resistance to oil, chemicals, ozone, weather, and flame.

Neoprene solvent solutions are prepared by dissolving neoprene in standard rubber solvents. These solutions can be formulated in a range of viscosities suitable for application by brush, spray, or roller. Major areas of application include coating for storage tanks, industrial equipment, and chemical processing equipment. These coatings protect the vessels from corrosion by acids, oils, alkalies, and most hydrocarbons. Neoprene roofing applied in liquid form is used to protect concrete, plywood, and metal decks. The solvent solution can be readily applied and will cure into a roofing membrane that is tough, elastic, and weather resistant.

Solvent based adhesives develop quick initial bonds and remain strong and flexible almost indefinitely. They can be used to join a wide variety of rigid and flexible materials.

Collapsible nylon containers coated with neoprene are used for transporting and/or storing liquids, pastes, and flowable dry solids. Containers have been designed to hold oils, fuels, molasses, and various bulk-shipped products.

Neoprene in its many forms has proven to be reliable and indeed indispensable as a substitute for natural rubber, possessing many of the advantageous properties of natural rubber while also overcoming many of its shortcomings.

See Refs. 1 and 2.

NEUTRAL SOLUTION

A neutral solution is one that contains an equal number of hydrogen and hydroxyl ions, making it neither acid nor basic. It will have a pH of 7.

NEXUS

Nexus is the trademark of Burlington Industries for their polyester surfacing veil material. See "Thermoset Reinforcing Material."

NICKEL

Commercially, nickel is available in two main forms, alloy 200 and alloy 201. Alloy 200 is limited in operating temperatures to 600°F (315°C) while alloy 201 is used in applications having temperatures in excess of 600°F (315°C). Refer to Table N.7 for the physical and mechanical properties.

The outstanding characteristic of nickel 200 and 201 is its resistance to caustic soda and other alkalies. (Ammonium hydroxide is an exception.) These alloys are not attacked by anhydrous ammonia or ammonium hydroxide in concentrations of 1% or less. Stronger concentrations can cause rapid attack.

Nickel 200 shows excellent resistance to all concentrations of caustic soda at temperatures up to and including the molten state. Below 50% the corrosion rates are negligible, usually being less than 0.2 mpy even in boiling solutions. As concentrations and temperatures increase, corrosion rates increase very slowly.

Nickel 200 is not subject to stress corrosion cracking in any of the chloride salts, and it has excellent general resistance to nonoxidizing halides. The oxidizing acid chlorides, such as ferric, cupric, and mercuric are very corrosive and should not be used. Nickel 201 is extremely useful in handling dry chlorine and hydrogen chloride at elevated temperatures.

If aeration is not high, nickel 200 has excellent resistance to most organic acids, particularly fatty acids such as stearic and oleic.

Nickel is also used in handling food and synthetic fibers for maintaining purity. For food products, the presence of nickel ions is not detrimental to the flavor and is not toxic.

Table N.7 Mechanical and Physical Properties of Nickel 200 and Nickel 201

Property	Nickel 200	Nickel 201
Modulus of elasticity × 10^6, psi	28	30
Tensile strength × 10^3, psi	27000	58500
Yield strength 0.2% offset × 10^3, psi	21500	15000
Elongation in 2 in., %	47	50
Hardness, Brinell	105	87
Density, lb/in.^{3}	0.321	0.321
Specific gravity	8.89	8.89
Specific heat, Btu/lb °F	0.109	0.109
Thermal conductivity, Btu/h/ft^2/°F/in.		
at 0–70°F	500	569
at 70–200°F	465	512
at 70–400°F	425	460
at 70–600°F	390	408
at 70–800°F	390	392
at 70–1000°F	405	410
at 70–1200°F	420	428
Thermal expansion coefficient, in./in./°F × 10^{-6}		
at 0–70°F	6.3	
at 70–200°F	7.4	7.3
at 70–400°F	7.7	
at 70–600°F	8.0	
at 70–800°F	8.3	
at 70–1000°F	8.5	
at 70–1200°F	8.7	

For many organic chemicals such as phenol and viscose rayon, nickel does not discolor the solutions, unlike iron and copper.

The corrosion resistances of nickel 200 and nickel 201 are the same. Refer to Table N.8 for the compatibility of these alloys with selected corrodents. Reference 2 provides a more comprehensive listing of the compatibility of nickel with a variety of corrodents.

See Refs. 3–8.

NICKEL COATINGS

Nickel coatings are available in three types, bright, semibright, and dull bright. A bright nickel layer offers less protection from corrosion than the semibright and dull nickel layers.

High corrosion resistance is provided by a duplex nickel coating system in which the ratio of a semibright nickel deposit to a bright nickel deposit is 3:1, and a thickness of 20–25 μm is achieved.

In order to reduce production costs a nickel–iron alloy coating was developed. The nickel–iron alloy provides full brightness, high leveling, excellent ductility, and good receptivity for chromium. This coating is only suitable for use in mild atmospheres, since a formation of red rust will develop when the coating is immersed in water.

See Ref. 9.

Table N.8 Compatibility of Nickel 200 and Nickel 201 with Selected Corrodents[a]

Chemical	Maximum temp. °F	Maximum temp. °C	Chemical	Maximum temp. °F	Maximum temp. °C
Acetaldehyde	200	93	Barium sulfate	210	99
Acetamide			Barium sulfide	110	43
Acetic acid 10%	90	32	Benzaldehyde	210	99
Acetic acid 50%	90	32	Benzene	210	99
Acetic acid 80%	120	49	Benzene sulfonic acid 10%	190	88
Acetic acid, glacial	x	x	Benzoic acid	400	204
Acetic anhydride	170	77	Benzyl alcohol	210	99
Acetone	190	88	Benzyl chloride	210	99
Acetyl chloride	100	38	Borax	200	93
Acrylic acid			Boric acid	210	99
Acrylonitrile	210	99	Bromine gas, dry	60	16
Adipic acid	210	99	Bromine gas, moist	x	x
Allyl alcohol	220	104	Bromine liquid		
Allyl chloride	190	88	Butadiene	80	27
Alum	170	77	Butyl acetate	80	27
Aluminum acetate			Butyl alcohol	200	93
Aluminum chloride, aqueous	300	149	*n*-Butylamine		
Aluminum chloride, dry	60	16	Butyl phthalate	210	99
Aluminum fluoride	90	32	Butyric acid	x	x
Aluminum hydroxide	80	27	Calcium bisulfide		
Aluminum nitrate			Calcium bisulfite	x	x
Aluminum oxychloride			Calcium carbonate		
Aluminum sulfate	210	99	Calcium chlorate	140	60
Ammonia gas	90	32	Calcium chloride	80	27
Ammonium bifluoride			Calcium hydroxide 10%	210	99
Ammonium carbonate	190	88	Calcium hydroxide, sat.	200	93
Ammonium chloride 10%	230	110	Calcium hypochlorite	x	x
Ammonium chloride 50%	170	77	Calcium nitrate		
Ammonium chloride, sat.	570	299	Calcium oxide	90	32
Ammonium fluoride 10%	210	99	Calcium sulfate	210	99
Ammonium fluoride 25%	200	93	Caprylic acid[b]	210	99
Ammonium hydroxide 25%	x	x	Carbon bisulfide	x	x
Ammonium hydroxide, sat.	320	160	Carbon dioxide, dry	210	99
Ammonium nitrate	90	32	Carbon dioxide, wet	200	93
Ammonium persulfate	x	x	Carbon disulfide	x	x
Ammonium phosphate 30%	210	99	Carbon monoxide	570	290
Ammonium sulfate 10–40%	210	99	Carbon tetrachloride	210	99
Ammonium sulfide			Carbonic acid	80	27
Ammonium sulfite	x	x	Cellosolve	210	99
Amyl acetate	300	149	Chloracetic acid, 50% water		
Amyl alcohol			Chloracetic acid	210	99
Amyl chloride	90	32	Chlorine gas, dry	200	93
Aniline	210	99	Chlorine gas, wet	x	x
Antimony trichloride	210	99	Chlorine, liquid		
Aqua regia 3:1	x	x	Chlorobenzene	120	49
Barium carbonate	210	99	Chloroform	210	99
Barium chloride	80	27	Chlorosulfonic acid	80	27
Barium hydroxide	90	32	Chromic acid 10%	100	38

Table N.8 Continued

Chemical	Maximum temp. °F	Maximum temp. °C	Chemical	Maximum temp. °F	Maximum temp. °C
Chromic acid 50%	x	x	Methyl chloride	210	99
Chromyl chloride	210	99	Methyl ethyl ketone		
Citric acid 15%	210	99	Methyl isobutyl ketone	200	93
Citric acid, concentrated	80	27	Muriatic acid	x	x
Copper acetate	100	38	Nitric acid 5%	x	x
Copper carbonate	x	x	Nitric acid 20%	x	x
Copper chloride	x	x	Nitric acid 70%	x	x
Copper cyanide	x	x	Nitric acid, anhydrous	x	x
Copper sulfate	x	x	Nitrous acid, concentrated	x	x
Cresol	100	38	Oleum		
Cupric chloride 5%	x	x	Perchloric acid 10%	x	x
Cupric chloride 50%	x	x	Perchloric acid 70%		
Cyclohexane	80	27	Phenol, sulfur free	570	299
Cyclohexanol	80	27	Phosphoric acid 50–80%	x	x
Dichloroacetic acid			Picric acid	80	27
Dichloroethane (ethylene dichloride)	x	x	Potassium bromide 30%		
Ethylene glycol	210	99	Salicylic acid	80	27
Ferric chloride	x	x	Silver bromide 10%		
Ferric chloride 50% in water	x	x	Sodium carbonate to 30%	210	99
Ferric nitrate 10–50%	x	x	Sodium chloride to 30%	210	99
Ferrous chloride	x	x	Sodium hydroxide 10%[c]	210	99
Ferrous nitrate			Sodium hydroxide 50%[c]	300	149
Fluorine gas, dry	570	290	Sodium hydroxide, concentrated	200	93
Fluorine gas, moist	60	16	Sodium hypochlorite 20%	x	x
Hydrobromic acid, dilute	x	x	Sodium hypochlorite, concentrated	x	x
Hydrobromic acid 20%	x	x	Sodium sulfide to 50%	x	x
Hydrobromic acid 50%	x	x	Stannic chloride	x	x
Hydrochloric acid 20%	80	27	Stannous chloride, dry	570	299
Hydrochloric acid 38%	x	x	Sulfuric acid 10%	x	x
Hydrocyanic acid 10%			Sulfuric acid 50%	x	x
Hydrofluoric acid 30%[c]	170	77	Sulfuric acid 70%	x	x
Hydrofluoric acid 70%[c]	100	38	Sulfuric acid 90%	x	x
Hydrofluoric acid 100%[c]	120	49	Sulfuric acid 98%	x	x
Hypochlorous acid	x	x	Sulfuric acid 100%	x	x
Iodine solution 10%			Sulfuric acid, fuming	x	x
Ketones, general	100	38	Sulfurous acid	x	x
Lactic acid 25%	x	x	Thionyl chloride	210	99
Lactic acid, concentrated	x	x	Toluene	210	99
Magnesium chloride	300	149	Trichloroacetic acid	80	27
Malic acid	210	99	White liquor		
Manganese chloride 37%	90	32	Zinc chloride to 80%	200	93

[a]The chemicals listed are in the pure state or in a saturated solution unless otherwise indicated. Compatibility is shown to the maximum allowable temperature for which data are available. Incompatibility is shown by an x. A blank space indicates that data are unavailable. When compatible, corrosion rate is < 20 mpy.
[b]Material subject to pitting.
[c]Material subject to stress cracking.
Source: PA Schweitzer. *Corrosion Resistance Tables,* 4th ed. Vols. 1–3. New York: Marcel Dekker, 1995.

NITRIDING

Also see "High Temperature Corrosion."

Nitriding is the absorption of nitrogen atoms by a metal to remain either dissolved or to form metal nitrides. It occurs when hot surfaces react with "active" nitrogen, such as produced by thermal cracking of ammonia. (Molecular nitrogen is relatively stable.)

NITRILE RUBBER (NBR, BUNA-N)

The nitrile rubbers are an outgrowth of German Buna-N or Perbunan. They are copolymers of butadiene and acrylonitrile ($CH_2 = CH - C = N$) and are one of the four most widely used elastomers. XNBR is a carboxylic acrylonitrile butadiene nitrile rubber with somewhat improved abrasion resistance over that of the standard NBR nitrile rubbers. The main advantages of the nitrile rubbers are their low cost, good oil and abrasion resistance, and good low temperature and swell characteristics. Their greater resistance to oils, fuel, and solvents compared to that of neoprene is their primary advantage. As with other elastomers, appropriate compounding will improve certain properties.

Physical and Mechanical Properties

The physical and mechanical properties of the nitrile rubbers are very similar to those of natural rubber (see Table N.9). These rubbers have the exceptional ability to retain both

Table N.9 Physical and Mechanical Properties of Nitrile Rubber (NBR, Buna-N)[a]

Specific gravity	0.99
Refractive index	1.54
Brittle point	–32°F to –40°F (–1°C to –40°C)
Swelling, % by volume	
in kerosene at 77°F (25°C)	9–10
in benzene at 77°F (25°C)	120
in acetone at 77°F (25°C)	60–50
in mineral oil at 100°F (70°C)	2–10
in air at 77°F (25°C)	30–50
Tensile strength, psi	500–4000
Elongation, % at break	400
Hardness, Shore A	40–95
Abrasion resistance	Excellent
Maximum temperature, continuous use	250°F (120°C)
Compression set	Good
Tear resistance	Excellent
Resilience, %	63–74
Machining qualities	Can be ground
Resistance to sunlight	Fair
Effect of aging	Highly resistant
Resistance to heat	Softens

[a]These are representative values since they may be altered by compounding.

their strength and elasticity at extreme low temperatures. It is this property that makes them valuable for use as hose used in operating the hydraulic controls of airplanes.

Buna-N does not have exceptional heat resistance. It has a maximum operating temperature of 250°F (120°C), but it has a tendency to harden at high temperatures. The nitrile rubbers will support combustion and burn. Their electrical properties are relatively poor, and consequently they do not find wide use in electrical applications, since there are so many other elastomeric materials with far superior electrical properties.

NBR has good compression set recovery from deformation and good abrasion resistance and tensile strength.

Resistance to Sun, Weather, and Ozone

The nitrile rubbers offer poor resistance to sunlight and ozone, and their weathering qualities are not good.

Chemical Resistance

The nitrile rubbers exhibit good resistance to solvents, oil, water, and hydraulic fluids. A very slight swelling occurs in the presence of aliphatic hydrocarbons, fatty acids, alcohols, and glycols. The deterioration of physical properties as a result of this swelling is small, making NBR suitable for gasoline and oil resistant applications. NBR has excellent resistance to water. The use of highly polar solvents such as acetone and methyl ethyl ketone, chlorinated hydrocarbons, ozone, nitro hydrocarbons, ether, or esters should be avoided, since these materials will attack the nitrile rubbers.

The XNBR rubbers are used primarily in nonalkaline service. Refer to Table N.10 for the compatibility of nitrile rubber with selected corrodents.

Applications

Because of its exceptional resistance to fuels and hydraulic fluids, Buna-N's major area of application is in the manufacture of aircraft hose, gasoline and oil hose, and self-sealing fuel tanks. Other applications include carburetor diaphragms, gaskets, cables, printing rolls, and machinery mountings.

See Refs. 1 and 2.

NOBLE METAL

A noble metal is any metal that has a standard electrode potential more noble (positive) than that of hydrogen. The noble metals are gold, silver, platinum, iridium, osmium, palladium, rhodium, and ruthenium. The term is often used synonomously for precious metals when referring to such as platinum and gold.

NORMALIZING

Normalizing is a heat treatment process in which carbon or low alloy steel is heated to approximately 1650°F (900°C) and is then air cooled. This process partially relieves

Table N.10 Compatibility of Nitrile Rubber with Selected Corrodents[a]

Chemical	Maximum temp. °F	Maximum temp. °C	Chemical	Maximum temp. °F	Maximum temp. °C
Acetaldehyde	x	x	Barium hydroxide		
Acetamide	180	82	Barium sulfate		
Acetic acid 10%	x	x	Barium sulfide		
Acetic acid 50%	x	x	Benzaldehyde		
Acetic acid 80%	x	x	Benzene	150	66
Acetic acid, glacial	x	x	Benzene sulfonic acid 10%		
Acetic anhydride	x	x	Benzoic acid	150	66
Acetone	x	x	Benzyl alcohol		
Acetyl chloride	x	x	Benzyl chloride		
Acrylic acid	x	x	Borax		
Acrylonitrile	x	x	Boric acid	150	66
Adipic acid	180	82	Bromine gas, dry		
Allyl alcohol	180	82	Bromine gas, moist		
Allyl chloride	x	x	Bromine liquid		
Alum	150	66	Butadiene		
Aluminum acetate			Butyl acetate		
Aluminum chloride, aqueous	150	66	Butyl alcohol		
Aluminum chloride, dry			*n*-Butylamine		
Aluminum fluoride			Butyl phthalate		
Aluminum hydroxide	180	82	Butyric acid		
Aluminum nitrate	190	88	Calcium bisulfide		
Aluminum oxychloride			Calcium bisulfite		
Aluminum sulfate	200	93	Calcium carbonate		
Ammonia gas	190	88	Calcium chlorate		
Ammonium bifluoride			Calcium chloride		
Ammonium carbonate	x	x	Calcium hydroxide 10%		
Ammonium chloride 10%			Calcium hydroxide, sat.		
Ammonium chloride 50%			Calcium hypochlorite	x	x
Ammonium chloride, sat.			Calcium nitrate		
Ammonium fluoride 10%			Calcium oxide		
Ammonium fluoride 25%			Calcium sulfate		
Ammonium hydroxide 25%			Caprylic acid		
Ammonium hydroxide, sat.			Carbon bisulfide		
Ammonium nitrate	150	66	Carbon dioxide, dry		
Ammonium persulfate			Carbon dioxide, wet		
Ammonium phosphate	150	66	Carbon disulfide		
Ammonium sulfate 10–40%			Carbon monoxide		
Ammonium sulfide			Carbon tetrachloride		
Ammonium sulfite			Carbonic acid	100	38
Amyl acetate			Cellosolve		
Amyl alcohol	150	66	Chloracetic acid, 50% water		
Amyl chloride			Chloracetic acid		
Aniline	x	x	Chlorine gas, dry		
Antimony trichloride			Chlorine gas, wet		
Aqua regia 3:1			Chlorine, liquid		
Barium carbonate			Chlorobenzene		
Barium chloride	125	52	Chloroform		

Table N.10 Continued

Chemical	Maximum temp. °F	Maximum temp. °C	Chemical	Maximum temp. °F	Maximum temp. °C
Chlorosulfonic acid			Manganese chloride		
Chromic acid 10%			Methyl chloride		
Chromic acid 50%			Methyl ethyl ketone		
Chromyl chloride			Methyl isobutyl ketone		
Citric acid 15%			Muriatic acid		
Citric acid, concentrated			Nitric acid 5%		
Copper acetate			Nitric acid 20%	x	x
Copper carbonate			Nitric acid 70%	x	x
Copper chloride			Nitric acid, anhydrous	x	x
Copper cyanide			Nitrous acid, concentrated		
Copper sulfate			Oleum		
Cresol			Perchloric acid 10%		
Cupric chloride 5%			Perchloric acid 70%		
Cupric chloride 50%			Phenol	x	x
Cyclohexane			Phosphoric acid 50–80%	150	66
Cyclohexanol			Picric acid		
Dichloroacetic acid			Potassium bromide 30%		
Dichloroethane (ethylene dichloride)			Salicylic acid		
Ethylene glycol	100	38	Silver bromide 10%		
Ferric chloride	150	66	Sodium carbonate	125	52
Ferric chloride 50% in water			Sodium chloride	200	93
Ferric nitrate 10–50%	150	66	Sodium hydroxide 10%	150	66
Ferrous chloride			Sodium hydroxide 50%		
Ferrous nitrate			Sodium hydroxide, concentrated		
Fluorine gas, dry			Sodium hypochlorite 20%	x	x
Fluorine gas, moist			Sodium hypochlorite, concentrated	x	x
Hydrobromic acid, dilute			Sodium sulfide to 50%		
Hydrobromic acid 20%			Stannic chloride	150	66
Hydrobromic acid 50%			Stannous chloride		
Hydrochloric acid 20%			Sulfuric acid 10%	150	66
Hydrochloric acid 38%			Sulfuric acid 50%	150	66
Hydrocyanic acid 10%			Sulfuric acid 70%	x	x
Hydrofluoric acid 30%			Sulfuric acid 90%	x	x
Hydrofluoric acid 70%	x	x	Sulfuric acid 98%	x	x
Hydrofluoric acid 100%	x	x	Sulfuric acid 100%	x	x
Hypochlorous acid			Sulfuric acid, fuming	x	x
Iodine solution 10%			Sulfurous acid		
Ketones, general			Thionyl chloride		
Lactic acid 25%			Toluene		
Lactic acid, concentrated			Trichloroacetic acid		
Magnesium chloride			White liquor		
Malic acid			Zinc chloride	150	66

[a]The chemicals listed are in the pure state or in a saturated solution unless otherwise indicated. Compatibility is shown to the maximum allowable temperature for which data are available. Incompatibility is shown by an x. A blank space indicates that data are unavailable.

Source: PA Schweitzer. *Corrosion Resistance Tables,* 4th ed. Vols. 1–3. New York: Marcel Dekker, 1995.

residual stresses produced by prior processing and reduces the grain size and makes the grain structure more homogenous. This produces a tougher and more ductile material.

Also see "Annealing."

NO_x

NO_x is a term used to describe nitrogen oxides that are emitted to the atmosphere as pollutants. These emissions contribute to atmospheric corrosion. No_x emissions are the result of energy production and road traffic. During the combustion process the nitrogen oxides are emitted as NO, which is oxidized to NO_2, which can be further oxidized to HNO_3 (nitric acid). This latter reaction has a very slow rate, so that in the immediate vicinity of the emissions, the predominant corrodent is NO_2. At further distances the concentration of nitric acid increases.

NYLON

See "Polyamides."

REFERENCES

1. PA Schweitzer, Corrosion Resistance of Elastomers. New York: Marcel Dekker, 1990.
2. PA Schweitzer. Corrosion Resistance Tables. 4th ed. Vol. 1–3. New York: Marcel Dekker, 1995.
3. HH Uhlig. Corrosion and Corrosion Control. New York: John Wiley and Sons, 1963.
4. CP Dillon. Corrosion Control in the Chemical Process Industries. 2nd ed. St. Louis: Materials Technology Institute of the Chemical Process Industries, 1994.
5. B. MacDougall and MJ Graham. Growth and stability of passive films. In: P Marcus and J Oudar, eds. Corrosion Mechanisms in Theory and Practice. New York: Marcel Dekker, 1995, pp. 143–167.
6. WJ Lorenz and KE Heusler. Anodic dissolution of iron group metals. In: F Mansfield, ed. Corrosion Mechanisms. New York: Marcel Dekker, 1987, pp. 61–73.
7. N Sridhar and G Hodge. Nickel and high nickel alloys. In: PA Schweitzer, ed. Corrosion and Corrosion Protection Handbook. 2nd ed. New York: Marcel Dekker, 1989, p. 95.
8. PA Schweitzer. Corrosion Resistant Piping Systems. New York: Marcel Dekker, 1994.
9. I Suzuki. Corrosion Resistant Coatings Technology. New York: Marcel Dekker, 1989.

O

OIL ASH CORROSION

Oil ash corrosion takes place on secondary superheater and reheater tube sections of boilers using bunker "C" oil high in vanadium content, or using residual oil as the fuel. Molten ash deposits from this fuel dissolve protective iron oxide layers on the tube surfaces, act as an oxidation catalyst, and allow oxygen and other combustion gases to diffuse rapidly to the metal surface. If sodium, chlorine, or sulfur are present in the ash, the corrosion will be accelerated.

OIL/GAS WELL CORROSION INHIBITORS

These inhibitors are used in oil/gas wells to reduce the corrosion resulting from the presence of water, salt, carbon dioxide, and hydrogen sulfide contained in the hydrocarbon mix. The inhibitors can be classified as to their solubility and dispersibility in the two phases (water/oil-gas) of the well. They are described as (a) oil and water insoluble, (b) oil soluble and water soluble, (c) oil soluble and water dispersible, (d) oil and water soluble, (e) volatile inhibitors. Organic nitrogen molecules with molecular weights exceeding 200 are the most commonly used inhibitors.

OIL PAINT

Oil paints are those paints containing a drying oil, oil varnish, or oil modified resin as the vehicle.

ORGANIC

Organic is the branch of chemistry which treats of the hydrocarbons and their derivatives, or the chemistry of the carbon compounds, whether natural or of synthetic origin.

OXIDANT

See "Oxidation."

OXIDATION

Oxidation is the increase in positive valence or decrease in negative valence of any element in a substance. On the basis of the electron theory, oxidation is a process in which an element loses electrons. In a narrow sense, oxidation means the chemical addition of oxygen to a substance.

An oxidizing agent (or oxidant) is a chemical agent that causes oxidation of other substances and is itself reduced.

It also refers to the corrosion of a metal exposed to an oxidizing gas at elevated temperatures.

OXIDIZING ACIDS

The cathodic reaction in an oxidizing acid is the reduction of the acidic anion rather than hydrogen evolution. The reactions of dilute to moderate concentrations of nitric acid on steel is to liberate the brown oxides of nitrogen. It should be noted that the nature of the metal also influences the metal/acid reaction. For example, boiling 55% sulfuric acid is a reducing acid to steel or type 300 stainless steel liberating hydrogen, while it is oxidizing to cast silicon–nickel alloy, the sulfate ion being reduced to sulfur dioxide and hydrogen sulfide (and elemental sulfur being formed as a consequence of their interaction). This is in accord with the generality that any "oxidizing agent" can be a reducing agent in the presence of a stronger oxidant.

Solutions of oxidizing acid salts (e.g., ammonium nitrate) act like dilute solutions of the parent acid.

In general, oxidizing acids tend to corrode metals that do not form a passive oxide film (e.g., copper, lead) rather than those that do form passive films such as chromium-bearing alloys, titanium, and aluminum.

Reducing acids at times are more aggressive to the normally passive metals than to the more active metals, because of the reaction of nascent hydrogen with the oxide film or the direct hydriding of the metal itself as with zirconium, titanium, and tantalum.

Some of the most important oxidizing acids are nitric acid, chromic acid, concentrated sulfuric acid, and oleum.

Nitric Acid

Nitric acid is a strong mineral acid and a powerful oxidizing agent, even in dilute solutions. Sixty percent concentration acid is produced by a process involving the oxidation of ammonia. To produce 70% concentration reagent grade acid (chemically pure), the 60% acid is purified and concentrated. Strong acid in the 90 to 100% range is produced by dehydrating weaker concentrations with concentrated sulfuric acid. Nitric acid above a concentration of 85% is known as fuming nitric acid because red or white fumes of oxides of nitrogen are evolved. Very strong concentrations of nitric acid have different corrosion properties than dilute concentrations.

Chromic Acid

Chromic acid is a very strong oxidizing acid but because of its nature more discrimination must be used when selecting materials of construction. Materials such as copper or nickel, which are attacked by oxidizing acids, are unsuitable.

Aluminum can handle a 10% concentration to approximately 150°F (66°C). Molybdenum free type 300 stainless steels may be used to a 30% concentration, but corrosion rates increase rapidly above 5% and 175°F (80°C).

Fluorinated plastics can handle chromic acid to their normal temperature limits, but conventional plastics (CPVC, PE, PP) are limited to not more than 50% acid at or below 160°F (70°C).

Concentrated Sulfuric Acid

Sulfuric acid in the 70 to 100% range is considered concentrated and as such is an oxidizing acid. Below 70% concentration it tends to act as a reducing acid. The oxidizing or reducing effect of sulfuric acid is also dependent upon the material to which it is exposed. For example, it starts to have an ozidizing effect at 25% when exposed to nickel or alloy 400 at the boiling point. At 60% and 175°F (80°C) it will carbonize PVDC over a prolonged period of time. At 95% and 77°F (25°C) it will carbonize FRP instantly.

Oleum

Oleum is 100% sulfuric acid with dissolved sulfur trioxide and is generally designated as for example 20% oleum, which is equivalent to 104.5% sulfuric acid if all of the sulfur trioxide is converted.

Compatibility of Various Materials of Construction with the Oxidizing Acids

Aluminum

Aluminum and its alloys are resistant to concentrations of nitric acid in excess of 80% at room temperature and to fuming acids in the 93 to 96% concentration range to a maximum temperature of 110°F (43°C). Higher temperatures can be employed above 96% concentration. Since weaker concentrations than 80% will rapidly attack aluminum, it is important that no moisture (moist air) be allowed to come into contact with the aluminum.

A 10% concentration of chromic acid to a temperature of approximately 150°F (66°C) can be handled satisfactorily by aluminum.

Sulfuric acid at concentrations greater than 96% can be handled in aluminum, even at elevated temperature. However, great care must be taken against dilution.

Iron and Steel

Dilute nitric acid attacks cast iron and steel rapidly. Even passivated iron and steel experience corrosion rates in excess of what would be considered acceptable for practical applications.

Conventional steel and cast irons are attacked by chromic acid solutions and should not be used.

Concentrations of sulfuric acid above 90% can be handled in carbon steel and cast iron. Care must be taken to prevent dilution.

Carbon steel is compatible with 20 to 30% concentration of oleum.

Silicon Cast Irons

Silicon cast irons form an adherent siliceous film when exposed to nitric acid above 45% concentration even up to the atmospheric boiling point. As the acid concentration increases, the corrosion resistance increases. In high concentrations at high temperatures, the corrosion rate is essentially nil.

Chromic acid up to concentrations of 50% can be handled at a temperature up to 200°F (93°C).

Silicon cast irons are resistant to concentrated sulfuric acid up to the boiling point. These materials would replace ordinary iron or steel above 122°F (50°C).

High silicon iron should not be exposed to oleum.

Stainless Steels

Because of the problem of intergranular attack, the low carbon type 304L or carbon stabilized grades Type 347 of the molybdenum free austenitic stainless steels should be used for handling all grades of nitric acid. If hexavalent chromium ions accumulate in the nitric acid to some concentration level, intergranular attack can take place regardless of the composition of the stainless steel.

Wrought type 316L stainless steel should not be used in nitric acid service because of the corrosion of the sigma phase.

Type 304 stainless steel is resistant to chromic acid in concentrations up to 50%. Concentrations up to 10% can be handled at temperatures up to 200°F (93°C). Concentrations exceeding 10% are limited to a maximum temperature of 175°F (80°C). However type 304 stainless steel is subject to pitting.

Type 316 stainless steel has a greater corrosion rate than type 304 at the comparable concentration and temperature and is subject to crevice corrosion.

Cold concentrated sulfuric acid can be handled by the 300 series austenitic stainless steels. The molybdenum free grades (type 304 and 304L) should be used in concentrations above 93% and in oleum. Higher alloys, such as type 309, are successfully used at elevated temperatures.

Titanium

Titanium is resistant to nitric acid below a concentration of 25% and in the 65 to 90% concentration range at the atmospheric boiling point. Boiling acid within the concentration range of 25 to 50% will corrode titanium at the rate of 10 mpy. Although titanium is resistant to fuming nitric acid, if the water content drops below 1.3% or the nitrogen dioxide concentration exceeds 6% there is the danger of a violent pyrophoric reaction.

Chromic acid can be handled up to 50% concentration to a temperature of 212°F (100°C).

Concentrated sulfuric acid will attack titanium.

Other Metals

With the exception of the chromium-bearing nickel alloys such as alloy 600, 625, or C-276, other nickel and copper alloys are rapidly attacked by even dilute nitric acid. Lead will also be attacked by nitric acid.

In the concentration range of 60 to 90% nitric acid, zirconium has a better resistance than titanium, but at concentrations greater than 70% it is subject to stress corrosion cracking and fails rapidly in boiling 94% nitric acid.

Of the noble metals gold and platinum are resistant, but silver will corrode.

Zirconium and tantalum are resistant to chromic acid up to 50% concentration at 212°F (100°C).

In the absence of chloride ion contamination, magnesium is highly resistant to chromic acid. A boiling 20% solution will not attack magnesium.

Tin will resist chromic acid to approximately 80% at 212°F (100°C).

Lead has a corrosion rate of <20 mpy in chromic acid at temperatures up to 200°F (93°C).

Tantalum will resist 95% sulfuric acid to 350°F (175°C) and lower concentrations to the atmospheric boiling point. Since tantalum is attacked by sulfur trioxide it cannot be used in the presence of oleum.

Zirconium will be attacked by concentrated sulfuric acid.

Nonmetallic Materials

PTFE plain or glass filled, and viton, are employed in nitric acid service. Carbon, if it is free of oxidizable binders, can also be used. Materials such as FEP, PVC, polypropylene, polyethylene, butyl rubber, and other elastomeric materials are also satisfactory with limitations. Refer to Ref. 1 for more complete details.

Most thermoplastic materials are compatible with chromic acid up to 50% concentration at a temperature of 160°F (70°C). Halogenated polyesters can also be used. Reference 1 provides more complete information on elastomers, thermoplastics, thermoset resins, and other nonmetallic materials.

Sulfuric acid concentrations up to 75% to 120°F (50°C) are compatible with thermoplastic materials such as polyethylene, polypropylene, and polyvinyl chloride, and 90% acid at 85°F (30°C). Carbonization occurs above these limits. Thermoset plastics with the proper resin will withstand 75% acid up to 77°F (25°C).

Fluorinated plastics such as PTFE, PFA, FEP, and PVDF are compatible with all concentrations of concentrated acid up to the maximum service temperature of the plastic. Oleum service is questionable because of permeability problems.

Pure carbon resists boiling 100% sulfuric acid and resists 115% acid to 160°F (70°C).

Elastomers other than the fluorinated variety are limited to a maximum of 75% sulfuric acid up to a temperature of 175°F (80°C).

Kalrez and viton can be used for oleum service.

See Ref. 1.

OXIDIZING AGENT

See "Oxidation."

OXYGEN CONCENTRATION CELL

Concentration cells occur when the concentration of identical ions differs from one region to another in the system. One of the common cells of this type is the oxygen concentration cell, which is a Galvanic cell caused by a difference in oxygen concentration at two points on a metal surface. The effect is the same as Galvanic action between two dissimilar metals, which can cause and accelerate pitting. Oxygen concentration cells are also known as differential aeration cells.

See Refs. 2 and 3.

OZHENNITE ALLOYS

Ozhennite alloys are zirconium alloys containing tin, iron, nickel, and niobium with a total alloy content of 0.5 to 1.5%. They were originally developed in the Soviet Union for use in pressurized water and steam in nuclear operations. Researchers at Atomic Energy of Canada Ltd. took a lead from the Russian zirconium–niobium alloys and developed the zirconium–2.5% niobium alloy. This alloy is strong and heat treatable.

OZONE

The ozone molecule consists of three atoms of oxygen (O_3) and is a powerful oxidizing allotropic form of oxygen.

REFERENCES

1. PA Schweitzer. Corrosion Resistance Tables. 4th ed. Vol. 1–3. New York: Marcel Dekker, 1995.
2. DM Berger. Fundamentals and prevention of metallic corrosion. In: PA Schweitzer, ed. Corrosion and Corrosion Protection Handbook. 2nd ed. New York: Marcel Dekker, 1989, pp. 1–22.
3. GT Murray. Introduction to Engineering Materials. New York: Marcel Dekker, 1993.

P

PAINT

See "Coatings."

PARTING

This is similar to dezincification in that one or more active components of the alloy corrode preferentially as in dezincification with the same results. Copper base alloys containing aluminum are subject to this reaction, with the aluminum corroding preferentially. Refer to "Dezincification."

See Ref. 1.

PASSIVATION

Metals, upon exposure to the atmosphere, form a protective oxide film on the exposed surface. This film, as long as it remains intact, will protect the metal from further corrosion. Some fabrication processes can prevent or retard the reformation of this protective film. To insure the reformation of the protective film the metal is subjected to "passivation" treatments.

For stainless steels the most common passivation treatment is to expose the metal to an oxidizing acid such as nitric and nitric–hydrofluoric acids.

See Refs. 2 and 3.

PASSIVE

State of a metal surface characterized by low corrosion rates in a potential region that is strongly oxidizing to the metal.

PASSIVE FILMS

All metals develop a diffusion barrier layer of reaction products on the surface, which is referred to as a passive film. These reaction products are either metal oxides or other compounds. The resistance of these films to dissolution is related to their physical and chemical nature, which determines the corrosion resistance of the metal. Other factors that influence the rate of metallic corrosion are pH, temperature, and anion content of the solution.

There are two theories regarding the formation of these films. The first theory states that the film formed is a metal oxide or other reaction compound. This is known as the "oxide-film theory." The second theory states that oxygen is adsorbed on the surface forming a chemisorbed film. However, all chemisorbed films react over a period of time with the underlying metal to form metal oxides. Oxide films are formed at room temperature. Metal oxides can be classified as network formers, intermediates, or modifiers. This division can be related to thin oxide films on metals. The metals that fall into the network forming or intermediate classes tend to grow protective oxides that support anion or mixed anion–cation movement. The network formers are noncrystalline, while the intermediates tend to be microcrystalline at low temperatures.

Passive Film on Iron

Iron in iron oxides can assume a valence of two or three. The former acts as a modifier and the latter as a network former. The iron is protected from the corrosive environment by a thin oxide film 1–4 mm in thickness with a composition of $\surd$ Fe_2O_3/Fe_3O_4. This is the same type of film formed by the reaction of clean iron with oxygen or dry air. The $\surd$ Fe_2O_3 layer is responsible for the passivity while the Fe_3O_4 provides the basis for the formation of a higher oxidizing state. Iron is more difficult to passivate than nickel, because with iron it is not possible to go directly to the passivation species $\surd$ Fe_2O_3. Instead, a lower oxidation state film of Fe_3O_4 is required, and this film is highly susceptible to chemical dissolution. The $\surd$ Fe_2O_3 layer will not form until the Fe_3O_4 phase has existed on the surface for a reasonable period of time. During this time the Fe_3O_4 layer continues to form.

Passive Film on Nickel

The passive film on nickel can be achieved quite readily in contrast to the formation of the passive film on iron. Differences in the nature of the oxide film on iron and nickel are responsible for this phenomenon. The film thickness on nickel is between 0.9 and 1.2 mm, while iron oxide film is between 1.5 and 4.5 mm. There are two theories as to exactly what the passive film on nickel is. It is either entirely NiO with a small amount of nonstoichiometry giving rise to Ni^{+3} and cation vacancies or it consists of an inner layer of NiO and an outer hydrous layer of $Ni(OH)_2$. The passive oxide film on nickel once formed cannot be easily removed by either cathodic treatment or chemical dissolution.

Passive Film on Austenitic Stainless Steel

The passive film formed on austenitic stainless steel is duplex in nature, consisting of an inner barrier oxide film and an outer deposit hydroxide or salt film. Passivation takes place by the rapid formation of surface-absorbed hydrated complexes of metals, which are sufficiently stable on the alloy surface that further reaction with water enables the formation of a hydroxide phase that rapidly deprotonates to form an insoluble surface oxide film. The three most commonly used austenite stabilizers, nickel, manganese, and nitrogen, all contribute to the passivity. Chromium, a major alloying ingredient, is in itself very corrosion resistant and is found in greater abundance in the passive film than iron, which is the majority element in the alloy.

See Refs. 1, 4–6.

PASSIVE METAL

A passive metal is one that substantially resists corrosion in a given environment resulting from marked anodic polarization.

PATENTING

See "Annealing."

PATINA

When exposed to the atmosphere over long periods of time, copper will form a coloration on the surface known as patina which in reality is a corrosion product that acts as a protective film against further corrosion. When first formed, the patina has a dark color that gradually turns green. The length of time required to form the patina depends upon the atmosphere, because the coloration is given by copper hydroxide compounds. In a marine atmosphere, the compound is a mixture of copper/hydroxide/chloride and in urban or industrial atmospheres copper/hydroxide/sulfate. These compounds will form in approximately seven years. When exposed in a clean rural atmosphere, tens or hundreds of years may be required to form the patina.

PEARLITE

Although most carbon steels contain enough carbon to allow hardening by heat treatment, they are not intended to be hardened by heat treatment. They are normally produced with a more ductile, lower strength microstructure that forms during cooling from austenitic temperatures. This microstructure is a mixture of ferrite and pearlite. As the carbon steel is cooled from the austenitic temperature ferrite starts to form. Since the ferrite contains essentially no carbon, it leaves behind an increasing concentration of carbon in the remaining austenite, which is eventually ejected. Under normal circumstances the excess carbon combines with iron to form iron carbide (Fe_3C), called cementite. If the austenite is cooled slowly in air, a binary mixture of ferrite and cementite is formed, which is called pearlite. The structure of pearlite is lamellar, consisting of very fine alternating layers of ferrite and cementite. Consequently the gross microstructure of normal carbon steel consists of a mixture of ferrite and pearlite.

Table P.1 Physical and Mechanical Properties of PFA

Specific gravity	2.12–2.17
Water absorption, 24 h at 73°F/23°C, %	< 0.03
Tensilde strength:	
at 73°F/23°C psi	4000
at 482°F/250°C psi	2000
Modulus of elasticity in tension, psi:	
at 73°F/23°C	40,000
at 482°F/250°C	6000
Compressive strengh, psi	
at 73°F/23°C psi	3500
at 320°F/196°C	60,000
Flexural modulus, psi	
at 73°F/23°C	90,000
at 482°F/250°C	10,000
Izod impact, notched at 73°F/23°C, ft-lb/in.	No break
Coefficient of linear thermal expansion, in./in. °F	
at 70–212°F/20–100°C	7.8×10^{-5}
at 212–300°F/100–150°C	9.8×10^{-5}
at 300–480°F/150–210°C	12.1×10^{-5}
Heat distortion temperature, °F/°C:	
at 66 psi	164/73
at 264 psi	118/48
Limiting oxygen index, %	< 95
Flame spread	10
Underwriters lab rating, Sub 94	94-V-O

PERFLUOROALKOXY (PFA)

PFA is a fully fluorinated thermoplast having the formula

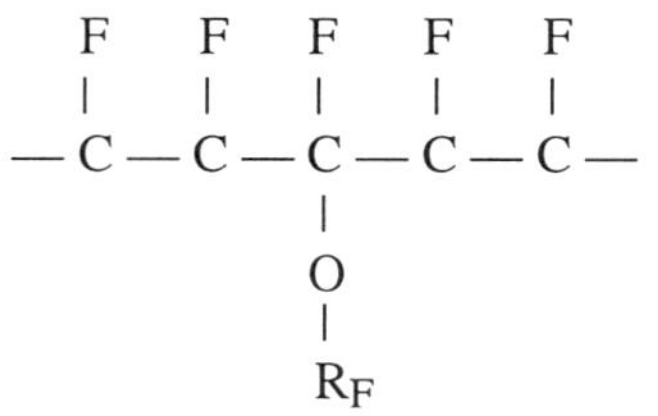

in which $R_F = C_nF_{2n} + 1$. Perfluoroalkoxy lacks the physical strength of PTFE at elevated temperatures but has somewhat better physical and mechanical properties than FEP above 300°F (149°C) and can be used up to 500°F (260°C). For example, PFA has reasonable tensile strength at 68°F (20°C), but its heat deflection temperature is the lowest of all the fluoroplastics. While PFA matches the hardness and impact strength of PTFE, it sustains only one quarter of the life of PTFE in flexibility tests. Refer to Table P.1 for the physical and mechanical properties of PFA.

Like PTFE, PFA is subject to permeation by certain gases and will absorb selected chemicals. Perfluoroalkoxy also performs well at cryogenic temperatures. Table P.2 compares the mechanical properties of PFA at room temperature and cryogenic temperatures.

Table P.2 Comparison of Mechanical Properties of PFA at Room Temperature and Cryogenic Temperature

	Temperature	
Property	73°F/23°C	320°F/190°C
Yield strength, psi	2100	No yield
Ultimate tensile strength, psi	2600	18,700
Elongation, %	260	8
Flexural modulus, psi	81,000	840,000
Izod impact strength, notched, ft-lb/in.	No break	12
Compressive strength, psi	3500	60,000
Compressive strain, %	20	35
Modulus of elasticity, psi	10,000	680,000

PFA is inert to strong mineral acids, organic bases, inorganic oxidizers, aromatics, some aliphatic hydrocarbons, alcohols, aldehydes, ketones, ethers, esters, chlorocarbons, fluorocarbons, and mixtures of these.

PFA will be attacked by certain halogenated complexes containing fluorine. This includes chlorine trifluoride, bromine trifluoride, iodine pentafluoride, and fluorine. It can also be attacked by such metals as sodium or potassium, particularly in their molten state. Refer to Table P.3 for the compatibility of PFA with selected corrodents.

Refer to Ref. 8 for the compatibility of PFA with a wide variety of selected corrodents.

PERFLUOROELASTOMERS (FPM)

Perfluoroelastomers provide the elastomeric properties of fluoroelastomers and the chemical resistance of PTFE. These compounds are true rubbers. Compared with other elastomeric compounds they are more resistant to swelling and embrittlement and retain their elastomeric properties over the long term. In difficult environments there are no other elastomers that can outperform the perfluoroelastomers. These synthetic rubbers provide the sealing force of a true elastomer and the chemical inertness and thermal stability of polytetrafluoroethylene.

As with other elastomers, perfluoroelastomers are compounded to modify certain of their properties. Such materials as carbon black, perfluorinated oil, and various fillers are used for this purpose.

The ASTM designation for these elastomers is FPM.

Physical and Mechanical Properties

One of the outstanding physical properties of the perfluoroelastomers is their thermal-stability. They retain their elasticity and recovery properties up to 600°F (316°C) in long term service and up to 650°F (343°C) in intermittant service. This is the highest temperature rating of any elastomer.

In general the physical properties of the perfluoroelastomers are similar to those of

Table P.3 Compatiblity of PFA with Selected Corrodents[a]

Chemical	Maximum temp. °F	°C	Chemical	Maximum temp. °F	°C
Acetaldehyde	450	232	Benzene[b]	450	232
Acetamide	450	232	Benzene sulfonic acid 10%	450	232
Acetic acid 10%	450	232	Benzoic acid	450	232
Acetic acid 50%	450	232	Benzyl alcohol[c]	450	232
Acetic acid 80%	450	232	Benzyl chloride[b]	450	232
Acetic acid, glacial	450	232	Borax	450	232
Acetic anhydride	450	232	Boric acid	450	232
Acetone	450	232	Bromine gas, dry[b]	450	232
Acetyl chloride	450	232	Bromine liquid[b,c]	450	232
Acrylonitrile	450	232	Butadiene[b]	450	232
Adipic acid	450	232	Butyl acetate	450	232
Allyl alcohol	450	232	Butyl alcohol	450	232
Allyl chloride	450	232	*n*-Butylamine[c]	450	232
Alum	450	232	Butyl phthalate	450	232
Aluminum chloride, aqueous	450	232	Butyric acid	450	232
Aluminum fluoride	450	232	Calcium bisulfide	450	232
Aluminum hydroxide	450	232	Calcium bisulfite	450	232
Aluminum nitrate	450	232	Calcium carbonate	450	232
Aluminum oxychloride	450	232	Calcium chlorate	450	232
Aluminum sulfate	450	232	Calcium chloride	450	232
Ammonia gas[b]	450	232	Calcium hydroxide 10%	450	232
Ammonium bifluoride[b]	450	232	Calcium hydroxide, sat.	450	232
Ammonium carbonate	450	232	Calcium hypochlorite	450	232
Ammonium chloride 10%	450	232	Calcium nitrate	450	232
Ammonium chloride 50%	450	232	Calcium oxide	450	232
Ammonium chloride, sat.	450	232	Calcium sulfate	450	232
Ammonium fluoride 10%[b]	450	232	Caprylic acid	450	232
Ammonium fluoride 25%[b]	450	232	Carbon bisulfide[b]	450	232
Ammonium hydroxide 25%	450	232	Carbon dioxide, dry	450	232
Ammonium hydroxide, sat.	450	232	Carbon dioxide, wet	450	232
Ammonium nitrate	450	232	Carbon disulfide[b]	450	232
Ammonium persulfate	450	232	Carbon monoxide	450	232
Ammonium phosphate	450	232	Carbon tetrachloride[b,c,d]	450	232
Ammonium sulfate 10–40%	450	232	Carbonic acid	450	232
Ammonium sulfide	450	232	Chloracetic acid, 50% water	450	232
Amyl acetate	450	232	Chloracetic acid	450	232
Amyl alcohol	450	232	Chlorine gas, dry	x	x
Amyl chloride	450	232	Chlorine gas, wet[b]	450	232
Aniline[c]	450	232	Chlorine, liquid[c]	x	x
Antimony trichloride	450	232	Chlorobenzene[b]	450	232
Aqua regia 3:1	450	232	Chloroform[b]	450	232
Barium carbonate	450	232	Chlorosulfonic acid[c]	450	232
Barium chloride	450	232	Chromic acid 10%	450	232
Barium hydroxide	450	232	Chromic acid 50%[c]	450	232
Barium sulfate	450	232	Chromyl chloride	450	232
Barium sulfide	450	232	Citric acid 15%	450	232
Benzaldehyde[c]	450	232	Citric acid, concentrated	450	232

Table P.3 Continued

Chemical	Maximum temp. °F	Maximum temp. °C	Chemical	Maximum temp. °F	Maximum temp. °C
Copper carbonate	450	232	Muriatic acid[b]	450	232
Copper chloride	450	232	Nitric acid 5%[b]	450	232
Copper cyanide	450	232	Nitric acid 20%[b]	450	232
Copper sulfate	450	232	Nitric acid 70%[b]	450	232
Cresol	450	232	Nitric acid, anhydrous[b]	450	232
Cupric chloride 5%	450	232	Nitrous acid 10%	450	232
Cupric chloride 50%	450	232	Oleum	450	232
Cyclohexane	450	232	Perchloric acid 10%	450	232
Cyclohexanol	450	232	Perchloric acid 70%	450	232
Dichloroacetic acid	450	232	Phenol[b]	450	232
Dichloroethane	450	232	Phosphoric acid 50–80%[c]	450	232
(ethylene dichloride)[b]			Picric acid	450	232
Ethylene glycol	450	232	Potassium bromide 30%	450	232
Ferric chloride	450	232	Salicylic acid	450	232
Ferric chloride 50%			Sodium carbonate	450	232
in water[c]	450	232	Sodium chloride	450	232
Ferric nitrate 10–50%	450	232	Sodium hydroxide 10%	450	232
Ferrous chloride	450	232	Sodium hydroxide 50%	450	232
Ferrous nitrate	450	232	Sodium hydroxide,		
Fluorine gas, dry	x	x	concentrated	450	232
Fluorine gas, moist	x	x	Sodium hypochlorite 20%	450	232
Hydrobromic acid, dilute[b,d]	450	232	Sodium hypochlorite,		
Hydrobromic acid 20%[b,d]	450	232	concentrated	450	232
Hydrobromic acid 50%[b,d]	450	232	Sodium sulfide to 50%	450	232
Hydrochloric acid 20%[b,d]	450	232	Stannic chloride	450	232
Hydrochloric acid 38%[b,d]	450	232	Stannous chloride	450	232
Hydrocyanic acid 10%	450	232	Sulfuric acid 10%	450	232
Hydrofluoric acid 30%[b]	450	232	Sulfuric acid 50%	450	232
Hydrofluoric acid 70%[b]	450	232	Sulfuric acid 70%	450	232
Hydrofluoric acid 100%[b]	450	232	Sulfuric acid 90%	450	232
Hypochlorous acid	450	232	Sulfuric acid 98%	450	232
Iodine solution 10%[b]	450	232	Sulfuric acid 100%	450	232
Ketones, general	450	232	Sulfuric acid, fuming[b]	450	232
Lactic acid 25%	450	232	Sulfurous acid	450	232
Lactic acid, concentrated	450	232	Thionyl chloride[b]	450	232
Magnesium chloride	450	232	Toluene[b]	450	232
Malic acid	450	232	Trichloroacetic acid	450	232
Methyl chloride[b]	450	232	White liquor	450	232
Methyl ethyl ketone[b]	450	232	Zinc chloride[c]	450	232
Methyl isobutyl ketone[b]	450	232			

[a]The chemicals listed are in the pure state or in a saturated solution unless otherwise indicated. Compatibility is shown to the maximum allowable temperature for which data are available. Incompatibility is shown by an x. A blank space indicates that the data are unavailable.
[b]Material will permeate.
[c]Material will be absorbed.
[d]Material will cause stress cracking.
Source: Schweitzer, PA. *Corrosion Resistance Tables,* 4th ed. Vols 1–3. New York: Marcel Dekker, 1995.

Table P.4 Physical and Mechanical Properties of Perfluoroelastomers[a]

Specific gravity	1.9–2.0
Specific heat at 122–302°F (50–150°C), cal/g	0.226–0.250
Brittle point	–9 to –58°F (–23 to –50°C)
Coefficient of friction (to steel)	0.25–0.60
Tear strength, psi	1.75–27
Coefficient of linear expansion	
in./°F	1.3×10^{-4}
in./°C	2.3×10^{-4}
Thermal conductivity, BTU-in./hr-°F-ft^2	
at 122°F (50°C)	1.3
at 212°F (100°C)	1.27
at 392°F (200°C)	1.19
at 572°F (300°C)	1.10
Dielectric constant, kV/mm	17.7
Dielectric constant at 1000 Hz	4.9
Dissipation factor at 1000 Hz	5×10^{-3}
Permeability ($\times 10^{-9}$ cm^3-cm/S-cm^2-cmHg P)	
to nitrogen, at room temperature	0.05
to oxygen, at room temperature	0.09
to helium, at room temperature	2.5
to hydrogen, at 199°F (93°C)	113
Tensile strength, psi	1850–3800
Elongation, % at break	20–190
Hardness, Shore A	65–95
Maximum temperature, continuous use	600°F (316°C)
Abrasion resistance, NBS	121
Compression set, %	
at room temperature	15–40
at 212°F (100°C)	32–54
at 400°F (204°C)	63–82
at 500°F (260°C)	63–79
Resistance to sunlight	Excellent
Effect of aging	Nil
Resistance to heat	Excellent

[a]These are representative values since they may be altered by compounding.

the fluoroelastomers. As with most compounding, the enhancement of one property usually has the opposite effect on another property. For example, as the coefficient of friction increases the hardness decreases. Because of these factors the physical and mechanical properties given in Table P.4 are in ranges where they are compound dependent. Special compounds are available for applications requiring thermal cycling, increased resistance to strong oxidizing environments, other pressures, different hardnesses, or other specific physical properties. Selection of specific compounds for special applications should be done in cooperation with the manufacturer.

These elastomers provide excellent performance in high vacuum environments. They exhibit negligible outgassing over a wide temperature range. This is an important property in any application where freedom from contamination of process streams is critical.

Typical applications include semiconductor manufacturing operations, aerospace applications, and use in analytical instruments.

The perfluoroelastomers have good mechanical properties. See Table P.4. Because these elastomers are based on expensive monomers and require complex processing, they cost more than other elastomeric materials. As a result these materials are limited to use in extremely hostile environments and/or applications where high heat will quickly attack other elastomers. Perfluoro rubbers retain their elastic properties in long-term service at elevated temperatures.

Resistance to Sun, Weather, and Ozone

The perfluoroelastomers provide excellent resistance to sun, weather, and ozone. Long-term exposure under these conditions has no effect on them.

Chemical Resistance

The perfluoroelastomers have outstanding chemical resistance. They are virtually immune to chemical attack at ambient and elevated temperatures. Typical corrodents to which the perfluoroelastomers are resistant include the following.

Polar solvents (ketones, esters, ethers)
Strong organic solvents (benzene, dimethyl formamide, perchloroethylene, tetrahydrofuran)
Inorganic and organic acids (hydrochloric, nitric, sulfuric, trichloroacetic) and bases (hot caustic soda)
Strong oxidizing agents (dinitrogen tetroxide, fuming nitric acid)
Metal halides (titanium tetrachloride, diethyl aluminum chloride)
Hot mercury
Chlorine, wet and dry
Inorganic salt solutions
Fuels (ASTM Reference Fuel C, JP-5 jet fuel, aviation gas, kerosene)
Hydraulic fluids
Heat transfer fluids
Oil well sour gas (methane, hydrogen sulfide, carbon dioxide, steam)
Steam

These perfluoroelastomers should not be exposed to molten or gaseous alkali metals such as sodium because a highly exothermic reaction may occur. Service life can be greatly reduced in fluids containing high concentrations of some diamines, nitric acid, and basic phenol when the temperature exceeds 212°F (100°C). Uranium hexafluoride and fully halogenated freons (F-11 and F-12) cause considerable swelling.

The corrosion resistance as given above is for the base polymer. Since the polymer is quite often compounded with fillers and curatives, these additives may interact with the environment even though the polymer is resistant. Therefore a knowlege of the additives present is essential in determining the material's suitability for a particular application. A corrosion testing program is the best method whereby this evaluation can be undertaken. Refer to Table P.33 for the compatibility of PTFE with selected corrodents.

Applications

Perfluoroelastomer parts are a practical solution wherever the sealing performance of rubber is desirable but not feasible because of severe chemical or thermal conditions. In the petrochemical industry FPM is widely used for O-ring seals on equipment. O-rings of FPM are employed in mechanical seals, pump housings, compressor casings, valves, rotameters, and other instruments. Custom-molded parts are also used as valve seats, packings, diaphragms, gaskets, and miscellaneous sealing elements including U-cups and V-rings.

Other industries where FPM contributes importantly are aerospace (versus jet fuels, hydrazine, N_2O_4 and other oxidizers, Freon-21 fluorocarbon, etc.); nuclear power (versus radiation, high temperatures); oil, gas, and geothermal drilling (versus sour gas, acidic fluids, amine-containing hydraulic fluids, extreme temperatures and pressures); and analytical and process instruments (versus high vacuum, liquid and gas chromotography exposures, high purity reagents, high temperature conditions).

The semiconductor industry makes use of FPM O-rings to seal the aggressive chemical reagents and specialty gases required for producing silicon chips. Also the combination of thermal stability and low outgassing characteristics are desirable in furnaces for growing crystals and in high vacuum applications.

The chemical transportation industry is also a heavy user of FPM components in safety relief and unloading valves to prevent leakage from tank trucks and trailers, rail cars, ships, and barges carrying hazardous and corrosive chemicals.

Other industries that also use FPM extensively include pharmaceuticals, agricultural chemicals, oil and gas recovery, and analytical and process control instrumentation.

Because of their cost, the perfluoroelastomers are used primarily as seals where their corrosion and/or heat resistance properties can be utilized and other elastomeric materials will not do the job or where high maintenance costs result if other elastomeric materials are used.

See Refs. 7 and 8.

PERMEATION

Also see "Sheet Linings."

All materials are somewhat permeable to chemical molecules, but plastic materials tend to be an order of magnitude greater in their permeability rates than metals. Gases, vapors, or liquids will permeate polymers.

Permeation is a molecular migration either through microvoids in the polymer (if the polymer is more or less porous) or between polymer molecules. In neither case is there any attack on the polymer. This action is strictly a physical phenomenon. In lined equipment permeation can result in

1. Failure of the substrate from corrosive attack.
2. Bond failure and blistering, resulting from accumulation of fluids at the bond when the substrate is less permeable than the liner, or from corrosion/reaction products if the substrate is attacked by the permeant.
3. Loss of contents through substrate and liner as a result of the eventual failure of

the substrate. In unbonded linings it is important that the space between the liner and support member be vented to the atmosphere, not only to allow minute quantities of permeant vapors to escape but also to prevent expansion of entrapped air from collapsing the liner.

Permeation is a function of two variables, one relating to diffusion between molecular chains and the other to the solubility of the permeant in the polymer. The driving force of diffusion is the partial pressure gradient for gases, and the concentration gradient for liquids. Solubility is a function of the affinity of the permeant for the polymer.

There is no relation between permeation and the passage of materials through cracks and voids, even though in both cases migrating chemicals travel through the polymer from one side to the other.

The user has some control over permeation, which is affected by

1. Temperature and pressure
2. Permeant concentration
3. Thickness of the polymer

Increasing the temperature will increase the permeation rate, since the solubility of the permeant in the polymer will increase, and as the temperature rises the polymer chain movement is stimulated, permitting more permeants to diffuse among the chains more easily. For many gases, the permeation rates increase linearly with the partial pressure gradient, and the same effect is experienced with the concentration gradients of liquids. If the permeant is highly soluble in the polymer, the permeability increase may be nonlinear. The thickness will generally decrease permeation by the square of the thickness. For general corrosion resistance, thicknesses of 0.010–0.020 inch are usually satisfactory.

The density of the polymer, as well as the thickness, will have an effect on the permeation rate. The greater the density of the polymer, the fewer voids through which permeation can take place. A comparison of the density of sheets produced from different polymers does not provide any indication of the relative permeation rates. However, a comparison of the density of sheets produced from the same polymer will provide an indication of the relative permeation rates. The denser the sheet, the lower the permeation rate.

Thickness of lining is a factor affecting permeation. For general corrosion resistance, thicknesses of 0.010–0.020 in. are usually satisfactory, depending upon the combination of elastomeric material and specific corrodent. When mechanical factors such as thinning due to cold flow, mechanical abuse, and permeation rates are a consideration, thicker linings may be required.

Increasing the lining thickness will normally decrease permeation by the square of the thickness. Although this would appear to be the approach to follow to control permeation, there are disadvantages. First, as the thickness increases, the thermal stresses on the boundary increase, which can result in bond failure. Temperature changes and large differences in coefficients of thermal expansion are the most common causes of bond failure. Thickness and modulus of elasticity of the elastomer are two of the factors that would influence these stresses. Second, as the thickness of the lining increases, installation becomes more difficult, with a resulting increase in labor costs.

The rate of permeation is also affected by temperature and temperature gradient in

Table P.5 Vapor Permeation of PTFE

Gases	Permeation (g/100 in.^{2}/24 h/mil) at 73°F (23°C)	86°F (30°C)
Carbon dioxide		0.66
Helium		0.22
Hydrogen chloride; anh.		< .01
Nitrogen		0.11
Acetophenone	0.56	
Benzene	0.36	0.80
Carbon tetrachloride	0.06	
Ethyl alcohol	0.13	
Hydrochloric acid 20%	< 0.01	
Piperdine	0.07	
Sodium hydroxide 50%	5×10^{-5}	
Sulfuric acid 98%	1.8×10^{-5}	

Based on PTFE having a specific gravity > 2.2.

the lining. Lowering these will reduce the rate of permeation. Lined vessels that are used under ambient conditions, such as storage tanks, provide the best service.

Other factors affecting permeation consisting of chemical and physiochemical properties are

1. Ease of condensation of the permeant: chemicals that condense readily will permeate at higher rates.
2. The higher the intermolecular chain forces (e.g., van der Waals hydrogen bonding) of the polymer the lower the permeation rate.
3. The higher the level of crystallinity in the polymer, the lower the permeation rate.
4. The greater the degree of cross-linking within the polymer, the lower the permeation rate.
5. Chemical similarity between the polymer and permeant. When the polymer and permeant both have similar functional groups, the permeation rate will increase.
6. The smaller the molecule of the permeant, the greater the permeation rate.

Vapor permeation of PTFE, FEP, and PFA are shown in Tables P.5 P.6, and P.7.

pH

pH is the negative logarithm of the hydrogen ion concentration of a solution. It is a means of expressing the hydrogen ion concentration in terms of the powers of 10. A value of 7 is neutral, the lower the number the greater is the acidity, the higher the number the greater is the alkalinity.

Table P.6 Vapor Permeation of FEP

	Permeation (g/100 in.^{2}/24 h/mil) at		
	73°F (23°C)	95°F (35°C)	122°F (50°C)
Gases			
Nitrogen	0.18		
Oxygen	0.39		
Vapors			
Acetic acid		0.42	
Acetone	0.13	0.95	3.29
Acetophenone	0.47		
Benzene	0.15	0.64	
N-Butyl ether	0.08		
Carbon tetrachloride	0.11	0.31	
Decane	0.72		1.03
Ethyl acetate	0.06	0.77	2.9
Ethyl alcohol	0.11	0.69	
Hexane		0.57	
Hydrochloric acid 20%	< 0.01		
Methanol			5.61
Sodium hydroxide 50%	4×10^{-5}		
Sulfuric acid 98%	8×10^{-6}		
Toluene	0.37		2.93
Water	0.09	0.45	0.89

PHENOL-FORMALDEHYDE RESIN

This is one of the oldest systhetic materials available, having been in existance for more than 50 years. In general it does not have the impact resistance of the polyesters or epoxies.

Corrosion Resistance

It is generally recommended for service with mineral acids, salts, and chlorinated aromatic hydrocarbons. Refer to Table P.8 for the compatibility of phenol-formaldehyde with selected corrodents. Since these resins possess little alkaline and bleach resistance, application in such services should be avoided.

Table P.7 Permeation of Various Gases in PFA at 77°F (25°C)

Gas	Permeation (cc mil thickness/100 in.^{2} 24 h atm)
Carbon dioxide	2260
Nitrogen	291
Oxygen	881

Table P.8 Compatiblity of Phenol-Formaldehyde with Selected Corrodents[a]

Chemical	Maximum temp. °F	Maximum temp. °C	Chemical	Maximum temp. °F	Maximum temp. °C
Acetaldehyde	x	x	Ferric chloride 50% in water	300	149
Acetamide			Ferric nitrate 10–50%	300	149
Acetic acid 10%	212	100	Ferrous chloride 40%	300	149
Acetic acid 50%	160	71	Hydrobromic acid, dilute	212	100
Acetic acid 80%	120	49	Hydrobromic acid 20%	212	100
Acetic acid, glacial	120	49	Hydrochloric acid 20%	300	149
Acetone	x	x	Hydrochloric acid 38%	300	149
Acetyl chloride	x	x	Hydrocyanic acid 10%	160	71
Acrylic acid 90%	80	27	Hydrofluoric acid 30%	x	x
Acrylonitrile	x	x	Hydrofluoric acid 70%	x	x
Aluminum acetate			Hydrofluoric acid 100%	x	x
Aluminum chloride, aqueous	300	149	Hypochlorous acid		
Aluminum chloride, dry	300	149	Iodine solution 10%	x	x
Aluminum sulfate	300	149	Lactic acid 25%	160	71
Ammonium hydroxide 25%	x	x	Lactic acid, concentrated	160	71
Ammonium hydroxide, sat.	x	x	Methyl chloride	300	149
Amyl alcohol	160	71	Methyl ethyl ketone	x	x
Aniline	x	x	Methyl isobutyl ketone	x	x
Aqua regia 3:1	x	x	Muriatic acid	300	149
Benzene	160	71	Nitric acid 5%	x	x
Benzene sulfonic acid 10%	160	71	Nitric acid 20%	x	x
Benzyl chloride	160	71	Nitric acid 70%	x	x
Boric acid	300	149	Nitric acid, anhydrous	x	x
Bromine liquid 3% max.	300	149	Phenol	x	x
Butyric acid	260	127	Phosphoric acid 50%	212	100
Calcium chloride	300	149	Sodium carbonate	x	x
Calcium hydroxide 10%	x	x	Sodium hydroxide 10%	x	x
Calcium hydroxide, sat.	x	x	Sodium hydroxide 50%	x	x
Calcium hypochlorite	x	x	Sodium hydroxide, concentrated	x	x
Carbon bisulfide	160	71			
Carbon tetrachloride	212	100	Sodium hypochlorite 15%	x	x
Chlorine gas, dry	160	71	Sodium hypochlorite, concentrated	x	x
Chlorine gas, wet	160	71			
Chlorobenzene			Sulfuric acid 10%	300	149
Chloroform	160	71	Sulfuric acid 50%	300	149
Chlorosulfonic acid	80	27	Sulfuric acid 70%	250	121
Chromic acid 10%	x	x	Sulfuric acid 90%	100	38
Chromic acid 50%	x	x	Sulfurous acid	160	71
Copper sulfate	300	149	Thionyl chloride	80	27
Cresol			Toluene	212	100
Cupric chloride 5%	300	149	Trichloroacetic acid 30%	80	27
Cupric chloride 50%	300	149	Zinc chloride	300	149
Ethylene glycol	80	27			

[a]The chemicals listed are in the pure state or in a saturated solution unless otherwise indicated. Compatibility is shown to the maximum allowable temperature for which data are available. Incompatibility is shown by an x. A blank space indicates that the data are unavailable.

Source: Schweitzer, PA. *Corrosion Resistance Tables,* 4th ed. Vols 1–3. New York: Marcel Dekker, 1995.

Furan resins are produced from furfuryl alcohol and furfural. These resins are more expensive than other thermoset set resins but are the most economical choice when the presence of solvents exists in a combination with acids and bases or when process changes may occur that result in exposure to solvents in oxidizing atmospheres.

The furan laminates have the ability to retain their physical properties at elevated temperatures.

See Refs. 8–10.

PHENOLIC RESINS

These are the oldest commercial classes of polymers in use today. Although first discovered in 1872 it was not until 1907 after the "heat and pressure" patent was applied for by Leo H. Bakeland that the development of and application of phenolic molding compounds became economical.

The phenolics display excellent resistance to most organic solvents, particuarly chlorinated and aromatic solvents. These resins are not suitable for use with caustics, strong mineral acids (such as nitric, chromic, and hydrochloric). Refer to Table P.9 for the compatibility of phenolic resins with selected corrodents.

See Refs. 8 and 11.

PHOSPHATING

See "Coatings."

PITTING

Pitting corrosion is a form of localized corrosion, with the rate of attack being greater in some areas than in others. The main factors that cause and accelerate pitting are electrical contact between dissimilar metals or between areas of the same metal where oxygen or conductive salt concentrations in water differ.

Pitting develops when the anodic (corroding) area is small in relation to the cathodic (protected) area. For example, it can be expected when large areas of the surface are generally covered by mill scale, applied coatings, or deposits of various kinds but breaks exist in the continuity of the protective material. Pitting may also develop on bare clean metal surfaces because of irregularities in the physical or chemical structure of the metal. Localized dissimilar soil conditions at the surfaces of steel can also create conditions that promote pitting.

If appreciable attack is confined to a small area of metal acting as an anode, the developed pits are described as deep. If the area of attack is relatively large the pits are called shallow. The ratio of deepest metal penetration to average metal penetration as determined by weight loss of the specimen is known as the pitting factor. A pitting factor of one represents uniform corrosion.

Performance in the area of pitting and crevice corrosion is often measured using critical pitting temperatures, (CPT), critical crevice temperatures, (CCT), and pitting resistance equivalent numbers (PREN). As a general rule the higher the PREN,

Table P.9 Compatiblity of Phenolics with Selected Corrodents[a]

Chemical	Maximum temp. °F	°C	Chemical	Maximum temp. °F	°C
Acetaldehyde			Barium hydroxide		
Acetamide			Barium sulfate		
Acetic acid 10%	212	100	Barium sulfide		
Acetic acid 50%			Benzaldehyde	70	21
Acetic acid 80%			Benzene	160	71
Acetic acid, glacial	70	21	Benzene sulfonic acid 10%	70	21
Acetic anhydride	70	21	Benzoic acid		
Acetone	x	x	Benzyl alcohol		
Acetyl chloride			Benzyl chloride	70	21
Acrylic acid			Borax		
Acrylonitrile			Boric acid		
Adipic acid			Bromine gas, dry		
Allyl alcohol			Bromine gas, moist		
Allyl chloride			Bromine liquid		
Alum			Butadiene		
Aluminum acetate			Butyl acetate	x	x
Aluminum chloride, aqueous	90	32	Butyl alcohol		
Aluminum chloride, dry			*n*-Butylamine		
Aluminum fluoride			Butyl phthalate		
Aluminum hydroxide			Butyric acid	160	71
Aluminum nitrate			Calcium bisulfide		
Aluminum oxychloride			Calcium bisulfite		
Aluminum sulfate	300	149	Calcium carbonate		
Ammonia gas	90	32	Calcium chlorate		
Ammonia bifluoride			Calcium chloride	300	149
Ammonium carbonate	90	32	Calcium hydroxide 10%		
Ammonium chloride 10%	80	27	Calcium hydroxinde, sat.		
Ammonium chloride 50%	80	27	Calcium hypochlorite	x	x
Ammonium chloride, sat.	80	27	Calcium nitrate		
Ammonium fluoride 10%			Calcium oxide		
Ammonium fluoride 25%			Calcium sulfate		
Ammonium hydroxide 25%	x	x	Caprylic acid		
Ammonium hydroxide, sat.	x	x	Carbon bisulfide		
Ammonium nitrate	160	71	Carbon dioxide, dry	300	149
Ammonium persulfate			Carbon dioxide, wet	300	149
Ammonium phosphate			Carbon disulfide		
Ammonium sulfate 10–40%	300	149	Carbon monoxide		
Ammonium sulfide			Carbon tetrachloride	200	93
Ammonium sulfite			Carbonic acid	200	93
Amyl acetate			Cellosolve		
Amyl alcohol			Chloracetic acid, 50% water		
Amyl chloride			Chloracetic acid		
Aniline	x	x	Chlorine gas, dry		
Antimony trichloride			Chlorine gas, wet	x	x
Aqua regia 3:1			Chlorine, liquid	x	x
Barium carbonate			Chlorobenzene	260	127
Barium chloride			Chloroform	160	71

Table P.9 Continued

Chemical	Maximum temp. °F	°C	Chemical	Maximum temp. °F	°C
Chlorosulfonic acid			Manganese chloride		
Chromic acid 10%	x	x	Methyl chloride	160	71
Chromic acid 50%	x	x	Methyl ethyl ketone	x	x
Chromyl chloride			Methyl isobutyl ketone	160	71
Citric acid 15%	160	71	Muriatic acid	300	149
Citric acid, concentrated	160	71	Nitric acid 5%	x	x
Copper acetate			Nitric acid 20%	x	x
Copper carbonate			Nitric acid 70%	x	x
Copper chloride			Nitric acid, anhydrous	x	x
Copper cyanide			Nitrous acid, concentrated		
Copper sulfate	300	149	Oleum		
Cresol			Perchloric acid 10%		
Cupric chloride 5%			Perchloric acid 70%		
Cupric chloride 50%			Phenol	x	x
Cyclohexane			Phosphoric acid 50–80%	212	100
Cyclohexanol			Picric acid		
Dichloroacetic acid			Potassium bromide 30%		
Dichloroethane (ethylene dichloride)			Salicylic acid		
Ethylene glycol	70	21	Silver bromide 10%		
Ferric chloride	300	149	Sodium carbonate		
Ferric chloride 50% in water	300	149	Sodium chloride	300	149
Ferric nitrate 10–50%			Sodium hydroxide 10%	x	x
Ferrous chloride 40%			Sodium hydroxide 50%	x	x
Ferrous nitrate			Sodium hydroxide, concentrated	x	x
Fluorine gas, dry			Sodium hypochlorite 15%	x	x
Fluorine gas, moist			Sodium hypochlorite, concentrated	x	x
Hydrobromic acid, dilute	200	93	Sodium sulfide to 50%		
Hydrobromic acid 20%	200	93	Stannic chloride		
Hydrobromic acid 50%	200	93	Stannous chloride		
Hydrochloric acid 20%	300	149	Sulfuric acid 10%	250	121
Hydrochloric acid 38%	300	149	Sulfuric acid 50%	250	121
Hydrocyanic acid 10%			Sulfuric acid 70%	200	93
Hydrofluoric acid 30%	x	x	Sulfuric acid 90%	70	21
Hydrofluoric acid 60%	x	x	Sulfuric acid 98%	x	x
Hydrofluoric acid 100%	x	x	Sulfuric acid 100%	x	x
Hypochlorous acid			Sulfuric acid, fuming		
Iodine solution 10%			Sulfurous acid	80	27
Ketones, general			Thionyl chloride	200	93
Lactic acid 25%	160	71	Toluene		
Lactic acid, concentrated			Trichloroacetic acid 30%		
Magnesium chloride			White liquor		
Malic acid 10%			Zinc chloride	300	149

[a]The chemicals listed are in the pure state or in a saturated solution unless otherwise indicated. Compatibility is shown to the maximum allowable temperature for which data are available. Incompatibility is shown by an x. A blank space indicates that the data are unavailable.

Source: Schweitzer, PA. *Corrosion Resistance Tables,* 4th ed. Vols 1–3. New York: Marcel Dekker, 1995.

the better the resistance. Alloys having similar values may differ in actual service. The pitting resistance number is determined by the chromium, molybdenum, and nitrogen contents.

PREN = %Cr + 3.3 × %Mo = 30 × %N.

The PREN for various austenitic stainless steels is listed in the table.

Alloy	PREN	Alloy	PREN
654	63.09	316LN	31.08
31	54.45	316	27.90
25-6Mo	47.45	20Cb3	27.26
Al-6XN	46.96	348	25.60
20Mo-6	42.81	347	19.0
317LN	39.60	331	19.0
904L	36.51	304N	18.3
20Mo-4	36.20	304	18.0
317	33.2		

Prevention can be accomplished first by proper material selection, followed by a design that prevents stagnation of material, and alternate wetting and drying of the surface. Also, if coatings are to be applied, care should be taken that they are continuous, without "holidays."

See refs. 12, 13, and 1.

PITTING POTENTIAL

The pitting potential of an alloy (specifically stainless steel alloys) indicates the relative susceptibility of that alloy to suffer pitting. Potential is measured using an electrochemical apparatus in a standard chloride solution. The more positive the potential the less likely the alloy is to suffer pitting.

PITTING RESISTANCE EQUIVALENT NUMBER

See "Pitting."

PLASTICS

See "Polymers."

POLARIZATION

Polarization is the extent of potential charge caused by net current to or from an electrode, measured in volts, and results in the retardation of an electrochemical reaction (corrosion) caused by various physical and chemical factors. There are three types of polarization, concentration polarization, activation polarization, and IR drop.

Concentration polarization refers to an electrochemical process controlled by the diffusion in the electrolyte. Activation polarization refers to an electrochemical process controlled by the reaction sequence at the metal–electrolyte interface. IR drop polarization is the change in voltage associated with effects of the environment and the circuit between the anode and cathode sites. It includes the resistivity of the media, surface films, corrosion products, etc.

See Refs. 1 and 14.

POLYAMIDES (PA)

The polyamides are more commonly known as the nylons. They are linear molecules with a high degree of crystallinity and have the formula

$$\text{Nylon 6} \quad -\overset{\overset{\displaystyle H}{|}}{N}-[CH_2]_5-\overset{\overset{\displaystyle O}{\|}}{C}-C-$$

$$\text{Nylon 6, 10} \quad -\overset{\overset{\displaystyle H}{|}}{N}-[CH_2]_4-\overset{\overset{\displaystyle H}{|}}{N}-\overset{\overset{\displaystyle O}{\|}}{C}-[CH_2]_8-\overset{\overset{\displaystyle O}{\|}}{C}-$$

They are a series of high strength thermoplasts. Average physical and mechanicl properties are shown in Table P.10.

The nylons are resistant to weak acids, strong and weak alkalies, most common solvents, hydrocarbons, esters, and ketones. They will be attacked by strong acids, Refer to Table P.11 for the compatibility of PA with selected corrodents.

See Ref. 8.

Table P.10 Physical and Mechanical Properties of PA

Specific gravity	1.01–1.17
Water absorption, 24 h at 73°F/23°C, %	0.4–1.8
Tensile strength at 73°F/23°C, psi × 10^3	8.3–12.5
Modulus of elasticity in tension at 73°F/23°C × 10^3 psi	2–17
Compressive stength, psi × 10^3	9.7–12.5
Flexural strength, psi × 10^3	12.5–14
Izod impact strength, notched at 73°F/23°C, ft-lb/in.	0.5–3.3
Coefficient of thermal expansion, in./in. °F × 10^{-5}	4.5–5
Thermal conductivity, Btu/h/ft^2/°F/in.	1.2–1.7
Heat distortion temperature, °F/°C	
at 66 psi	
Nylon 6	360/182
Nylon 6/6	470/243
Nylon 11	302/150
at 264 psi	
Nylon 6	155–160/68–71
Nylon 6/6	220/104
Nylon 11	131/55

Table P.11 Compatiblity of Polyamides with Selected Corrodents[a]

Chemical	Max. temp. °F	°C	Chemical	Max. temp. °F	°C
Acetaldehyde	x	x	Benzene sulfonic acid 10%	x	x
Acetamide	250	121	Benzoic acid	80	27
Acetic acid 10%	200	93	Benzyl alcohol	200	93
Acetic acid 50%	x	x	Benzyl chloride	250	121
Acetic acid 80%	x	x	Borax	200	93
Acetic acid, glacial	x	x	Boric acid	x	x
Acetic anhydride	200	93	Bromine gas, dry	x	x
Acetone	80	27	Bromine gas, moist	x	x
Acetyl chloride	x	x	Bromine liquid	x	x
Acrylonitrile	80	27	Butadiene	80	27
Allyl alcohol	80	27	Butyl acetate	250	121
Alum	x	x	Butyl alcohol	200	93
Aluminum chloride, aqueous	x	x	*n*-Butylamine	200	93
Aluminum chloride, dry	x	x	Butyl phthalate	240	116
Aluminum fluoride	80	27	Butyric acid	x	x
Aluminum hydroxide	250	121	Calcium bisulfite	140	60
Aluminym nitrate	80	27	Calcium carbonate	200	93
Aluminum sulfate	140	60	Calcium chloride	250	121
Ammonia gas	200	93	Calcium hydroxide 10%	150	66
Ammonium carbonate	240	160	Calcium hydroxide, sat.	150	66
Ammonium chloride 10%	200	93	Calcium hypochlorite	x	x
Ammonium chloride 50%	200	93	Calcium nitrate	x	x
Ammonium fluoride 10%	80	27	Calcium oxide	80	27
Ammonium fluoride 25%	80	27	Calcium sulfate	80	27
Ammonium hydroxide 25%	250	121	Caprylic acid	230	110
Ammonium hydroxide, sat.	250	121	Carbon bisulfide	80	27
Ammonium nitrate	190	88	Carbon dioxide, dry	80	27
Ammonium persulfate	x	x	Carbon disulfide	80	27
Ammonium phosphate	80	27	Carbon monoxide	80	27
Ammonium sulfite	80	27	Carbon tetrachloride	250	121
Amyl acetate	150	66	Carbonic acid	100	38
Amyl alcohol	200	93	Cellosolve	250	121
Amyl chloride	x	x	Chloracetic acid, 50% water	x	x
Aniline	x	x	Chloracetic acid	x	x
Antimony trichloride	x	x	Chlorine gas, dry	x	x
Aqua regia 3:1	x	x	Chlorine gas, wet	x	x
Barium carbonate	80	27	Chlorine, liquid	x	x
Barium chloride	250	121	Chlorobenzene	250	121
Barium hydroxide	80	27	Chloroform	130	54
Barium sulfate	80	27	Chlorosulfonic acid	x	x
Barium sulfide	80	27	Chromic acid 10%	x	x
Benzaldehyde	150	66	Chromic acid 50%	x	x
Benzene	250	121	Citric acid 15%	200	93

Table P.11 Continued

Chemical	Max. temp. °F	°C	Chemical	Max. temp. °F	°C
Citric acid, conc.	200	93	Nitric acid 5%	x	x
Copper cyanide	80	27	Nitric acid 20%	x	x
Copper sulfate	140	60	Nitric acid 70%	x	x
Cresol	x	x	Nitric acid, anhydrous	x	x
Cupric chloride 5%	x	x	Nitrous acid, concentrated	x	x
Cupric chloride 50%	x	x	Perchloric acid 10%	x	x
Cyclohexane	250	121	Perchloric acid 70%	x	x
Cyclohexanol	250	121	Phenol	x	x
Dichloroacetic acid	x	x	Phosphoric acid 50–80%	x	x
Dichloroethane			Picric acid	x	x
(ethylene dichloride)	80	27	Salicylic acid	80	27
Ethylene glycol	200	93	Sodium carbonate	240	116
Ferric chloride	x	x	Sodium chloride	230	110
Ferric chloride 50% in water	x	x	Sodium hydroxide 10%	250	121
Ferric nitrate 10–50%	x	x	Sodium hydroxide 50%	250	121
Ferrous chloride	x	x	Sodium hypochlorite 20%	x	x
Fluorine gas, dry	x	x	Sodium hypochlorite,		
Fluorine gas, moist	x	x	concentrated	x	x
Hydrobromic acid, dilute	x	x	Sodium sulfide to 50%	230	110
Hydrobromic acid 20%	x	x	Stannic chloride	80	27
Hydrobromic acid 50%	x	x	Stannous chloride	x	x
Hydrochloric acid 20%	x	x	Sulfuric acid 10%	x	x
Hydrochloric acid 38%	x	x	Sulfuric acid 50%	x	x
Hydrocyanic acid 10%	x	x	Sulfuric acid 70%	x	x
Hydrofluoric acid 30%	x	x	Sulfuric acid 90%	x	x
Hydrofluoric acid 70%	x	x	Sulfuric acid 98%	x	x
Hydrofluoric acid 100%	x	x	Sulfuric acid 100%	x	x
Ketones, general	150	66	Sulfuric acid, fuming	x	x
Lactic acid 25%	200	93	Sulfurous acid	x	x
Lactic acid, concentrated	200	93	Thionyl chloride	x	x
Magnesium chloride	240	116	Toluene	200	93
Malic acid	x	x	Trichloroacetic acid	x	x
Methyl chloride	80	27	White liquor	80	27
Methyl ethyl ketone	240	116	Zinc chloride	x	x
Methyl isobutyl ketone	110	42			

[a]The chemicals listed are in the pure state or in a saturated solution unless otherwise indicated. Compatibility is shown to the maximum allowable temperature for which data are available. Incompatibility is shown by an x. A blank space indicates that the data are unavailable.

Source: Schweitzer, PA. *Corrosion Resistance Tables,* 4th ed. Vols 1–3. New York: Marcel Dekker, 1995.

POLYAMIDE/ACRYLONITRILE–BUTADIENE–STYRENE ALLOY

See "Triax."

POLYAMIDE ELASTOMERS

Polyamides are produced under a variety of trade names, the most popular of which are the nylons (nylon 6, etc.) manufactured by DuPont. Although the polyamides find their greatest use as textile fibers, they can also be formulated into thermoplastics molding compounds with many attractive properties. Their relatively high price tends to restrict their use.

There are many varieties of polyamides in production, but the four major types are nylon 6, nylon 6/6, nylon 11, and nylon 12. Of these, nylon 11 and 12 find application as elastomeric materials.

Physical and Mechanical Properties

The polyamide materials have an unusual combination of high tensile strength, ductility, and toughness. They can withstand extremely high impact, even at extremely low temperatures. According to ASTM test D-746, the cold brittleness temperature is –94°F (–74°C). Nylons 11 and 12 have an extremely high elastic memory that permits them to withstand repeated stretching and flexing over long periods of time. They also exhibit good abrasion resistance. Their absorption of moisture is very low, which means that parts produced from these materials will have good dimensional stability regardless of the humidity of the environment. Moisture also has little effect on the mechanical properties, particularly the modulus of elasticity.

The electrical insulation properties of polyamides grade 11 and grade 12 are good. They have a stable volume resistivity and offer excellent resistance to tracking. These materials are also self-extinguishing.

The physical and mechanical properties of the polyamides are given in Table P.12.

Resistance to Sun, Weather, and Ozone

The polyamides are resistant to sun, weather, and ozone. Many metals are coated with polyamide to provide protection from harsh weather.

Chemical Resistance

The polyamides exhibit excellent resistance to a broad range of chemicals and harsh environments. They have good resistance to most inorganic alkalies, particularly ammonium hydroxide and ammonia, even at elevated temperatures, and to sodium and potassium hydroxide at ambient temperatures. They also display good resistance to almost all inorganic salts and almost all hydrocarbons and petroleum based fuels.

They are also resistant to organic acids, such as citric, lactic, oleic, oxalic, stearic,

Table P.12 Physical and Mechanical Properties of the Polyamide Elastomers[a]

	Grade 11	Grade 12
Specific gravity	1.03–1.05	1.01–1.02
Specific heat, Btu/lb-°F	0.3–0.5	
Thermal conductivity		
10^{-4} cal/cm-s-°C	5.2	5.2
Btu/h-ft^2-°F-in.	1.5	1.5
Coefficient of linear therml expansion		
10^{-5}/°C	10.0	10.0
10^{-5}/°F	5.1	5.3
Continuous service temperature		
°F	180–300	180–250
°C	82–149	82–121
Brittle point		
°F	–94	—[b]
°C	–70	—[b]
Volume resistivity, ohm-cm	10^{13}	10^{13}
Dielectric strength (3 mm thick)	17	18
Dielectric constant at 10^3–10^6 Hz	3.2–3.7	3.8
Dissipation factor	0.05	5×10^{12}
Water absorption, %	0.3	0.25
Tensile strength, psi	8000	8000–9500
Elongation, % at break	300	300
Hardness, Rockwell	R108	R106–R109
Abrasion resistance	Good	Good
Impact resistance, kg-cm/cm	9.72	10.88–29.92
Resistance to compression set	Good	Good
Tear resistance	Good	Good
Machining qualities	Can be machined	
Resistance to sunlight	Good	Good
Effect of aging	Nil	Nil
Resistance to heat	Good	Good

[a]These are representative values since they may be altered by compounding.
[b]Data not available.

tartaric, and uric, and most aldehydes and ketones at normal temperatures. They display limited resistance to hydrochloric, sulfonic, and phosphoric acids at ambient temperatures Refer to Table P.11 for the compatibility of the polyamides with selected corrodents.

Applications

The polyamides find many diverse applications resulting from their many advantageous properties. A wide range of flexibility permits material to be produced that is soft enough for high quality bicycle seats and other materials whose strength and rigidity are comparable to those of many metals. Superflexible grades are also available that are used for shoe soles, gaskets, diaphragms, and seals. Because of the high elastic memory of the polyamides, these parts can withstand repeated stretching and flexing over long periods of time.

Since the polyamides can meet specification SAE J844 for airbrake hose, coiled tubing is produced for this purpose for use on trucks. Coiled airbrake hose produced from this material has been used on trucks that have traveled over 2 million miles without a single reported failure. High pressure hose and fuel lines are also produced from this material.

Corrosion resistant and wear resistant coverings for aircraft control cables, automotive cables, and electrical wire are also produced.

See Refs. 7 and 8.

POLYBUTADIENE RUBBER (BR)

Butadiene ($CH_2 = CH — CH = CH_2$) has two unsaturated linkages and can be polymerized readily. When butadiene or its derivatives become polymerized, the units link together to form long chains that contain over 1000 units. Simple butadiene does not yield a good grade of rubber, apparently because the chains are too smooth. Better results are obtained by introducing side groups into the chains either by modifying butadiene or by making a copolymer of butadiene and some other compound.

Physical and Mechanical Properties

Polybutadiene, designated BR, is very similar to the butadiene–styrene rubbers but is extremely difficult to process. As a result it is widely used as an admixture with Buna-S and other elastomers. It is rarely used in an amount larger than 75% of the total polymer in a compound.

It has outstanding properties of resilience and hysteresis, almost equivalent to that of natural rubber, excellent abrasion resistance, and good resistance to water adsorption and heat aging. It also possesses good electrical properties. Its tensile strength, tear tolerance, and impermeability are all good. Polybutadiene has an operating temperature range only slightly greater than that of natural rubber, ranging from –150 to 200°F (–101 to 93°C).

Polybutadiene will burn and has poor flame resistance. Its physical and mechanical properties are given in Table P.13.

Resistance to Sun, Weather, and Ozone

Although polybutadiene has good weather resistance, it will deteriorate when exposed to sunlight for prolonged periods of time. It also exhibits poor resistance to ozone.

Table P.13 Physical and Mechanical Properties of Polybutadiene (BR)[a]

Specific gravity	0.94
Specific heat, cal/g	0.45
Brittle point	–68°F (–56°C)
Insulation resistance, ohms/cm	10^{17}
Coefficient of linear expansion at 32–140°F, in./in.-°F	0.000036
Dielectric constant at 50 Hz	2.9
Power factor at 50 Hz	7×10^{-4}
Swelling, % by volume	
in kerosene at 77°F (25°C)	200
in benzene at 77°F (25°C)	200
in acetone at 77°F (25°C)	25
in mineral oil at 100°F (70°C)	120
Tensile strength, psi	2000–3000
Elongation, % at break	700–750
Hardness, Shore A	45–80
Abrasion resistance	Excellent
Maximum temperature, continuous use	200°F (93°C)
Impact resistance	Excellent
Compression set	Good
Resilience, %	90
Tear resistance, psi	1600
Machining qualities	Can be ground
Resistance to sunlight	Deteriorates
Resistance to heat	Poor

[a]These are representative values since they may be altered by compounding.

Chemical Resistance

The chemical resistance of polybutadiene is similar to that of natural rubber. It shows poor resistance to aliphatic and aromatic hydrocarbons, oil, and gasoline, but displays fair to good resistance in the presence of mineral solids and oxygenated compounds. Refer to Table P.14 for the compatibility of polybutadiene with selected corrodents.

Applications

Very rarely is polybutadiene used by itself. It is generally used as a blend with other elastomers to impart better resiliency, abrasion resistance, and/or low temperature properties, particuarly in the manufacture of automobile tire treads, shoe heels and soles, gaskets, seals, and belting.

See Refs. 7 and 8.

POLYBUTYLENE

Polybutylene is a semicrystalline polyolefin thermoplastic based on poly-1-butene and includes homopolymers and a series of copolymers (butene/ethylene). This thermoplast

Table P.14 Compatiblity of Polybutadiene (BR) with Selected Corrodents[a]

Chemical	Max. temp. °F	Max. temp. °C	Chemical	Max. temp. °F	Max. temp. °C
Alum	90	32	Nitric acid 10%	80	27
Alum ammonium	90	32	Nitric acid 20%	80	27
Alum ammonium sulfate	90	32	Nitric acid 30%	80	27
Alum chrome	90	32	Nitric acid 40%	x	x
Alum potassium	90	32	Nitric acid 50%	x	x
Aluminum chloride, aqueous	90	32	Nitric acid 70%	x	x
Aluminum sulfate	90	32	Nitric acid, anhydrous	x	x
Ammonia gas	90	32	Nitrous acid, concentrated	80	27
Ammonium chloride 10%	90	32	Ozone	x	x
Ammonium chloride 28%	90	32	Phenol	80	27
Ammonium chloride 50%	90	32	Sodium bicarbonate 20%	90	32
Ammonium chloride, saturated	90	32	Sodium bisulfate	80	27
Ammonium nitrate	90	32	Sodium bisulfite	90	32
Ammonium sulfate 10–40%	90	32	Sodium carbonate	90	32
Calcium chloride, saturated	80	27	Sodium chlorate	80	27
Calcium hypochlorite, saturated	90	32	Sodium hydroxide 10%	90	32
Carbon dioxide, wet	90	32	Sodium hydroxide 15%	90	32
Chlorine gas, wet	x	x	Sodium hydroxide 30%	90	32
Chrome alum	90	32	Sodium hydroxide 50%	90	32
Chromic acid 10%	x	x	Sodium hydroxide 70%	90	32
Chromic acid 30%	x	x	Sodium hydroxide, concentrated	90	32
Chromic acid 40%	x	x	Sodium hypochlorite to 20%	90	32
Chromic acid 50%	x	x	Sodium nitrate	90	32
Copper chloride	90	32	Sodium phosphate, acid	90	32
Copper sulfate	90	32	Sodium phosphate, alkaline	90	32
Fatty acids	90	32	Sodium phosphate, neutral	90	32
Ferrous chloride	90	32	Sodium silicate	90	32
Ferrous sulfate	90	32	Sodium sulfide to 50%	90	32
Hydrochloric acid, dilute	80	27	Sodium sulfite 10%	90	32
Hydrochloric acid 20%	90	32	Sodium dioxide, dry	x	x
Hydrochloric acid 35%	90	32	Sulfur trioxide	90	32
Hydrochloric acid 38%	90	32	Sulfuric acid 10%	80	27
Hydrochloric acid 50%	90	32	Sulfuric acid 30%	80	27
Hydrochloric acid fumes	90	32	Sulfuric acid 50%	80	27
Hydrogen peroxide 90%	90	32	Sulfuric acid 60%	80	27
Hydrogen sulfide, dry	90	32	Sulfuric acid 70%	90	32
Nitric acid 5%	80	27	Toluene	x	x

[a]The chemicals listed are in the pure state or in a saturated solution unless otherwise indicated. Compatibility is shown to the maximum allowable temperature for which data are available. Incompatibility is shown by an x. A blank space indicates that the data are unavailable.

Source: Schweitzer, PA. *Corrosion Resistance Tables,* 4th ed. Vols 1–3. New York: Marcel Dekker, 1995.

possesses the combination of stress cracking resistance, chemical resistance, and abrasion resistance. Its structural formula is

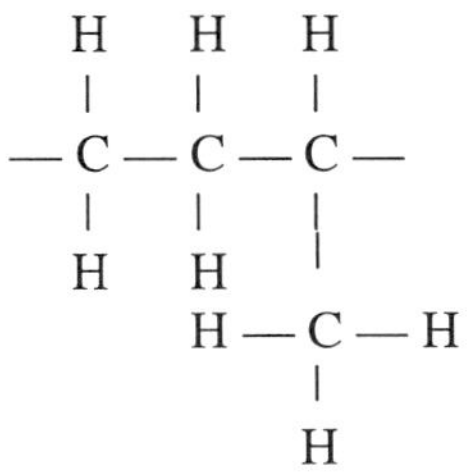

Polybutylene maintains its mechanical properties at elevated temperatures. Its long-term strength is greater than that of high density polyethylene. Polybutylene has an upper temperature limit of 200°F (93°C).

It is resistant to acids, bases, soaps, and detergents but not to aliphatic solvents at room temperature, it is partially soluble in aromatic and chlorinated hydrocarbons. Chlorinated water will cause pitting attack.

POLYCARBONATE (PC)

This thermoplast is produced under the tradename of Lexan by GE Plastics. Its structure is

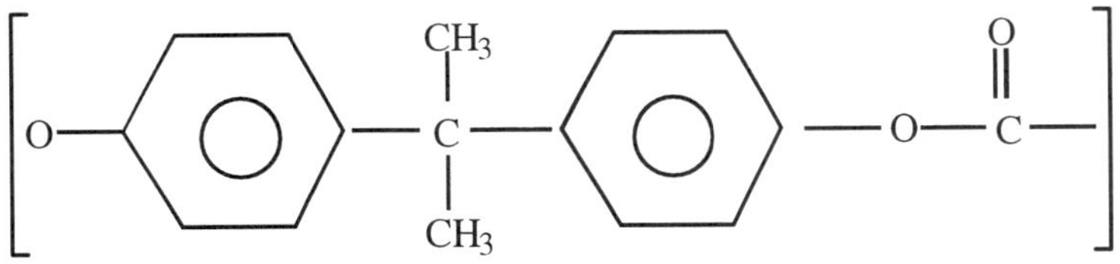

Because of its extremely high impact resistance and good clarity it is widely used for windows in chemical equipment and glazing in chemical plants. The exceptional weatherability, corrosion resistance, and high impact strength gives it wide use in outdoor energy management devices, network interfaces, electrical wiring blocks, telephone equipment, lighting diffusers, globes, and housings. Table P.15 provides the physical and mechanical properties of PC.

Polycarbonate is resistant to aliphatic hydrocarbons and weak acids and has limited resistance to weak alkalies. It is resistant to most oils and greases. PC will be attacked by strong alkalies and strong acids and is soluble in ketones and esters and aromatic and chlorinated hydrocarbons.

See Refs. 15 and 16.

POLYCARBONATE/ACRYLONITRILE–BUTADIENE–STYRENE ALLOY

See "Cycoloy."

Table P.15 Physical and Mechanical Properties of PC

Specific gravity	1.2
Water absorption, 24 h at 73°F/23°C, %	0.15–0.2
Tensile strength at 73°F/23°C, psi × 10^3	8–9.5
Modulus of elasticity at 73°F/23°C × 10^5 psi	3.2–4.5
Compressive strengh, psi × 10^3	10–14
Flexural strength, psi × 10^3	11.5–15
Izod impact strength, notched at 73°F/23°C, ft-lb/in.	4–16
Coefficient of linear thermal expansion, in./in. °F × 10^{-5}	1.79–3.9
Thermal conductivity, Btu/h/ft^2/°F/in.	1.33–1.41
Heat distortion temperature, °F/°C:	
at 66 psi	285/140
at 264 psi	265/129
Limiting oxygen index, %	25–31.5
Underwriters lab rating, Sub 94	SEO-SEI

POLYCARBONATE/POLYBUTYLENE–TEREPHTHALATE ALLOY

See "Xenoy."

POLYCHLOROPRENE

See "Neoprene."

POLYESTER (PE) ELASTOMER

This elastomer combines characteristics of thermoplastics and elastomers. It is structurally strong, resilient, and resistant to impact and flexural fatigue. Physical and mechanical properties vary depending upon the hardness of the elastomer. Hardnesses range from 40 to 72 on the Shore D scale. The standard hardnesses to which PE elastomers are formulated are 40, 55, 63, and 72 Shore D. Combined are such features as resilience, high resistance to deformation under moderate strain conditions, outstanding flex-fatigue resistance, good abrasion resistance, retention of flexibility at low temperatures, and good retention of properties at elevated temperatures. Polyester elastomers can successfully replace other thermoset rubbers at lower cost in many applications by taking advantage of their higher strength and using thinner cross-sections.

Physical and Mechanical Properties

When modulus is an important design consideration, thinner sections of PE elastomer can be used, since its tensile stress at low strain shows higher modulus values than other elastomeric materials. Polyester elastomers yield at approximately 25% strain and beyond the yield point elongate to 300–500% with some permanent set. To take full advantage of the elastomeric properties of PE, parts should be designed to function with deformations that do not exceed the yield point (up to approximately 20% strain).

Table P.16 Physical and Mechanical Properties of Polyester Elastomer[a]

Specific gravity	1.17–1.25
Brittle point	–94°F (–70°C)
Resilience, %	42–62
Coefficient of linear expansion, in./in.-°C	2×10^{-5} –21×10^{-5}
Dielectric strength, V/mil	645–900
Dielectric constant at 1 kHz	4.16–6
Permeability to air	Low
Tear strength	
lb/in.	631–1055
kn/m	110–185
Water absorption, %/24 h	0.6–.03
Tensile strength, psi	3700–5700
Elongation, % at break	350–450
Hardness, Shore D	40–72
Abrasion resistance	Good
Maximum temperature, continuous use	302°F (150°C)
Impact resistance	Good
Compression set, %	
at 73°F (23°C)	1–11
at 158°F (70°C)	2–27
at 212°F (100°C)	4–44
Resistance to sunlight	Good
Effect of aging	Nil
Resistance to heat	Excellent
Flexural modulus, psi	7,000–75,000

These are representative values since they may be altered by compounding.

The Bashore resilience of PE exceeds 60% for the 40D hardness grade. Even as the hardness approaches that of plastics (63D), these materials still have a resilience of 40%.

The load-bearing properties of PE elastomers under compression are good. As the hardness increases, so does the compression modulus. The compression modulus of PE elastomers is 50–100% greater than that of other rubbers of comparable hardness.

Table P.16 lists the compression set of PE elastomers at constant load (1350 psi) and at varying temperatures. These conditions were chosen because they very closely simulate the conditions under which parts fabricated from PE elastomers operate. The compression set of these rubbers measured for 22 h at 158°F (70°C) with a constant deflection of 25% is 60% for materials having a hardness of 40D and 56% for materials of 55D hardness. Because of the high load-bearing capability of these elastomers, the majority of compression applications employ deflections well below 25%. At 25% deflection, most formulations of PE elastomers are deformed beyond their yield point, or at least beyond the limits of good design practice. Under these conditions PE elastomers are generally comparable to urethane elastomers. Annealing of PE parts will reduce the compression set values to 40% or less with a 25% deflection.

Very little hysteresis loss is exhibited by PE elastomers when they are stressed below the yield point. Applications at low strain levels can usually be expected to exhibit complete recovery and continue to do so in cyclic applications with little heat buildup.

Because of the high resilience and low heat buildup of these elastomers, their resistance to cut growth is outstanding. This means that a part made of PE, engineered to operate at low strain levels, can usually be expected to exhibit complete recovery from deformation and to continue to do so under repeated cycling for extremely long periods of time without heat buildup or distortion.

The mechanical properties of these elastomers are maintained up to 302°F (150°C), better than many rubbers, particularly the harder polymers. Above 248°F (120°C) their tensile strengths far exceed those of other rubbers. Their hot strength, and good resistance to hot fluids, make PE elastomers suitable for many applications involving fluid containment.

All of the PE formulations have brittle points below –94°F (–70°C) and exhibit high impact strength and resistance to stiffening at temperatures down to –40°F (–40°C). As would be expected, the softer members exhibit the better low temperature flexibility.

The physical and mechanical properties of polyester elastomers are given in Table P.16.

Resistance to Sun, Weather, and Ozone

The polyester rubbers possess excellent resistance to ozone. When formulated with appropriate additives, their resistance to sunlight aging is very good. Resistance to general weathering is good.

Chemical Resistance

In general, the fluid resistance of polyester rubbers increases with increasing hardness. Since these rubbers contain no plasticizers, they are not susceptible to the solvent extraction or heat volatilization of such additives. Many fluids and chemicals will extract plasticizers from elastomers, causing a significant increase in stiffness (modulus) and volume shrinkage.

Overall, PE elastomers are resistant to the same classes of chemicals and fluids as the polyurethanes are. However, PE has better high temperature properties than the polyurethanes and can be used satisfactorily at higher temperatures in the same fluids.

Polyester elastomers have excellent resistance to non polar materials such as oils and hydraulic fluids, even at elevated temperatures. At room temperature, elastomers are resistant to most polar fluids, such as acids, bases, amines, and glycols. Resistance is very poor at temperatures of 158°F (70°C) or above. These rubbers should not be used in applications requiring continuous exposure to polar fluids at elevated temperatures.

Polyester elastomers also have good resistance to hot moist atmospheres. Their hydrolytic stability can be further improved by compounding.

Applications

Applications for PE elastomers are varied. Large quantities of PE materials are used for liners for tanks, ponds, swimming pools, and drums. Because of their low permeability to air, they are also used for inflatables. Their chemical resistance to oils and hydraulic fluids coupled with their high heat resistance make PE elastomers very suitable for automotive hose applications.

Since the PE elastomers do not contain any plasticizers, hoses and tubing produced from them do not stiffen with age. Other PE products include seals, gaskets, specialty belting, noise damping devices, low pressure tires, industrial solid tires, wire and cable jacketing, pump parts, electrical connectors, flexible shafts, sports equipment, piping clamps and cushions, gears, flexible couplings, and fasteners.

See Refs. 7 and 8.

POLYESTER FIBERS

Polyester fibers are used primarily as a surfacing mat for the resin rich inner surfaces of filament wound or contact molded thermoset structures. See "Thermoset Reinforcing Materials."

POLYETHERETHERKETONE (PEEK)

PEEK is a thermoplast suitable for applications that require mechanical strength with the need to resist difficult thermal and chemical environments. Its chemical structure is

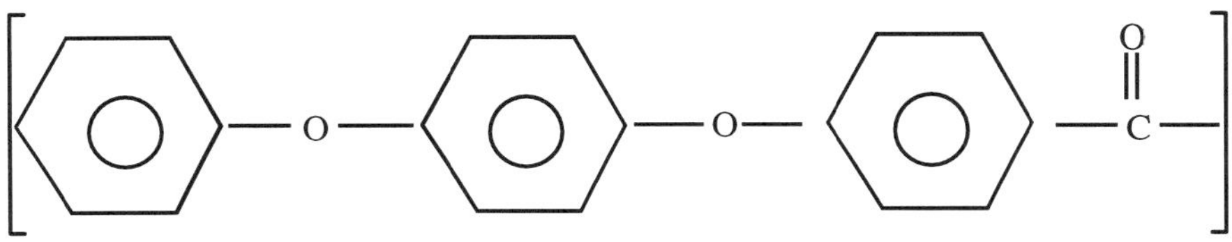

PEEK has a continuous maximum service temperature of 480°F (260°C) with excellent mechanical properties retained to temperatures over 570°F (300°C). Table P.17 gives the physical and mechanical properties of PEEK.

PEEK is insoluble in all common solvents and has excellent resistance to a wide range of organic and inorganic liquids. Refer to Table P.18 for the compatibility of PEEK with selected corrodents.

This thermoplast is not subject to hydrolysis at ambient or elevated temperatures in a continuous cycled environment, permitting the material to be steam sterilized using conventional sterilization equipment.

Table P.17 Physical and Mechanical Properties of PEEK

Specific gravity	1.32
Water absorption, 24 h at 73°F/23°C, %	0.5
Tensile strength at 73°F/23°, psi	14,500
Modulus of elasticity in tension at 73°F/23°C $\times 10^{-5}$ psi	4.9
Compressive strength, psi	17,100
Flexural strength, psi	24,650
Izod impact strength, notched at 73°F/23°C, ft-lb/in.	1.57
Coefficient of thermal expansion in./in. °F $\times 10^{-5}$	
at 0–290°F	2.6
at 290–500°F	6.1
Thermal conductivity, Btu/h/ft^2/°F/in.	1.75
Heat distortion temperature, °F/°C	
at 264 psi	320/160
Limiting oxygen index, %	24
Underwriters lab rating, Sub UL 94	V-O (1.45)

Table P.18 Compatiblity of PEEK with Selected Corrodents[a]

Chemical	Max. temp. °F	Max. temp. °C	Chemical	Max. temp. °F	Max. temp. °C
Acetaldehyde	80	27	Hydrobromic acid 50%	x	x
Acetic acid 10%	80	27	Hydrochloric acid 20%	100	38
Acetic acid 50%	140	60	Hydrochloric acid 38%	100	38
Acetic acid 80%	140	60	Hydrofluoric acid 30%	x	x
Acetic acid, glacial	140	60	Hydrofluoric acid 70%	x	x
Acetone	210	99	Hydrofluoric acid 100%	x	x
Acrylic acid	80	27	Hydrogen sulfide, wet	200	93
Acrylonitrile	80	27	Ketones, general	80	27
Aluminum sulfate	80	27	Lactic acid 25%	80	27
Ammonia gas	210	99	Lactic acid, concentrated	100	38
Ammonium hydroxide, sat.	80	27	Magnesium hydroxide	100	38
Aniline	200	93	Methyl alcohol	80	27
Aqua regia 3:1	x	x	Methyl ethyl ketone	370	188
Benzaldehyde	80	27	Naphtha	200	93
Benzene	80	27	Nitric acid 5%	200	93
Benzoic acid	170	77	Nitric acid 20%	200	93
Boric acid	80	27	Nitrous acid 10%	80	27
Bromine gas, dry	x	x	Oxalic acid 5%	x	x
Bromine gas, moist	x	x	Oxalic acid 10%	x	x
Calcium carbonate	80	27	Oxalic acid, saturated	x	x
Calcium chloride	80	27	Phenol	140	60
Calcium hydroxide, 10%	80	27	Phosphoric acid 50–80%	200	93
Calcium hydroxide, sat.	100	38	Potassium bromide 30%	140	60
Carbon dioxide, dry	80	27	Sodium carbonate	210	99
Carbon tetrachloride	80	27	Sodium hydroxide 10%	220	104
Carbonic acid	80	27	Sodium hydroxide 50%	180	82
Chlorine gas, dry	80	27	Sodium hydroxide,		
Chlorine, liquid	x	x	concentrated	200	93
Chlorobenzene	200	93	Sodium hypochlorite 20%	80	27
Chloroform	80	27	Sodium hypochlorite,	80	27
Chlorosulfonic acid	80	27	concentrated		
Chromic acid 10%	80	27	Sulfuric acid 10%	80	27
Chromic acid 50%	200	93	Sulfuric acid 50%	200	93
Citric acid, conc.	170	77	Sulfuric acid 70%	x	x
Cyclohexane	80	27	Sulfuric acid 90%	x	x
Ethylene glycol	160	71	Sulfuric acid 98%	x	x
Ferrous chloride	200	93	Sulfuric acid 100%	x	x
Fluorine gas, dry	x	x	Sulfuric acid, fuming	x	x
Fluorine gas, moist	x	x	Toluene	80	27
Hydrobromic acid, dilute	x	x	Zinc chloride	100	38
Hydrobromic acid 20%	x	x			

[a]The chemicals listed are in the pure state or in a saturated solution unless otherwise indicated. Compatibility is shown to the maximum allowable temperature for which data are available. Incompatibility is shown by an x. A blank space indicates that the data are unavailable.

Source: Schweitzer, PA. *Corrosion Resistance Tables,* 4th ed. Vols 1–3. New York: Marcel Dekker, 1995.

Table P.19 Physical and Mechanical Properties of PES

Specific gravity	1.51
Tensile strength at 68°F/20°C, psi	12,200
Flexural strength, psi	18,700
Izod impact strength, notched at 73°F/23°C, ft-lb/in.	1.57
Coefficient of thermal expansion in./in. °F × 10^{-5}	5.5
Thermal conductiity, Btu/h/ft^2/°F/in.	1.25
Heat distortion temperature, °F/°C t 264 psi	397/203
Limiting oxygen index, %	34
Underwriters Lab rating, Sub UL 94	V-O at 0.46

Like most linear polyaromatics, PEEK suffers from the effects of degradation during outdoor weathering. Painting or pigmenting the polymer will provide protection.

PEEK exhibits excellent resistance to hard (gamma) irradiation, absorbing over 1000 M rads of irradiation without suffering significant damage.

See Refs. 8 and 15.

POLYETHERSULFONE (PES)

PES is a thermoplast having a continuous maximum operating temperature of 390°F (200°C). At room temperature PES is tough, rigid, and strong with outstanding long-term load bearing properties. Most of these properties are retained at the maximum operating temperature. Table P.19 lists the physical and mechanical properties of PES.

It has the chemical structure

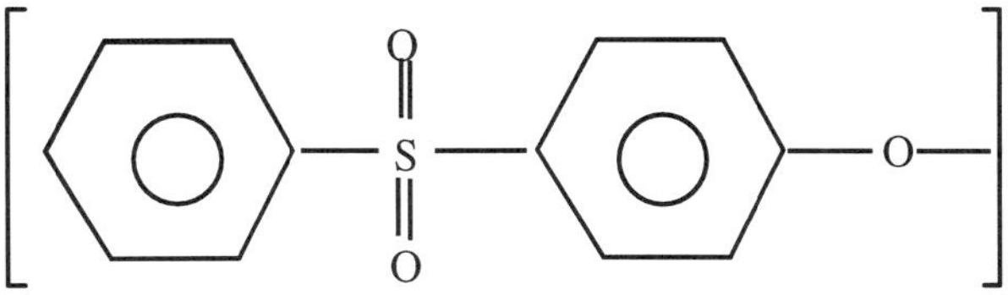

PES has excellent resistance to aliphatic hydrocarbons, some chlorinated hydrocarbons, and aromatics. It is also resistant to most inorganic chemicals. Hydrocarbon and mineral oils, greases, and transmission fluids have no effect of PES.

PES will be attacked by strong oxidizing acids, but glass fiber reinforced grades are resistant to more dilute acids. It is soluble in highly polar solvents and is subject to stress cracking in ketones and esters.

PES is not resistant to outdoor weathering and is not recommended for outdoor applications unless stabilized by incorporating carbon black or unless painted.

Refer to Table P.20 for the compatibility of PES with selected corrodents.

Refer to Ref. 8 for a wide range of compatibility of PES with selected corrodents. See also Ref. 15.

POLYETHYLENE (PE)

See also "Polymers."

Table P.20 Compatiblity of PES with Selected Corrodents[a]

Chemical	Max. temp. °F	Max. temp. °C	Chemical	Max. temp. °F	Max. temp. °C
Acetic acid 10%	80	27	Nitric acid 5%	80	27
Acetic acid 50%	140	60	Nitric acid 20%	x	x
Acetic acid 80%	200	93	Oxalic acid 5%	80	27
Acetic acid, glacial	200	93	Oxalic acid 29%	80	27
Acetone	x	x	Oxalic acid, saturated	80	27
Ammonia, gas	80	27	Phenol	x	x
Aniline	x	x	Phosphoric acid 50–80%	200	93
Benzene	80	27	Potassium bromide 30%	140	60
Benzene sulfonic acid 10%	100	38	Sodium carbonate	80	27
Benzoic acid	80	27	Sodium chloride	80	27
Carbon tetrachloride	80	27	Sodium hydroxide 20%	80	27
Chlorobenzene	x	x	Sodium hydroxide 50%	80	27
Chlorsulfonic acid	x	x	Sodium hypochlorite 20%	80	27
Chromic acid 10%	x	x	Sodium hypochlorite, concentrated	80	27
Chromic acid 50%	x	x	Sulfuric acid 10%	80	27
Citric acid, concentrated	80	27	Sulfuric acid 50%	x	x
Ethylene glycol	100	38	Sulfuric acid 70%	x	x
Ferrous chloride	100	38	Sulfuric acid 90%	x	x
Hydrochloric acid 20%	140	60	Sulfuric acid 98%	x	x
Hydrochloric acid 38%	140	60	Sulfuric acid 100%	x	x
Hydrogen sulfide, wet	80	27	Sulfuric acid, fuming	x	x
Methyl ethyl ketone	x	x	Toluene	80	27
Naphtha	80	27			

[a]The chemicals listed are in the pure state or in a saturated solution unless otherwise indicated. Compatibility is shown to the maximum allowable temperature for which data are available. Incompatibility is shown by an x.
Material extracted from PA Schweitzer. *Corrosion Resistance Tables*, 4th ed. Vols 1–3. New York: Marcel Dekker, 1995.

Polyethylene is produced in various types that differ in molecular structure, crystallinity, molecular weight, and molecular weight distribution. The basic formula is

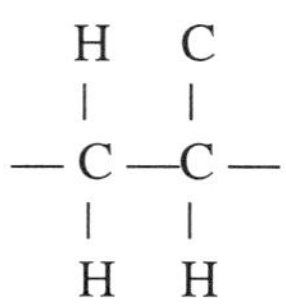

PE is produced by polymerizing ethylene gas obtained from petroleum hydrocarbons. Changes in the polymerizing conditions are responsible for the various types of PE.

The terms low, high, and medium density refer to the ASTM designation based on the unmodified PE. Low density PE has a specific gravity of 0.91–0.925; medium density PE has a specific gravity of 0.926–0.940; and high density PE has a specific gravity of

Table P.21 Physical and Mechanical Properties of PE UHMW

Specific gravity	0.94–0.96
Water absorption 24 h at 73°F (23°C), %	< 0.01
Tensile strength at 73°F (23°C), psi	3100–3500
Modulus of elasticity in tension at 73°F (23°C) × 10^5	1.18
Flexural modulus, psi × 10^5	1.33
Izod impact strength, notched at 73°F (23°C)	0.4–6.0
Coefficient of thermal expansion	
in./in.-°F × 10^{-5}	11.1
in./10°F/100 ft	0.111
Thermal conductivity, Btu/h/sq ft/°F/in.	0.269
Heat distortion temperature, °F/°C	
at 66 psi	150/66
at 264 psi	250/121
Resistance to heat at continuous drainage, °F/°C	180/82
Flame spread	slow burning

0.941–0.959. The densities, being related to the molecular structure, are indications of the properties of the final product.

The two grades of polyethylene primarily used for corrosion resistance are the high molecular weight (HMW) and the ultra high molecular weight (UHMW). The HMW material has an average molecular weight of 200,000–500,000, while the UHMW material has an average molecular weight of at least 3.1 million. Table P.21 gives the physical and mechanical properties of UHMW polyethylene.

If exposed to ultraviolet radiation, from the sun or other source, photo or light oxidation will occur. To prevent this it is necessary to incorporate carbon black into the resin to stabilize it. Other stabilizers will not provide complete protection. PE does not support biological growth.

Polyethylene is resistant to a wide variety of acids, bases, inorganic salts, and many fertilizer solutions.

The compatibility of EHMW polyethylene with selected corrodents is given in Table P.22.

See Refs. 8 and 10.

POLYMERS

Also see "Thermoplasts," "Thermosets," "Elastomers," and individual compounds. Polymers are better known as "plastics." This is really a misnomer since the term plastic as defined in the dictionary is "a material capable of being molded," such as putty or wet clay. Many of the so-called plastic materials available today are not capable of being molded or once formed of being reshaped. Consequently the term polymer is more descriptive.

Polymers can be classified into three categories: thermoplastic polymers, commonly called thermoplasts; thermosetting polymers, commonly called thermosets; and elastomers, commonly called rubbers.

Table P.22 Compatibility of EHMW PE with Selected Corrodents[a]

Chemical	Maximum temp. °F	°C	Chemical	Maximum temp. °F	°C
Acetaldehyde 40%	90	32	Barium sulfate	140	60
Acetamide			Barium sulfide	140	60
Acetic acid 10%	140	60	Benzaldehyde	x	x
Acetic acid 50%	140	60	Benzene	x	x
Acetic acid 80%	80	27	Benzene sulfonic acid 10%	140	60
Acetic acid, glacial			Benzoic acid	140	60
Acetic anhydride	x	x	Benzyl alcohol	170	77
Acetone	120	49	Benzyl chloride		
Acetyl chloride			Borax	140	60
Acrylic acid			Boric acid	140	60
Acrylonitrile	150	66	Bromine gas, dry	x	x
Adipic acid	140	60	Bromine gas, moist	x	x
Allyl alcohol	140	60	Bromine liquid	x	x
Allyl chloride	80	27	Butadiene	x	x
Alum	140	60	Butyl acetate	90	32
Aluminum acetate			Butyl alcohol	140	60
Aluminum chloride, aqueous	140	60	*n*-Butylamine	x	x
Aluminum chloride, dry	140	60	Butyl phthalate	80	27
Aluminum fluoride	140	60	Butyric acid	130	54
Aluminum hydroxide	140	60	Calcium bisulfide	140	60
Aluminum nitrate			Calcium bisulfite	80	27
Aluminum oxychloride			Calcium carbonate	140	60
Aluminum sulfate	140	60	Calcium chlorate	140	60
Ammonia gas	140	60	Calcium chloride	140	60
Ammonium bifluoride			Calcium hydroxide 10%	140	60
Ammonium carbonate	140	60	Calcium hydroxide, sat.	140	60
Ammonium chloride 10%	140	60	Calcium hypochlorite	140	60
Ammonium chloride 50%	140	60	Calcium nitrate	140	60
Ammonium chloride, sat.	140	60	Calcium oxide	140	60
Ammonium fluoride 10%	140	60	Calcium sulfate	140	60
Ammonium fluoride 25%	140	60	Caprylic acid		
Ammonium hydroxide 25%	140	60	Carbon bisulfide	x	x
Ammonium hydroxide, sat.	140	60	Carbon dioxide, dry	140	60
Ammonium nitrate	140	60	Carbon dioxide, wet	140	60
Ammonium persulfate	140	60	Carbon disulfide	x	x
Ammonium phosphate	80	27	Carbon monoxide	140	60
Ammonium sulfate 10–40%	140	60	Carbon tetrachloride	x	x
Ammonium sulfide	140	60	Carbonic acid	140	60
Ammonium sulfite			Cellosolve		
Amyl acetate	140	60	Chloracetic acid, 50% in water	x	x
Amyl alcohol	140	60	Chloracetic acid	x	x
Amyl chloride	x	x	Chlorine gas, dry	80	27
Aniline	130	54	Chlorine gas, wet, 10%	120	49
Antimony trichloride	140	60	Chlorine, liquid	x	x
Aqua regia 3:1	130	54	Chlorobenzene	x	x
Barium carbonate	140	60	Chloroform	80	27
Barium chloride	140	60	Chlorosulfonic acid	x	x
Barium hydroxide	140	60			

Table P.22 Continued

Chemical	Maximum temp. °F	Maximum temp. °C	Chemical	Maximum temp. °F	Maximum temp. °C
Chromic acid 10%	140	60	Methyl chloride	x	x
Chromic acid 50%	90	32	Methyl ethyl ketone	x	x
Chromyl chloride			Methyl isobutyl ketone	80	27
Citric acid 15%	140	60	Muriatic acid	140	60
Citric acid, conc.	140	60	Nitric acid 5%	140	60
Copper acetate			Nitric acid 20%	140	60
Copper carbonate			Nitric acid 70%	x	x
Copper chloride	140	60	Nitric acid, anhydrous	x	x
Copper cyanide	140	60	Nitrous acid, concentrated		
Copper sulfate	140	60	Oleum		
Cresol	80	27	Perchloric acid 10%	140	60
Cupric chloride 5%	80	27	Perchloric acid 70%	x	x
Cupric chloride 50%			Phenol	100	38
Cyclohexane	130	54	Phosphoric acid 50–80%	100	38
Cyclohexanol	170	77	Picric acid	100	38
Dichloroacetic acid	73	23	Potassium bromide 30%	140	60
Dichloroethane			Salicylic acid		
(ethylene dichloride)	x	x	Silver bromide 10%		
Ethylene glycol	140	60	Sodium carbonate	140	60
Ferric chloride	140	60	Sodium chloride	140	60
Ferric chloride 50% in water	140	60	Sodium hydroxide 10%	170	77
Ferric nitrate 10–50%	140	60	Sodium hydroxide 50%	170	77
Ferrous chloride	140	60	Sodium hydroxide,		
Ferrous nitrate	140	60	concentated		
Fluorine gas, dry	x	x	Sodium hypochlorite 20%	140	60
Fluorine gas, moist	x	x	Sodium hypochlorite,		
Hydrobromic acid, dilute	140	60	concentrated	140	60
Hydrobromic acid 20%	140	60	Sodium sulfide to 50%	140	60
Hydrobromic acid 50%	140	60	Stannic chloride	140	60
Hydrochloric acid 20%	140	60	Stannous chloride	140	60
Hydrochloric acid 38%	140	60	Sulfuric acid 10%	140	60
Hydrocyanic acid 10%	140	60	Sulfuric acid 50%	140	60
Hydrofluoric acid 30%	80	27	Sulfuric acid 70%	80	27
Hydrofluoric acid 70%	x	x	Sulfuric acid 90%	x	x
Hydrofluoric acid 100%	x	x	Sulfuric acid 98%	x	x
Hypochlorous acid			Sulfuric acid 100%	x	x
Iodine solution 10%	80	27	Sulfuric acid, fuming	x	x
Ketones, general	x	x	Sulfurous acid	140	60
Lactic acid 25%	140	60	Thionyl chloride	x	x
Lactic acid, concentrated	140	60	Toluene	x	x
Magnesium chloride	140	60	Trichloroacetic acid	140	60
Malic acid	100	38	White liquor		
Manganese chloride	80	27	Zinc chloride	140	60

[a]The chemicals listed are in the pure state or in a saturated solution unless otherwise indicated. Compatibility is shown to the maximum allowable temperature for which data are available. Incompatibility is shown by an x. A blank space indicates that the data are unavailable.

Source: Schweitzer, PA. *Corrosion Resistance Tables*, 4th ed. Vols. 1–3. New York: Marcel Dekker, 1995.

Thermoplasts are long-chain linear molecules that can easily be formed by heat and pressure at temperatures above a critical temperature known as the "glass temperature." Since this critical temperature is below room temperature for many polymers, these polymers are brittle at room temperature. However these polymers can be reheated and re-formed into new shapes and can therefore be recycled.

Thermosets are polymers that take on a permanent shape or "set" when heated, although some will set at room temperature. An example of the latter are epoxies, which result from combining an epoxy polymer with a curing agent or catalyst at room temperature. Thermosets will decompose on heating and therefore cannot be re-formed or recycled.

An elastomer is generally considered to be any material, either natural or synthetic, that is elastic or resilient and in general resembles natural rubber in feeling and appearance. A more technical definition is provided by ASTM, which states, "An elastomer is a polymeric material which at room temperature can be stretched to at least twice its original length and upon immediate release of the stress will return quickly to its original length."

In general elastomers must be cooled to below room temperature to be made brittle.

Metallic materials undergo a specific corrosion rate as a result of an electrochemical reaction. Because of this it is possible to predict the life of a metal when in contact with a specific corrodent under a given set of conditions. This is not the case with polymeric materials. Plastic materials do not experience a specific corrosion rate. They are usually completely resistant to chemical attack or they deteriorate rapidly. They are attacked either by chemical reaction or by solvation. Solvation is the penetration of the plastic by a corrodent, which causes softening, swelling, and ultimate failure. Corrosion of plastics can be classified in the following ways as to the attack mechanism:

1. Disintegration or degradation of a physical nature due to absorption, permeation, solvent action, or other factors
2. Oxidation, where chemical bonds are attacked
3. Hydrolysis, where ester linkages are attacked
4. Radiation
5. Thermal degradation involving depolymerization and possibly repolymerization
6. Dehydration (rather uncommon)
7. Any combination of the above

Results of such attacks will appear in the form of softening, charring, crazing, delamination, embrittlement, discoloration, dissolving, or swelling.

The corrosion resistance of polymer matrix composites is also affected by two other factors, the nature of the laminate and, in the case of the thermoset resins, the cure. Improper or insufficient cure time will adversely affect the corrosion resistance, while proper cure time and procedures will generally improve the corrosion resistance.

All of the polymers are compounded. The final product is produced to certain specific properties for a specific application. When the corrosion resistance of a polymer is discussed, the data referred to are that of the pure polymer. In many instances other ingredients are blended with the polymer to enhance certain properties, which in many cases will reduce the ability of the polymer to resist the attack of some media. Therefore it is essential to know the makeup of any polymer prior to its use.

Thermoplasts

A general rule as to the differences in the corrosion resistance of the thermoplasts may be derived from the periodic table. In the periodic table the basic elements of nature are organized by atomic structure as well as by chemical nature. The elements are placed into classes with similar properties, i.e., elements and compounds that exhibit similar behavior. These classes are the alkali metals, alkaline earth metals, transition metals, rare earth series, other metals, nonmetals, and noble (inert) gases.

The category known as halogens is of particular importance and interest in the case of thermoplasts. These elements include fluorine, chlorine, bromine, and iodine. They are the most electronegative elements in the periodic table, making them the most likely to attract an electron from another element and become a stable structure. Of all the halogens, fluorine is the most electronegative, permitting it to bond strongly with carbon and hydrogen atoms, but not well with itself. The carbon–fluorine bond is predominant in PVDF and is responsible for the important properties of these materials. These are among the strongest known organic compounds. The fluorine acts like a protective shield for other bonds of lesser strength within the main chain of the polymer. The carbon–hydrogen bond, of which such plastics as PE and PP are composed, is considerably weaker. The carbon–chlorine bond, a key bond in PVC, is still weaker.

The arrangement of the elements in the molecule, the symmetry of the structure, and the degree of branching of the polymer chains are as important as the specific elements contained in the molecule. Plastics containing the carbon–hydrogen bonds, such as PP and PE, and carbon–chlorine bonds, such as PVC, ECTFE, and CTFE are different in the important property of chemical resistance from a fully fluorinated plastic such as PTFE.

The fluoroplastic materials are divided into two groups: fully fluorinated fluorocarbon polymers, such as PTFE, FEP, and PFA, called perfluoropolymers; and the partially fluorinated polymers such as ETFE, PVDF, and ECTFE, which are called fluoropolymers. The polymeric characteristics within each group are similar, but there are important differences between the groups.

The abbreviations used for the common thermoplasts are given in Table P.23, while Table P.24 gives the heat distortion temperatures. Table P.25 provides the tensile strength of the thermoplasts.

Refer to individual thermoplasts for additional information.

Polyvinyl Chloride (PVC)

There are two basic types of PVC produced, Type 1, which is a rigid unplasticized PVC that has optimum chemical resistance, and type 2, which has optimum impact resistance but reduced chemical resistance. Unplasticized PVC resists attack by most acids and strong alkalies as well as gasoline, kerosene, aliphatic alcohols, and hydrocarbons. It is particularly useful in the handling of inorganic materials such as hydrochloric acid. It has been approved by the National Sanitation Foundation for the handling of potable water. Type 2 PVC's resistance to oxidizing and highly alkaline mediums is reduced.

PVC can be attacked by aromatics, chlorinated organic compounds, and lacquer solvents. Refer to "Polyvinyl Chloride" for additional information.

Table P.23 Abbreviations Used for Thermoplastics

ABS	Acrylonitrile–butadiene–styrene
CPVC	Chlorinated polyvinyl chloride
CR	Chloroprene rubber (Neoprene)
CSM	Chlorine sulfonyl polyethylene (Hypalon)
EP	Epoxide epoxy
ECTFE	Ethylene-chlorotrifluorethylene
EPDM	Ethylene propylene rubber
ETFE	Ethylene tetrafluoroethylene
FEP	Perfluorethyleneproplylene
FPM	Fluorine rubber (Viton[a])
HDPE	High density polyethylene
HP	Isobutene isoprene (butyl) rubber
NBR	Nitrile (butadiene) rubber
NR	Natural rubber
PFA	Perfluoralkoxy resin
PA	Polyamide
PB	Polybutylene
PC	Polycarbonate
PCTFE	Polychlorotrifluoroethylene
PF	Phenol-formaldehyde
PP	Polypropylene
PTFE	Polytetrafluoroethylene (Teflon[a])
PVC	Polyvinyl chloride
PVDC	Polyvinylidene chloride
PVDF	Polyvinylidene fluoride

[a]Registered trademark of E. I. DuPont.

Chlorinated Polyvinyl Chloride (CPVC)

Although the corrosion resistance of CPVC is similar to that of PVC, there are enough differences that prevent CPVC from being used in all environments where PVC is used. In general CPVC cannot be used in the presence of most polar organic materials, including chlorinated or aromatic hydrocarbons, esters, and ketones.

It is compatible for use with most acids, alkalies, salts, halogens, and many corrosive wastes. Refer to "Chlorinated Polyvinyl Chloride" for additional information.

Polypropylene

Polypropylene is available as a homopolymer or a copolymer. The homopolymers are generally long-chain high molecular weight molecules with a minimum of random molecular orientation, thus optimizing their chemical, thermal, and physical properties. For maximum corrosion resistance, homopolymers should be used.

Polypropylene is resistant to salt water, crude oil, sulfur-bearing compounds, caustic, solvents, acids, and other organic chemicals. It is not recommended for use with oxidizing type acids, detergents, low boiling hydrocarbons, alcohols, aromatics, and some chlorinated organic materials. Unpigmented polypropylene is degraded by ultraviolet light. Refer to "Polypropylene" for additional information.

Table P.24 Heat Distortion Temperature of the Common Thermoplasts

Polymer	Pressure 66 psi	264 psi	Melt point
PTFE	250°F/121°C	132°F/56°C	620°F/327°C
PVC	135°F/57°C	140°F/60°C	285°F/141°C
LDPE	—	104°F/40°C	221°F/105°C
UHMWPE	155°F/68°C	110°F/43°C	265°F/129°C
PP	225°F/107°C	120°F/49°C	330°F/166°C
PFA	164°F/73°C	118°F/48°C	590°F/310°C
FEP	158°F/70°C	124°F/51°C	554°F/290°C
PVDF	298°F/148°C	235°F/113°C	352°F/178°C
ECTFE	240°F/116°C	170°F/77°C	464°F/240°C
CTFE	258°F/126°C	167°F/75°C	424°F/218°C
ETFE	220°F/104°C	165°F/74°C	518°F/270°C

Polyethylene (PE)

Polyethylene material varies from type to type depending upon the molecular structure and its crystallinity, molecular weight, and molecular weight distribution. The terms low, high, and medium density refer to the ASTM designations based on unmodified polyethylene. The densities, being related to the molecular structure, are indicators of the properties of the final product. High molecular weight (HMW) and ultra high molecular weight (UHMW) are the two forms most often used for corrosion resistance. When PE is exposed to ultraviolet radiation, usually from the sun, photo or light oxidation will occur. In order to protect against this it is necessary to incorporate carbon black into the resin to stabilize it. Other types of stabilizers will not provide complete protection.

PE exhibits a wide range of corrosion resistance ranging from potable water to

Table P.25 Tensile Strength of Thermoplastics 83°F(25°C) at Break

Plastic	Strength (psi)
PVDF	8000
ETFE	6500
CTFE	4500–6000
PFA	4000–4300
ECTFE	7000
PTFE	2500–6000
FEP	2700–3100
PVC	6000–7500
PE	1200–4550
PP	4500–6000
UHMW PE	5600

corrosive wastes. It is resistant to most mineral acids, including sulfuric up to 70% concentration, inorganic salts, including chlorides, alkalies, and many organic acids.

It is not resistant to bromine, aromatics, or chlorinated hydrocarbons. Refer to "Polyethylene" for additional information.

Polybutylene

The combination of stress cracking resistance, chemical resistance, and abrasion resistance makes this polymer extremely useful. It is resistant to acids, bases, soaps, and detergents up to 200°F (93°C). It is not resistant to aliphatic solvents at room temperatures and is partially soluble in aromatic and chlorinated hydrocarbons. Chlorinated water will cause pitting attack. Refer to "Polybutylene" for additional information.

Polyphenylene Sulfide (PPS) (Ryton)

This thermoplast exhibits good resistance to aqueous inorganic salts and bases and is inert to many organic solvents. It also finds applications in oxidizing environments. Refer to "Polyphenylene Sulfide" for additional information.

Polycarbonate

Polycarbonate has exceptional weatherability and good corrosion resistance to mineral acids. Organic solvents will attack the polymer. Strong alkalies will decompose it. It is sold under the tradename of Lexan. Refer to "Polycarbonate" for additional information.

Polyetheretherketone (PEEK)

PEEK exhibits excellent corrosion resistance to a wide range of organic and inorganic chemicals. It is resistant to acetic, nitric, hydrochloric, phosphoric, and sulfuric acids among others. Refer to "Polyetheretherketone" for additional information.

Polyethersulfone (PES)

PES resists most inorganic chemicals but is attacked by strong oxidizing acids. PES has excellent resistance to aliphatic hydrocarbons and aromatics. It is soluble in highly polar solvents and is subject to stress cracking in certain solvents, notably ketones and esters.

Hydrocarbons and mineral oils, greases, and transmission fluids have no effect on PES. Refer to "Polyethersulfone" for additional information.

Phenolics

The phenolics are relatively inert to acids but have little alkaline or bleach resistance. They exhibit a wider range of corrosion resistance as a composite material with a glass filling. Refer to "Phenolic Resins" for additional information.

ABS

This thermoplastic resin is resistant to aliphatic hydrocarbons but not resistant to aromatic and chlorinated hydrocarbons. Refer to "Acrylonitrile–butadiene–styrene" for additional information.

Vinylidene Fluoride (PVDF)

Vinylidene fluoride is chemically resistant to most acids, bases, and organic solvents. It also has the ability to handle wet or dry chlorine, bromine, and other halogens.

PVDF is not suitable for use with strong alkalies, fuming acids, polar solvents, amines, ketones, and esters. When used with strong alkalies it is subject to stress cracking. Refer to "Vinylidene Fluoride" for additional information.

Ethylene Chlorotrifluorethylene (ECTFE)

The chemical resistance of ECTFE is outstanding. It is resistant to strong mineral and oxidizing acids, alkalies, metal etchants, liquid oxygen, and practically all organic solvents except hot amines such as aniline, dimethylamine, etc. Severe stress tests have shown that ECTFE is not subject to chemically induced stress cracking from strong acids, bases, or solvents. Some halogenated solvents can cause ECTFE to become slightly plasticized when it comes into contact with them. Upon removal of the solvent from contact, and upon drying, the mechanical properties of the ECTFE return to their original values, indicating that no chemical attack has taken place.

ECTFE will be attacked by metallic sodium and potassium and fluorine. Refer to "Ethylene Chlorotrifluorethylene" for additional information.

Ethylene–tetrafluoroethylene (ETFE)

ETFE is inert to strong mineral acids, inorganic bases, halogens, and metal salt solutions. Even carboxylic acids, anhydrides, aromatic and aliphatic hydrocarbons, alcohols, aldehydes, ketones, ethers, esters, chlorocarbons, and classic polymer solvents have little effect on ETFE.

Very strong oxidizing acids, such as nitric, near their boiling points at high concentration, will attack ETFE in varying degrees, as will organic bases such as amines and sulfonic acids. Refer to "Ethylene–Tetrafluoroethylene" for additional information.

Polytetrafluoroethylene (PTFE)

PTFE is unique in its corrosion resistant properties. There are very few chemicals that will attack PTFE within normal use temperature. Elemental sodium in intimate contact removes fluorine from the polymer molecule. The other alkali metals (potassium, lithium, etc.) react in a similar manner.

Fluorine and related compounds (e.g., chlorine trifluoride) are absorbed into PTFE resin with such intimate contact that the mixture becomes sensitive to a source of ignition such as impact. These potent oxidizers should only be handled with great care and a recognition of the potential hazards.

The handling of 80% sodium hydroxide, aluminum chloride, ammonia, and certain amines at high temperatures may produce the same effect as elemental sodium. Also, slow oxidative attack can be produced by 70% nitric acid under pressure and at 480°F (250°C). Refer to "Polytetrafluoroethylene" for additional information.

Fluorinated Ethylene Propylene (FEP)

FEP basically exhibits the same corrosion resistance as PTFE, with a few exceptions, but at a lower operating temperature. It is resistant to practically all chemicals, exceptions being the extremely potent oxidizing agents such as chlorine trifluoride and related compounds. Some chemicals will attack FEP when present in high concentrations at or near the service temperature limit of 400°F (200°C). Refer to "Fluorinated Ethylene Propylene" for additional information.

Perfluoroalkoxy (PFA)

PFA is inert to strong mineral acids, inorganic bases, inorganic oxidizers, aromatics, some aliphatic hydrocarbons, alcohols, aldehydes, ketones, ethers, esters, chlorocarbons, fluorocarbons, and mixtures of these.

PFA will be attacked by certain halogenated complexes containing fluorine. These include chlorine trifluoride, bromine trifluoride, iodine pentachloride, and fluorine. It can also be attacked by such metals as sodium or potassium, particularly in their molten state. Refer to "Perfluoroalkoxy" for additional information.

Thermosets

Thermoset resins are "families" of compounds rather than unique individual compounds. Similar to the thermoplast resins, they can be formulated to improve certain specific properties but often at the expense of another property. In the chemical corrosion field there are four families of thermoset resins that are of importance:

The polyesters
The epoxies
The vinyl esters
The furans

Refer to the specific resins for additional information.

Polyester Resins

The members of this family of resins that are of greatest importance for their corrosion resistance are

Isophthalic resins
Bisphenol A–fumarate resins
Hydrogenated–bisphenol A–bisphenol A resins
Halogenated resins
Terephthalate resins

Elastomers

See also individual elastomers. Elastomeric materials fail in the same manner as other polymeric materials. When an elastomer is used as a lining material for a tank or piping, additional consideration must be given to the problems of permeation and absorption.

Physical properties of the elastomer determine the reaction of the elastomer to such physical actions as permeation and absorption. If a lining material is subject to permeation by a corrosive chemical, it is possible for the base metal to be attacked and corroded even though the lining material itself is unaffected. Because of this, permeation and absorption must be taken into account when specifying a lining material. See "Permeation and Absorption."

Environmental Stress Cracking

When a tough polymer is stressed for a long period of time under loads that are small relative to the polymer's yield point, stress cracks develop. Crystallinity is an important factor affecting stress corrosion cracking. The less the crystallization that takes place, the less the likelihood of stress cracking. Unfortunately the lower the crystallinity, the greater the likelihood of permeation.

The presence of contaminants in the fluid may act as an acceleration for stress corrosion cracking. For example, polypropylene can safely handle sulfuric and hydrochloric acids. However, iron or copper contamination in concentrated sulfuric or hydrochloric acid can result in stress cracking of polypropylene.

Outdoor Use

Elastomers in outdoor use can be subject to degradation as a result of the action of ozone, oxygen, and sunlight. These three weathering agents can greatly affect the properties and appearance of a large number of elastomeric materials. Surface cracking, discoloration of colored stocks, and serious loss of tensile strength, elongation, and serious loss of tensile strength, elongation, and other rubber-like properties are the result of this attack.

Natural Rubber (NR)

Cold water preserves natural rubber, but if exposed to air, particularly in sunlight, rubber tends to become hard and brittle. It has only fair resistance to ozone. In general it has poor weathering and aging properties.

Natural rubber offers excellent resistance to most inorganic salt solutions, alkalies, and nonoxidizing acids. Hydrochloric acid will react with rubber to form rubber hydrochloride. Strong oxidizing media such as nitric acid, concentrated sulfuric acid, permanganates, dichromates, chlorine dioxide, and sodium hypochlorite will severely attack rubber. Mineral and vegetable oils, gasoline, benzene, toluene, and chlorinated hydrocarbons also affect rubber. Natural rubber offers good resistance to radiation and alcohols. Refer to "Natural Rubber" for additional information.

Isoprene Rubber (IR)

This is the synthetic form of natural rubber and as such can be used in the same applications as natural rubber. For additional information refer to "Isoprene Rubber."

Neoprene (CR)

Neoprene possesses excellent resistance to sun, weather, and ozone. Because of its low rate of oxidation, neoprene has a high resistance to both outdoor and indoor aging. If severe ozone is to be expected, as for example around electrical equipment, neoprene can be compounded to resist thousands of parts per million of ozone for hours without surface cracking. Natural rubber will crack within minutes when exposed to ozone concentrations of only 50 ppm.

Neoprene provides excellent resistance to attack from solvents, fats, waxes, oils, greases, and many other petroleum based products. A minimum amount of swelling and relatively little loss of strength is experienced when in contact with aliphatic compounds (methyl and ethyl alcohols, ethylene glycols, etc.), aliphatic hydrocarbons, and most refrigerants. Neoprene is also resistant to dilute mineral acids, inorganic salt solutions, and alkalies.

Neoprene has only limited serviceability when exposed to chlorinated and aromatic hydrocarbons, organic esters, aromatic hydroxy compounds, and certain ketones. Highly oxidizing acid and salt solutions cause surface deterioration and loss of strength. This includes such materials as nitric acid and concentrated sulfuric acid. Refer to "Neoprene" for additional information.

Butadiene–Styrene Rubber (SBR, BUNA-S, GR-S)

Buna-S has poor weathering and aging properties. Sunlight will cause it to deteriorate. It does have better water resistance than natural rubber.

The chemical resistance of Buna-S is similar to that of natural rubber. It is resistant to water and exhibits fair to good resistance to dilute acids, alkalies, and alcohol. It is not resistant to oils, gasoline, hydrocarbons, or oxidizing agents. Refer to "Butadiene–styrene" rubber for additional information.

Butyl Rubber (IRR) and Chlorobutyl Rubber (CIIR)

Butyl rubber has excellent resistance to sun, weather, and ozone. Its resistance to water absorption and its weathering qualities are outstanding.

Butyl rubber is resistant to dilute mineral acids, alkalies, phosphate ester oils, acetone, ethylene, ethylene glycol, and water. It is resistant to swelling by vegetable and animal oils.

Butyl rubber is not resistant to concentrated nitric and sulfuric acids, petroleum oils, gasoline, and most solvents, except oxygenated solvents.

Chlorobutyl rubber exhibits the same general resistance as natural rubber but can be used at a higher temperature. It cannot be used with hydrochloric acid even though butyl rubber is suitable. Refer to "Butyl Rubber" and "Chlorobutyl Rubber" for additional information.

Chlorosulfonated Polyethylene Rubber (Hypalon)

Hypalon is one of the most weather resistant elastomers available. Sunlight and ultraviolet light have little if any adverse effect on its physical properties. Many elastomers are degraded by ozone concentrations of 1 ppm parts in air, while Hypalon is unaffected by concentrations as high as 1 ppm per 100 parts of air.

Hypalon is capable of resisting attack by hydrocarbon oils and fats, to such oxidizing chemicals as sodium hypochlorite, sodium peroxide, ferric chloride, and sulfuric, chromic, and hydrofluoric acids. Concentrated hydrochloric acid (37%) at temperatures above 158°F (70°C) will attack Hypalon, but it can be handled in all concentrations below this temperature. Nitric acid at room temperature and up to 60% concentration can also be handled without adverse effects. Hypalon is also resistant to salt solutions, alcohols, and both weak and concentrated alkalies and is generally unaffected by soil chemicals, moisture, and other deteriorating factors associated with burial in the earth.

Hypalon has poor resistance to aliphatic, aromatic, and chlorinated hydrocarbons, aldehydes, and ketones. Refer to "Chlorinated Polyethylene Rubber" for additional details.

Polybutadiene Rubber (BR)

Polybutadiene has good weather resistance but will deteriorate when exposed to sunlight for extended periods of time. It also has poor resistance to ozone.

In general the chemical resistance of BR is similar to that of natural rubber. For additional information refer to "Polybutadiene."

Ethylene-Acrylic Rubber (EA)

The EA elastomers have extremely good resistance to sun, weather, and ozone. Its resistance to water absorption is very good.

The EA elastomers exhibit good resistance to hot oils and hydrocarbon or glycol based lubricants, and to transmission and power steering fluids. Good resistance is also displayed to dilute acids, aliphatic hydrocarbons, gasoline, and animal and vegetable oils.

The EA elastomers will be attacked by esters, ketones, highly aromatic hydrocarbons, and concentrated acids. Refer to "Ethylene-Acrylic Rubber" for additional information.

Acrylate–Butadiene Rubber (ABR) and Acrylic Ester–Acrylic Halide Rubbers (ACM)

These rubbers exhibit good resistance to sun, weather, and ozone.

They have excellent resistance to aliphatic hydrocarbons (gasoline, kerosene) and offer good resistance to water, acids, synthetic lubricants, and silicate hydraulic fluids.

These rubbers will be attacked when exposed to alkalies, aromatic hydrocarbons (benzene, toluene), halogenated hydrocarbons, alcohol, and phosphate hydraulic fluids. Refer to "Butadiene Rubber" and "Acrylic Ester–Acrylic Halide Rubber" for additional information.

Ethylene Propylene Rubber (EPDM and EPT)

Ethylene-propylene rubbers are particularly resistant to sun, weather, and ozone attack. Ozone resistance is inherent in the polymer, and for all practical purposes it can be considered immune to ozone attack. It is not necessary to add any compounding ingredients to produce the immunity.

Ethylene propylene rubbers are resistant to oxygenated solvents, such as acetone, methyl ethyl ketone, ethyl acetate, weak acids and alkalies, detergents, phosphate esters, alcohols, and glycols.

The elastomer will be attacked by hydrocarbon solvents and oils, chlorinated hydrocarbons, and turpentine. EPT rubbers, in general, are resistant to most of the same corrodents as EPDM. Refer to "Ethylene Propylene Rubbers" for additional information.

Styrene–Butadiene–Styrene Rubber (SBS)

The SBS rubbers are not resistant to sun, weather, or ozone. Their chemical resistance is similar to that of natural rubber. They have excellent resistance to water, acids, and bases. Refer to "Styrene–Butadiene–Styrene Rubber" for additional information.

Styrene–Ethylene–Butylene–Styrene Rubber (SEBS)

The SEBS rubbers possess excellent resistance to ozone. For prolonged outdoor exposure the addition of an ultraviolet light absorber or carbon black pigment or both is required. The chemical resistance of the SEBS rubbers is similar to that of natural rubber while possessing excellent resistance to water, acids, and bases. Refer to "Styrene–Ethylene–Butylene–Styrene Rubber" for additional information.

Polysulfide Rubbers (ST and FA)

FA polysulfide rubbers possess excellent resistance to ozone, weathering, and ultraviolet light. ST polysulfide rubber, compounded with carbon black, is resistant to ultraviolet light and sunlight. It also has satisfactory weather resistance.

The polysulfide rubbers exhibit excellent resistance to oils, gasoline, and aliphatic and aromatic hydrocarbon solvents, good water and alkali resistance, and fair acid resistance. The FA polysulfide rubbers are more resistant to solvents than the ST polysulfide rubbers. The ST rubbers exhibit better resistance to chlorinated organics than the FA rubbers.

The polysulfide rubbers are not resistant to strong concentrated inorganic acids such as sulfuric, nitric, and hydrochloric. Refer to "Polysulfide Rubbers" for additional information.

Urethane Rubbers (AU)

The urethane rubbers exhibit excellent resistance to ozone attack and have good resistance to weathering. Extended exposure to ultraviolet light will cause the rubbers to darken and reduce their physical properties. The addition of pigments or ultraviolet-screening agents will prevent this.

The urethane rubbers are resistant to most mineral and vegetable oils, grease and fuels, and aliphatic, aromatic and chlorinated hydrocarbons.

Aromatic hydrocarbons, polar solvents, esters, ethers, and ketones will attack the urethane rubbers. The urethane rubbers have limited service in weak acid solutions and cannot be used in concentrated acids. Neither are they resistant to caustic or steam. Refer to "Urethane Rubbers" for additional information.

Polyamides

Of the many varieties of polyamides (Nylons) produced, only grades 11 and 12 find application as elastomeric materials. They are resistant to sun, weather, and ozone.

The polyamides are resistant to most inorganic alkalies, particularly ammonium hydroxide and ammonia at elevated temperatures, and sodium and potassium hydroxide at ambient temperatures. They are also resistant to almost all inorganic salts and almost all hydrocarbons and petroleum based fuels. At normal temperatures they are also resistant to organic acids (citric, lactic, oleic, oxalic, stearic, tartaric, and uric) and most aldehydes and ketones.

The polyamides have limited resistance to hydrochloric, sulfonic, and phosphoric acids at ambient temperatures.

Refer to "Polyamides" for additional information.

Polyester Elastomer (PE)

The polyesters exhibit excellent resistance to ozone and good resistance to weathering. When formulated with proper additives they are capable of exhibiting very good resistance to sunlight aging.

Polyester elastomers have excellent resistance to nonpolar materials such as oils and hydraulic fluids, even at elevated temperatures. At room temperature they are resistant to most polar fluids such as acids, bases, amines, and glycols. Resistance is very poor at temperatures of 158°F (70°C) or higher. Refer to "Polyester Elastomer" for additional information.

Thermoplastic Elastomers-Olefinic Type (TPE)

The TPEs exhibit good resistance to sun, weather, and ozone. Their water resistance is excellent, showing essentially no property changes after prolonged exposure to water at elevated temperatures.

The TPEs display reasonably good resistance to oils and automotive fluids, comparable to that of neoprene. However, they do not have the outstanding oil resistance of the polyester elastomers. Refer to "Thermoplasic Elastomers" for additional information.

Silicone (SI) and Fluorosilicone (FSI) Rubbers

The SI and FSI rubbers show excellent resistance to sun, weathering, and ozone, even after long-term exposure.

The silicone rubbers are resistant to dilute acids and alkalies, alcohols, animal and vegetable oils, and lubricating oils and aliphatic hydrocarbons. Aromatic solvents such as benzene, toluene, gasoline, and chlorinated solvents, and high temperature steam, will attack SI rubbers.

The FSI rubbers have better chemical resistance than the SI rubbers. They possess excellent resistance to aliphatic hydrocarbons and good resistance to aromatic hydrocarbons, oil and gasoline, animal and vegetable oils, dilute acids and alkalies, and alcohols. Refer to "Silicone" and "Fluorosilicone Rubbers" for additional information.

Vinylidene Fluoride (PVDF)

PVDF is highly resistant to the chlorinated solvents, aliphatic solvents, weak bases and salts, strong acids, halogens, strong oxidants, and aromatic solvents.

Strong bases will attack PVDF. Sodium hydroxide can cause stress cracking. Refer to "Vinylidene Fluoride Elastomers" for additional information.

Ethylene–Tetrafluoroethylene Elastomer (ETFE)

Because of ETFE's outstanding resistance to sunlight, ozone, and weather, coupled with its wide range of corrosion resistance, it is ideally suited for outdoor applications subject to atmospheric corrosion.

ETFE is inert to strong mineral acids, inorganic bases, halogens, and metal salt solutions. Carboxylic acids, anhydrides, aromatic and aliphatic hydrocarbons, alcohols, aldehydes, ketones, esters, ethers, chlorocarbons, and classic polymer solvent have little effect on ETFE.

Strong oxidizing acids near their boiling points, such as nitric acid at high concentrations, organic bases, such as amines, and sulfonic acids will have a deleterious effect on ETFE. Refer to "Ethylene–Tetrafluorethylene Elastomer" for additional information.

Ethylene–Chlorotrifluoroethylene Elastomer (ECTFE)

ECTFE is extremely resistant to sun, weather, and ozone attack.

Ethylene–chlorotrifluoroethylene is resistant to strong mineral and oxidizing acids, alkalies, metal etchants, liquid oxygen, and practically all organic solvents except hot amines (aniline, dimethylamine, etc.).

ECTFE will be attacked by metallic sodium and potassium. Refer to "Ethylene–Chlorotrifluoroethylene Elastomer" for additional information.

Perfluoroelastomers (FPM)

The FPM elastomers have excellent resistance to sun, weather, and ozone, even after long term exposure.

The perfluoroelastomers are resistant to polar solvents (ketones, esters, ethers), strong organic solvents (benzene, dimethyl formamide), inorganic and organic acids (hydrochloric, nitric, sulfuric) and bases, strong oxidizing agents (fuming nitric acid), metal halides, chlorine wet and dry, inorganic salt solutions, hydraulic fluids, and heat transfer fluids. Refer to "Perfluoroelastomers" for additional information.

See Refs. 9, 8, 10, 7, 17, and 15.

POLYMER CONCRETES

Polymer concretes are totally chemical resistant synthetic resin compounds. They pass the total immersion test at varying temperatures for extended periods of time. Polymer concretes are not to be confused with polymer modified portland cement concrete. Polymer modified portland cement concrete can use some of the same generic resins as used in polymer concretes, but with different results.

The polymer concretes are usually formulated from the resins

1. Furan
2. Epoxy
3. Polyester
4. Acrylics
5. Sulfur
6. Vinyl ester

The major advantage of the polymer concretes are their wide range of corrosion resistance. The polymer modified portland cement concretes offer the advantages of

1. Thinner concrete cross-sections can be applied.
2. They lower absorption of concrete.
3. They improve impact resistance.
4. They provide improved adhesion for pours onto existing concrete.
5. They provide improved corrosion resistance to salt, but not to aggressive corrosive chemicals such as hydrochloric acid.

Polymer concretes are usually installed at thicknesses of greater than 1/2 inch.

A monolithic surfacing or topping is usually installed at thicknesses of 1/16 to 1/2 inch. The most popular monolithic surfacings are formulated from the resins

1. Epoxy
2. Polyester
3. Vinyl ester
4. Acrylic
5. Urethane

Monolithic surfacings are versatile materials used primarily as flooring systems.

The chemical resistance of monolithic surfacings and polymer concretes is the same as the chemical resistance of their mortar counterparts. Refer to "Mortars."

See Refs. 18, 8, 19, and 20.

POLYPHENYLENE SULFIDE (PPS)

See also "Polymers."

Polyphenylene sulfide is a thermoplastic capable of being used at high temperatures. It has a maximum service rating of 450°F (230°C). As the temperature increases, there is a corresponding increase in toughness. Table P.26 provides the physical and mechanical properties of PPS.

The structure is

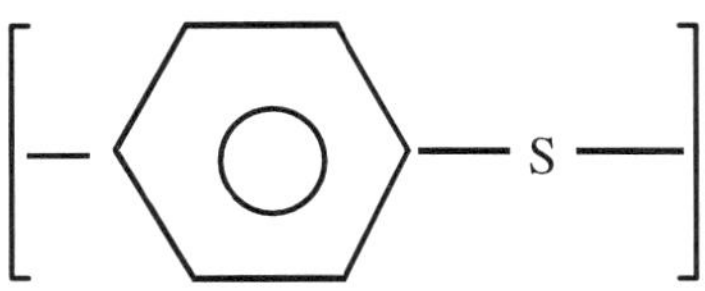

Polyphenylene sulfide offers excellent resistance to aqueous inorganic salts and bases, and many organic solvents. It can also be used under highly oxidizing conditions. Relatively few materials react with PPS at high temperatures. PPS is resistant to organic solvents except for chlorinated solvents, some halogenated gases, and alkyl amines. It stress cracks

Table P.26 Physical and Mechanical Properties of PPS

Specific gravity	1.34
Water absorption 24 h at 73°F/23°C, %	0.01
Tensile strength at 73°F/23°C, psi	10,800
Modulus of elasticity in tension at 73°F/23°C × 10^5 psi	4.8–6.3
Compressive strength, psi	16,000
Flexural modulus, psi × 10^5	11–20
Izod impact strength, notched at 73°F/23°C, ft-lb/in.	0.03
Coefficient of thermal expansion in./in. °F × 10^{-5}	2.7–3.0
Thermal conductivity, Btu/h/ft^2/°F/in.	2.0
Heat distortion temperature, °F/°C at 264 psi	275/135
Limiting oxygen index, %	47
Underwriters lab rating Sub 94	SEO

in chlorinated solvents. Weak and strong alkalies have no effect on PPS. Polyphenylene sulfide is resistant to weak acids with the exception of hydrochloric. Strong oxidizing acids such as sulfuric, nitric, chromic, and 10% perchloric will attack PPS. Refer to Table P.27 for the compatibility of PPS (Ryton) with selected corrodents.

POLYPROPYLENE (PP)

See also "Polymers."

Polypropylene is one of the most common and versatile thermoplastics. It is closely related to polyethylene, both of which are members of a group known as "polyolefins." The polyolefins are composed of only carbon and hydrogen. When unmodified, PP is the lightest of the common thermoplastics, having a specific gravity of 0.91. In addition to its light weight it has the advantages of high heat resistance, stiffness, and a wide range of chemical resistance. Within the chemical structure of PP a distinction is made between isotactic PP and atactic PP; the isotactic form accounts for 97% of the PP. This form is highly ordered:

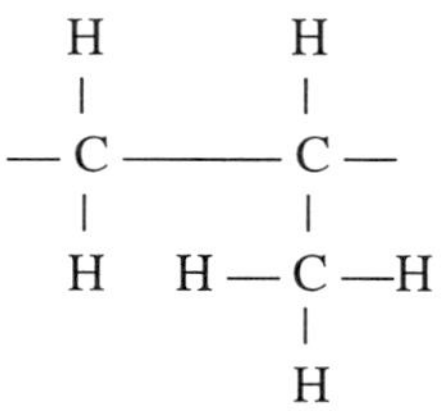

Atactic PP is a viscous liquid type PP having a PP polymer matrix.

Polypropylene can be produced either as a homopolymer or as a copolymer with polyethylene. The copolymer is less brittle than the homopolymer and is able to withstand impact forces down to –20°F (–29°C), while the homopolymer is extremely brittle below 40°F (4°C). The physical and mechanical properties are shown in Table P.28.

Although the copolymers have increased impact resistance, their tensile strength and stiffness are considerably lower, increasing the potential for distortion and cold flow, particuarly at elevated temperatures.

The homopolymers, being long-chain, high molecular weight molecules with a minimum of random orientation, have optimum chemical, thermal, and physical properties.

Table P.27 Compatibility of Polyphenylene Sulfide with Selected Corrodents[a]

Chemical	Max. temp. °F	°C	Chemical	Max. temp. °F	°C
Acetaldehyde	230	110	Bromine liquid	x	x
Acetamide	250	121	Butadiene	100	38
Acetic acid 10%	250	121	Butyl acetate	250	121
Acetic acid 50%	250	121	Butyl alcohol	200	93
Acetic acid 80%	250	121	*n*-Butaylamine	200	93
Acetic acid, glacial	190	88	Butyric acid	240	116
Acetic anhydride	280	138	Calcium bisulfite	200	93
Acetone	260	127	Calcium carbonate	300	149
Acrylic acid 25%	100	38	Calcium chloride	300	149
Acrylonitrile	130	54	Calcium hydroxide 10%	300	149
Adipic acid	300	149	Calcium hydroxide, sat.	300	149
Alum	300	149	Carbon bisulfide	200	93
Aluminum acetate	210	99	Carbon dioxide, dry	200	93
Aluminum chloride, aqueous	300	149	Carbon disulfide	200	93
Aluminum chloride, dry	270	132	Carbon tetrachloride	120	49
Aluminum hydroxide	250	121	Cellosolve	220	104
Aluminum nitrate	250	121	Chloracetic acid	190	88
Aluminum oxychloride	460	238	Chlorine gas, dry	x	x
Ammonia gas	250	121	Chlorine gas, wet	x	x
Ammonium carbonate	460	238	Chlorine, liquid	200	93
Ammonium chloride 10%	300	149	Chloroform	150	66
Ammonium chloride 50%	300	149	Chlorosulfonic acid	x	x
Ammonium chloride, sat.	300	149	Chromic acid 10%	200	93
Ammonium hydroxide 25%	250	121	Chromic acid 50%	200	93
Ammonium hydroxide, sat.	250	121	Citric acid 15%	250	121
Ammonium nitrate	250	121	Citric acid, conc.	250	121
Ammonium phosphate 65%	300	149	Copper acetate	300	149
Ammonium sulfate 10–40%	300	149	Copper chloride	220	104
Amyl acetate	300	149	Copper cyanide	210	99
Amyl alcohol	210	99	Copper sulfate	250	121
Amyl chloride	200	93	Cresol	200	93
Aniline	300	149	Cupric chloride 5%	300	149
Barium carbonate	200	93	Cyclohexane	190	88
Barium chloride	200	93	Cyclohexanol	250	121
Barium hydroxide	200	93	Dichloroethane		
Barium sulfate	220	104	(ethylene dichloride)	210	99
Barium sulfide	200	93	Ethylene glycol	300	149
Benzaldehyde	250	121	Ferric chloride	210	99
Benzene	300	149	Ferric chloride 50% in water	210	99
Benzene sulfonic acid 10%	250	121	Ferric nitrate 10–50%	210	99
Benzoic acid	230	110	Ferrous chloride	210	99
Benzyl alcohol	200	93	Ferrous nitrate	210	99
Benzyl chloride	300	149	Fluorine gas,dry	x	x
Borax	210	99	Hydrobromic acid, dilute	200	93
Boric acid	210	99	Hydrobromic acid 20%	200	93
Bromine gas, dry	x	x	Hydrobromic acid 50%	200	93
Bromine gas, moist	x	x	Hydrochloric acid 20%	230	110

Table P.27 Continued

Chemical	Max. temp. °F	Max. temp. °C	Chemical	Max. temp. °F	Max. temp. °C
Hydrochloric acid 38%	210	99	Sodium hydroxide 10%	210	99
Hydrocyanic acid 10%	250	121	Sodium hydroxide 50%	210	99
Hydrofluoric acid 30%	200	93	Sodium hypochlorite 5%	200	93
Lactic acid 25%	250	121	Sodium hypochlorite, concentrated	250	121
Lactic acid, concentrated	250	121	Sodium sulfide to 50%	230	110
Magnesium chloride	300	149	Stannic chloride	210	99
Methyl ethyl ketone	200	93	Sulfuric acid 10%	250	121
Methyl isobutyl ketone	250	121	Sulfuric acid 50%	250	121
Muriatic acid	210	99	Sulfuric acid 70%	250	121
Nitric acid 5%	150	66	Sulfuric acid 90%	220	104
Nitric acid 20%	100	38	Sulfuric acid, fuming	80	27
Oleum	80	27	Sulfurous acid 10%	200	93
Phenol 88%	300	149	Thionyl chloride	x	x
Phosphoric acid 50–80%	220	104	Toluene	300	149
Potassium bromide 30%	200	93	Zinc chloride 70%	250	121
Sodium carbonate	300	149			
Sodium chloride	300	149			

[a]The chemicals listed are in the pure state or in a saturated solution unless otherwise indicated. Compatibility is shown to the maximum allowable temperature for which data are available. Incompatibility is shown by an x. A blank space indicates that the data are unavailable.
Source: Schweitzer, PA. *Corrosion Resistance Tables*, 4th ed. Vols 1–3. New York: Marcel Dekker, 1995.

For this reason, homopolymer material is preferred for difficult chemical, thermal, and physical conditions.

Polypropylene is subject to degradation by ultraviolet light. Therefore if it is exposed to sunlight, an ultraviolet absorber or screening agent must be used to protect the material. It is not affected by most inorganic chemicals, except the halogens and severe oxidizing conditions. PP can be used with sulfur-bearing compounds, caustics, solvents, acids, and other organic chemicals.

It should not be used with oxidizing type acids, detergents, low boiling hydrocarbons, alcohols, aromatics, and some chlorinated organic materials. Refer to Table P.29 for the compatibility of polypropylene with selected corrodents. Reference 8 provides a wider range of corrodents and the compatibility of PP with them.

See Refs. 8 and 10.

POLYSILOXANE RUBBER

See "Silicone Rubber."

Table P.28 Physical and Mechanical Properties of Copolymer and Homopolymer PP

Property	Homopolymer	Copolymer
Specific gravity	0.905	0.91
Water absorption, 24 h at 73°F/23°C, %	0.02	0.03
Tensile strength at 73°F/23°C, psi	5000	4000
Modulus of elasticity in tension at 73°F/23°C $\times 10^5$ psi	1.7	1.5
Compressive strength, psi	9243	8500
Flexural strength, psi	7000	—
Izod impact strength, notched at 73°F/23°C, ft-lb/in.	1.3	8
Coefficient of thermal expansion in./in. °F $\times 10^{-5}$	5.0	6.1
Thermal conductivity, Btu/h/ft^2/°F/in.	1.2	1.3
Heat distortion temperature °F/°C		
at 66 psi	220/107	220/107
at 264 psi	140/60	124/49
Limiting oxygen index, %	17	—
Flame spread	Slow burning	
Underwriters lab rating Sub 94	94 HB	

POLYSULFIDE RUBBERS (ST AND FA)

Polysulfide rubbers are manufactured by combining ethylene ($CH_2 = CH_2$) with an alkaline polysulfide. The sulfur forms a part of the polymerized molecule. They are also known as Thiokol rubbers. In general these elastomers do not have great elasticity, but they do have good resistance to most solvents. Compared to nitrile rubber they have poor tensile strength, a pungent odor, poor rebound, high creep under strain, and poor abrasion resistance.

Modified organic polysulfides are made by substituting other unsaturated compounds for ethylene, which results in compounds that have little objectionable odor.

Physical and Mechanical Properties

ST polysulfide rubber is prepared from bis(2-chloroethyl) formal and sodium polysulfide. Products made from these rubbers have good low temperature properties and exhibit outstanding resistance to oils and solvents, to gas permeation, and to weathering. Pure gum vulcanizates possess poor physical properties, and as a result all practical compounds contain reinforcing fillers, usually carbon blacks.

Compounds can be used continuously at temperatures of 212°F (100°C), and intermittently at temperatures up to 300°F (149°C). The polymer does not melt or soften at elevated temperatures, but a gradual shortening in elongation and drop in tensile strength do occur. As the temperature increases, the effect becomes more pronounced. With the addition of plasticizers, ST compounds can be made to remain flexible at temperatures below –60°F (–51°C).

Compression set values for ST elastomers are shown in Table P.30. The use of ST

Table P.29 Compatibility of PP with Selected Corrodents[a]

Chemical	Maximum temp. °F	°C	Chemical	Maximum temp. °F	°C
Acetaldehyde	120	49	Benzaldehyde	80	27
Acetamide	110	43	Benzene	140	60
Acetic acid 10%	220	104	Benzene sulfonic acid 10%	180	82
Acetic acid 50%	200	93	Benzoic acid	190	88
Acetic acid 80%	200	93	Benzyl alcohol	140	60
Acetic acid, glacil	190	88	Benzyl chloride	80	27
Acetic anhydride	100	38	Borax	210	99
Acetone	220	104	Boric acid	220	104
Acetyl chloride	x	x	Bromine gas, dry	x	x
Acrylic acid	x	x	Bromine gas, moist	x	x
Acrylonitrile	90	32	Bromine liquid	x	x
Adipic acid	100	38	Butadiene	x	x
Allyl alcohol	140	60	Butyl acetate	x	x
Allyl chloride	140	60	Butyl alcohol	200	93
Alum	220	104	*n*-Butylamine	90	32
Aluminum acetate	100	38	Butyl phthalate	180	82
Aluminum chloride, aqueous	200	93	Butyric acid	180	82
Aluminum chloride, dry	220	104	Calcium bisulfide	210	99
Aluminum fluoride	200	93	Calcium bisulfite	210	99
Aluminum hydroxide	200	93	Calcium carbonate	210	99
Aluminum nitrate	200	93	Calcium chlorate	220	104
Aluminum oxychloride	220	104	Calcium chloride	220	104
Aluminum sulfate			Calcium hydroxide 10%	200	93
Ammonia gas	150	66	Calcium hydroxide, sat.	220	104
Ammonium bifluoride	200	93	Calcium hypochlorite	210	99
Ammonium carbonate	220	104	Calcium nitrate	210	99
Ammonium chloride 10%	180	82	Calcium oxide	220	104
Ammonium chloride 50%	180	82	Calcium sulfate	220	104
Ammonium chloride, sat.	200	93	Caprylic acid	140	60
Ammonium fluoride 10%	210	99	Carbon bisulfide	x	x
Ammonium fluoride 25%	200	93	Carbon dioxide, dry	220	104
Ammonium hydroxide 25%	200	93	Carbon dioxide, wet	140	60
Ammmonium hydroxide, sat.	200	93	Carbon disulfide	x	x
Ammonium nitrate	200	93	Carbon monoxide	220	104
Ammonium persulfate	220	104	Carbon tetrachloride	x	x
Ammonium phosphate	200	93	Carbonic acid	220	104
Ammonium sulfate 10–40%	200	93	Cellosolve	200	93
Ammonium sulfide	220	104	Chloracetic acid, 50% water	80	27
Ammonium sulfite	220	104	Chloracetic acid	180	82
Amyl acetate	x	x	Chlorine gas, dry	x	x
Amyl alcohol	200	93	Chlorine gas, wet	x	x
Amyl chloride	x	x	Chlorine, liquid	x	x
Aniline	180	82	Chlorobenzene	x	x
Antimony trichloride	180	82	Chlroform	x	x
Aqua regia 3:1	x	x	Chlorosulfonic acid	x	x
Barium carbonate	200	93	Chromic acid 10%	140	60
Barium chloride	220	104	Chromic acid 50%	150	66
Barium hydroxide	200	93	Chromyl chloride	140	60
Barium sulfate	200	93	Citric acid 15%	220	104
Barium sulfide	200	93	Citric acid, conc.	220	104

Table P.29 Continued

Chemical	Maximum temp. °F	°C	Chemical	Maximum temp. °F	°C
Copper acetate	80	27	Muriatic acid	200	93
Copper carbonate	200	93	Nitric acid 5%	140	60
Copper chloride	200	93	Nitric acid 20%	140	60
Copper cyanide	200	93	Nitric acid 70%	x	x
Copper sulfate	200	93	Nitric acid, anhydrous	x	x
Cresol	x	x	Nitrous acid, concentrated	x	x
Cupric chloride 5%	140	60	Oleum	x	x
Cupric chloride 50%	140	60	Perchloric acid 10%	140	60
Cyclohexane	x	x	Perchloric acid 70%	x	x
Cyclohexanol	150	66	Phenol	180	82
Dichloroacetic acid	100	38	Phosphoric acid 50–80%	210	99
Dichloroethane			Picric acid	140	60
(ethylene dichloride)	80	27	Potassium bromide 30%	210	99
Ethylene glycol	210	99	Salicylic acid	130	54
Ferric chloride	210	99	Silver bromide 10%	170	77
Ferric chloride 50% in water	210	99	Sodium carbonate	220	104
Ferric nitrate 10–50%	210	99	Sodium chloride	200	93
Ferrous chloride	210	99	Sodium hydroxide 10%	220	104
Ferrous nitrate	210	99	Sodium hydroxide 50%	220	104
Fluorine gas, dry	x	x	Sodium hydroxide,		
Fluorine gas, moist	x	x	concentrated	140	60
Hydrobromic acid, dilute	230	110	Sodium hypochlorite 20%	120	49
Hydrobromic acid 20%	200	93	Sodium hypochlorite,		
Hydrobromic acid 50%	190	88	concentrated	110	43
Hydrochloric acid 20%	220	104	Sodium sulfide to 50%	190	88
Hydrochloric acid 38%	200	93	Stannic chloride	150	66
Hydrocyanic acid 10%	150	66	Stannous chloride	200	93
Hydrofluoric acid 30%	180	82	Sulfuric acid 10%	200	93
Hydrofluoric acid 70%	200	93	Sulfuric acid 50%	200	93
Hydrofluoric acid 100%	200	93	Sulfuric acid 70%	180	82
Hypochlorous acid	140	60	Sulfuric acid 90%	180	82
Iodine solution 10%	x	x	Sulfuric acid 98%	120	49
Ketones, general	110	43	Sulfuric acid 100%	x	x
Lactic acid 25%	150	66	Sulfuric acid, fuming	x	x
Lactic acid, concentrated	150	66	Sulfurous acid	180	82
Magnesium chloride	210	99	Thionyl chloride	100	38
Malic acid	130	54	Toluene	x	x
Manganese chloride	120	49	Trichloroacetic acid	150	66
Methyl chloride	x	x	White liquor	220	104
Methyl ethyl ketone	x	x	Zinc chloride	200	93
Methyl isobutyl ketone	80	27			

[a]The chemicals listed are in the pure state or in a saturated solution unless otherwise indicated. Compatibility is shown to the maximum allowable temperature for which data are available. Incompatibility is shown by an x. A blank space indicates that the data are unavailable.

Source: Schweitzer, PA. *Corrosion Resistance Tables,* 4th ed. Vols 1–3. New York: Marcel Dekker, 1995.

Table P.30 Physical and Mechanical Properties of Polysulfide ST Rubber[a]

Specific gravity		1.27	
Brittle point		–60°F (–51°C)	
Permeability			
to helium gas, cm^3/sec-in.^{2}, 0.1-in. thick film		1.5	
to solvents, fl oz/in.-24 h-ft^2	75°F (24°C)		180°F (82°C)
methanol	0.005		0.140
tetrachloride	0.042		0.210
acetate	0.150		0.510
benzene	0.540		1.800
diisobuytlene	0.0		0.001
methyl ethyl ketone	0.28		0.65
Tensile strength, psi		500–1750	
Elongation, % at break		230–450	
Hardness, Shore A		30–90	
Abrasion resistance		Good	
Maximum temperature, continuous use		212°F (100°C)	
Compression set, % after 22 h			
at 158°F (70°C)		15	
at 212°F (100°C)		75	
Machining qualities		Excellent	
Resistance to sunlight		Excellent	
Effect of aging		None	
Resistance to heat		Fair	

[a]These are representative values since they may be altered by compounding.

rubber in compressive applications should be limited to those applications whose service temperature does not exceed 200°F (93°C).

ST polysulfide rubber is blended with nitrile rubber (NBR) and neoprene to obtain a balance of properties unattainable with either polymer alone. High ratios of ST to NBR or neoprene W decrease the swelling from aromatics, fuels, ketones, and esters. The low temperature flexibility is also improved. Higher ratios of NBR or neoprene W to ST result in improvements in physical properties, tear strength, and compressive set resistance before and after heat aging.

The electrical insulating properties of ST are poor, as is its flame resistance.

Table P.30 lists the physical and mechanical properties of ST polysulfide.

FA polysulfide is prepared by reacting a mixture of bis (2-chloroethyl) formal and ethylene dichloride with sodium polysulfide. Cured compounds are particularly noted for outstanding volume swell resistance to aliphatic and aromatic solvents; resistance to alcohols, ketones, and esters; exceptional low permeability to gases, water, and organic liquids; and excellent low temperature flexibility.

FA polysulfide rubbers that contain no plasticizers are flexible at temperatures as low as –50°F (–45°C), and they retain their excellent flexing characteristics at subnormal temperatures even when oil and solvent are present. With the addition of plasticizers or reinforcing pigments, a wide range of hardness can be achieved. Depending on the hardness, tensile strengths up to 1500 psi can be developed. When the materials are immersed in solvents, they retain a very high percentage of this tensile strength. The FA

Table P.31 Physical and Mechanical Properties of Polysulfide FA Rubber[a]

Specific gravity	1.34
Refractive index	1.65
Brittle point	–30°F (–35°C)
Dielectric constant	
at 1 kHz	7.3
at 1 MHz	6.8
Dissipation factor	
at 1 kHz	5.3×10^{-3}
at 1MHz	5.2×10^{13}
Volume resistivity, ohm-cm	
at 73°F (23°C)	5×10^{13}
at 140°F (60°C)	2×10^{12}
Surface resistivity, ohms	
at 73°F (23°C)	7×10^{14}
at 140°F (60°C)	2×10^{14}
Permeability at 75°F (24°C)[b]	
to methanol	0.001
to carbon tetrachloride	0.01
to ethyl acetate	0.04
to benzene	0.14
Permeability at room temperature[c]	
to hydrogen	14.8×10^{-6}
to helium	9.4×10^{-6}
Swelling, % by volume	
in kerosene at 77°F (25°C)	4
in benzene at 77°F (25°C)	50
in acetone at 77°F (25°C)	25
in mineral oil at 158°F (70°C)	1
Tensile strength, psi	150–1200
Elongation, % at break	210–700
Hardness, Shore A	25–90
Abrasion resistance	Fair
Maximum temperture, continuous use	250°F (121°C)
Machining qualities	Excellent
Resistance to sunlight	Excellent
Effect of aging	None

[a]These are representative values since they may be altered by compounding.
[b]$^1/_{16}$-in. (1.6 mm), sheet, in.-oz.-in.^{2} (24 h)-ft.
[c]0.25-mm thick sheet in cm^3/cm^2-min.

polysulfide rubbers have a wider operating temperature range than the ST elastomers. The FA series remain serviceable over a range of –50 to 250°F (–45 to 121°C).

The electrical properties of the FA polysulfide rubbers are good, but their flame resistance is poor. Table P.31 lists the physical and mechanical properties of FA polysulfide rubber.

ST polysulfide rubber compounded with carbon black is resistant to ultraviolet light and sunlight. Its resistance to ozone is good but can be improved by the addition of NBC,

though this addition can degrade the material's compression set. ST polysulfide rubber also possesses satisfactory weather resistance.

Chemical Resistance

The polysulfide rubbers possess outstanding resistance to solvents. They exhibit excellent resistance to oils, gasoline, and aliphatic and aromatic hydrocarbon solvents, very good water resistance, good alkali resistance, and fair acid resistance. FA polysulfide rubbers are somewhat more resistant to solvents than the ST rubbers. Compounding of the FA polymers with NBR will provide high resistance to aromatic solvents and improve the physical properties of the blend. For high resistance to esters and ketones, neoprene W is compounded with FA polysulfide rubber to produce improved physical properties.

ST polysulfide rubbers exhibit better resistance to chlorinated organics than the FA polysulfide rubbers. Contact of either rubber with strong, concentrated inorganic acids such as sulfuric, nitric, or hydrochloric should be avoided. Refer to Table P.32 for the compatibility of polysulfide ST rubber with selected corrodents.

Applications

FA polysulfide rubber is one of the elastomeric materials commonly used for the fabrication of rubber rollers for printing and coating equipment. The major reason for this is its high degree of resistance to the many types of solvents, including ketones, esters, aromatic hydrocarbons, and plasticizers, that are used as vehicles for the various printing inks and coatings.

Applications are also found in the fabrication of hose and hose liners for the handling of aromatic solvents, esters, ketones, oils, fuels, gasolines, paints, lacquers, and thinners. Large amounts of material are also used to produce caulking compounds, cement, paint can gaskets, seals, and flexible mountings.

The impermeability of the polysulfide rubbers to air and gas has promoted the use of these materials for inflatable products such as life jackets, life rafts, balloons, and other items.

Resistance to Sun, Weather, and Ozone

FA polysulfide rubber compounds display excellent resistance to ozone, weathering, and exposure to ultraviolet light. Their resistance is superior to that of the ST polysulfide rubbers. If high concentrations of ozone are to be present, the use of 0.5 parts of nickel dibutyldithiocarbamate (NBC) per 100 parts of FA polysulfide rubber will improve the ozone resistance.

ST polysulfide rubber compounded with carbon black is resistant to ultraviolet light and sunlight. Its resistance to ozone is good but can be improved by the addition of NBC, though this addition can degrade the material's compression set. ST polysulfide rubber also possesses satisfactory weather resistance.

See Refs. 7 and 8.

Table P.32 Compatibility of Polysulfide ST Rubber with Selected Corrodents[a]

Chemical	Max. temp. °F	Max. temp. °C	Chemical	Max. temp. °F	Max. temp. °C
Acetamide	x	x	Citric acid 15%	x	x
Acetic acid 10%	80	27	Citric acid, conc.	x	x
Acetic acid 50%	80	27	Copper sulfate	x	x
Acetic acid 80%	80	27	Corn oil	90	32
Acetic acid, glacial	80	27	Cottonseed oil	90	32
Acetone	80	27	Cresol	x	x
Ammonia gas	x	x	Diacetone alcohol	80	27
Ammonium chloride 10%	150	66	Dibutyl phthalate	80	27
ammonium chloride 28%	150	66	Ethers, general	90	32
Ammonium chloride 50%	150	66	Ethyl acette	80	27
Ammonium chloride, sat.	90	32	Ethyl alcohol	80	27
Ammonium hydroxide 10%	x	x	Ethylene chloride	80	27
Ammonium hydroxide 25%	x	x	Ethylene glycol	150	66
Ammonium hydroxide, sat.	x	x	Formaldehyde dilute	80	27
Ammonium sulfate 10–40%	x	x	Formaldehyde 37%	80	27
Amyl alcohol	80	27	Formaldehyde 50%	80	27
Aniline	x	x	Glycerine	80	27
Benzene	x	x	Hydrochloric acid dilute	x	x
Benzoic acid	150	66	Hydrochloric acid 20%	x	x
Bromine water, dilute	80	27	Hydrochloric acid 35%	x	x
Bromine water, sat.	80	27	Hydrochloric acid 38%	x	x
Butane	150	66	Hydrochloric acid 50%	x	x
Butyl acetate	80	66	Hydrofluoric acid, dilute	x	x
Butyl alcohol	80	66	Hydrofluoric acid 30%	x	x
Calcium chloride dilute	150	66	Hydrofluoric acid 40%	x	x
Calcium chloride, sat.	150	66	Hydrofluoric acid 50%	x	x
Calcium hydroxide 10%	x	x	Hydrofluoric acid 70%	x	x
Calcium hydroxide 20%	x	x	Hydrofluoric acid 100%	x	x
Calcium hydroxide 30%	x	x	Hydrogen peroxide,		
Calcium hydroxide, sat.	x	x	all concentrations	x	x
Cane sugar liquors	x	x	Lactic acid,		
Carbon bisulfide	x	x	all concentrations	x	x
Carbon tetrachloride	x	x	Methyl ethyl ketone	150	66
Carbonic acid	150	66	Methyl isobutyl ketone	80	27
Castor oil	80	27	Monochlorobenzene	x	x
Cellosolve	80	66	Muriatic acid	x	x
Chlorine water, sat.	x	x	Nitric acid, all concentrations	x	x
Chlorobenzene	x	x	Oxalic acid, all concentrations	x	x
Chloroform	x	x	Phenol, all concentrations	x	x
Chromic acid 10%	x	x	Phosphoric acid,		
Chromic acid 30%	x	x	all concentrations	x	x
Chromic acid 40%	x	x	Potassium hydroxide to 50%	80	27
Chromic acid 50%	x	x	Potassium sulfate 10%	90	32
Citric acid 5%	x	x	Propane	150	66
Citric acid 10%	x	x	Silicone oil	x	x

Table P.32 Continued

Chemical	Max. temp. °F	Max. temp. °C	Chemical	Max. temp. °F	Max. temp. °C
Sodium carbonate	x	x	Toluene	x	x
Sodium chloride	80	27	Trichloroethylene	x	x
Sodium hydroxide, all concentrations	x	x	Water, demineralized	80	27
			Water, distilled	80	27
Sodium hypochlorite, all concentrations	x	x	Water, salt	80	27
			Water, sea	80	27
Sulfuric acid, all concentrations	x	x	Whiskey	x	x
			Zinc chloride	x	x

[a]The chemicals listed are in the pure state or in a saturated solution unless otherwise indicated. Compatibility is shown to the maximum allowable temperature for which data are available. Incompatibility is shown by an x. A blank space indicates that the data are unavailable.
Source: Schweitzer, PA. *Corrosion Resistance Tables*, 4th ed. Vols 1–3. New York: Marcel Dekker, 1995.

POLYTETRAFLUORETHYLENE (PTFE)

Also see "Permeation." PTFE is a fully fluorinated thermoplastic having the formula

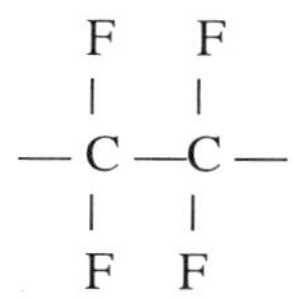

It has an operating temperature range of from –20°F (–29°C) to 450°F (232°C). This temperature range is based on the physical/mechanical properties of PTFE. When handling certain aggressive chemicals it may be necessary to reduce the upper temperature limit. PTFE is a relatively weak material and tends to creep under stress at elevated temperatures. The physical and mechanical properties are shown in Table P.33.

PTFE is unique in its corrosion resistant properties. It is chemically inert in the presence of most materials. There are very few materials that will attack PTFE within normal use temperatures. Among materials that will attack PTFE are the most violent oxidizing and reducing agents known. Elemental sodium removes fluorine from the polymer molecule. The other alkali metals (potassium, lithium, etc.) act in a similar manner.

Fluorine and related compounds (e.g., chlorine trifluoride) are absorbed into the PTFE resin with such intimate contact that the mixture becomes sensitive to a source of ignition such as impact. These potent oxidizers should only be handled with great care and a recognition of the potential hazards.

The handling of 80% sodium hydroxide, aluminum chloride, ammonia, and certain amines at high temperatures have the same effect as elemental sodium. Slow oxidative

Table P.33 Physical and Mechanical Properties of PTFE

Specific gravity	2.13–2.2
Water absorption, 24 h at 73°F/23°C, %	0.01
Tensile strength at 73°F/23°C, psi	2000–6500
Compressive strength, psi	1700
Flexural strength, psi	No break
Flexural modulus, psi $\times 10^{-5}$	0.7–1.1
Izod impact strength, notched at 73°F/23°C, ft-lb/in.	3
Coefficient of thermal expansion, in./in. °F $\times 10^{-5}$	5.5
Heat distortion temperature at 66 psi, °F/°C	250/121
Low-temperature embrittlement, °F/°C	–450/–268

attack can be produced by 70% nitric acid under pressure at 480°F (250°C). Refer to Table P.34 for the compatibility of PTFE with selected corrodents.

See Ref. 8.

POLYVINYL CHLORIDE (PVC)

See also "Polymers." Polyvinyl chloride is the most widely used of any of the thermoplasts. PVC is polymerized vinyl chloride, which is produced from acetylene and anhydrous hydrochloric acid. The structure is

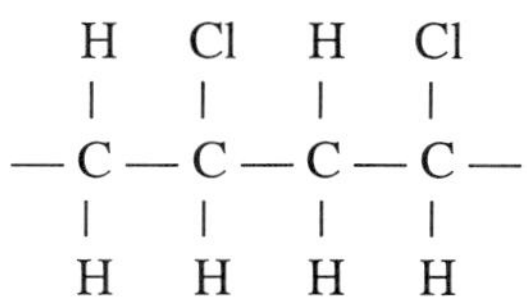

PVC is stronger and more rigid than other thermoplastic materials. It has a high tensile strength and modulus of elasticity. Additives are used to further specific end uses, such as thermal stabilizers, lubricity, impact modifiers, and pigmentation.

Two types of PVC are produced, normal impact (type 1) and high impact (type 2). Type 1 is a rigid unplasticized PVC having normal impact with optimum chemical resistance. Type 2 has optimum impact resistance and reduced chemical resistance. Table P.35 lists the physical and mechanical properties of PVC.

Type 1 (unplasticized PVC) resists attack by most acids and strong alkalies, gasoline, kerosene, aliphatic alcohols, and hydrocarbons. It is particularly useful in the handling of hydrochloric acid.

The chemical resistance of type 2 PVC to oxidizing and highly alkaline materials is reduced.

PVC may be attacked by aromatics, chlorinated organic compounds, and lacquer solvents. Refer to Table P.36 for the compatibility of PVC with selected corrodents. Reference 8 provides a wider listing of the compatibility of PVC with a variety of corrodents.

See Refs. 8 and 10.

Table P.34 Compatibility of PTFE with Selected Corrodents[a]

Chemical	Maximum temp. °F	Maximum temp. °C	Chemical	Maximum temp. °F	Maximum temp. °C
Acetaldehyde	450	232	Benzene[b]	450	232
Acetamide	450	232	Benzene sulfonic acid 10%	450	232
Acetic acid 10%	450	232	Benzoic acid	450	232
Acetic acid 50%	450	232	Benzyl alcohol	450	232
Acetic acid 80%	450	232	Benzyl chloride	450	232
Acetic acid, glacial	450	232	Borax	450	232
Acetic anhydride	450	232	Boric acid	450	232
Acetone	450	232	Bromine gas, dry[b]	450	232
Acetyl chloride	450	232	Bromine liquid[b]	450	232
Acrylonitrile	450	232	Butadiene[b]	450	232
Adipic acid	450	232	Butyl acetate	450	232
Allyl alcohol	450	232	Butyl alcohol	450	232
Allyl chloride	450	232	*n*-Butylamine	450	232
Alum	450	232	Butyl phthalate	450	232
Aluminum chloride, aqueous	450	232	Butyric acid	450	232
Aluminum fluoride	450	232	Calcium bisulfide	450	232
Aluminum hydroxide	450	232	Calcium bisulfite	450	232
Aluminum nitrate	450	232	Calcium carbonate	450	232
Aluminum oxychloride	450	232	Calcium chlorate	450	232
Aluminum sulfate	450	232	Calcium chloride	450	232
Ammonia gas[b]	450	232	Calcium hydroxide 10%	450	232
Ammonium bifluoride	450	232	Calcium hydroxide, sat.	450	232
Ammonium carbonate	450	232	Calcium hypochlorite	450	232
Ammonium chloride 10%	450	232	Calcium nitrate	450	232
Ammonium chloride 50%	450	232	Calcium oxide	450	232
Ammonium chloride, sat.	450	232	Calcium sulfate	450	232
Ammonium fluoride 10%	450	232	Caprylic acid	450	232
Ammonium fluoride 25%	450	232	Carbon bisulfide[b]	450	232
Ammonium hydroxide 25%	450	232	Carbon dioxide, dry	450	232
Ammonium hydroxide, sat.	450	232	Carbon dioxide, wet	450	232
Ammonium nitrate	450	232	Carbon disulfide	450	232
Ammonium persulfate	450	232	Carbon monoxide	450	232
Ammonium phosphate	450	232	Carbon tetrachloride[c]	450	232
Ammonium sulfate 10–40%	450	232	Carbonic acid	450	232
Ammonium sulfide	450	232	Chloracetic acid, 50% water	450	232
Amyl acetate	450	232	Chloracetic acid	450	232
Amyl alcohol	450	232	Chlorine gas, dry	x	x
Amyl chloride	450	232	Chlorine gas, wet[b]	450	232
Aniline	450	232	Chlorine, liquid	x	x
Antimony trichloride	450	232	Chlorobenzene[b]	450	232
Aqua regia 3:1	450	232	Chloroform[b]	450	232
Barium carbonate	450	232	Chlorosulfonic acid	450	232
Barium chloride	450	232	Chromic acid 10%	450	232
Barium hydroxide	450	232	Chromic acid 50%	450	232
Barium sulfate	450	232	Chromyl chloride	450	232
Barium sulfide	450	232	Citric acid 15%	450	232
Benzaldehyde	450	232	Citric acid, concentrated	450	232

Table P.34 Continued

Chemical	Maximum temp. °F	Maximum temp. °C	Chemical	Maximum temp. °F	Maximum temp. °C
Copper carbonate	450	232	Muriatic acid[b]	450	232
Copper chloride	450	232	Nitric acid 5%[b]	450	232
Copper cyanide 10%	450	232	Nitric acid 20%[b]	450	232
Copper sulfate	450	232	Nitric acid 70%[b]	450	232
Cresol	450	232	Nitric acid, anhydrous[b]	450	232
Cupric chloride 5%	450	232	Nitrous acid 10%	450	232
Cupric chloride 50%	450	232	Oleum	450	232
Cyclohexane	450	232	Perchloric acid 10%	450	232
Cyclohexanol	450	232	Perchloric acid 70%	450	232
Dichloroacetic acid	450	232	Phenol[b]	450	232
Dichloroethane (ethylene dichloride)[b]	450	232	Phosphoric acid 50–80%	450	232
Ethylene glycol	450	232	Picric acid	450	232
Ferric chloride	450	232	Potassium bromide 30%	450	232
Ferric chloride 50% in water	450	232	Salicylic acid	450	232
Ferric nitrate 10–50%	450	232	Sodium carbonate	450	232
Ferrous chloride	450	232	Sodium chloride	450	232
Ferrous nitrate	450	232	Sodium hydroxide 10%	450	232
Fluorine gas, dry	x	x	Sodium hydroxide 50%	450	232
Fluorine gas, moist	x	x	Sodium hydroxide, concentrated	450	232
Hydrobromic acid, dilute[bc]	450	232	Sodium hypochlorite 20%	450	232
Hydrobromic acid 20%[c]	450	232	Sodium hypochlorite, concentrated	450	232
Hydrobromic acid 50%[c]	450	232	Sodium sulfide to 50%	450	232
Hydrochloric acid 20%[c]	450	232	Stannic chloride	450	232
Hydrochloric acid 38%[c]	450	232	Stannous chloride	450	232
Hydrocyanic acid 10%	450	232	Sulfuric acid 10%	450	232
Hydrofluoric acid 30%[b]	450	232	Sulfuric acid 50%	450	232
Hydrofluoric acid 70%[b]	450	232	Sulfuric acid 70%	450	232
Hydrofluoric acid 100%[b]	450	232	Sulfuric acid 90%	450	232
Hypochlorous acid	450	232	Sulfuric acid 98%	450	232
Iodine solution 10%[b]	450	232	Sulfuric acid 100%	450	232
Ketones, general	450	232	Sulfuric acid, fuming[b]	450	232
Lactic acid 25%	450	232	Sulfurous acid	450	232
Lactic acid, concentrated	450	232	Thionyl chloride	450	232
Magnesium chloride	450	232	Toluene[b]	450	232
Malic acid	450	232	Trichloroacetic acid	450	232
Methyl chloride[b]	450	232	White liquor	450	232
Methyl ethyl ketone[b]	450	232	Zinc chloride[d]	450	232
Methyl isobutyl ketone[c]	450	232			

[a]The chemicals listed are in the pure state or in a saturated solution unless otherwise indicated. Compatibility is shown to the maximum allowable temperature for which data are available. Incompatibility is shown by an x. A blank space indicates that the data are unavailable.
[b]Material will permeate.
[c]Material will cause stress cracking.
[d]Material will be absorbed.
Source: Schweitzer, PA. *Corrosion Resistance Tables*, 4th ed. Vols 1–3. New York: Marcel Dekker, 1995.

Table P.35 Physical and Mechanical Properties of PVC

Property	Type 1	Type 2
Specific gravity	1.45	1.38
Water absorption (24 h at 73° F (23°C), %	0.04	0.05
Tensile strength at 73°F (23°C), psi	6,800	5,500
Modulus of elasticity in tension at 73°F (23°C) $\times 10^5$	5.0	4.2
Compressive strength, psi	10,000	7,900
Flexural strength, psi	14,000	11,000
Izod impact strength, notched at 73°F (23°C)	0.88	12.15
Coefficient of thermal expansion		
in./in.-°F $\times 10^{-5}$	4.0	6.0
in./10°F/100 ft	0.40	0.60
Thermal conductivity Btu/h/sq ft/°F/in	1.33	1.62
Heat distortion temperature, °F (°C)		
at 66 psi	130/54	135/57
at 264 psi	155/68	160/71
Resistance to heat, °F (°C) at continuous drainage	150/66	140/60
Limiting oxygen index, %		43
Flame spread		15–20
Underwriters lab rating (Sub 94)		94V-O

POLYVINYLIDENE CHLORIDE (SARAN)

Saran (polyvinylidene chloride) is manufactured by Dow Chemical. The resin is a proprietary product of Dow. It has found wide application in the plating industry and for handling deionized water, pharmaceuticals, food processing, and other applications where stream purity protection is critical. The material complies with FDA regulations for food processing and potable water and also with regulations prescribed by the Meat Inspection Division of the Department of Agriculture for transporting fluids used in meat production. In applications such as plating solutions, chlorines, and certain other chemicals, Saran is superior to polypropylene and finds many applications in the handling of municipal water supplies and waste waters. Refer to Table P.37 for the physical and mechanical properties. Refer to Table P.38 for the compatibility of Saran with selected corrodents.

POLYVINYLIDENE FLUORIDE (PVDF)

Vinylidene fluoride is a crystalline, high molecular weight thermoplastic polymer containing 59% fluorine. It is similar in chemical structure to PTFE except that it is not fully fluorinated. The chemical structure is

Table P.36 Compatibility of Type 2 PVC with Selected Corrodents[a]

Chemical	Maximum temp. °F	°C	Chemical	Maximum temp. °F	°C
Acetaldehyde	x	x	Barium chloride	140	60
Acetamide	x	x	Barium hydroxide	140	60
Acetic acid 10%	100	38	Barium sulfate	140	60
Acetic acid 50%	90	32	Barium sulfide	140	60
Acetic acid 80%	x	x	Benzaldehyde	x	x
Acetic acid, glacial	x	x	Benzene	x	x
Acetic anhydride	x	x	Benzene sulfonic acid 10%	140	60
Acetone	x	x	Benzoic acid	140	60
Acetyl chloride	x	x	Benzyl alcohol	x	x
Acrylic acid	x	x	Borax	140	60
Acrylonitrile	x	x	Boric acid	140	60
Adipic acid	140	60	Bromine gas, dry	x	x
Allyl alcohol	90	32	Bromine gas, moist	x	x
Allyl chloride	x	x	Bromine liquid	x	x
Alum	140	60	Butadiene	60	16
Aluminum acetate	100	38	Butyl acetate	x	x
Aluminum chloride, aqueous	140	60	Butyl alcohol	x	x
Aluminum fluoride	140	60	*n*-Butylamine	x	x
Aluminum hydroxide	140	60	Butyric acid	x	x
Aluminum nitrate	140	60	Calcium bisulfide	140	60
Aluminum oxychloride	140	60	Calcium bisulfite	140	60
Aluminum sulfate	140	60	Calcium carbonate	140	60
Ammonia gas	140	60	Calcium chlorate	140	60
Ammonium bifluoride	90	32	Calcium chloride	140	60
Ammonium carbonate	140	60	Calcium hydroxide 10%	140	60
Ammonium chloride 10%	140	60	Calcium hydroxide, sat.	140	60
Ammonium chloride 50%	140	60	Calcium hypochlorite	140	60
Ammonium chloride, sat.	140	60	Calcium nitrate	140	60
Ammonium fluoride 10%	90	32	Calcium oxide	140	60
Ammonium fluoride 25%	90	32	Calcium sulfate	140	60
Ammonium hydroxide 25%	140	60	Calcium bisulfide	x	x
Ammonium hydroxide, sat.	140	60	Carbon dioxide, dry	140	60
Ammonium nitrate	140	60	Carbon dioxide, wet	140	60
Ammonium persulfate	140	60	Carbon disulfide	x	x
Ammonium phosphate	140	60	Carbon monoxide	140	60
Ammonium sulfate 10–40%	140	60	Carbon tetrachloride	x	x
Ammonium sulfide	140	60	Carbonic acid	140	60
Amyl acetate	x	x	Cellosolve	x	x
Amyl alcohol	x	x	Chloracetic acid	105	40
Amyl chloride	x	x	Chlorine gas, dry	140	60
Aniline	x	x	Chlorine gas, wet	x	x
Antimony trichloride	140	60	Chlorine, liquid	x	x
Aqua regia 3:1	x	x	Chlorobenzene	x	x
Barium carbonate	140	60	Chloroform	x	x

Table P.36 Continued

Chemical	Maximum temp. °F	Maximum temp. °C	Chemical	Maximum temp. °F	Maximum temp. °C
Chlorosulfonic acid	60	16	Nitric acid 5%	100	38
Chromic acid 10%	140	60	Nitric acid 20%	140	60
Chromic acid 50%	x	x	Nitric acid 70%	70	140
Citric acid 15%	140	60	Nitric acid, anhydrous	x	x
Citric acid, con.	140	60	Nitrous acid, concentrated	60	16
Copper carbonate	140	60	Oleum	x	x
Copper chloride	140	60	Perchloric acid 10%	60	16
Copper cyanide	140	60	Perchloric acid 70%	60	16
Copper sulfate	140	60	Phenol	x	x
Cresol	x	x	Phosphoric acid 50–80%	140	60
Cyclohexanol	x	x	Picric acid	x	x
Dichloroacetic acid	120	49	Potassium bromide 30%	140	60
Dichloroethane (ethylene dichloride)	x	x	Salicylic acid	x	x
Ethylene glycol	140	60	Silver bromide 10%	105	40
Ferric chloride	140	60	Sodium carbonate	140	60
Ferric nitrate 10–50%	140	60	Sodium chloride	140	60
Ferrous chloride	140	60	Sodium hydroxide 10%	140	60
Ferrous nitrate	140	60	Sodium hydroxide 50%	140	60
Fluorine gas, dry	x	x	Sodium hydroxide, concentrated	140	60
Fluorine gas, moist	x	x	Sodium hypochlorite 20%	140	60
Hydrobromic acid, dilute	140	60	Sodium hypochlorite, concentrated	140	60
Hydrobromic acid 20%	140	60	Sodium sulfide to 50%	140	60
Hydrobromic acid 50%	140	60	Stannic chloride	140	60
Hydrochloric acid 20%	140	60	Stannous chloride	140	60
Hydrochloric acid 38%	140	60	Sulfuric acid 10%	140	60
Hydrocyanic acid 10%	140	60	Sulfuric acid 50%	140	60
Hydrofluoric acid 30%	120	49	Sulfuric acid 70%	140	60
Hydrofluoric acid 70%	68	20	Sulfuric acid 90%	x	x
Hypochlorous acid	140	60	Sulfuric acid 98%	x	x
Ketones, general	x	x	Sulfuric acid 100%	x	x
Lactic acid 25%	140	60	Sulfuric acid, fuming	x	x
Lactic acid, concentrated	80	27	Sulfurous acid	140	60
Magnesium chloride	140	60	Thionyl chloride	x	x
Malic acid	140	60	Toluene	x	x
Methyl chloride	x	x	Trichloroacetic acid	x	x
Methyl ethyl ketone	x	x	White liquor	140	60
Methyl isobutyl ketone	x	x	Zinc chloride	140	60
Muriatic acid	140	60			

[a]The chemicals listed are in the pure state or in a saturated solution unless otherwise indicated. Compatibility is shown to the maximum allowable temperature for which data are available. Incompatibility is shown by an x. A blank space indicates that the data are unavailable.
Source: Schweitzer, PA. *Corrosion Resistance Tables,* 4th ed. Vols 1–3. New York: Marcel Dekker, 1995.

Table P.37 Physical and Mechanical Properties of Polyvinylidene Chloride

Specific gravity	1.75–1.85
Water absorption 24 h at 73°F (23°C), %	nil
Tensile strength at 73°F (23°C), psi	2700–3700
Coefficient of thermal expansion	
in./in.-°F (°C) × 10^{-5}	3.9 to 5
in./10°F/100 ft	0.039 to 0.05
Thermal conductivity, Btu/h/sq ft/°F/in.	1.28
Flame spread	self-extinguishing

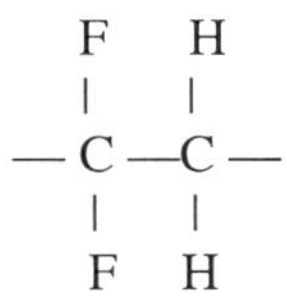

Much of the strength and chemical resistant properties are maintained through an operating range of –40 to 320°F (–40 to 160°C). It has high tensile strength and heat deflection temperature and is resistant to the permeation of gases. Approval has been granted by the Food and Drug Administration for repeated use in contact with food, as in food handling and processing equipment. The physical and mechanical properties can be found in Table P.39.

PVDF is chemically resistant to most acids, bases, and organic solvents. It is also resistant to wet or dry chlorine, bromine, and other halogens.

It should not be used with strong alkalies, fuming acids, polar solvents, amines, ketones, or esters. When used with strong alkalies it stress cracks. Refer to Table P.40 for the compatibility of PVDF with selected corrodents.

PVDF is manufactured under the trade name of Kynar by Elf Atochem, Solef by Solvay, Hylar by Ausimont USA, and Super Pro 230 and ISO by Asahi/America.

See Refs. 8 and 15.

POTENTIAL–pH DIAGRAMS (POURBAIX DIAGRAMS)

Potential–pH diagrams represent graphically the stability of a metal and its corrosion products as a function of the potential and pH (acidity or alkalinity) of an aqueous solution. The pH is shown on the horizontal axis and the potential in the vertical axis.

In order to trace such a diagram the concentration of the dissolved material must be fixed. The thermodynamic possibilities for reaction between zinc and the atmosphere with its various constituents can be determined in potential–pH diagrams.

Zinc is a relatively base metal. The stability areas of various zinc-containing species in the system Zn—CO_2—H_2O at 77°F (25°C) are shown in Fig. P.1. The diagram takes into account the formation of zinc hydroxide, of Zn^{2+}, and of the zincate ions $HZnO_2$ and ZnO^{2-}. The numbers indicate the H_2CO_3 content in the moisture film in mole/liter. The broken lines indicate the area of thermodynamic stability of water.

Table P.38 Compatibility of Polyvinylidene Chloride (Saran) with Selected Corrodents[a]

Chemical	Maximum temp. °F	Maximum temp. °C	Chemical	Maximum temp. °F	Maximum temp. °C
Acetaldehyde	150	66	Benzene	x	x
Acetic acid 10%	150	66	Benzene sulfonic acid 10%	120	49
Acetic acid 50%	130	54	Benzoic acid	120	49
Acetic acid 80%	130	54	Benzyl chloride	80	27
Acetic acid, glacial	140	60	Boric acid	170	77
Acetic anhydride	90	32	Bromine liquid	x	x
Acetone	90	32	Butadiene	x	x
Acetyl chloride	130	54	Butyl acetate	120	49
Acrylonitrile	90	32	Butyl alcohol	150	66
Adipic acid	150	66	Butyl phthalate	180	82
Allyl alcohol	80	27	Butyric acid	80	27
Alum	180	82	Calcium bisulfite	80	27
Aluminum chloride, aqueous	150	66	Calcium carbonate	180	82
Aluminum fluroide	150	66	Calcium chlorate	160	71
Aluminum hydroxide	170	77	Calcium chloride	180	82
Aluminum nitrate	180	82	Calcium hydroxide 10%	160	71
Aluminum oxychloride	140	60	Calcium hydroxide, sat.	180	82
Aluminum sulfate	180	82	Calcium hypochlorite	120	49
Ammonia gas	x	x	Calcium nitrate	150	66
Ammonium bifluoride	140	60	Calcium oxide	180	82
Ammonium carbonate	180	82	Calcium sulfate	180	82
Ammonium chloride, sat.	160	71	Caprylic acid	90	32
Ammonium fluoride 10%	90	32	Carbon bisulfide	90	32
Ammonium fluoride 25%	90	32	Carbon dioxide, dry	180	82
Ammonium hydroxide 25%	x	x	Carbon dioxide, wet	80	27
Ammonium hydroxide, sat.	x	x	Carbon disulfide	80	27
Ammonium nitrate	120	49	Carbon monoxide	180	82
Ammonium persulfate	90	32	Carbon tetrachloride	140	60
Ammonium phosphate	150	66	Carbonic acid	180	82
Ammonium sulfate 10–40%	120	49	Cellosolve	80	27
Ammonium sulfide	80	27	Chloracetic acid, 50% water	120	49
Amyl acetate	120	49	Chloracetic acid	120	49
Amyl alcohol	150	66	Chlorine gas, dry	80	27
Amyl chloride	80	27	Chlorine gas, wet	80	27
Aniline	x	x	Chlorine, liquid	x	x
Antimony trichloride	150	66	Chlorobenzene	80	27
Aqua regia 3:1	120	49	Chloroform	x	x
Barium carbonate	180	82	Chlorosulfonic acid	x	x
Barium chloride	180	82	Chromic acid 10%	180	82
Barium hydroxide	180	82	Chromic acid 50%	180	82
Barium sulfate	180	82	Citric acid 15%	180	82
Barium sulfide	150	66	Citric acid, concentrated	180	82
Benzaldehyde	x	x	Copper carbonate	180	82

Table P.38 Continued

Chemical	Maximum temp. °F	Maximum temp. °C	Chemical	Maximum temp. °F	Maximum temp. °C
Copper chloride	180	82	Nitric acid 5%	90	32
Copper cyanide	130	54	Nitric acid 20%	150	66
Copper sulfate	180	82	Nitric acid 70%	x	x
Cresol	150	66	Nitric acid, anhydrous	x	x
Cupric chloride 5%	160	71	Oleum	x	x
Cupric chloride 50%	170	77	Perchloric acid 10%	130	54
Cyclohexane	120	49	Perchloric acid 70%	120	49
Cyclohexanol	90	32	Phenol	x	x
Dichloroacetic acid	120	49	Phosphoric acid 50–80%	130	54
Dichloroethane			Picric acid	120	49
(ethylene dichloride)	80	27	Potassium bromide 30%	110	43
Ethylene glycol	180	82	Salicylic acid	130	54
Ferric chloride	140	60	Sodium carbonate	180	82
Ferric chloride 50% in water	140	60	Sodium chloride	180	82
Ferric nitrate 10–50%	130	54	Sodium hydroxide 0%	90	32
Ferrous chloride	130	54	Sodium hydroxide 50%	150	66
Ferrous nitrate	80	27	Sodium hydroxide,		
Fluorine gas, dry	x	x	concentrated	x	x
Fluorine gas, moist	x	x	Sodium hypochlorite 10%	130	54
Hydrobromic acid, dilute	120	49	Sodium hypochlorite,		
Hydrobromic acid 20%	120	49	concentrated	120	49
Hydrobromic acid 50%	130	54	Sodium sulfide to 50%	140	60
Hydrochloric acid 20%	180	82	Stannic chloride	180	82
Hydrochloric acid 38%	180	82	Stannous chloride	180	82
Hydrocyanic acid 10%	120	49	Sulfuric acid 10%	120	49
Hydrofluoric acid 30%	160	71	Sulfuric acid 50%	x	x
Hydrofluoric acid 100%	x	x	Sulfuric acid 70%	x	x
Hypochlorous acid	120	49	Sulfuric acid 90%	x	x
Ketones, general	90	32	Sulfuric acid 98%	x	x
Lactic acid, concentrated	80	27	Sulfuric acid 100%	x	x
Magnesium chloride	180	82	Sulfuric acid, fuming	x	x
Malic acid	80	27	Sulfurous acid	80	27
Methyl chloride	80	27	Thionyl chloride	x	x
Methyl ethyl ketone	x	x	Toluene	80	27
Methyl isobutyl ketone	80	27	Trichloroacetic acid	80	27
Muriatic acid	180	82	Zinc chloride	170	77

[a]The chemicals listed are in the pure state or in a saturated solution unless otherwise indicated. Compatibility is shown to the maximum allowable temperature for which data are available. Incompatibility is shown by an x. A blank space indicates that the data are unavailable.

Source: Schweitzer, PA. *Corrosion Resistance Tables,* 4th ed. Vols 1–3. New York: Marcel Dekker, 1995.

Table P.39 Physical and Mechanical Properties of PVDF

Specific gravity	1.76
Water absorption 24 h at 73° (23°C), %	< 0.04
Tensile strength at 73°F (23°C), psi	6000
Modulus of elasticity in tension at 73°F (23°C) $\times 10^5$	2.1
Compressive strength, psi	11600
Flexural strength, psi	10750
Izod impact sttrength, notched at 73°F (23°C)	3.8
Coefficient of thermal expansion	
in./in.-°F $\times 10^{-5}$	7.9
in./10°F/100 ft	0.079
Thermal conductivity, Btu/h/sq ft/°F/in.	0.79
Heat distortion temperature, °F/°C	
at 66 psi	284/140
at 264 psi	194/90
Resistance to heat at continuous drainage, °F/°C	280/138
Limiting oxygen index, %	44
Flame spread	0
Underwriters lab rating (Sub94)	94V-O

Table P.40 Compatibility of PVDF with Selected Corrodents[a]

Chemical	Maximum temp. °F	Maximum temp. °C	Chemical	Maximum temp. °F	Maximum temp. °C
Acetaldehyde	150	66	Aluminum hydroxide	260	127
Acetamide	90	32	Aluminum nitrate	300	149
Acetic acid 10%	300	149	Aluminum oxychloride	290	143
Acetic acid 50%	300	149	Aluminum sulfate	300	149
Acetic acid 80%	190	88	Ammonia gas	270	132
Acetic acid, glacial	190	88	Ammonium bifluoride	250	121
Acetic anhydride	100	38	Ammonium carbonate	280	138
Acetone	x	x	Ammonium chloride 10%	280	138
Acetyl chloride	120	49	Ammonium chloride 50%	280	138
Acrylic acid	150	66	Ammonium chloride, sat.	280	138
Acrylonitrile	130	54	Ammonium fluoride 10%	280	138
Adipic acid	280	138	Ammonium fluoride 25%	280	138
Allyl alcohol	200	93	Ammonium hydroxide 25%	280	138
Allyl chloride	200	93	Ammonium hydroxide, sat.	280	138
Alum	180	82	Ammonium nitrate	280	138
Aluminum acetate	250	121	Ammonium persulfate	280	138
Aluminum chloride, aqueous	300	149	Ammonium phosphate	280	138
Aluminum chloride, dry	270	132	Ammonium sulfate 10–40%	280	138
Aluminum fluoride	300	149	Ammonium sulfide	280	138

Table P.40 Continued

Chemical	Maximum temp. °F	Maximum temp. °C	Chemical	Maximum temp. °F	Maximum temp. °C
Ammonium sulfite	280	138	Carbon disulfide	80	27
Amyl acetate	190	88	Carbon monoxide	280	138
Amyl alcohol	280	138	Carbon tetrachloride	280	138
Amyl chloride	280	138	Carbonic acid	280	138
Aniline	200	93	Cellosolve	280	138
Antimony trichloride	150	66	Chloracetic acid, 50% water	210	99
Aqua regia 3:1	130	54	Chloracetic acid	200	93
Barium carbonate	280	138	Chlorine gas, dry	210	99
Barium chloride	280	138	Chlorine gas, wet, 10%	210	99
Barium hydroxide	280	138	Chlorine, liquid	210	99
Barium sulfate	280	138	Chlorobenzene	220	104
Barium sulfide	280	138	Chloroform	250	121
Benzaldehyde	120	49	Chlorosulfonic acid	110	43
Benzene	150	66	Chromic acid 10%	220	104
Benzene sulfonic acid 10%	100	38	Chromic acid 50%	250	121
Benzoic acid	250	121	Chromyl chloride	110	43
Benzyl alcohol	280	138	Citric acid 15%	250	121
Benzyl chloride	280	138	Citric acid, concentrated	250	121
Borax	280	138	Copper acetate	250	121
Boric acid	280	138	Copper carbonate	250	121
Bromine gas, dry	210	99	Copper chloride	280	138
Bromine gas, moist	210	99	Copper cyanide	280	138
Bromine liquid	140	60	Copper sulfate	280	138
Butadiene	280	138	Cresol	210	99
Butyl acetate	140	60	Cupric chloride 5%	270	132
Butyl alcohol	280	138	Cupric chloride 50%	270	132
n-Butylamine	x	x	Cyclohexane	250	121
Butyl phthalate	80	27	Cyclohexanol	210	99
Butyric acid	230	110	Dichloroacetic acid	120	49
Calcium bisulfide	280	138	Dichlorethane		
Calcium bisulfite	280	138	(ethylene dichloride)	280	138
Calcium carbonate	280	138	Ethylene glycol	280	138
Calcium chlorate	280	138	Ferric chloride	280	138
Calcium chloride	280	138	Ferric chloride 50% in water	280	138
Calcium hydroxide 10%	270	132	Ferrous nitrate 10–50%	280	138
Calcium hydroxide, sat.	280	138	Ferrous chloride	280	138
Calcium hypochlorite	280	138	Ferrous nitrate	280	138
Calcium nitrate	280	138	Fluorine gas, dry	80	27
Calcium oxide	250	121	Fluorine gas, moist	80	27
Calcium sulfate	280	138	Hydrobromic acid, dilute	260	127
Caprylic acid	220	104	Hydrobromic acid 20%	280	138
Carbon bisulfide	80	27	Hydrobromic acid 50%	280	138
Carbon dioxide, dry	280	138	Hydrochloric acid 20%	280	138
Carbon dioxide, wet	280	138	Hydrochloric acid 38%	280	138

Table P.40 Continued

Chemical	Maximum temp. °F	Maximum temp. °C	Chemical	Maximum temp. °F	Maximum temp. °C
Hydrocyanic acid 10%	280	138	Salicylic acid	220	104
Hydrofluoric acid 30%	260	127	Silver bromide 10%	250	121
Hydrofluoric acid 70%	200	93	Sodium carbonate	280	138
Hydrofluoric acid 100%	200	93	Sodium chloride	280	138
Hypochlorous acid	280	138	Sodium hydroxide 10%	230	110
Iodine solution 10%	250	121	Sodium hydroxide 50%	220	104
Ketones, general	110	43	Sodium hydroxide, concentrated[b]	150	66
Lactic acid 25%	130	54	Sodium hypochlorite 20%	280	138
Lactic acid, concentrated	110	43	Sodium hypochlorite, concentrated	280	138
Magnesium chloride	280	138	Sodium sulfide to 50%	280	138
Malic acid	250	121	Stannic chloride	280	138
Manganese chloride	280	138	Stannous chloride	280	138
Methyl chloride	x	x	Sulfuric acid 10%	250	121
Methyl ethyl ketone	x	x	Sulfuric acid 50%	220	104
Methyl isobutyl ketone	110	43	Sulfuric acid 70%	220	104
Muriatic acid	280	138	Sulfuric acid 90%	210	99
Nitric acid 5%	200	93	Sulfuric acid 98%	140	60
Nitric acid 20%	180	82	Sulfuric acid 100%	x	x
Nitric acid 70%	120	49	Sulfuric acid, fuming	x	x
Nitric acid, anhydrous	150	66	Sulfurous acid	220	104
Nitrous acid, concentrated	210	99	Thionyl chloride	x	x
Oleum	x	x	Toluene	x	x
Perchloric acid 10%	210	99	Trichloroacetic acid	130	54
Perchloric acid 70%	120	49	White liquor	80	27
Phenol	200	93	Zinc chloride	260	127
Phosphoric acid 50–80%	220	104			
Picric acid	80	27			
Potassium bromide 30%	280	138			

[a]The chemicals listed are in the pure state or in a saturated solution unless otherwise indicated. Compatibility is shown to the maximum allowable temperature for which data are available. Incompatibility is shown by an x. A blank space indicates that the data are unavailable.
Source: Schweitzer, PA. *Corrosion Resistance Tables*, 4th ed. Vols 1–3. New York: Marcel Dekker, 1995.

Pourbaix diagrams are widely used in corrosion because they permit easy identification of the predominant materials at equilibrium for a given potential and pH. However they do not give any information as to the possible rate of corrosion.

POULTICE CORROSION

Poultice corrosion is a special form of crevice corrosion. It is the gradual collection of hydroscopic particulate matter on ledges and the like. This is a typical type of corrosion

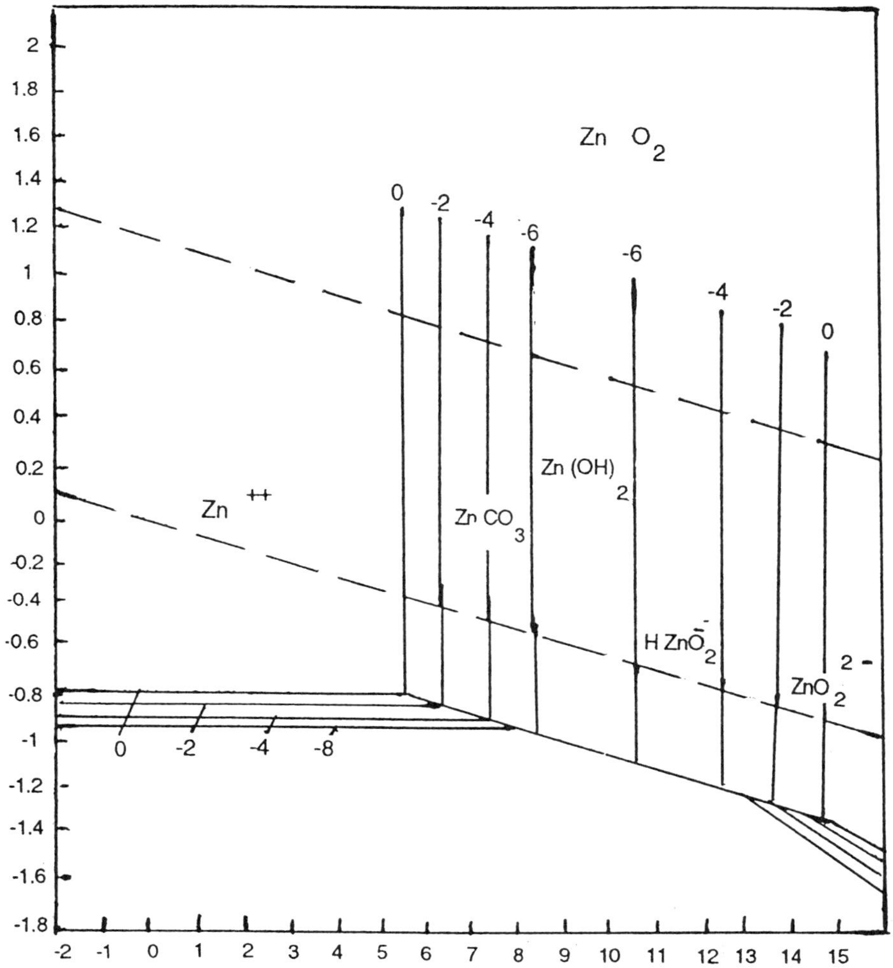

Figure P.1 Potential pH diagram for the system Zn–CO_2–H_2O at 77°F (25°C).

experienced on vehicle body parts due to the collection of road salts and debris on ledges and in pockets that are kept moist by weather and washing.

Poultice corrosion is also early initiation of corrosion occurring beneath an hygroscopic attachment or insert. Poultice corrosion is also referred to as deposit corrosion and deposit attack.

PRECIPITATION HARDENABLE STAINLESS STEELS

This is a family of stainless steel alloys that utilize a thermal treatment to precipitate phases intentionally that cause a strengthening of the alloy. The precipitation hardenable stainless steels are divided into three alloy types; martensitic, austenitic, and semiaustenitic.

Martensitic and austenitic precipitation hardenable stainless steels are directly hardened by thermal treatment. The semiaustenitic steels are supplied as an unstable austenite, which is the workable condition, and must be transformed to martensite before aging.

See Ref. 2.

PYREX

This is the trade name of Corning Glass's borosilicate glass. See "Borosilicate Glass."

PYROLYSIS

Pyrolysis is a chemical decomposition caused by heat, or severe oxidation. The molecule of all plastic materials is an organic molecule that is subject to pyrolysis. This can occur by fire or it can occur, for example, by contact with very strong sulfuric acid (e.g., 93%), which completely destroys the molecule, and severe oxidation takes place. The resin turns black and loses all physical strength, so that the structure is destroyed. Service life is measured in a matter of hours. The thermoset resins are particularly subject to pyrolysis from concentrated sulfuric or nitric acids. Polyesters can withstand up to 78% sulfuric acid for short periods and 70% sulfuric acid almost indefinitely at room temperature. Above room temperature, oxidation becomes so severe that the molecule is destroyed. Some of the thermoplasts such as polyethylene, polypropylene, and polyvinyl chloride are exceptions and can tolerate higher concentrations.

REFERENCES

1. HH Uhlig. Corrosion and Corrosion Control. New York: John Wiley, 1963.
2. PH Whitcraft, Corrosion of stainless steels. In: PA Schweitzer, ed. Corrosion Engineering Handbook. New York: Marcel Dekker, 1996, pp. 53–77.
3. CP Dillon. Corrosion Resistance of Stainless Steels. New York: Marcel Dekker, 1995.
4. FP Fehlner and MJ Graham. Thin oxide film formation on metals. In: P Marcus and J Oudar, eds, Corrosion Mechanisms in Theory and Practice. New York: Marcel Dekker, 1995, pp. 123–141.
5. B MacDougall and MJ Graham. Growth and stability of passive films. In: P Marcus and J Oudar, eds. Corrosion Mechanisms in Theory and Practice. New York: Marcel Dekker, 1995, pp. 143–173.
6. CR Clayton and I Olefjord, Passivity of austentic stainless steels. In: P Marcus and J Oudar, eds. Corrosion Mechanisms in Theory and Practice. New York: Marcel Dekker, 1995, pp. 175–199.
7. PA Schweitzer, Corrosion Resistance of Elastomers. New York: Marcel Dekker, 1990.
8. PA Schweitzer, Corrosion Resistance Tables. 4th ed. Vol. 1–3. New York: Marcel Dekker, 1995.
9. JH Mallinson. Corrosion Resistant Plastic Composites in Chemical Plant Design. New York: Marcel Dekker, 1988.
10. PA Schweitzer, Corrosion Resistant Piping Systems. New York: Marcel Dekker, 1994.
11. DL Pletcher. Corrosion of thermoset plastics. In: PA Schweitzer, ed. Corrosion Engineering Handbook, New York: Marcel Dekker, 1996, pp. 372–373.
12. HH Strehblow. Mechanisms of pitting corrosion. In: P Marcus and J Oudar, eds. Corrosion Mechanisms in Theory and Practice. New York: Marcel Dekker, 1995, pp. 201–238.

13. B Baroux. Further insights on the pitting of stainless steels. In: P Marcus and J Oudar, eds. Corrosion Mechanisms in Theory and Practice. New York: Marcel Dekker, 1995, pp. 265–310.
14. PK Whitcraft. Fundamentals of metallic corrosion. In: PA Schweitzer, ed. Corrosion Engineering Handbook. New York: Marcel Dekker, 1996, pp. 1–11.
15. GT Murray. Introduction to Engineering Materials. New York: Marcel Dekker, 1993.
16. JH Mallinson. Development and applications of plastic materials. In: PA Schweitzer, ed. Corrosion and Corrosion Protection Handbook. 2nd ed. New York: Marcel Dekker, 1988, pp. 323–393.
17. PA Schweitzer. Mechanisms of chemical attack, corrosion resistance, and failure of plastic materials. In: PA Schweitzer, ed. Corrosion Engineering Handbook. New York: Marcel Dekker, 1996, pp. 297–346.
18. AA Boova. Chemical-resistant mortars, grouts, and monolithic surfacings. In: PA Schweitzer, ed. Corrosion Engineering Handbook. New York: Marcel Dekker, 1996, pp. 459–487.
19. WL Sheppard Jr. Chemically Resistant Masonry. 2nd ed. New York: Marcel Dekker, 1982.
20. GW Read Jr, CE Zimmer, and GR Hall. Cements and mortars. In: PA Schweitzer, ed. Corrosion and Corrosion Protection Handbook. New York: Marcel Dekker, 1989, pp. 521–531.

Q

QUENCH

To quench is to cool rapidly a metal or alloy from an elevated temperature by means of immersion in water, salt solution, or oil, or by forced air cooling, to prevent it from reaching a stable condition and in so doing impart special properties to the metal or alloy.

QUENCH ANNEALING

Quench annealing is a high temperature solution heat treatment followed by a water quenching. This procedure dissolves chromium carbide, thereby forming a more homogeneous alloy. It is a procedure used to minimize intergranular corrosion of austenitic stainless steels. It is also called solution quenching.

QUENCHING AND TEMPERING (HARDENING AND TEMPERING)

The quenching of steel into water or oil, converting it to martensite and increasing internal stress, which is relieved by heating (tempering) up to 212°F–392°F (100–200°C) to diffuse carbon and precipitate small carbides to improve the material's strength and ductility.

R

RADIATION CORROSION

Metals that are exposed to intense radiation in the form of neutrons or other energetic particles undergo lattice changes resembling those resulting from severe cold work. Lattice vacancies, interstitial atoms, and dislocations are produced, all of which increase the diffusion rate of specific impurities or alloyed components.

RADIOGRAPHY

Radiography is the use of x-rays, gamma rays, or other forms of radiation to measure the thickness or imperfection within solid materials.

REBAR CORROSION

Rebar corrosion is the corrosion of the steel reinforcing bars in concrete. Reinforcement for concrete, both regular and prestressed, must be protected to prevent corrosion. The amount of cover required for protection under various conditions are shown in the table.

Condition	Cover, inches
Concrete surface poured against the ground	3
Concrete surface to come into contact with the ground after casting:	
reinforcement larger than No. 5	2
reinforcement smaller than No. 5	1½
Beams and girders not exposed to weather:	
main steel	1½
stirrups and ties	1
Joists, slabs, and walls not exposed to weather	¾
Columns, spirals and ties	1½

The cover must be at least equal to the bar diameter except for joists and slabs. For columns the cover must be 1½ times the maximum size of the coarse aggregate.

The above recommendations for cover are minimums. When corrosive conditions are to be encountered the cover should be increased. In all cases the concrete in the cover should be made as impermeable as possible.

RED BRASS

See "Copper–Tin Alloys."

REDUCING ACIDS

Reducing or nonoxidizing acids are those inorganic and organic acids that evolve gaseous hydrogen during the corrosion of active metals. They are corrosive to metals above hydrogen in the electromotive series when in the presence of oxygen or oxidizing agents, whose reduction takes the place of hydrogen evolution.

The behavior of passive metals cannot always be predicted because behavior is dependent upon the acid concentration, temperature, dissolved oxygen, and specific contaminants.

Inorganic Acids

The inorganic acids include hydrochloric acid, hydrofluoric, acid, phosphoric acid, and dilute and intermediate concentrations of sulfuric acid.

Hydrochloric Acid

Hydrochloric acid is produced by dissolving hydrogen chloride gas in water. The concentrated acid is 36%. It is a highly corrosive acid whose corrosive behavior can be drastically modified as a result of contaminants. Muriatic acid is a 30% commercial acid contaminated with dissolved ferric iron salts. The presence of trace amounts of chlorinated solvents or aromatic solvents has a bearing on the resistance of plastics and elastomers to what would nominally be called hydrochloric acid.

Hydrofluoric Acid

Hydrogen fluoride gas dissolved in water produces hydrofluoric acid, a very toxic and dangerous acid. The laboratory grade is usually a 48% concentration.

Phosphoric Acid

Commercial grade phosphoric acid contains fluorides and other impurities such as chlorides, sulfates, and metal ions, particuarly if manufactured by the "wet" process of digestion of phosphate rock with sulfuric acid. Food grade phosphoric acid is free of these contaminants. Consequently the corrosion characteristics of the commercial grade of acid can be unpredictable. In any case it is quite corrosive.

Sulfuric Acid

Sulfuric acid reacts both as an oxidizing and as a reducing acid. Concentrated acid, above 70% concentration, reacts as an oxidizing acid (see "Oxidizing Acids"), while dilute and intermediate concentrations below 70% react as a reducing acid. The corrosive effect is also affected by temperature and contaminants as well as dilution.

Organic Acids

There are two reducing organic acids that are of particular concern. They are formic and acetic acids.

Formic Acid

Formic acid is a strong organic acid similar in corrosiveness to the dilute mineral acids. Its greatest corrosive properties are exhibited when the acid is hot and anaerobic. When anhydrous it is a powerful dehydrating agent.

Acetic Acid

Of the organic acids, acetic acid is the most commercially important. The corrosion properties of chemically pure acid are very predictable. Crude or unrefined acetic acid may contain peracids or peroxides that affect the corrosive properties and at times are unpredictable.

Other Organic Acids

The higher molecular weight acids are generally less corrosive than acetic acid. Material recommendations for acetic acid can usually be used for butyric, propionic, and other acids and will be conservative.

Specific Materials

Following is a general analysis of the behavior of various materials of construction in the presence of the inorganic and organic reducing acids. More specific and expanded information can be gotten from Ref. 1, and the specific material of construction.

Light Metals

Aluminum and magnesium are severely attacked by hydrochloric acid.

Magnesium is resistant to hydrofluoric acid up to a 2% concentration due to an insoluble film of corrosion products. Aluminum and its alloys should not be exposed to hydrofluoric acid, even in dilute concentrations.

Aluminum and magnesium have no practical application in the handling of phosphoric acid.

Dilute sulfuric acid will attack aluminum and magnesium.

Aluminum is resistant to formic acid in concentrations above 30%. Below 30% aluminum is rapidly attacked at temperatures only slightly above ambient.

Concentrated (glacial) acetic acid is compatible with aluminum up to a temperature of 200°F (93°C). As the concentration decreases the corrosivity increases, even in cold solutions.

Iron and Steel

Iron and steel are not compatible with hydrochloric acid. They are rapidly attacked.

Although steels are resistant to concentrated hydrofluoric acid < 64% up to approximately 90°F (32°C) because of their protective film of corrosion products, they should not be used, since hydrogen blistering may occur, and welds may be preferentially attacked. Hardened steels are subject to environmental cracking.

Cast irons and alloy irons are not compatible with hydrofluoric acid and should not be used. High silicon iron is also not compatible.

Phosphoric acid above 70% concentration will form a protective film on steel. However, because of iron contamination, steel is not used to handle phosphoric acid.

High silicon iron is resistant to all concentrations of phosphoric acid to the atmospheric boiling point providing no fluorides are present.

Cast iron and steel are attacked by sulfuric acid in concentrations less than 70%.

High silicon cast iron is resistant to all concentrations of sulfuric acid up to 70% and 200°F (93°C).

Formic acid will corrode iron and steel. High silicon iron is resistant to all aqueous concentrations but is attacked by anhydrous acid.

Acetic acid will attack iron and steel in all concentrations, even cold. High silicon iron is resistant to all concentrations of acetic acid to the atmospheric boiling point.

Stainless Steels

All grades of stainless steel will be attacked by hydrochloric acid.

Hydrofluoric acid is not compatible with any stainless steel. Martensitic grades are subject to hydrogen cracking and other grades are subject to pitting and/or stress corrosion cracking.

Above 5% concentration of phosphoric acid, type 304 and type 316 stainless steels are subject to intergranular corrosion. Type 316L can be used to handle uncontaminated phosphoric acid up to 85% to approximately 175°F (80°C). Type 317L or alloy 20Cb3 are better choices. Alloy 20Cb3 is resistant to all concentrations of phosphoric to 210°F (99°C). In the absence of contaminants alloy 825 is resistant to all concentrations of phosphoric acid to the atmospheric boiling point.

Martensitic and ferritic grades of stainless steel will be attacked by all concentrations of sulfuric acid to 70% with varying temperature limits based on concentration. Refer to Ref. 1.

Formic acid will cause pitting in type 316 stainless steel and intergranular corrosion in type 304. Alloy 20Cb3 is resistant to all concentrations of formic acid, including anhydrous acid to a temperature of 210°F (99°C).

Type 304 and Type 316 stainless steels are resistant to all concentrations of acetic acid. However type 316L can be used up to a temperature of 400°F (204°C). Type 304 stainless is limited to a maximum temperature of 160°F (70°C).

Copper Alloys

In the presence of dissolved oxygen or oxidizing ions copper and its alloys will be attacked. Since the cupric ion is an oxidant, few practical applications can be found. Dealloying is also a potential problem.

The amount of oxygen or oxidants contained in hydrofluoric acid will determine the degree of corrosion to copper and its alloys. Normally copper and its alloys are not considered for this service.

Copper and copper alloys can be used in sulfuric acid concentrations to 70% providing there are absolutely no oxidizing species present, including dissolved oxygen.

Copper is resistant to formic acid up to 5% concentration in the absence of other oxidants.

Copper is not recommended for the handling of acetic acid, since even a slight attack will discolor the refined acid.

Lead

Corrosion products formed on lead by hydrochloric acid are soluble and are easily washed away by flow, removing any protective film.

Hydrofluoric acid is compatible with lead up to a 60% concentration at 77°F (25°C), or with 25% acid up to 175°F (80°C). Attack will increase with acid strength and temperature.

Lead is resistant to 80% pure phosphoric and/or 85% impure phosphoric acid up to

390°F (200°C). The resistance is due to the formation of films of insoluble lead phosphates. Erosion or impingement effects will cause corrosion.

Lead and its alloys are resistant to sulfuric acid up to 70% concentration below the atmospheric boiling point. However, the protective sulfide film may be removed by organic contaminants (organic acids, alkyl sufates).

Lead is attacked by formic acid and acetic acid.

Nickel Alloys

Alloy B-2 resists boiling hydrochloric acid providing there are no traces of oxidizing agents such as ferric ions present. Alloy C-276 will resist dilute acid containing ferric chloride but only to intermediate temperatures 140°F (60°C). Alloy 600 is compatible with acid up to 15% at 80°F (26°C).

Alloy 400 is compatible with all concentrations of hydrofluoric acid up to 250°F (120°C). However it is susceptible to stress corrosion cracking in the vapors when air is present. Nickel 200 is resistant to anhydrous HF up to 300°F (150°C). In aqueous solutions it is limited to nonoxidizing conditions below 175°F (85°C). Alloy 400 is a much better choice. Alloy 600 is compatible with dilute hydrofluoric acid but has a great tendency to pit at higher concentrations.

Alloy 400 is resistant to all concentrations of phosphoric acid to approximately 200°F (90°C) providing there are no stronger oxidants present than dissolved oxygen and the cupric ion corrosion products do not accumulate.

Alloy B-2 is resistant to all concentrations of phosphoric acid to approximately 150°F (65°C) and 50% acid to the atmospheric boiling point.

Alloy 200, alloy 400, and alloy B-2 are resistant to nonaerated sulfuric acid, free of oxidants in all concentrations to 70%.

Formic acid is compatible with alloy 200 and alloy 400. Alloy C-276 is resistant to formic acid that is contaminated at elevated temperatures.

Reactive Metals

Titanium is not resistant to hydrochloric acid. Zirconium is compatible with all concentrations of hydrochloric acid to 225°F (107°C) providing the concentration of oxidizing species is less than 50 ppm. If the oxidizing species exceed 50 ppm, pyrophoric corrosion products may be formed. Tantalum is resistant to all concentrations of hydrochloric acid to 345°F (175°C) but is subject to attack by hydrogen chloride vapors at 265°F (130°C).

Hydrofluoric acid, or even small traces of fluorides, will severely attack titanium, zirconium, and tantalum.

In the absence of fluorides, zirconium will resist phosphoric acid to approximately 60% concentration, and tantalum is resistant to any concentration up to 345°F (175°C) providing the fluoride contamination is less than 10 ppm. Titanium will be attacked by phosphoric acid unless the acid contains oxidizing contaminants.

Zirconium is resistant to all concentrations of sulfuric acid to 70% concentration at the atmospheric boiling point. Unless oxidizing contaminants are present, titanium will be attacked by dilute sulfuric acid. Tantalum is resistant to all concentrations of sulfuric acid to 70% and at the boiling point providing no fluorides are present.

Formic acid will attack titanium unless strongly oxidizing contaminants are present. Zirconium is resistant to formic acid up to 90% concentration up to 210°F (99°C). Tantalum is resistant to all concentrations of formic acid, including anhydrous, up to a temperature of 300°F (149°C).

Nonmetallic Materials

Ceramic and glass materials are extremely resistant to hydrochloric acid. Hydrochloric acid can be handled in rubber lined equipment up to 175°F (80°C) providing organic solvent contaminants are not present. Contaminents such as chlorinated hydrocarbons or aromatic solvents (benzene, toluene, etc.) will be preferentially absorbed and concentrated to cause failure of the rubber or of its adhesive. Halogenated or hydrogenated polyesters are also compatible with all concentrations of hydrochloric acid to various temperatures depending upon the concentrations. Refer to Ref. 1.

Hydrofluoric acid will attack plastics with hydroxyl groups. Polyethylene, polystyrene, methacrylates, and vinyls are resistant up to 60% concentration at 120°F (50°C). Fluorinated plastics are compatible up to their operating temperature limits but are subject to permeation. Polyesters are rapidly attacked, as are glass and other siliceous ceramics. Synthetic soft rubbers such as butyl, neoprene, and hypalon are compatible with 60% acid to 160°F (70°C).

Fluoride free phosphoric acid can be handled by glass up to 60% concentration and 212°F (100°C).

Carbon is resistant to all concentrations of phosphoric acid to 700°F (350°C). Various plastic materials and elastomers are resistant to all concentrations of phosphoric acid to the temperature limitations shown in the table.

Resistance of Plastic and Elastomeric Materials to Phosphoric Acid

	Temperature Limits	
Material	°F	°C
CPVC	180	82
E-CTFE, PVDF	240	116
ETFE, furan	260	127
FEP	400	204
PFA, PEEK, PES, PP	200	93
Polyesters	220	104
PTFE	500	260
Hypalon	200	93
Viton, EPDM	300	149
Natural rubber	100	38
Butyl rubber	140	60
PE, PVC	140	60

As long as there are no fluorides present, glass and other ceramics will resist all concentrations of sulfuric acid up to 70%. Dilute and intermediate concentrations of sulfuric acid can also be handled by carbon and graphite to the boiling point. Plastic and elastomeric materials can handle all concentrations of sulfuric acid to 70% to the temperature limitations shown in the table.

Resistance of Plastic and Elastomeric Materials to Sulfuric Acid up to 70%

Material	Temperature Limits °F	Temperature Limits °C
CPVC	180	82
E-CTFE, furan, PFA, PVDF	240	116
ETFE, polysulfone, EPDM	300	149
FEP	400	204
Polyesters, PP, CIIR	200	93
PE, PVC	140	60
PPFE	500	260
Hypalon	240	116
Viton	340	171
Neoprene	200	93
Natural Rubber	140	60

Glass, carbon and ceramic ware are fully resistant to formic acid including anhydrous. Because of the solvent nature of formic acid, plastics can be subject to attack, depending upon temperature. Temperature limitations of plastics and elastomeric materials are shown in the table.

Resistance of Plastic and Elastomeric Materials to Formic Acid

Material	Temperature Limits °F	Temperature Limits °C
CPVC, PE	140	60
E-CTFE, PVDF	200	93
ETFE, furan	260	127
FEP, PTFE	400	204
PFA, PP	200	93
PVC	100	38
Butyl rubber, CIIR	140	60
Hypalon	200	93
EPDM	300	149
Viton	180	82
Neoprene	160	70

Glass, carbon, and graphite are fully resistant to acetic acid. Because of the solvent effects of organic acids, the applications of plastics and elastomeric materials are limited. However, the fully fluorinated grades of plastics are completely resistant. Temperature limitations of plastics and elastomeric materials are shown in the table.

Resistance of Plastic and Elastomeric Materials to Acetic Acid

Material	Temperature Limits °F	Temperature Limits °C
E-CTFE, PES	200	93
ETFE	220	104
FEP	400	204
Furan	260	127
PFA	240	116
PEEK	140	60
PP, PVDF	180	82
PVC	120	49
PTFE	460	238
CIIR	160	71
EPDM	300	148
Butyl rubber, Buna-N	80	26

See Ref. 1.

REDUCING ATMOSPHERE CORROSION

Reducing atmosphere corrosion occurs in coal and oil-fired boilers resulting from direct reaction of the water wall tubes with a substoichiometric gaseous environment containing sulfur, or with deposited, partially burned char containing iron pyrites. The reducing conditions affect corrosion in two ways. They tend to lower the melting temperature of any deposited slag, which increases its ability to dissolve the normal protective oxide scale on the tubes, and the stable gaseous sulfur compounds, including hydrogen sulfide, which react to form iron sulfide. The iron sulfide scale formed provides less protection than the normal iron oxide scale, thereby allowing corrosion to take place.

RIDDICK'S CORROSION INDEX

Riddick's corrosion index is used to determine the corrosivity of soft waters. Characteristics of the water such as calcium carbonate solubility, chloride ion concentration, dissolved oxygen, silica concentration, and noncarbonate hardness are taken into account when making the determination.

RIMMED STEEL

See "Killed Carbon Steel."

RUST

Rust is the corrosion product of iron or iron based alloys consisting largely of hydrous ferric oxides. It is an electrochemical process that takes place only in the presence of acids or other electrolytes in the water. The hydrous ferric oxide is orange to reddish brown in color. It consists of nonmagnetic $\propto Fe_2O_3$ (hematite) or the magnetic$\surd Fe_2O_3$. Nonferrous metals, therefore, corrode but do not rust.

REFERENCES

1. PA Schweitzer. Corrosion Resistance Tables. 4th ed. New York: Marcel Dekker, 1995.

S

SACRIFICIAL ANODE

See also "Cathodic Protection."

Use of a sacrificial anode is a means of cathodic protection. By selecting an anode constructed of a metal more active in the galvanic series than the metal to be protected a galvanic cell may be established with the current flowing such that the metal to be protected becomes the cathode and is protected from corrosion while the anode is corroded (sacrificed). These sacrificial anodes are usually composed of magnesium or magnesium based alloys. Occasionally, zinc and aluminum have been used.

Most sacrificial anodes in use in the United States are of magnesium construction. Approximately 10 million pounds of magnesium are used annually for this purpose. These anodes are used to protect buried pipe lines and metal structures. Magnesium rods have also been placed in steel hot-water tanks to increase the life of these tanks.

See Ref. 1.

SARAN

See "Polyvinylidene Chloride"

SATURATED SOLUTION

A saturated solution is one in which the maximum amount of solute has been dissolved in a solvent (usually water) for a specific set of conditions, e.g., temperature, pressure.

SCAB CORROSION

Scab corrosion is the condition where the paint film remains intact but corrosion has taken place under the paint film as a result of external damage and the paint film is nonadherent to the metal substrate.

SEASON CRACKING

This is a form of stress corrosion cracking that is usually applied to the stress corrosion cracking of brass. The term originates from early in the twentieth century when cartridge shells made of 70% copper and 30% zinc were found to crack over a period of time. It was later realized that ammonia from decaying organic matter in combination with residual stresses in the brass was responsible for the crack of these shells. This phenomenon was called "season cracking" because the presence of high humidity during warm moist climates (or seasons) promoted the stress corrosion cracking.

Stress corrosion cracking of brass commonly occurs when brass is subjected to an applied or residual tensional stress while in contact with a trace of ammonia or amine in the presence of moisture and oxygen.

SELECTIVE CORROSION

See "Dezincification." Also called selective leaching.

SELECTIVE LEACHING

Removal of one element from a solid alloy by corrosion process. See "Dezincification."

SEMIKILLED STEEL

See "Killed Steel."

SENSITIZATION

Sensitization is the term applied to the precipitation of chromium carbides in the grain boundaries of both austenitic 300 series alloys and the straight chromium grades such as types 405 and 410 stainless steel, as a result of exposure to temperatures in the range of 800 to 1600°F (425 to 870°C). This results in intergranular corrosion or cracking.

As the chromium carbides develop, the adjacent metal becomes depleted in dissolved chromium, creating a zone adjacent to the grain boundary of locally corrosion susceptible iron–nickel alloy. The local composition of the straight chromium grades may approach that of carbon steel. This chromium depleted zone has less corrosion resistance than that of the adjacent unaffected alloy and can react Galvanically with the unaffected zone accelerating corrosion rates.

There is not complete agreement over the lower threshold temperature that causes sensitization. It has been reported that cold worked austenitic stainless steels sensitize at temperatures as low as 700°F (370°C). Solution annealed austenitic stainless steels require extremely long exposure times to sensitize at temperatures below 850°F (455°C). In general, sensitization takes place most rapidly when the temperature is in the range of 1500°F (815°C).

Sensitization can cause two types of corrosion problems, weld rusting and intergranular corrosion.

S-GLASS

S-glass is used as a reinforcing material for thermosetting resins. See "Thermoset Reinforcing Materials."

SHEET LININGS

Designers of tanks and process vessels are faced with the problem of choosing the most reliable material of construction at reasonable cost. When handling corrosive materials a choice must often be made between the use of an expensive metallic material of construction or the use of a low cost material from which to fabricate the shell and then install a corrosion resistant lining. Carbon steel has been and still is the material predominantly selected, although there has been a tendency over the past few years to use a fiberglass reinforced plastic shell. This latter choice has the advantage of providing atmospheric corrosion protection of the shell exterior.

For many years vessels have been successfully lined with various rubber formulations, both natural and synthetic. Many such vessels have given over 20 years of reliable service.

With the development of newer synthetic elastomeric and plastic materials, the variety of lining materials available has greatly increased.

As with any material, the corrosion resistance, allowable operating values, and cost varies with each. Care must be taken when selecting the lining material that it is compatible with the corrodent being handled at the operating temperatures and pressures required.

Shell Design

In order for a lining to perform satisfactorily, the vessel shell must meet certain design configurations. Although these details may vary slightly depending upon the specific lining material to be used, there are certain basic principles that apply in all cases.

1. The vessel must be of butt welded construction.
2. All internal welds must be ground flush.
3. All weld spatter must be removed.
4. All sharp corners must be ground to a minimum of 1/8 in. radius.
5. All outlets must be of flanged or pad type. Certain lining materials require that nozzles be not less than 2 in. (55 mm) in diameter.
6. No protrusions are permitted inside of the vessel. Once the lining has been installed, there should be no welding permitted on the exterior of the vessel.

After the fabrication has been completed, the interior surface of the vessel must be prepared to accept the lining. This is a critical step. Unless the surface is properly prepared, proper bonding of the lining to the shell will not be achieved. The basic requirement is that the surface be absolutely clean. To insure proper bonding, all surfaces to be lined should be abrasive blasted to white metal in accordance with SSPC specification Tp5-63 or NACE specification NACE-1. A white-metal surface condition is defined as being one from which all rust, scale, paint and the like has been removed and the surface has a uniform gray-white appearance. Streaks or stains of rust or other contaminants are not allowed. A near white blast cleaned finish equal to SSPC SP-10 is allowed on occasion. This is a more economical finish. In any case it is essential that the finish be as the lining contractor has specified. Some lining contractors will fabricate the vessel as well as preparing the surface. When the total responsibility is placed on the lining contractor the problem is simplified, and usually a better quality product will be the result.

When a vessel shell is fabricated from a reinforced thermosetting plastic (RTP) several advantages are realized. The RTPs themselves generally have a wider range of corrosion resistance but relatively low allowable operating temperatures. When a fluoropolymer type lining is applied to an RTP shell, the temperature to which the backup RTP is exposed has been reduced, in addition to preventing the RTP from becoming exposed to the chemicals in the process system. An upper temperature limit for using RTP dual laminates is 350°F (177°C).

The dual laminate construction lessens the problem of permeation through liners. If there is permeation it is believed to pass through the RTP structure at a rate equal to or greater than through the fluoropolymer itself, resulting in no potential for collection of permeate at the thermoplastic-to-thermoset interface. If delamination does not occur, permeation is not a problem.

Considerations in Liner Selection

Before a lining material is selected, careful consideration should be given to several broad categories, specifically materials being handled, operating conditions, conditions external to the vessel.

The following questions must be answered about the materials being handled.

1. What are the primary chemicals being handled and at what concentrations?
2. Are there any secondary chemicals, and if so at what concentrations?
3. Are there any trace impurities or chemicals?
4. Are there any solids present, and if so what is their particle size and concentration?
5. If a vessel, will there be agitation, and to what degree? If a pipeline, what are the flow rates, maximum and minimum?
6. What are the fluid purity requirements?

The answers will narrow the selection to those materials that are compatible. This next set of questions will narrow the selection still further by eliminating those materials that do not have the required physical and/or mechanical properties.

1. What is the normal operating temperature and temperature range?
2. What peak temperatures can be reached during shutdown, startup, process upset, etc.?
3. Will any mixing areas exist where exothermic or heat of mixing temperatures can develop?
4. What is the normal operating pressure?
5. What vacuum conditions and range are possible during operation, startup, shutdown, or upset conditions?
6. Will there be temperature cycling?
7. What cleaning methods will be used?

Finally, consideration should be given to the conditions external to the vessel or pipe.

1. What are the ambient temperature conditions?
2. What is the maximum surface temperature during operation?
3. What are the insulation requirements?
4. What is the nature of the external environment? This can dictate finish requirements and/or affect the selection of the shell material.
5. What are the external heating requirements?
6. Is grounding necessary?

With the answers to these questions an intelligent selection of liner and shell can be made.

Design Considerations

In addition to selecting a lining material that is resistant to the corrodent being handled, there are three other factors that must be considered in the design, namely permeation, absorption, and environmental stress cracking. Permeation and absorption can cause

1. Bond failure and blistering, resulting from the accumulation of fluids at the bond when the substrate is less permeable than the liner, or from corrosion/reaction products if the substrate is attacked by the permeant.
2. Failure of the substrate from corrosive attack.
3. Loss of contents through the substrate and liner as a result of the eventual failure of the substrate. In unbonded linings it is important that the space between the liner and substrate be vented to the atmosphere, not only to allow minute quantities of permeant vapor to escape but also to prevent entrapped air from collapsing the liner.

Permeation

Also see "Permeation."

All materials are some what permeable to chemical molecules, but plastic materials tend to be an order of magnitude greater in their permeability rates than metals. Polymers can be permeated by gases, vapors, or liquids. Permeation is strictly a physical phenomenon: there is no chemical attack on the polymer. It is a molecular migration either through microvoids in the polymer (if the structure is more or less porous) or between polymer molecules.

Permeation is a function of two variables, one relating to diffusion between molecular chains and the other to the solubility of the permeant in the polymer. The driving forces of diffusion are the concentration gradients in liquids and the partial pressure gradient for gases. Solubility is a function of the affinity of the permeant for the polymer.

Material passing through cracks and voids is not related to permeation. These are two distinct happenings. They are not related in any way.

Permeation is affected by the factors

1. Temperature and pressure
2. Permeant concentration
3. Thickness of the polymer

An increase in temperature will increase the permeation rate, since the solubility of the permeant in the polymer will increase, and as the temperature rises, the polymer chain movement is stimulated, permitting more permeants to diffuse among the chain more easily. For many gases the permeant rates increase linearly with the partial pressure gradient, and the same effect is experienced with concentration gradients of liquids. If the permeant is highly soluble in the polymer, the permeability increase may not be linear.

The thickness of the polymer affects the permeation. An increase in thickness will generally decrease permeation by the square of the thickness. However there are disadvantages to this approach. First, as the lining thickness is increased, thermal stresses on the bond are increased, resulting in bond failure. Temperature changes and large differences in coefficients of thermal expansion are the most common causes of bond failure. The thickness and modulus of elasticity of the lining material are two of the factors that influence these stresses. In addition, as the thickness of the sheet lining material increases, it becomes more difficult to form, and heat may have to be applied. Also, the thicker sheets are more difficult to weld. Third is cost. As the thickness of the material increases, not only does the material cost more but also the labor cost increases because of the greater difficulty of working with the material. If polymers such as FEP, PTFE, or PVDF are being used, the cost may become prohibitive.

The density of the polymer, in addition to the thickness, will also have an effect on the permeation rate. The higher the specific gravity of the sheet, the fewer will be the voids present through which permeation can take place. A comparison, of the specific gravity between two different polymers will not give an indication as to the relative permeation rates. However, a comparison between two liners of the same polymer will provide the difference in the relative permeation rates. The liner having the greater density will have the lower permeation rate.

Other factors affecting permeation consisting of chemical and physicochemical properties are

1. Ease of condensation of the permeant. Chemicals that condense readily will permeate at a higher rate.
2. The higher the intermolecular chain forces (e.g., Van der Waals hydrogen bonding) of the polymer, the lower the permeation rate.
3. The higher the level of crystallinity in the polymer, the lower the permeation rate.
4. The greater the degree of cross-linking within the polymer, the lower the permeation rate.
5. Chemical similarity between the polymer and the permeant. When the permeant and polymer both have similar functional groups, the permeant rate will increase.
6. The smaller the molecule of the permeant, the greater the permeation rate.

The magnitude of any of the effects will be a function of the combination of the polymer and the permeant in actual service.

Absorption

Also see "Absorption."

Polymers have the potential to absorb varying amounts of corrodents they come into contact with, particuarly organic liquids. This can result in swelling, cracking, and penetration to the substrate. Swelling can cause softening of the polymer, introduce high stresses, and cause failure of the bond. If the polymer has a high absorption rate, permeation will probably take place. An approximation of the expected permeation and/or absorption of a polymer can be based on the absorption of water. These data are usually available. Table A.1 provides the water absorption rates for the more common polymers.

The failure due to absorption can best be understood by considering the "steam cycle" test described in the ASTM standards for lined pipe. A section of lined pipe is subjected to thermal and pressure fluctuations. This is repeated for 100 cycles. The steam creates a temperature and pressure gradient through the liner, causing absorption of a small quantity of steam, which condenses to water within the inner wall. Upon pressure release, or on reintroduction of steam, the entrapped water can expand to vapor, causing an original micropore. The repeated pressure and thermal cycling enlarges the micropores, ultimately producing visible water filled blisters within the liner.

In an actual process, the polymer may absorb process fluids, and repeated temperature or pressure cycling can cause blisters. Eventually corrodent may find its way to the substrate.

Related effects can occur when process chemicals are absorbed, which may later react, decompose, or solidify within the structure of the plastic. Prolonged retention of the chemicals may lead to their decomposition within the polymer. Although it is unusual, it is possible for absorbed monomers to polymerize.

Several steps can be taken to reduce absorption. Thermal insulation of the substrate will reduce the temperature gradient across the vessel, thereby preventing condensation and subsequent expansion of the absorbed fluids. This also reduces the rate and magnitude of temperature changes, keeping blisters to a minimum. The use of operating procedures or devices that limit the ratio of process pressure reductions or temperature increases will provide additional protection.

Environmental Stress Cracking

Stress cracks develop when a tough polymer is stressed for an extended period of time under loads that are small relative to the polymer's yield point. Cracking will occur with little elongation of the material. The higher the molecular weight of the polymer, the less likelihood of environmental stress cracking, other things being equal. Molecular weight is a function of length of individual chains that make up the polymer. Longer-chain polymers tend to crystallize less than polymers of lower molecular weight or shorter chains and also have greater load-bearing capacity.

Crystallinity is an important factor affecting stress corrosion cracking. The less the crystallization that takes place, the less the likelihood of stress cracking. Unfortunately the lower the crystallinity, the greater the likelihood of permeation.

Resistance to stress cracking can be reduced by the absorption of substances that chemically resemble the polymer and will plasticize it. In addition, the mechanical strength will be reduced. Halogenated chemicals, particuarly those consisting of small molecules containing fluorine or chlorine, are especially likely to be similar to the fluoropolymers and should be tested for their effect.

The presence of contaminants in a fluid may act as an accelerator. For example, polypropylene can safely handle sulfuric or hydrochloric acids, but iron or copper contamination in concentrated sulfuric or hydrochloric acids can result in the stress cracking of polypropylene.

Elastomeric Linings

Elastomers, sometimes referred to as rubbers, have given many years of service in providing protection to steel vessels. Each of these materials can be compounded to improve certain of its properties. Because of this it is necessary that a complete specification for a lining using these materials include specific properties that are required for the application. These include resilience, hysteresis, static or dynamic shear and compression modulus, flex fatigue and cracking, creep resistance to oils and chemicals, permeability, and brittle point, all in the temperature range to be encountered in service. This will permit a competent manufacturer to propose lining material for the application.

Elastomeric linings are sheet applied and bonded to the steel substrate. The bonding material to be used is dependent upon the specific elastomer to be installed. Repair of these linings is relatively simple. Many older vessels with numerous repair patches are still operating satisfactorily.

Natural Rubber

The maximum temperature for continuous use of natural rubber is 175°F (80°C). The degree of curing to which natural rubber is subjected will determine whether it is classified as soft, semihard, or hard. Soft rubber is the form primarily used for lining material, although some hard linings are produced.

It provides excellent resistance to most inorganic salt solutions, alkalies, and nonoxidizing acids. Hydrochloric acid will react with soft rubber to form rubber hydrochloride, and therefore it is not recommended that natural (soft) rubber be used in contact with this acid. Strong oxidizing media such as nitric acid, concentrated sulfuric acid, permangates, dichromates, chlorine dioxide, and sodium hypochlorite will severely attack rubber. Mineral and vegetable oils, gasoline, benzene, toluene, and chlorinated hydrocarbons also affect rubber. Cold water tends to preserve natural rubber. Natural rubber offers good resistance to radiation and alcohols. Refer to Table S.1 for the compatibility of multiple ply (soft/hard/soft) natural rubber.

See also "Natural Rubber."

Isoprene

Natural rubber chemically is a natural *cis*-polyisoprene. Isoprene (IR) is a synthetic *cis*-polyisoprene. The corrosion resistance of IR is the same as that of natural rubber. The major difference is that isoprene has no odor and therefore can be used in the handling of certain food products.

Neoprene

The maximum temperature under which neoprene (CR) can be used is 180–200°F (82–93°C). Excellent service is experienced in contact with aliphatic compounds (methyl and ethyl alcohols, ethylene glycols, etc.), aliphatic hydrocarbons, and most freon refrigerants. However, the outstanding property of neoprene is its resistance to attack from solvents, waxes, fats, oils, greases, and many other petroleum based products. Dilute mineral acids, inorganic salt solutions, and alkalies can also be handled successfully.

Chlorinated and aromatic hydrocarbons, organic esters, aromatic hydroxy compounds, and certain ketones will attack neoprene, as will highly oxidizing acid and salt solutions such as nitric and concentrated sulfuric acids.

See also "Neoprene."

Butyl Rubber

The maximum temperature to which butyl rubber (IIR) can be exposed on a continuous basis is 250–300°F (120–148°C). Butyl rubber is very nonpolar. It has exceptional

Table S.1 Compatibility of Multiply Ply (Soft/Hard/Soft) Natural Rubber with Selected Corrodents[a]

	Maximum temp.			Maximum temp.	
Chemical	°F	°C	Chemical	°F	°C
Acetaldehyde	x	x	Barium hydroxide	160	71
Acetamide	x	x	Barium sulfate	160	71
Acetic acid 10%	x	x	Barium sulfide	160	71
Acetic acid 50%	x	x	Benzaldehyde	x	x
Acetic acid 80%	x	x	Benzene	x	x
Acetic acid, glacial	x	x	Benzene sulfonic acid 10%	x	x
Acetic anhydride	x	x	Benzoic acid	160	71
Acetone	140	60	Benzyl alcohol	x	x
Acetyl chloride			Benzyl chloride	x	x
Acrylic acid			Borax	160	71
Acrylonitrile			Boric acid	140	60
Adipic acid			Bromine gas, dry		
Allyl alcohol			Bromine gas, moist		
Allyl chloride			Bromine liquid		
Alum	160	71	Butadiene		
Aluminum acetate			Butyl acetate	x	x
Aluminum chloride, aqueous	160	71	Butyl alcohol	160	71
Aluminum chloride, dry	160	71	*n*-Butylamine		
Aluminum fluoride	x	x	Butyl phthalate		
Aluminum hydroxide			Butyric acid	x	x
Aluminum nitrate	x	x	Calcium bisulfide		
Aluminum oxychloride			Calcium bisulfite	160	71
Aluminum sulfate	160	71	Calcium carbonate	160	71
Ammonia gas			Calcium chlorate	140	60
Ammonium bifluoride			Calcium chloride	140	60
Ammonium carbonate	160	71	Calcium hydroxide 10%	160	71
Ammonium chloride 10%	160	71	Calcium hydroxide, sat.	160	71
Ammonium chloride 50%	160	71	Calcium hypochlorite	x	x
Ammonium chloride, sat.	160	71	Calcium nitrate	x	x
Ammonium fluoride 10%	x	x	Calcium oxide	160	71
Ammonium fluoride 25%	x	x	Calcium sulfate	160	71
Ammonium hydroxide 25%	100	38	Caprylic acid		
Ammonium hydroxide, sat.	100	38	Carbon bisulfide	x	x
Ammonium nitrate	160	71	Carbon dioxide, dry		
Ammonium persulfate			Carbon dioxide, wet		
Ammonium phosphate	160	71	Carbon disulfide	x	x
Ammonium sulfate 10–40%	160	71	Carbon monoxide	x	x
Ammonium sulfide	160	71	Carbon tetrachloride	x	x
Ammonium sulfite			Carbonic acid	160	71
Amyl acetate	x	x	Cellosolve	x	x
Amyl alcohol	100	38	Chloracetic acid, 50% water	x	x
Amyl chloride	x	x	Chloracetic acid	x	x
Aniline	x	x	Chlorine gas, dry	x	x
Antimony trichloride			Chlorine gas, wet	x	x
Aqua regia 3:1	x	x	Chlorine, liquid	x	x
Barium carbonate	160	71	Chlorobenzene	x	x
Barium chloride	160	71	Chloroform	x	x

Table S.1 Continued

Chemical	Maximum temp. °F	Maximum temp. °C	Chemical	Maximum temp. °F	Maximum temp. °C
Chlorosulfonic acid	x	x	Manganese chloride		
Chromic acid 10%	x	x	Methyl chloride	x	x
Chromic acid 50%	x	x	Methyl ethyl ketone	x	x
Chromyl chloride			Methyl isobutyl ketone	x	x
Citric acid 15%	x	x	Muriatic acid	140	60
Citric acid, concentrated	x	x	Nitric acid 5%	x	x
Copper acetate			Nitric acid 20%	x	x
Copper carbonate	x	x	Nitric acid 70%	x	x
Copper chloride	x	x	Nitric acid, anhydrous	x	x
Copper cyanide	160	71	Nitrous acid, concentrated	x	x
Copper sulfate	160	71	Oleum		
Cresol	x	x	Perchloric acid 10%		
Cupric chloride 5%	x	x	Perchloric acid 70%		
Cupric chloride 50%	x	x	Phenol	x	x
Cyclohexane	x	x	Phosphoric acid 50–80%	160	71
Cyclohexanol			Picric acid		
Dichloroacetic acid			Potassium bromide 30%	160	71
Dichloroethane (ethylene dichloride)	x	x	Salicylic acid		
Ethylene glycol	160	71	Silver bromide 10%		
Ferric chloride	160	71	Sodium carbonate	160	71
Ferric chloride 50% in water	160	71	Sodium chloride	160	71
Ferric nitrate 10–50%	x	x	Sodium hydroxide 10%	160	71
Ferrous chloride	140	60	Sodium hydroxide 50%	x	x
Ferrous nitrate	x	x	Sodium hydroxide, concentrated	x	x
Fluorine gas, dry	x	x	Sodium hypochlorite 20%	x	x
Fluorine gas, moist			Sodium hypochlorite, concentrated	x	x
Hydrobromic acid, dilute	160	71	Sodium sulfide to 50%	160	71
Hydrobromic acid 20%	160	71	Stannic chloride	160	71
Hydrobromic acid 50%	160	71	Stannous chloride	160	71
Hydrochloric acid 20%	x	x	Sulfuric acid 10%	160	71
Hydrochloric acid 38%	160	71	Sulfuric acid 50%	x	x
Hydrocyanic acid 10%			Sulfuric acid 70%	x	x
Hydrofluoric acid 30%	x	x	Sulfuric acid 90%	x	x
Hydrofluoric acid 70%	x	x	Sulfuric acid 98%	x	x
Hydrofluoric acid 100%	x	x	Sulfuric acid 100%	x	x
Hypochlorous acid			Sulfuric acid, fuming	x	x
Iodine solution 10%			Sulfurous acid	x	x
Ketones, general			Thionyl chloride		
Lactic acid 25%	x	x	Toluene		
Lactic acid, concentrated	x	x	Trichloroacetic acid		
Magnesium chloride	160	71	White liquor		
Malic acid	100	38	Zinc chloride	160	71

[a]The chemicals listed are in the pure state or in a saturated solution unless otherwise indicated. Compatibility is shown to the maximum allowable temperature for which data are available. Incompatibility is shown by an x. A blank space indicates that data are unavailable.

Source: PA Schweitzer. *Corrosion Resistance Tables,* 4th ed. Vols. 1–3. New York: Marcel Dekker, 1995.

resistance to dilute mineral acids, alkalies, phosphate ester oils, acetone, ethylene glycol, ethylene, and water. Resistance to concentrated acids, except nitric and sulfuric, is good. It will be attacked by petroleum oils, gasoline, and most solvents (except oxygenated solvents) but is resistant to swelling by vegetable and animal oils. Refer to "Butyl Rubber" for more details.

Chlorsulfonated Polyethylene

The maximum temperature that Hypalon can be exposed to on a continuous basis is 250°F (121°C). Hypalon is highly resistant to attack by hydrocarbon oils and fuels, even at elevated temperatures. It is also resistant to such oxidizing chemicals as sodium hypochlorite, sodium peroxide, ferric chloride, and sulfuric, chromic, and hydrofluoric acids. Concentrated hydrochloric acid (37%) can be handled below 158°F (70°C). Below this temperature Hypalon can handle all concentrations without adverse effect. Nitric acid up to 60% concentration at room temperature can also be handled without adverse effect. Hypalon is also resistant to salt solutions, alcohols, and both weak and concentrated alkalies.

Aliphatic, aromatic, and chlorinated hydrocarbons, aldehydes, and ketones will attack Hypalon. Refer to Table C.6 for compatibility of Hypalon with selected corrodents.

Urethane Rubber

The maximum temperature to which the urethane rubbers (AU) can be exposed on a continuous basis is 250°F (121°C). The urethane rubbers are resistant to most mineral and vegetable oils, greases, and fuels. They have limited service in weak acid solutions and cannot be used in concentrated acids. Neither are they resistant to steam or caustic, aromatic hydrocarbons, polar solvents, esters, or ethers. Ketones will attack urethane. Alcohols will soften and swell the urethane rubbers.

Polyester Elastomer

The maximum temperature to which polyester elastomers (PE) can be exposed continuously is 302°F (150°C). The elastomers have excellent resistance to nonpolar materials such as oils and hydraulic fluids, even at elevated temperatures. At room temperature they are resistant to most polar fluids, such as acids, bases, amines, and glycols. Resistance is very poor at temperatures of 158°F (70°C) or above.

These elastomers are not resistant to polar fluids at elevated temperatures.

Perfluoroelastomers

The maximum temperature to which PFEs may be subjected on a continuous basis is 400°F (205°C). These materials are sold under various trade names. Three typical brands are

Trade Name	Manufacturer
Viton	DuPont
Technoflon	Ausimont
Fluorel	3M

The fluoroelastomers provide excellent resistance to oils, fuels, lubricants, most mineral acids, many aliphatic and aromatic hydrocarbons (carbon tetrachloride, benzene, toluene, xylene), gasoline, naphtha, chlorinated solvents, and pesticides.

These materials are not suitable for use with low molecular weight esters, ethers, ketones, certain amines, or hot anhydrous hydrofluoric or chlorosulfonic acids. Refer to Table E.3 for the compatibility of Viton with selected corrodents.

Thermoplastic Linings

Thermoplastic linings are used more extensively than any other type of lining. The most common thermoplasts used are

Polyvinyl chloride (PVC)
Chlorinated polyvinyl chloride (CPVC)
Polyvinylidene fluoride (PVDF)
Polypropylene (PP)
Polyethylene (PE)
Tetrafluoroethylene (PTFE)
Fluorinated ethylene propylene (FEP)
Perfluoralkoxy (PFA)
Ethylene-Tetrafluoroethylene (ETFE)
Ethylene-Chlorotrifluorethylene (ECTFE)

These materials are capable of providing a wide range of corrosion resistance. Table S.2 lists the general area of corrosion resistance for each of the thermoplasts. This table is only a general guide. The resistance of a lining material to a specific corrodent should be checked.

These materials are used to line vessels as well as pipe and fittings. When vessels are lined, the linings, with the exception of plasticized PVC, are fabricated from sheet stock that must be cut, shaped, and joined. Joining is usually accomplished by hot gas welding. Several problems exist when a thermoplastic lining is bonded to a metal shell, primarily the large differences in the coefficient of thermal expansion and the difficulty of adhesion. If the vessel is to be used under ambient temperature, such as a storage vessel, the problem of thermal expansion differences is eliminated. However, the problem of adhesion is still present. Because of this many linings are installed as loose linings in the vessel.

Table S.2 General Corrosion Resistance of Thermoplastic Lining Materials

Material	Strong acids	Strong bases	Chlorinated solvents	Esters and ketones	Strong oxidants
PVC (type 1)	F	F	P	P	P
CPVC	F	F	F	P	P
PVDF	E	P	E	P	E
Polypropylene	G	E	P	F	P
Polyethylene	F	G	P	P	E
PTFE	E	E	E	E	E
FEP	E	E	E	E	E
ECTFE	E	E	E	E	E

E, excellent; G, good; F, fair; P, poor.

Techniques have been developed to overcome the problem of adhesion, making use of an intermediate bond. One approach is to heat the thermoplastic sheet, then impress into one surface a fiber cloth or nonwoven web. This provides half of the bond. Bonding to the metal surface is accomplished by the use of an epoxy adhesive that will bond to both the fiber and the metal.

Welding techniques for joining the thermoplastic sheets are critical. Each weld must be continuous, leak tight, and mechanically strong. It is important that only qualified welders be used for this operation. Poor welding is a common cause of lining failure.

Polyvinyl Chloride

The maximum allowable temperature for continuous operation of Type 1 PVC is 140°F (60°C). Plasticized PVC can be bonded directly to a metal substrate. Unplasticized PVC cannot be bonded directly. It is bonded to a plasticized material forming a dual laminate that is then bonded to a metal substrate.

Plasticized PVC does not have the same range of corrosion resistance as the unplasticized PVC (Type 1). Unplasticized PVC is resistant to attack by most acids and strong alkalies as well as gasoline, kerosene, and aliphatic alcohols and hydrocarbons. It is subject to attack by aromatics, chlorinated organic compounds, and lacquer solvents.

See also "Polyvinyl Chloride."

Chlorinated Polyvinyl Chloride

CPVC can be operated continuously at a maximum temperature of 200°F (93°C). Chlorinated polyvinyl chloride is very similar in properties to PVC, except that it has a higher allowable operating temperature and a somewhat better resistance to chlorinated solvents. It is extremely difficult to hot weld the joints in a sheet lining. This factor should be considered when selecting this material.

CPVC is resistant to most acids, alkalies, salts, halogens, and many corrosive waters. In general it should not be used to handle most polar organic materials including chlorinated or aromatic hydrocarbons, esters, and ketones.

See also "Chlorinated Polyvinyl Chloride."

Polyvinylidene Fluoride

This material may be operated continuously at a maximum temperature of 275°F (135°C). PVDF is one of the most popular lining materials because of its range of corrosion resistance and high allowable operating temperature. Unless a dual laminate is used, linings of PVDF will be loose.

PVDF is resistant to most acids, bases, and organic solvents. It also has the ability to handle wet or dry chlorine, bromine, and other halogens.

It is not resistant to strong alkalies, fuming acids, polar solvents, amines, ketones, and esters. When used with strong alkalies it stress cracks.

See also "Vinylidene Fluoride."

Polypropylene

The maximum allowable temperature under which PP can be operated continuously is

180°F (82°C). PP must be attached to a backing sheet in order to be secured as a lining. If a backing sheet is not used, the lining will be loose.

PP is resistant to sulfur bearing compounds, caustics, solvents, acids, and other organic chemicals. It is not resistant to oxidizing type acids, detergents, low boiling hydrocarbons, alcohols, aromatics, and some chlorinated organic materials.

See also "Polypropylene."

Polyethylene

The maximum allowable temperature, at continuous contact, to which polyethylene (PE) can be used is 120°F (49°C). PE is the least expensive of all the plastic materials.

PE has a wide range of corrosion resistance ranging from potable water to corrosive wastes. It exhibits excellent resistance to strong oxidizing chemicals, alkalies, acids, and salt solutions.

See also "Polyethylene."

Tetrafluorethylene

PTFE can be used continuously at 500°F (260°C). It cannot be bonded directly to metal substrates but can be bonded by using a laminated fiberglass sheet as a backing.

The material tends to creep under stress at elevated temperatures. When the vessel is designed, provisions should be made to retain the PTFE. PTFE liners are subject to permeation by some corrodents. Table P.5 provides the vapor permeation of PTFE by selected materials.

PTFE is chemically inert in the presence of most corrodents. There are very few chemicals that will attack it within normal use temperatures. These reactants are among the most violent oxidizers and reducing agents known. Elemental sodium in intimate contact with fluorocarbons removes fluorine from the polymer molecule. The other alkali metals (potassium, lithium, etc.) react in a similar manner.

Fluorine and related compounds (e.g., chlorine trifluoride) are absorbed into the PTFE resin with such intimate contact that the mixture becomes sensitive to a source of ignition such as impact.

The handling of 80% sodium hydroxide, aluminum chloride, ammonia, and certain amines at high temperatures may produce the same effect as elemental sodium. Also slow oxidative attack can be produced by 70% nitric acid under pressure at 480°F (250°C).

See also "Permeation and Polytetrafluorethylene."

Fluorinated Ethylene Propylene

FEP has a lower maximum operating temperature than PTFE. It exhibits changes in physical strength after prolonged exposure above 400°F (204°C), so the recommended maximum continuous operating temperature is 375°F (190°C).

Permeation of FEP liners can pose a problem. Table P.6 provides some permeation data relating to the more common chemicals. There is also some absorption of chemicals by FEP. This absorption can also lead to problems. Table A.2 is a listing of the absorption of selected liquids by FEP.

FEP basically has the same corrosion resistant properties as PTFE but at a lower maximum temperature. It is resistant to practically all chemicals, the exception being extremely potent oxidizers such as chlorine trifluoride and related compounds. Some chemicals will attack FEP when present in high concentrations, at or near the service temperature limit.

See also "Fluorinated Ethylene Propylene."

Perfluoralkoxy

PFA can be used continuously at 500°F (260°C). It is subject to permeation by certain gases and will absorb liquids. Table P.7 illustrates the permeability of PFA and Table A.3 lists the absorption of representative liquids by PFA.

Perfluoralkoxy is inert to strong mineral acids, inorganic bases, inorganic oxidizers, aromatics, some aliphatic hydrocarbons, alcohols, aldehydes, ketones, esters, ethers, chlorocarbons, fluorocarbons, and mixtures of these compounds.

PFA will be attacked by certain halogenated complexes containing fluorine. These include chlorine trifluoride, bromine trifluoride, iodine pentafluoride, and fluorine. PFA can also be attacked by such metals as sodium or potassium, particularly in the molten state.

See also "Perfluoralkoxy."

Ethylene-Tetrafluoroethylene

ETFE has a maximum continuous service temperature of 300°F (140°C). It is fairly inert to strong mineral acids, halogens, inorganic bases, and metal salt solutions. Under most conditions ETFE is resistant to alcohols, ketones, ethers, and chlorinated hydrocarbons.

Strong oxidizers (e.g., nitric acid), organic bases (e.g., amines), and sulfonic acid will attack FEP.

See also "Ethylene-Tetrafluoroethylene."

Ethylene-Chlorotrifluorethylene

The maximum service temperature of ECTFE is 340°F (170°C). ECTFE is very similar to PTFE as far as corrosion resistance is concerned, but it does not have the permeation problems associated with PTFE. It is resistant to strong mineral and oxidizing acids, alkalies, metal etchants, liquid oxygen, and essentially all organic solvents *except* hot amine (e.g., aniline, dimethylamine). ECTFE will be attacked by metallic sodium and potassium.

See also "Ethylene-Chlorotrifluoroethylene."

Causes of Lining Failure

Linings, if properly selected, installed, and maintained, and if the vessel has been properly designed, fabricated, and prepared to accept the lining, will usually give many useful years of service. However, on occasion there have been lining failures, which can be attributed to one or more of the following causes.

Liner Selection

This is the first step. Essential to this step is a careful analysis of the materials to be handled, their concentrations, and operating conditions as outlined in the beginning of this section. Consideration must also be given to the physical and mechanical properties of the liner to insure that they meet the specified operating conditions. If there is any doubt, corrosion testing should be undertaken to guarantee the resistance of the liner material.

Inadequate Surface Preparation

Surface preparation is extremely important. All specifications for surface preparation must be followed. If surface preparation is not done properly, poor bonding can result and/or mechanical damage to the liner is possible.

Thermal Stresses

If sheet linings are not properly designed, thermal stresses produced by thermal cycling can eventually result in bond failure.

Permeation

Certain lining materials are subject to permeation when in contact with specific corrodents. When the possibility of permeation exists, an alternate lining material should be selected. Permeation can result in debonding resulting from corrosion products of fluids accumulating at the interface between the liner and the substrate. In addition, corrosion of the substrate can result, leading to leakage problems and eventual failure of the substrate.

Absorption

As with permeation, absorption of the corrodent by the liner material can result in swelling of the liner, cracking, and eventual penetration to the substrate. This can lead to high stresses and debonding.

Welding Flaws

It is essential that qualified personnel perform the welding and that only qualified and experienced contractors be used to install the linings. A welding flaw is a common cause of lining failure.

Debonding

Debonding can also occur as a result of the use of the wrong bonding agent. Care should be taken that the proper bonding agent is employed for the specific lining being used.

Operation

Lined vessels should be properly identified when installed, with the allowable operating characteristics of the liner posted to avoid damage to the liner during cleaning or repair operations. Most failures from this cause result while vessels are being cleaned or repaired. If live steam is used to clean the vessel, allowable operating temperatures may be exceeded. If the vessel is solvent cleaned, chemical attack may occur.

See Ref. 2.

SHELTERED CORROSION

Corrosion taking place in locations where moisture condenses or accumulates and does not dry out for long periods of time is known as sheltered corrosion. Typical locations are the inside surfaces of automobile doors, storage tanks, etc.

SHOT PEENING

Since cracks will not initiate or propagate in an area of compressive stress, stress corrosion cracking cannot occur. After all manufacturing operations on metals are completed, residual stresses remain in the finished part. These stresses can be either tensile or compressive. High residual tensile stresses will be present in the heat-affected zone adjacent to a welded joint, while compressive stresses may be present on the surface of induction hardened components.

Shot peening is a cold working process in which the surface of a part is bombarded with small spherical media called shot. Each piece of shot upon striking the metal part imparts a small indentation or dimple on the surface. For the dimple to form, the surface fibers of the material must be yielded in tension. The fibers below the surface try to restore the surface to its original shape, thereby producing, below the dimple, an area of cold worked material highly stressed in compression. By overlapping the dimples an even area of residual compressive stress is produced.

Materials typically used for shot peening are small spheres of cast steel, conditioned cut wire (both carbon and stainless steel), and ceramic and glass beads. Normally cast steel is used, but in cases where iron contamination on the surface is a concern, stainless steel cut wire, glass, or ceramic beads are employed.

The effect of the compressive stress and the cold working induced by shot peening are responsible for the benefits derived. Compressive stresses are beneficial in increasing resistance to fatigue failure, corrosion fatigue, stress corrosion cracking, hydrogen assisted cracking, fretting, galling, and erosion caused by cavitation. Benefits received due to cold working include work hardening, intergranular corrosion resistance, surface texturing, closing of porosity, and testing the bond of coatings.

Fretting can develop when the relative motion of microscopic amplitude occurs between two metal surfaces. Fine abrasive oxides form as a result of this rubbing. These oxides contribute to the scoring of the surfaces. Fretting gives rise to one or more forms of damage, such as fretting corrosion, fretting wear, or fretting fatigue. (See "Fretting.") Shot peening has proven to be successful in retarding fretting by increasing the surface hardness and providing residual compressive stresses at the fretting surfaces.

Stress corrosion cracking is the failure of metal by cracking as a result of the combined action of corrosion and static tensile stress (either externally applied or internal i.e. residual). Cracking may be either intergranular or transgranular depending on the metal and the corrodent. Refer to "Stress Corrosion Cracking." Most metals such as aluminum, copper, magnesium, nickel, steel, and stainless steel alloys are susceptible to stress corrosion cracking when a tensile stress at or above their threshold limits exists and they are exposed to specific corrosive environments.

The primary sources of stress contributing to SCC are applied as a result of pressure, poor fit up, residual resulting from heat treatment, welding, machining, or forming. Shot peening produces compressive stresses that retard and in many cases prevent SCC.

Hydrogen assisted cracking, or hydrogen embrittlement, is the result of atomic hydrogen penetrating and reacting with the metal, reducing the metal's ductility and ability to withstand cyclic loads. (See "Hydrogen Embrittlement.") Shot peening retards the migration of hydrogen through the metal. It not only increases the time it takes hydrogen to migrate through the metal but also lowers the steady state permeation rate of hydrogen by as much as 24%.

When austenitic stainless steels are subjected to heating in the range of 900 to 1500°F (482 to 815°C), such as in welding, there is a preferential precipitation of chromium carbides in the grain boundaries. This results in a depletion of chromium in the areas adjacent to these grain boundaries, reducing the corrosion resistance of such regions and making the alloy susceptible to intergranular corrosion. (See "Intergranular Corrosion.")

Shot peening prior to exposure to this sensitization temperature range breaks up grains and grain boundaries and provides many nucleation sites for chromium carbide precipitation. Since the chromium carbides precipitate randomly in the peened surface rather than preferentially along the grain boundaries, there is no continuous path for corrosion to follow, and intergranualr corrosion does not occur.

See Refs. 3–10.

SILICON CARBIDE

Silicon carbide fibers are used as a reinforcing material for thermosetting resins. See "Thermoset Reinforcing Materials."

SILICON CARBIDE FIBERS

Silicon carbide fibers are produced by a chemical vapor deposition process in which a heated carbon monofilament approximately 33 μm in diameter reacts with a mixture of hydrogen and chlorinated alkyl silicones. These fibers have been used on a development basis for reinforcing aluminum and titanium and to some extent for ceramic and organic matrices.

SILICONE AND FLUOROSILICONE RUBBERS

The silicone rubbers (SI), also known as polysiloxanes, are a series of compounds whose polymer structure consists of silicone and oxygen atoms rather than the carbon structures of most other elastomers. The silicones are derivatives of silica, SiO_2 or $O = Si = O$. When the atoms are combined so that the double linkages are broken and methyl groups enter the linkages, silicone rubber is produced:

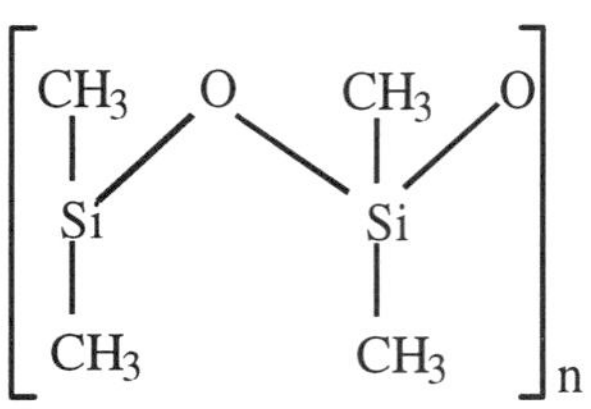

Silicone is in the same chemical group as carbon but is a more stable element, and therefore more stable compounds are produced from it. The basic structure can be modified with vinyl or fluoride groups, which improve such properties as tear resistance, oil resistance, and chemical resistance. This results in a family of silicones that covers a wide range of physical and environmental requirements.

Physical and Mechanical Properties

The silicones are some of the most heat resistant elastomers available and the most flexible at low temperatures. Their effective operating temperature range is from –60 to 450°F (–51 to 232°C). They exhibit excellent properties even at the lowest temperature.

The fluorosilicones have an effective operating temperature range of –100 to 375°F (–73 to 190°C).

Silicone rubbers possess outstanding electrical properties, superior to those of most elastomers. The decomposition product of carbon based elastomers is conductive carbon black, which can sublime and thus leave nothing for insulation, whereas the decomposition product of the silicone rubbers is an insulating silicone dioxide. This property is taken advantage of in the insulation of electric motors. The polysiloxanes have poor abrasion resistance, tensile strength, and tear resistance, but they exhibit good compression set resistance and rebound properties in both cold and hot environments. Their resistance to flame is good.

The physical and mechanical properties of silicone rubbers are given in Table S.3.

The fluorosilicones (FSIs) have essentially the same physical and mechanical properties as the silicones but with some improvement in adhesion to metals and impermeability.

Table S.4 lists the physical and mechanical properties of the fluorosilicones.

Resistance to Sun, Weather, and Ozone

The silicone and fluorosilicone rubbers display excellent resistance to sun, weathering, and ozone. Their properties are virtually unaffected by long-term exposure.

Chemical Resistance

Silicone rubbers can be used in contact with dilute acids and alkalies, alcohols, animal and vegetable oils, and lubricating oils. They are also resistant to aliphatic hydrocarbons, but

Table S.3 Physical and Mechanical Properties of Silicone (SI) Rubbers[a]

Specific gravity	1.05–1.94
Brittle point	–75°F (–6°C)
Water absorption, %/24 h	0.02–0.6
Dielectric strength, V/mil	350–590
Dissipation (power) factor	
at 60 Hz	0.0007
at 1 MHz	8.5×10^{-3}–2.6×10^{-3}
Dielectric constant	
at 60 Hz	2.91
at 1 MHz	2.8–3.94
Volume resistivity, ohm-cm	1×10^{14}–1×10^{16}
Water absorption, %/24 h	0.02–0.1
Tensile strength, psi	1200–6000
Elongation, % at break	800
Hardness, Shore A	20–90
Abrasion resistance	Poor
Maximum temperature, continuous use	450°F (232°C)
Tear resistance	Fair to good
Compression set, %	10–15
Impact resistance, notch 1/8-in. specimen, ft-lb/in.	0.25–0.30
Resistance to sunlight	Excellent
Effect of aging	Nil
Resistance to heat	Excellent

[a]These are representative values since they may be altered by compounding.

Table S.4 Physical and Mechanical Properties of Fluorosilicones[a]

Specific gravity	1.4
Hardness, Shore A	40–75
Tensile strength, psi	1000–5400
Elongation, % at break	100–500
Compression set, %	15
Tear resistance	Poor to fair
Maximum temperature, continuous use	375°F (190°C)
Abrasion resistance	Poor
Resistance to sunlight	Excellent
Effect of aging	Nil
Resistance to heat	Excellent

[a]These are representative values since they may be altered by compounding.

aromatic solvents such as benzene, toluene, gasoline, and chlorinated solvents will cause excessive swelling. Although they have excellent resistance to water and weathering, they are not resistant to high pressure, high temperature steam.

The fluorosilicone rubbers have better chemical resistance than the silicone rubbers. They have excellent resistance to aliphatic hydrocarbons and good resistance to aromatic hydrocarbons, oil and gasoline, animal and vegetable oils, dilute acids and alkalies, and alcohols; and fair resistance to concentrated alkalies. Table S.5 provides the compatibility of the silicone rubbers with selected corrodents.

Applications

Because of their unique thermal stability and or their insulating values, the silicone rubbers find many uses in the electrical industries, primarily in appliances, heaters, furnaces, aerospace devices, and automotive parts.

Their excellent weathering qualities and wide temperature range have also resulted in their employment as caulking compounds.

When silicone or fluorosilicone rubbers are infused with a high density conductive filler, an electric path is created. These conductive elastomers are used as part of an EMI/RFI/EMP shielding process in forms such as O-rings and gaskets to provide

1. Shielding for containment to prevent the escape of EMI internally generated by the device
2. Shielding for exclusion to prevent the intrusion of EMI/RFI/EMP created by outside sources into the protected device
3. Exclusion or containment plus pressure or vacuum sealing to provide EMI/EMP attenuation and pressure containment and/or weatherproofing
4. Grounding and contacting to provide a dependable low impedance connection to conduct electric energy to ground, often used where mechanical mating is imperfect or impractical

The following equipment is either capable of generating EMI or susceptible to EMI.

Aircraft and aerospace electronics

Table S.5 Compatibility of Silicone Rubbers with Selected Corrodents[a]

Chemical	Maximum temp. °F	°C	Chemical	Maximum temp. °F	°C
Acetamide	80	27	Ethyl chloride	x	x
Acetic acid 10%	90	32	Ethylene chloride	x	x
Acetic acid 20%	90	32	Ethylene diamine	400	204
Acetic acid 50%	90	32	Ethylene glycol	400	204
Acetic acid 80%	90	32	Ferric chloride	400	204
Acetic acid, glacial	90	32	Fluosilicic acid	x	x
Acetic acid vapors	90	32	Formaldehyde, all concentrations	200	93
Acetone	110	43	Fuel oil	x	x
Acetone, 50% water	110	43	Gasoline	x	x
Acetophenone	x	x	Glucose (corn syrup)	400	204
Acrylic acid 75%	80	27	Glycerine	410	210
Acrylonitrile	x	x	Green liquor	400	204
Aluminum acetate	x	x	Hexane	x	x
Aluminum phosphate	400	204	Hydrobromic acid	x	x
Aluminum sulfate	410	210	Hydrochloric acid dilute	90	32
Ammonia gas	x	x	Hydrochloric acid 20%	90	32
Ammonium chloride 10%	80	27	Hydrochloric acid 35%	x	x
Ammonium chloride 28%	80	27	Hydrofluoric acid	x	x
Ammonium chloride, sat.	80	27	Hydrogen peroxide, all concentrations	200	93
Ammonium hydroxide 10%	210	99	Lactic acid, all concentrations	80	27
Ammonium hydroxide, sat.	400	204	Lead acetate	x	x
Ammonium nitrate	80	27	Lime sulfur	400	204
Amyl acetate	x	x	Linseed oil	x	x
Amyl alcohol	x	x	Magnesium chloride	400	204
Aniline	80	27	Magnesium sulfate	400	204
Aqua regia 3:1	x	x	Mercury	80	27
Barium sulfide	400	204	Methyl alcohol	410	210
Benzene	x	x	Methyl cellosolve	x	x
Benzyl chloride	x	x	Methyl chloride	x	x
Boric acid	390	189	Methyl ethyl ketone	x	x
Butyl alcohol	80	27	Methylene chloride	x	x
Calcium acetate	x	x	Mineral oil	300	149
Calcium bisulfite	400	204	Naphtha	x	x
Calcium chloride, all concentrations	300	149	Nickel acetate	x	x
Calcium hydroxide to 30%	210	99	Nickel chloride	400	204
Calcium hydroxide, sat.	400	204	Nickel sulfate	400	204
Carbon bisulfide	x	x	Nitric acid 5%	80	27
Carbon monoxide	400	204	Nitric acid 10%	80	27
Carbonic acid	400	204	Nitric acid 20%	x	x
Chlorobenzene	x	x	Nitric acid, anhydrous	x	x
Chlorosulfonic acid	x	x	Nitrous acid–sulfuric acid 50:50	x	x
Dioxane	x	x	Nitrobenzene	x	x
Ethane	x	x	Nitrogen	400	204
Ethers, general	x	x	Nitromethane	x	x
Ethyl acetate	170	77	Oils, vegetable	400	204
Ethyl alcohol	400	204	Oleic acid	x	x

Table S.5 Continued

Chemical	Maximum temp. °F	Maximum temp. °C	Chemical	Maximum temp. °F	Maximum temp. °C
Oleum	x	x	Sodium borate	400	204
Oxalic acid to 50%	80	27	Sodium carbonate	300	149
Ozone	400	204	Sodium chloride 10%	400	204
Palmitic acid	x	x	Sodium hydroxide, all concentrations	90	32
Parrafin	x	x	Sodium peroxide	x	x
Peanut oil	400	204	Sodium sulfate	400	204
Perchloric acid	x	x	Sodium thiosulfate	400	204
Phenol	x	x	Stannic chloride	80	27
Phosphoric acid	x	x	Styrene	x	x
Picric acid	x	x	Sulfite liquors	x	x
Potassium chloride 30%	400	204	Sulfuric acid	x	x
Potassium cyanide 30%	410	210	Sulfurous acid	x	x
Potassium dichromate	410	210	Tartaric acid	400	204
Potassium hydroxide to 50%	210	99	Tetrahydrofuran	x	x
Potassium hydroxide 90%	80	27	Toluene	x	x
Potassium nitrate to 80%	400	204	Tributyl phosphate	x	x
Potassium sulfate 10%	400	204	Turpentine	x	x
Potassium sulfate, pure	400	204	Vinegar	400	204
Propane	x	x	Water, acid mine	210	99
Propyl acetate	x	x	Water, demineralized	210	99
Propyl alcohol	400	204	Water, distilled	210	99
Propyl nitrate	x	x	Water, salt	210	99
Pyridine	x	x	Water, sea	210	99
Silver nitrate	410	210	Xylene	x	x
Sodium acetate	x	x	Zinc chloride	400	204
Sodium bisulfite	410	210			

[a]The chemicals listed are in the pure state or in a saturated solution unless otherwise indicated. Compatibility is shown to the maximum allowable temperature for which data are available. Incompatibility is shown by an x.
Extracted from PA Schweitzer. *Corrosion Resistance Tables,* 4th ed. Vols. 1–3. New York: Marcel Dekker, 1995.

Digital instrumentation and process control systems
Analog instrumentation
Automotive electronics
Communication systems
Radio-frequency instrumentation and radar
Medical electronics
Security systems (military and commercial)
Home appliances
Business machines
Military and marine electronics

Table S.6 lists some typical properties of these infused elastomers.
See Refs.11 and 2.

Table S.6 Typical Properties of Conductive Elastomers[a]

Property	SI	FSI
Volume resistivity, ohm-cm	0.002–0.01	0.002–0.01
Shielding effectiveness, dB		
at 200 kHz (H field)	30–75	60–75
at 100 kHz (E field)	70–120	95–120
at 500 kHz (E field)	60–120	90–120
at 2 GHz (plane wave)	40–120	80–115
at 10 GHz (plane wave)	30–120	75–120
Heat aging, ohm-cm	0.006–0.012	0.006–0.015
Electrical stability after break, ohm-cm	0.003–0.015	0.003–0.015
Vibration resistance, ohm-cm		
during	0.004–0.008	0.004–0.008
after	0.002–0.010	0.002–0.005
EMP survivability, kA/in. perimeter	0.9	0.9
Specific gravity	1.9–4.5	2.1–4.1
Hardness, Shore A	50–85	70–85
Tensile strength, psi	175–600	200–500
Elongation, % at break	20–300	70–300
Tear strength, psi	20–75	40–50
Compression set, %	22–40	24–29
Operating temperature		
min, °F/°C	–85/–65	–85/–65
max, °F/°C	392/200	392/200

[a]These are typical values since they may be altered by compounding. SI = silicone; FSI = fluorosilicone.

SILOXIRANE

Siloxirane is the registered trademark for Tankinetics homopolymerized polymer with an ether cross-linking (carbon–oxygen–carbon) having a very dense, highly cross-linked molecular structure. The end products are extremely resistant to material abrasion, have a wide range of chemical resistance, and have an operating temperature range of –80 to 500°F (–62 to 260°C). The specific maximum operating temperature will be dependent upon the material being handled.

Following is a list of some of the chemicals with which Siloxirane is compatible.

Acetamide	Bromine water
Acetic acid, glacial	Calcium hypochlorite
Acetic anhydride	Carbon tetrachloride
Acetone	Chloric acid
Aluminum chloride	Chlorine water
Ammonium chloride	Chloroacetic acid
Ammonium hydroxide	Chlorobenzene
Aqua regia	Chromic acid 10%
Benzene	Chromic acid 50%
Benzene sulfonic acid	Dibutylphthalate
Black liquor (paper)	Dichlorobenzene

Dimethyl formamide	Molten sulfur
Ethanol	Monochloroacetic acid
Ethyl acetate	Nickel plating
Ferric chloride	Nitrous oxide
Formaldehyde	Phosphoric acid
Furan	Phosphoric acid 85%
Furfural alcohol	Sodium chloride
Gasohol	Sodium dichromate
Gasoline	Sodium hydroxide
Green liquor	Sodium hypochlorite 17%
Hydraulic oil	Sodium hypochlorite, aged
Hydrazine	Sulfite liquor (paper)
Hydrochloric acid 1%	Sulfur trioxide
Hydrochloric acid 0–37%	Sulfuric acid 1–70%
Hydrofluoric acid 40%	Sulfuric acid 70–90%
Hydrofluoric acid 52%	Sulfuric acid 90–98%
Iodine	Sulfuric acid, fuming oleum
Jet fuel	Tallow
Kerosene	Thionyl chloride
Ketones	Toluene
Latex	Trichloroethylene
Methanol	Tricresylphosphate
Methyl ethyl ketone	Water, deionized
Methyl isobutyl ketone	Water, salt
Methylene chloride	White liquor (paper)

SOIL CORROSION

When soil is dry it is not corrosive. It becomes corrosive as a result of its water content and related water-soluble salts that permit it to act as an electrolyte. Moisture received by the soil from the atmosphere contains specific contaminants and picks up specific water-soluble materials from the soil. Such materials include salts of aluminum, calcium, magnesium, and sodium, sulfates, chlorides, carbonates, phosphates, and silicates. This subjects construction materials to a wide variety of corrosive conditions.

In addition to the problem of localized cells, buried pipe lines are also subject to macrocell action. Because of the differences in soil chemistry, soil compaction, thermal effects, and bacterial action, long sections of pipeline may become anodic to other long sections. Stray DC currents will cause electrolysis. Corrosion of buried structures can also be caused by telluric currents, which are a result of fluctuations in the earth's magnetic field as well as stray currents from AC power lines.

The corrosivity of soil is influenced by three factors: resistivity, chemistry, and physical characteristics.

Resistivity

Resistivity is the property of a material, as opposed to resistance, which is the resultant property of a physical entity. For example, copper has a certain resistivity, but a piece of copper wire has resistance.

Many physical and chemical aspects of the soil give it its resistivity. It is determined by passing a known current I through a known volume of soil (volume = depth × width × length) and measuring the difference in voltage E due to current flow. From Ohm's Law we have

$$R = \frac{E}{I}$$

and resistivity is

$$P = R \times \frac{W \times d}{L}$$

where W is width, d is depth, and L is length.

Resistivity is expressed in ohm-centimeters (Ω cm) when R is expressed in ohms (Ω), area ($W \times d$) in square centimeters, and length (L) in centimeters. For example, seawater has a resistivity of 25 to 75 Ω cm, while a specific volume of seawater has so many ohms resistance. The lower the value of the resistivity the more highly corrosive a material is.

Soil is considered highly corrosive if its resistivity is 2000 Ω cm or less. It is considered mildly corrosive when its resistivity is between 2000 and 10,000 Ω cm, while at 10,000 to 20,000 Ω cm it is considered only slightly corrosive. If the resistivity of the soil is below 25,000 Ω cm cathodic protection is recommended.

Soil Chemistry

Different types of soils can contain a wide variety of chemical species. These not only influence the electrolytic nature of the soil but will also have specific ion effects (for example, chlorides for pitting or stress corrosion cracking or sulfides for stress corrosion cracking). Varying amounts of water may be available from the water table or atmospheric ingress.

Composition and Condition

Sand, clay, loam, and rock are four major categories of soil. With the exception of nonporous rock, any of these categories can have varying degrees of moisture and varying chemical contaminants. Dry desert sand is totally different from wet salty sand.

Ordinary earth will have different degrees of compaction. Freshly excavated and filled trenches will be more apt to absorb moisture and oxygen than undisturbed soil.

Overall Corrosivity

Resistivity alone is not the only criterion for judging a soil's corrosivity. Besides resistivity, the other factors that must be taken into account are pH, chloride content, redox potential, and type of soil. Overall rating of soil as to its corrosivity can be made by use of the different arithmetical point values from the following table.

Factor	Points	Factor	Points
1. pH		4. Soil type	
0–2	5	clay (blue gray)	10
2–4	3	clay/stone	5
4–8.5	0	clay	3
8.5+	3	silt	2
		clean sand	0
2. Chlorides, ppm		5. Soil resistivity	
1000+	10	< 1000	10
500–1000	6	1000–1500	8
200–500	4	1500–2500	6
50–200	2	2500–5000	4
0–50	0	5000–10000	2
		10000 +	0
3. Redox, mV (vs. copper; copper sulfate)			
negative	5		
0–50	4		
50–100	3.5		
100+	0		

Based on the above factors the overall rating of the corrosivity of the soil is as follows.

Point Total	Rating
15+	Severe
10–15	Appreciable
5–10	Moderate
0–5	Mild

Reaction of Specific Materials

Depending upon the specific chemistry of the soil, different materials will react in different ways. However, some generalizations can be made and specific problems highlighted.

Concrete

Portland cement type concrete is suitable for soil contact if a type appropriate to the sulfate concentration and chlorides and other chemical variants is used. Most important are the pH and sulfate effects.

Steels

Steel should not be exposed to soil unless some type of corrosion control is employed, such as a barrier coating, cathodic protection, or environmental control (mixing of extraneous materials in the soil such as lime, sand, or water repellents to reduce the corrosivity). The most effective approach is to use a combination of a coating and cathodic protection.

Cast Iron

Cast iron is more resistant than steel because of the adherent nature of the rust formed under normal conditions. In soils of low pH or those saturated with soft aggressive waters it can be subject to graphitic corrosion. Corrosion can also be aggravated by bacterial action. Underground cast iron pipe is usually protected by a barrier coating. Because of the difficulty of establishing continuity across the mechanical joints of cast iron pipe, cathodic protection is not often used.

Zinc

Specific soil chemistry will determine the suitability of zinc. In general, galvanized steel is not recommended for underground service since the zinc coating is very thin and anodically active to every metallic structure in the area. However, galvanized steel structured legs for power lines and such can be used provided they are protected by a coating.

Aluminum

If protected from galvanic demands and specific ion effects (chlorides), aluminum can provide satisfactory underground service. Because of the possibility of generating high alkali concentrations when using cathodic protection, care must be taken, since such alkali concentrations can corrode aluminum.

Stainless Steels

High chloride concentrations and/or oxygen concentration cells will pit all stainless steels. Austenitic stainless steel underground piping should be coated and cathodically protected, since corrosive conditions can arise, and localized attack can lead to rapid penetration.

Lead

Lead is generally resistant but can be corroded by stray electrical currents and specific chemical contaminants.

Copper Alloys

Sulfate-bearing bacteria in soils can cause corrosion by sulfides. If ammonia is present from the rotting of nitrogenous compounds, corrosion or stress corrosion cracking can take place. However, copper is usually satisfactory. Copper pipe was used by the Egyptian Pharaoh Cheops to transport water to the royal bath. Several years ago a remnant of this pipe was unearthed still in usable condition, a testimony to copper's durability and resistance to corrosion.

Plastics

Plastic is an ideal material for underground service, providing temperature and pressure otherwise permit. Major concerns are mechanical damage from rock fill and other foreign objects, due to ground subsidence or surface traffic imposing a loading force against the pipe or vessel wall.

SOLEF

See "Polyvinylidene Fluoride."

SOLUTE

A solute is a material that is dissolved in a solvent to form a solution. Water is the most common solvent.

SOLUTION QUENCHING

See "Quench Annealing."

SOLVENT

A solvent is a liquid in which another substance may be dissolved; also the liquid part that dissolves the solute.

SPHERADIZING

See "Annealing."

STAINLESS STEELS

Probably the most widely known and most widely used metallic material for corrosion resistance is stainless steel. For many years this was the only material available.

Stainless steel is not a single material as its name might imply but rather a broad group of alloys, each of which exhibits its own physical and corrosion resistant properties. Stainless steels are alloys of iron to which a minimum of 11% chromium has been added to provide a passive film to resist "rusting" when the material is exposed to weather. This film is self-forming and self-healing in environments where the stainless steel is resistant. As more chromium is added to the alloy, improved corrosion resistance results. Consequently, there are stainless steels with chromium contents of 15, 17, and 20% and even higher. Chromium provides resistance to oxidizing environments such as nitric acid and also provides resistance to pitting and crevice attack.

Other alloying ingredients are added further to improve the corrosion resistance and mechanical strength. Molybdenum is extremely effective in improving pitting and crevice corrosion resistance.

By the addition of copper, improved resistance to general corrosion in sulfuric acid is obtained. This will also strengthen some precipitation hardenable grades. In sufficient amounts, though, copper will reduce the pitting resistance of some alloys.

The addition of nickel will provide improved resistance to reducing environments and stress corrosion cracking. Nitrogen can also be added to improve corrosion resistance to pitting and crevice attack, and to improve strength.

Columbium and titanium are added to stabilize carbon. They form carbides and reduce the amount of carbon available to form chromium carbides, which can be deleterious to corrosion resistance.

As a result of these alloying possibilities, more than 70 stainless steels are available.

These can be divided into four major categories depending upon their microstructure. The classifications are

Austenitic
Ferritic
Martensitic
Duplex

Refer to these headings for specific information about each type.

See Refs. 12–14.

STRESS CORROSION CRACKING (SCC)

Stress corrosion cracking occurs at points of stress. Usually the metal or alloy is virtually free of corrosion over most of its surface, yet fine cracks penetrate through the surface at the points of stress. This type of attack takes place in certain mediums. All metals are potentially subject to SCC. The conditions necessary for stress corrosion cracking are

1. A suitable environment
2. A tensile stress
3. A sensitive metal
4. Appropriate temperature and pH values

An ammonia-containing environment can induce SCC in copper-containing alloys, while with low alloy austenitic stainless steels a chloride-containing environment is necessary. It is not necessary to have a concentrated solution to cause SCC. A solution containing only a few parts per million of the critical ion is all that is necessary. Temperature and pH are also factors. There is usually a threshold temperature below which SCC will not occur and a plus or minus pH value before cracking will start.

Normally stress corrosion cracking will not occur if the part is in compression. Failure is triggered by a tensile stress that must approach the yield strength of the metal. These stresses may be induced by faulty installation, or they may be residual stresses from welding, straightening, bending, or accidental denting of the component. Pits, which act as stress concentration sites, will often initiate SCC.

Alloy content of stainless steels, particularly nickel, determine the sensitivity of the metal to SCC. Ferritic stainless steels, which are nickel free, and the high nickel alloys, are not subject to stress corrosion cracking. An alloy with a nickel content greater than 30% is immune to SCC. The most common grades of stainless steel (304, 304L, 316, 316L, 321, 347, 303, 302, and 301) have nickel contents in the range of 7–10% and are the most susceptible to stress corrosion cracking.

Examples of stress corrosion cracking include the cracking of austenitic stainless steels in the presence of chlorides, caustic embrittlement cracking of steel in caustic solutions, cracking of cold-formed brass in ammonia environments, and cracking of Monel in hydrofluorosilicic acid. Table S.7 contains a partial listing of alloy systems subject to stress corrosion cracking.

It is important that any stresses that may have been induced during fabrication be removed by an appropriate stress-relief operation. The design should also avoid stagnant areas that could lead to pitting and the initiation of stress concentration sites.

See Refs. 15, 14, 13, and 16.

Table S.7 Alloy Systems Subject to Stress Corrosion Cracking

Alloy	Environment
Carbon steel	Anhydrous liquid ammonia, HCN, ammonium nitrate, sodium nitrite, sodium hydroxide
Aluminum base	Air, seawater, salt and chemical combination
Magnesium base	Nitric acid, caustic, HF solutions, salts, coastal atmospheres
Copper base	Primarily ammonia and ammonium hydroxide, amines, mercury
Martensitic and precipitation hardening stainless steels	Seawater, chlorides, H_2S solutions
Austenitic stainless steels	Chlorides (organic and inorganic), caustic solutions, sulfurous and polythiuric acids
Nickel base	Caustic above 600°F (315°C) fused caustic, hydrofluoric acid
Titanium	Seawater, salt atmosphere, fused salt
Zirconium	$FeCl_3$ or $CuCl_2$ solutions

STRESS RELIEF

Fabrication processes, such as rolling or forging, uneven heating or cooling, or welding are all capable of inducing residual stresses into a metal. The magnitude of these stresses is usually on the order of the yield strength of the metal, but in some cases it approaches the tensile strength of the metal.

Carbon and low alloy steels are heated to a temperature in the range of 1000 to 1350°F (595 to 730°C) to be stress relieved. They are held at this temperature for a period of time, then air cooled. The minimum holding time is specified by the appropriate engineering code. The holding temperature must be less than the lower transformation temperature of the steel, which is the lowest temperature at which austenite forms. For plain carbon steels this temperature is 1333°F (720°C).

As metals and alloys are heated, their yield strengths decrease. Residual stresses in excess of the reduced yield strength are eliminated. Upon cooling, the maximum residual stress possible is the yield strength at the holding temperature. For carbon steels, this heat treatment process reduces residual stresses by approximately two-thirds.

Austenitic stainless steels are not usually stress relieved or post weld heat treated. When such treatments are given to austenitic stainless steels they are held at a temperature of 1600 to 1650°F (870 to 900°C) followed by rapid cooling. The rapid cooling is necessary to prevent sensitization. If austenitic stainless steels are exposed to heat treating operation at less than 1600°F (870°C), they can be sensitized. For this reason, local stress relief of unstabilized austenitic stainless steel is usually impractical, since the runout areas immediately adjacent to the region being heat treated will be sensitized.

STYRENE–BUTADIENE–STYRENE (SBS) RUBBER

Styrene–butadiene–styrene (SBS) rubbers are either pure or oil modified block copolymers. They are most suitable as performance modifiers in blends with thermoplastics or as a base rubber for adhesive, sealant, or coating formulations. SBS compounds are formulations containing block copolymer rubber and other suitable ingredients. These compounds have a wide range of properties and provide the benefits of rubberiness and easy processing on standard thermoplastic processing equipment.

Physical and Mechanical Properties

Since the physical and mechanical properties vary greatly depending upon the formulation, this discussion of these properties is based on the fact that proper formulation will provide a material having the desired combination of properties.

The degree of hardness will determine the flexibility of the final product. With a Shore A hardness range from 37 to 74, SBS rubbers offer excellent impact resistance and low temperature flexibility. Their maximum service temperature is 150°F (65°C). They also exhibit good abrasion resistance and good resistance to water absorption and heat aging. Their resistance to compression set and tear is likewise good, as is their tensile strength.

The electrical properties of SBS rubber are only fair, and resistance to flame is poor.

The physical and mechanical properties of SBS rubber are given in Table S.8.

Resistance to Sun, Weather, and Ozone

The SBS rubbers are not resistant to ozone, particularly when they are in a stressed condition. Neither are they resistant to prolonged exposure to sun or weather.

Chemical Resistance

The chemical resistance of SBS rubbers is similar to that of natural rubber. They have excellent resistance to water, acids, and bases. Prolonged exposure to hydrocarbon solvents and oils will cause deterioration; however, short exposures can be tolerated.

Table S.8 Physical and Mechanical Properties of Styrene–Butadiene–Styrene (SBS) Rubber[a]

Specific gravity	0.92–1.09
Tear resistance	Good
Tensile strength, psi	625–4600
Elongation, % at break	500–1400
Hardness, Shore A	37–74
Abrasion resistance	Good
Maximum temperature, continuous use	150°F (65°C)
Impact resistance	Good
Compression set at 74°F (23°C), %	10–15
Machining qualities	Can be ground
Resistance to sunlight	Poor
Effect of aging	Little
Resistance to heat	Fair

[a]These are representative values since they may be altered by compounding.

Applications

The specific formulation will determine the applicability of various products. Applications include a wide variety of general-purpose rubber items and use in the footwear industry. These rubbers are used primarily in blends with other thermoplastic materials and as performance modifiers.

See Refs. 11 and 2.

STYRENE–ETHYLENE–BUTYLENE–STYRENE (SEBS) RUBBER

Styrene–ethylene–butylene–styrene (SEBS) rubbers are either pure or oil modified block copolymer rubbers. These rubbers are used as performance modifiers in blends with thermoplastics or as the base rubber for adhesive, sealant, or coating formulations. Formulations of SEBS compounds provide a wide range of properties with the benefits of rubberiness and easy processing on standard thermoplastic processing equipment.

Physical and Mechanical Properties

The SEBS rubbers offer excellent impact resistance and low temperature flexibility, are highly resistant to oxidation and ozone, and do not require vulcanization. Their degree of flexibility is a function of hardness, with their Shore A hardness ranging from 37 to 95. These rubbers are serviceable from –120 to 220°F (–75 to 105°C). They have excellent resistance to very low temperature impact and bending. Their thermal life and aging properties are excellent, as is their abrasion resistance.

The electrical properties of SEBS rubbers are extremely good. Although the SEBS rubbers have poor flame resistance, compounding can improve their flame retardancy. This compounding reduces the operating temperature slightly.

The physical and mechanical properties of SEBS rubbers are given in Table S.9.

Resistance to Sun, Weather, and Ozone

The SEBS rubbers and compounds exhibit excellent resistance to ozone. For prolonged outdoor exposure the addition of an ultraviolet absorber or carbon black pigment or both is recommended.

Chemical Resistance

The chemical resistance of the SEBS rubbers is similar to that of natural rubber. They have excellent resistance to water, acids, and bases. Soaking in hydrocarbon solvents and oils will deteriorate the rubber, but short exposures can be tolerated.

Applications

The SEBS rubbers find applications for a wide variety of general-purpose rubber items as well as in automotive, sporting goods, and other products. Many applications are found in

Table S.9 Physical and Mechanical Properties of Styrene–Ethylene–Butylene–Styrene (SEBS) Rubbers[a]

Specific gravity	0.885–1.17
Brittle point	–58 to –148°F (–50 to –100°C)
Tear strength, psi	275–470
Moisture absorption, mg/in.^{2}	1.2–3.3
Dielectric strength, at 77°F (25°C)	
at 60 Hz	2.1–2.8
at 1 kHz	2.1–2.8
at 1 MHz	2.1–2.35
Dissipation factor at 77°F (25°C)	
at 60 Hz	0.0001–0.002
at 1 kHz	0.0001–0.003
at 1 MHz	0.0001–0.01
Dielectric strength, V/mil	625–925
Volume resistivity, ohm/cm	9×10^5–9.1×10^{16}
Surface resistivity, ohm	2×10^{16}–9.5×10^{16}
Insulation resistance constant at 60°F (15.6°C) and 500 V dc	6.8×10^4–2.5×10^6
Insulation resistance at 60°F (15.6°C), megohms/1000 ft	2.1×10^4–1×10^6
Tensile strength, psi	1600–2700
Elongation, % at break	500–675
Hardness, Shore A	65–95
Abrasion resistance	Good
Maximum temperature, continuous use	220°F (105°C)
Impact resistance	Good
Resistance to sunlight	Good
Effect of aging	Small
Resistance to heat	Good

[a]These are representative values since they may be altered by compounding.

the electrical industry for such items as flexible cords, welding and booster cables, flame resistant appliance wiring materials, and automotive primary wire insulation.

See Refs. 11 and 2.

SULFATE REDUCING BACTERIA

These bacteria are widespread in seawater, freshwater, soil, and muddy sediments. When they are present and there is an abundance of sulfate, and the surface of the substrate is between a pH of 5.5 and 8.5, they may cause anaerobic corrosion of iron and steel. The usual by-product of this metabolic process is hydrogen sulfide, which tends to retard cathodic reactions, particularly hydrogen evolution, and to accelerate anodic dissolution, thereby increasing corrosion. The corrosion product is iron sulfide, which precipitates when ferrous and sulfide ions are in contact.

SULFIDATION

Also see "High Temperature Corrosion." Sulfidation is oxidation by sulfur forming a sulfide film on the metal surface similar to an oxide film but less protective than a corresponding oxide film.

SULFIDIC CORROSION

Sulfidic corrosion is most often found in petroleum refining. It is caused by a variety of sulfur compounds originating in the crude oils, including such as hydrogen sulfide, aliphatic sulfides, mercaptans, disulfides, polysulfides, and thiophenes. Corrosion takes place at process temperatures between 500 and 1000°F (260 and 540°C). The corrosion products are metal sulfides.

SULFIDE STRESS CRACKING

Sulfide stress cracking is a form of hydrogen-assisted cracking and refers to the cracking of metals when hydrogen sulfide environments generate the hydrogen. The term is used primarily in the petroleum industry.

SUPER PRO 230

See "Polyvinylidene Fluoride."

REFERENCES

1. PA Schweitzer. Cathodic protection. In: PA Schweitzer, ed. Corrosion and Corrosion Protection Handbook. 2nd ed. New York: Marcel Dekker, 1989, pp. 33–45.
2. PA Schweitzer. Corrosion Resistance Tables. 4th ed. Vols. 1–3. New York: Marcel Dekker, 1995.
3. PB Waterhouse and DA Saunders. The effect of shot peening on the fretting fatigue behavior of austenitic stainless steel and a mild steel. Wear 53:381–386, 1979.
4. WH Friske. Shot peening to prevent corrosion of austenitic stainless steels. Rockwell International Report No. A1-75-52.
5. DO Sprowls and RH Brown. What every engineer should know about stress corrosion of aluminum. Metal Progress, April/May, 1962.
6. FP Vaccaro et al. Effect of shot peening on transition stress corrosion cracking of alloy 600 steam generator tubing. Corrosion 87, paper No. 87.
7. RD Gillespie. Controlled shot peening can help prevent stress corrosion cracking. Third International Conference of Shot Peening, Garmisch-Partenkirchen, Germany, October 12–16, 1987.
8. CS Lin et al. Stress corrosion cracking of high strength bolting. 69th Annual Meeting of the American Society for Testing and Materials. Atlantic City, June 27–July 1.
9. JJ Daly. Controlled shot peening prevents stress corrosion cracking. Chemical Engineering, Feb. 16, 1976.
10. Stress corrosion cracking prevented by shot peening. Chemical Processing, March 1976.

11. PA Schweitzer. Corrosion Resistance of Elastomers. New York: Marcel Dekker, 1990.
12. PA Schweitzer. Stainless steels. In: PA Schweitzer, ed. Corrosion and Corrosion Protection Handbook. 2nd ed. New York: Marcel Dekker, 1989, pp. 69–85.
13. CP Dillon. Corrosion Resistance of Stainless Steels. New York: Marcel Dekker, 1995.
14. CP Dillon. Corrosion Control in the Chemical Process Industry. 2nd ed. St. Louis: Materials Technology Institute of the Chemical Process Industries, 1994.
15. RC Newman. Stress corrosion cracking mechanisms. In: P Marcus and J Oudar, eds. Corrosion Mechanisms in Theory and Practice. New York: Marcel Dekker, 1995, pp. 311–372.
16. HH Uhlig. Corrosion and Corrosion Control. New York: John Wiley, 1963.

T

TAFEL SLOPE

The slope of the linear portion of a plot of potential versus the log of the current density for a particular electrode in a corroding system.

See Ref. 1.

TANTALUM

Tantalum is not a new material. Its first commercial use at the turn of the century was as filaments in light bulbs. Later, when it became apparent that tantalum was practically inert to attack by most acids, applications in the laboratory and in the chemical and medical industries were developed. The rise of the electronics industry accelerated the development of many new applications.

Much of this growth can be attributed to a broader range of tantalum powders and mill products available from the producers: high melting point ability to form a dielectric oxide film and chemical inertness. With these applications, new reduction, melting, and fabrication techniques have led to higher purities, higher reliabilities, and improved yields to finished products.

Source of Tantalum

The earth's crust is made up of 92 naturally occurring elements, but these elements are not all present in equal amounts. Eight elements, oxygen, silicon, aluminum, iron, calcium, sodium, potassium, and magnesium make up 96.5% of the crust. The remaining 88 elements make up only 3.5%, with tantalum amounting to only 0.0002%.

If the tantalum were equally distributed in the rocks of the earth it would be uneconomical to recover, and there would be no tantalum industry today. However, the tantalum is concentrated in a few unusual rocks in sufficicent quantity to permit economical mining and refining. The most important tantalum minerals, tantalite microlites and wodginite, are found in rock formations known as pegmatites.

Pegmatites are coarse-grained rocks formed when molten rock material was cooled slowly. They range in size from 1 in. to many feet in diameter. Also found in the pegmatites are many rare elements such as tantalum, niobium, tin, lithium, and beryllium.

The only operating mine in North America is located at Bernic Lake in Manitoba, Canada. The other important mine in the Americas is found in Brazil. In the humid tropics the rocks weather and rot to great depths. Many times the rocks in which the tantalum minerals were formed have completely weathered and have been carried away by running water. The heavy tantalum minerals tend to be concentrated in deposits called placers.

These can be panned or washed with machinery, much as gold was recovered during the gold rush. One important placer type deposit is found in Greenbushes in western Australia.

Because tin and tantalum are often found together, tantalum is a by-product of the tin industry. Because most of the tantalum deposits are small, hard to find, and very expensive to mine, the result is a high priced ore and a correspondingly high priced metal.

Tantalum Manufacture

Ingot Consolidation

The first production route for tantalum was by powder metallurgy. Tantalum powder, produced by one of several reduction techniques, is pressed into suitably sized bars and then sintered in vacuum at temperatures in excess of 3800°F (2100°C). When completed, the pressed and sintered bars are ready for processing into mill shapes. Forging, rolling, swaging, and drawing of tantalum is performed at room temperature on standard metal-working equipment with relatively few modifications.

The powder metallurgy route, although still in use and adequate for many applications, has two major limitations: (1) the size of the bar capable of being pressed and sintered to a uniform density limits the size of the finished shape available, and (2) the amount of residual interstitial impurities, such as oxygen, carbon, and nitrogen, remaining after sintering adversely affects weldability.

The use of vacuum melting, either by consumable arc or electron beam process, overcomes these limitations. Either melting technique is capable of producing ingots that are big enough and high enough in purity to meet most requirements of product size and specifications adequately, provided that starting materials are selected with care.

Quality Description

The greatest volume of tantalum is supplied as powder for the manufacture of solid electolytic tantalum capacitors.

Because it is necessary to distinguish between capacitor-grade powder and melting-grade powder, manufacturers of electronic components tend to use the term capacitor-grade when ordering forms such as wire, foil, and sheet to identify end use and desired characteristics.

The term capacitor-grade means that the material should have the ability to form an anodic oxide film of certain characteristics. Capacitor-grade in itself does not mean an inherently higher purity, cleaner surface, or different type of tantalum. It does mean that the material should be tested using carefully standardized procedures for electrical properties. If certain objective standards, such as formation voltage and leakage current, are not available against which to test the material, the use of the phrase capacitor-grade is not definitive.

Metallurgical-grade coud be simply defined as non-capacitor-grade.

Properties of Tantalum

Alloys Available

Pure tantalum has a body centered, cubic crystal lattice. There is no allotropic transformation to the melting point, which means that unalloyed tantalum cannot be hardened by heat treatment. Additions of oxygen, carbon, or nitrogen above normal levels, either purposefully or accidentally, are considered as alloying ingredients no matter what the concentration.

Tantalum–niobium alloys containing more than about 5 to 10% niobium are much less corrosion resistant than tantalum itself.

Tantalum–tungsten alloys containing more than 18% tungsten are inert to 20% hydrofluoric acid at room temperature. Few data are available on the 90Ta-10W alloy. It is known to be somewhat more oxidation resistant (e.g., to air at higher temperatures) than tantalum. The indication is that it has about the same corrosion resistance to acids as tantalum itself.

Mechanical Properties

The room-temperature mechanical properties of tantalum are dependent on chemical purity, amount of reduction in cross-sectional area, and temperature of final annealing. Annealing time apparently is not critical. Close control over the many factors that affect mechanical properties is mandatory to ensure reproducible mechanical behavior. Typical mechanical properties for tantalum are shown in Table T.1.

Tantalum can be strengthened only by cold work, with a resulting loss in ductility. Because certain residual impurities have pronounced effects on ductility levels and metallurgical behavior, the purpose of most consolidation techniques is to make the material as pure as possible. Cold-working methods are used almost without exception to preclude the possibility of embrittlement by exposure to oxygen, carbon, nitrogen, and hydrogen at even moderate temperatures. Temperatures in excess of 800°F (425°C) should be avoided. Physical properties are shown in Table T.2.

Corrosion Resistance

Tantalum is inert to practically all organic and inorganic compounds at temperatures under 302°F (150°C); it is inert to all concentrations of hydrochloric acid, to all concentrations of nitric acid (including fuming), to 98% sulfuric acid, to 85% phosphoric acid, and to aqua regia.

Table T.1 Mechanical and Physical Properties of Tantalum

Modulus of elasticity $\times 10^6$, psi	27
Tensile strength $\times 10^3$, psi	
Grade VM	30
Grade PM	40
Yield strength 0.2% offset $\times 10^3$, psi	
Grade VM	20
Grade PM	30
Elongation in 2 in., %	30+
Hardness, Rockwell	
Grade VM	B-55
Grade PM	B-65
Density, lb/in.^{3}	0.6
Specific gravity	16.6
Specific heat, Btu/lb °F	0.036
Thermal conductivity at 68°F, Btu/h/ft^2/°F/in.	377
Coefficient of thermal expansion, in./in./°F $\times 10^{-6}$	3.6

Table T.2 Physical Properties of Tantalum

Atomic weight	180.9
Density	16.6 g/cm^3 (0.601 lb/in.^{3})
Melting point	2996°C (5432°F)
Vapor pressure at 1727°C	9.525×10^{-11} mmHg
Linear coefficient of expansion	1135K; 5.76×10^{-6}/°C
	1641K; 9.53×10^{-6}/°C
	2030K; 12.9×10^{-6}/°C
	2495K; 16.7×10^{-6}/°C
Thermal conductivity	20°C; 0.130 cal/cm-sec °C
	100°C; 0.131 cal/cm-sec °C
	1430°C; 0.174 cal/cm-sec °C
	1630°C; 0.186 cal/cm-sec °C
	1830°C; 0.198 cal/cm-sec °C
Specific heat	100°C; 0.03364 cal/g
Electrical conductivity	13.9% IACS
Electrical resistivity	–73°C; 9.0 μΩ/cm
	75°C; 12.4 μΩ/cm
	127°C; 18.0 μΩ/cm
	1000°C; 54.0 μΩ/cm
	1500°C; 71.0 μΩ/cm
	2000°C; 87.0 μΩ/cm

Materials to which tantalum is completely inert, up to at least 302°F (150°C), are shown in Table T.3.

Corrosion is first noticed at about 375°F (190°C) for 70% nitric acid, at about 345°F (175°C) for 98% sulfuric acid, and at about 355°F (180°C) for 85% phosphoric acid. Fuming sulfuric acid attacks tantalum even at room temperature. Similarly, hydrofluoric acid, anhydrous HF, or any acid medium containing fluoride ion will rapidly attack the metal. Commercial phosphoric acid may attack tantalum because of the presence of small amounts of fluoride impurity. One exception to fluoride attack appears to be in chromium-plating baths. Hot oxalic acid is the only organic acid known to attack tantalum. The corrosion rates of tantalum in various acid media are given in Table T.4.

Fused sodium and potassium hydroxides and pyrosulfates dissolve tantalum. It is attacked by concentrated alkaline solutions at room temperature; it is fairly resistant to dilute solutions. Tantalum is completely inert to body fluids.

Tantalum's resistance to oxidation by various gases is very good at low temperatures, but it reacts rapidly at high temperatures. Only HF and SO_3 attack the metal under 212°F (100°C); most gases begin to react with it at 570 to 750°F (300 to 400°C). As the temperature and concentration of such gases as oxygen, nitrogen, chlorine, hydrogen chloride, and ammonia are increased, oxidation becomes more rapid; the usual temperature for rapid failure is 930 to 1290°F (500 to 700°C).

Refer to Tables T.5 and T.13 for compatibility of tantalum with selected corrodents.

Tantalum Based Alloys

The use of tantalum based alloys provides certain advantages.

1. Alloying with a less expensive metal reduces the cost of the material while still retaining essentially all of the corrosion resistant properties.

Table T.3 Materials to Which Tantalum Is Completely Inert, Up to at Least 150°C (302°F)

Acetic acid	Hydrogen
Acetic anhydride	Hydrogen chloride
Acetone	Hydrogen iodide
Acids, mineral (except HF)	Hydrogen peroxide
Acid salts	Hydrogen sulfide
Air	Iodine
Alcohols	Hypochlorous acid
Aluminum chloride	Lactic acid
Aluminum sulfate	Magnesium chloride
Amines	Magnesium sulfate
Ammonium chloride	Mercury salts
Ammonium hydroxide	Methyl sulfuric acid
Ammonium phosphate	Milk
Ammonium sulfate	Mineral oils
Amyl acetate	Motor fuels
Amyl chloride	Nitric acid, industrial fuming
Aqua regia	Nitric oxides
Barium hydroxide	Nitrogen
Body fluids	Nitrosyl chloride
Bromine, wet or dry	Nitrous oxides
Butyric acid	Organic chlorides
Calcium bisulfate	Oxalic acid
Calcium chloride	Oxygen
Calcium hydroxide	Perchloric acid
Calcium hypochlorite	Petroleum products
Carbon tetrachloride	Phenols
Carbonic acid	Phosphoric acid, < 4 ppm F
Carbon dioxide	Phosphorus
Chloric acid	Phosphorus chlorides
Chlorinated hydrocarbons	Phosphorus oxychloride
Chlorine oxides	Phthalic anhydride
Chlorine water and brine	Potassium chloride
Chlorine wet or dry	Potassium dichromate
Chloroacetic acid	Potassium iodide, iodine
Chrome-plating solutions	Potassium nitrate
Chromic acid	Refrigerants
Citric acid	Silver nitrate
Cleaning solutions	Sodium bisulfate, aqueous
Copper salts	Sodium bromide
Ethyl sulfate	Sodium chlorate
Ethylene dibromide	Sodium chloride
Fatty acids	Sodium hypochlorite
Ferric chloride	Sodium nitrate
Ferrous sulfate	Sodium sulfate
Foodstuffs	Sodium sulfite
Formaldehyde	Sugar
Formic acid	Sulfamic acid
Fruit products	Sulfur
Hydriodic acid	Sulfur dioxide
Hydrobromic acid	Sulfuric acid, under 98%
Hydrochloric acid	Water

Table T.4 Corrosion Rates of Tantalum in Selected Media

Medium	Temperature (°C)	Temperature (°F)	Corrosion rate (mpy)
Acetic acid	100	212	Nil
$AlCl_3$ (10% soln.)	100	212	Nil
NH_4Cl (10% soln.)	100	212	Nil
HCl 20%	21	70	Nil
	100	212	Nil
Conc.	21	70	Nil
HNO_3 20%	100	212	Nil
70%	100	212	Nil
65%	170	338	1
H_3PO_4, 85%	25	76	Nil
	100	212	Nil
H_2SO_4			
10%	25	76	Nil
40%	25	76	Nil
98%	25	76	Nil
98%	50	122	Nil
98%	100	212	Nil
98%	200	392	3
98%	250	482	Rapid
H_2SO_4, fuming (15% SO_3)	23	73	0.5
	70	158	Rapid
Aqua regia	25	78	Nil
Chlorine wet	75	167	Nil
H_2O			
Cl_2 sat.	25	76	Nil
Sea	25	76	Nil
Oxalic acid	21	70	Nil
	96	205	0.1
NaOH 5%	21	70	Nil
	100	212	0.7
10%	100	212	1
40%	80	176	Rapid
HF, 40%	25	76	Rapid

2. The use of light material will reduce the overall weight.
3. Depending upon the alloying ingredient, the physical strength of the tantalum may be improved.

The formation of these alloys is a relatively new development and consequently there are few data available on their properties.

Tantalum–Tungsten Alloys

The tantalum–tungsten alloys are probably the most common. The addition of 2 to 3% tungsten will raise the strength of the tantalum by 30 to 50%. By also adding 0.15% columbium, a marked increase in the corrosion resistance to concentrated sulfuric acid at

Table T.5 Temperatures at Which Various Media Attack Tantalum

Medium	State	Remarks
Air	Gas	At temperatures over 300°C (572°F)
Alkaline solutions	Aqueous	At pH > 9, moderate temperature, some corrosion
Ammonia	Gas	Pits at high temperature and pressures
Bromine	Gas	At temperatures over 300°C (572°F)
Chlorine, wet	Gas	At temperatures over 250°C (482°F)
Fluorides, acid media	Aqueous	All temperatures and concentrations
Fluorine	Gas	At all temperatures
HBr, 25%	Aqueous	Begins to corrode at temperatures over 190°C (374°F)
Hydrocarbons	Gas	React at temperatures around 1500°C (2732°F)
HCl 25%	Aqueous	Begins to corrode at temperatures over 190°C (374°F)
HF	Aqueous	Corrodes at all temperatures and pressures
Hydrogen	Gas	Causes embrittlement, especially at temperatures over 400°C (752°F)
HBr	Gas	At temperatures over 400°C (752°F)
HCl	Gas	At temperatures over 350°C (662°F)
HF	Gas	At all temperatures
Iodine	Gas	At temperatures over 300°C (572°F)
Nitrogen	Gas	At temperatures over 300°C (572°F)
Oxalic acid, sat. soln.	Aqueous	At temperatures of about 100°C (212°F)
Oxygen	Gas	At temperatures over 350°C (662°F)
H_3PO_4, 85%	Aqueous	Corrodes at temperatures over 180°C (356°F), at higher temperatures for lower concentrations
Potassium carbonate	Aqueous	Corrodes at moderate temperatures depending on concentration
Sodium carbonate	Aqueous	Corrodes at moderate temperatures depending on concentration
NaOH, 10%	Aqueous	Corrodes at about 100°C (212°F)
NaOH	Molten	Dissolves metal rapidly (over 320°C) (608°F)
Sodium pyrosulfate	Molten	Dissolves metal rapidly (over 400°C) (752°F)
H_2SO_4, 98%	Aqueous	Begins to corrode at temperatures over 175°C (347°F); lower concentrations begin to corrode at higher temperatures
H_2SO_4 (oleum) (over 98% H_2SO_4)	Fuming	Corrodes at all temperatures
Sulfuric trioxide	Gas	At all temperatures
Water	Aqueous	Corrodes at pH > 9, reacts at high temperatures

392°F (200°C) is noted. In the lower temperature ranges, 347°F (175°C) or less, the resistance of the alloy is equal to that of pure tantalum.

When the tungsten concentration is increased to 18% or higher, the alloys exhibit essentially no corrosion rate in 20% hydrofluoric acid. This is a definite advantage over pure tantalum.

Tantalum–Titanium Alloys

The tantalum–titanium alloys are receiving a great deal of study because this series of alloys shows considerable promise of providing a less expensive, lower weight alloy having a corrosion resistance almost comparable with that of tantalum. Preliminary tests indicate that tantalum–titanium alloys show excellent resistance in nitric acid at 374°F (190°C) and at the boiling point.

Tantalum–Molybdenum Alloys

When these alloys are exposed to concentrated sulfuric acid and concentrated hydrochloric acid, they are extremely resistant, and the properties of tantalum are retained, as long as the tantalum concentration is higher than 50%.

Suggested Applications

Because of its relatively high price, tantalum can be recommended only for use in extremely corrosive media, in areas where no corrosion of the part can be tolerated, or where very high purity materials are being processed. Although some plastics, and even glass, fill these requirements to a large extent, tantalum is a structurally sound material of construction, can take considerable mechanical abuse, and has a much higher heat transfer coefficient. Tantalum should be used as a material of construction in locations and for equipment where hot concentrated hydrochloric, sulfuric, or phosphoric acids will be present. Tantalum is used by the medical profession for instruments and for metal implants in the body. In the manufacture of high purity chemicals and pharmaceuticals, tantalum ensures that no impurities are introduced from the container or reactor.

See Refs. 2–5.

TARNISH

Tarnish is a surface discoloration of a metal caused by a thin film of corrosion product. This is quite common on silver surfaces.

TECHNOFLON

See "Fluorelastomers."

TEFLON

See "Polytetrafluoroethylene."

TEFZEL

See "Ethylene tetrafluoroethylene."

TEMPERING

Tempering is usually performed after a quenching operation. Quenching produces a hard and strong but brittle phase called martensite. Tempering is performed to promote some carbon diffusion from the martensite, thereby greatly improving the toughness and ductility of the quenched steel.

Tempering is usually done at 1000 to 1300°F (595 to 705°C). Thick sections of many ferritic steels cannot be cooled quickly enough in air to obtain a normalized structure. In order to hasten the cooling rate the material is quenched. The objective is to produce the same type of microstructure that would be obtained from normalizing a thinner section of the same material. Quenching of very thick sections does not generate the cooling rates necessary to develop martensite. In these cases tempering is used primarily for stress relief rather than for softening of the martensite.

At times tempering is done in conjunction with other heat treatments such as normalizing. The purpose is usually to promote carbon diffusion with the intention of softening and/or toughening the steel. Stress relief may be a secondary or even a primary objective.

TEREPHTHALIC POLYESTERS

This family of thermoset resins is based on terephthalic acid, which is the paraisomer of phthalic acid. The corrosion resistance of these polyesters is similar to that of the isophthalics. Testing has shown that the benzene resistance of comparable formulated resin was lower for terephthalic versus isophthalic polyesters. At the same level of unsaturation as isophthalic polyester, the terephthalic's loss of properties in gasoline is greater, but as unsaturation increases the terephthalic polyesters perform better than the isophthalic polyesters.

THERMOGALVANIC CORROSION

Thermogalvanic corrosion is a form of corrosion resulting from temperature differences at two points of a common metal surface in the presence of an electrolyte.

THERMOPLASTIC ALLOYS

The alloying of several thermoplastic systems often produces better physical properties than any system can produce by itself. These include impact resistance, flame retardancy, and thermal stability. Alloying is generally done with intensive mixing or screw extruders. Polyvinyl chloride and ABS are frequently blended to provide rigidity, toughness, flame retardancy, and chemical resistance. Polycarbonate can be blended with ABS to provide better heat resistance and toughness. Polyurethane improves the abrasion resistance and toughness of ABS while retaining the advantage of reduced cost. The impact strength of polypropylene is increased by alloying with polyisobutylene. Alloying is generally done on a relatively small scale, varying from as low as 0.01% up to as high as 9%.

Also see "Xenoy," "Cycoloy," and "Triax."

THERMOPLASTIC COMPOSITES

See "Composite Laminates."

THERMOPLASTIC ELASTOMERS (TPE), OLEFINIC TYPE

This family of elastomers is based on cross-linked polyolefin alloys compounded with common fillers (pigments), plasticizers, stabilizers, and cross-linking agents. The plasticizers and fillers are used to tailor properties to specific applications. These materials can be injection molded, blow molded, and extruded on conventional thermoplastic equipment. Vulcanization is not required. See Table T.6.

Resistance to Sun, Weather, and Ozone

The TPEs possess good resistance to sun and ozone and have excellent weatherability. Their water resistance is excellent, showing essentially no property changes after prolonged exposure to water at elevated temperatures.

Chemical Resistance

The thermoplastic elastomers display reasonably good resistance to oils and automotive fluids, comparable to that of neoprene. However, they do not have the outstanding oil resistance of the polyester elastomers. They do have excellent water resistance, even at elevated temperatures.

Table T.6 Physical and Mechanical Properties of Thermoplastic Elastomers (TPE), Olefinic Type[a]

Specific gravity	0.88–1.0
Brittle point	–67 to –76°F (–55 to –60°C)
Tensile strength, psi	
at 73°F (23°C)	1700–2150
at 212°F (100°C)	1040–1170
at 277°F (136°C)	730–770
at 302°F (150°C)	640–750
Elongation, % at break	
at 73°F (23°C)	210–300
at 212°F (100°C)	290–670
at 277°F (136°C)	380–870
at 302°F (150°C)	300–750
Hardness, Shore	92A–54D
Abrasion resistance	Good
Maximum temperature, continuous use	277°F (136°C)
Compression set, method A, %	
after 22 h at 73°F (23°C)	8–18
after 22 h at 212°F (100°C)	47–48
Tear resistance	Good
Resistance to sunlight	Good
Resistance to heat	Good
Effect of aging	Small
Electrical properties	Excellent

[a]These are representative values since they may be altered by compounding.

Applications

These elastomeric compounds are found in a variety of applications including reinforced hose, seals, gaskets and profile extrusions, flexible and supported tubing, automotive trim, functional parts and under-hood components, mechanical goods, and wire and cable jacketing.

See Refs. 6 and 4.

THERMOPLASTIC POLYMERS

See "Thermoplasts."

THERMOPLASTS

See also "Polymers" and "Individual Thermoplasts." Thermoplasts are thermoplastic polymers that can be repeatedly reformed by the application of heat, similar to metallic materials. They are long-chain linear molecules that are easily formed by the application of heat and pressure at temperatures above a critical temperature referred to as the "glass temperature." Because of the ability to be reformed by heat, these materials can be recycled.

The most common thermoplasts are shown in Table T.7 along with their abbreviations. Table T.8 lists the heat distortion temperatures of the common thermoplasts, while the tensile strengths are given in Table T.9, and the maximum operating temperatures are shown in Table T.10.

Table T.7 Abbreviations Used for Plastics

ABS	Acrylonitrile–butadiene–styrene
CPVC	Chlorinated polyvinyl chloride
ECTFE	Ethylene–chlorotrifluorethylene
FEP	Perfluoroethylenepropylene
HDPE	High density polyethylene
PEEK	Polyetheretherketone
PES	Polyethersulfone
PFA	Perfluoralkoxy
PA	Polyamide
PB	Polybutylene
PC	Polycarbonate
PF	Phenol formaldehyde
PP	Polypropylene
PPS	Polyphenyl sulfide
PTFE	Polytetrafluoroethylene
PVC	Polyvinyl chloride
PVDC	Polyvinylidene chloride
PVDF	Polyvinylidene fluoride
UHMWPE	Ultrahigh molecular weight polyethylene

Table T.8 Heat Distortion Temperature of the Common Plastics (°F/°C)

	Pressure (psi)		
Polymer	66	264	Melt point
PTFE	250/121	132/56	620/327
PVC	135/57	140/60	285/141
LDPE	—	104/40	221/105
UHMW PE	155/68	110/43	265/129
PP	225/107	120/49	330/160
PFA	164/73	118/48	590/310
FEP	158/70	124/51	554/290
PVDF	248/148	235/113	352/178
ECTFE	240/116	170/77	464/240
ETFE	220/104	165/74	518/270
PEEK	—	320/160	644/340
PES	—	410/210	—
PC	280/138	265/129	—

THERMOSET COMPOSITES

Three general types of construction are used to produce thermoset composites to resist corrosion. They are hand laid up, filament winding, and chop-hoop construction.

In hand laid up construction a corrosion barrier of 100 mills (one layer C glass plus two layers of $1\frac{1}{2}$ oz mat) is followed by a structural laminate of alternate layers of rove mat to the desired thickness. This is to provide the optimum in corrosion resistance, but it is not the lowest in cost.

Table T.9 Tensile strength of Plastics at 73°F (25°C) at Break

Plastic	Strength (psi)
PVDF	8000
ETFE	6500
PFA	4000–4300
ECTFE	7000
PTFE	2500–6000
FEP	2700–3100
PVC	6000–7500
PE	1200–4550
PP	4500–6000
UHMW PE	5600
PEEK	13,200–23,800
PES	12,200–20,300
PC	10,000

Table T.10 Maximum Operating Temperature of the Common Thermoplasts

Thermoplast	°F/°C
PVC	140/160
CPVC	180/82
HMW PE	140/60
UHMW PE	180–220/82–104
ABS	140/60
PP	180–220/82–104
PB	220/104
ECTFE	300/149
ETFE	300/149
FEP	375/190
PEEK	480/250
PES	390/200
PFA	500/260
PA	212–250/100–121
PC	210–265/99–128
PPS	450/230
PTFE	450/230
PVDF	320/160

Filament winding construction starts with a 100 mil corrosion barrier as in hand laid up construction, followed by a structural laminate of filament winding to the desired thickness.

With chop-hoop construction a corrosion barrier of 100 mils is followed by alternate layers of filament winding and chopped glass. The advantage is that the chopped layers provide extra axial strength.

All three methods rely on the corrosion barrier to provide protection for the structural laminate. Only the hand laid up construction lends itself to easy repair. When splits occur in the latter two constructions, they invariably follow the wind angle, making field repair difficult, since the corrosion barrier has been breached.

There are several corrosion resistant laminate surfacing systems that are used. These are shown in the table.

Type of surface layer	Thickness, mils	Description	Specific uses
Gel coat	10–20, at times 60	Unreinforced layer of resin	Commonly used on epoxy piping systems
Type C glass surface mat	10–20	10 mil type C mat	Used with reinforcement for chemical plant applications
Organic veil	10–20	Dacron, acrylic, polypropylene, orlon	Good weathering properties, standard for hydrogen fluoride or caustic applications
Carbon mat	3	0.2 oz/yd^2	Provides high surface electrical conductivity

See Ref. 7.

THERMOSET LAMINATES

See "Thermoset Composites."

THERMOSET POLYMERS

Also see "Polymers."

Thermoset polymers, unlike the thermoplasts, once formed, cannot be heated to change their shape. Consequently they cannot be recycled. These resins are initially liquid at room temperature, then by adding a catalyst or accelerator they are changed into a rigid product that sets or cures into its final shape. The thermoset resins are high molecular weight polymers that are reinforced with glass or other suitable material to provide mechanical strength. The most commonly used resins are the vinyl esters, epoxies, polyesters, and furans.

For reinforcing these polymers, fibrous glass in the E and C grades are the most commonly used. Other reinforcing materials used include boron nitrade, carbon fiber, ceramic fibers, graphite jute, Kelvar, metallic wire or sheet, monacrylic fiber, polyester fiber, polypropylene fiber, quartz, sapphire whiskers, and S grade glass.

The advantages of the thermosets are many. They

1. Are less expensive than the stainless steels
2. Have a wide range of corrosion resistance
3. Are light in weight
4. Possess exceptional strength
5. Do not require painting
6. May be formulated to be fire retardant

See Refs. 7–9.

THERMOSET REINFORCING MATERIALS

The purpose of adding reinforcing to the thermoset resins is to provide mechanical strength and dimensional stability, which is not possible with the resin alone. Physical and chemical characteristics of the structure can be modified by changing the quantity and/or type of reinforcing. The most widely used reinforcing material for use with thermosetting resins is fibrous glass (fiberglass). Although glass itself is over 3000 years old, fiberglass was not commercialized until 1939 at the New York World's Fair. Other reinforcing materials that also find application are

Boron nitride
Silicon carbide
Carbon fiber
Ceramic fibers
Graphite
Polyester fibers

Aramid fibers
Polypropylene fibers
Acrylic fibers

Most composite materials manufactured use fiberglass for reinforcing. One of the main advantages in their use is that the fibers remain largely intact. The resin in liquid form can be made to flow around the fibers at room temperature and pressure. Any desired size or shape can be produced by building up layer by layer. Glass retains its strength only to a temperature of 752°F (400°C). Above this temperature it is necessary to use another type of reinforcing material, such as boron, carbon, or silicon carbide. The dangers of asbestos (mesothelioma etc.) have eliminated the use of asbestos as a reinforcing material, although it was widely used during the 1960s and 1970s.

Properties of Reinforcing Materials

As mentioned previously, there is a relatively wide choice of materials that can be used for reinforcing. Of these materials glass is the most often used, where corrosion resistance is required. It is available in several grades.

E-Glass

This is a boroaluminosilicate glass having excellent water resistance, strength, low elongation, and reasonable cost. Practically all glass mat, continuous filaments, and woven rovings come from this source.

C-Glass

C-glass is a calcium aluminosilicate glass used for surfacing mats, flakes, or flake glass linings and for acid resistant cloths. It has poor water resistance and carries a premium cost. Until the development of synthetic veils, 10 mil C-glass surfacing mats were widely used. It is available in 10, 15, 20, and 30 mil thicknesses.

S-Glass

S-glass has excellent resistance to acids and water with exceptional strength, comparable to an aramid fiber. Because of its cost, which is several times greater than that of E-glass, it is not used in the corrosion industry.

Boron Carbide

Boron carbide is used for applications at elevated temperatures or for those requiring high strength. It is used with epoxy resin to produce helicopter rotor blades that turn at high speed.

Silicon Carbide

Silicon carbide is used as a reinforcing material for applications above 752°F (400°C). It is resistant to hydrochloric, sulfuric, hydrofluoric, and nitric acids but will be attacked by mixtures of hydrofluoric and nitric acids.

Carbon Fibers

Carbon fibers are inert to most chemicals. Their use also imparts surface conductivity to FRP laminates. It is also used at temperatures above 752°F (400°C) to impart strength. Carbon fibers in epoxy resin provide compressor blades in lightweight jet engines.

In-depth grounding systems and static control in hazardous areas where static sparks may result in fires or explosions are provided when carbon fiber mat, either alone or in conjunction with graphite or ground carbon, is used for reinforcing.

Graphite Fibers

Graphite fibers, like carbon fibers, are inert to most chemicals and produce an inert conductive pathway. Applications are the same as for carbon fibers.

Polyester Fibers

Polyester is used primarily as a surfacing mat for the resin rich inner surfaces of filament wound or contact molded structures. The Nexus veil (registered trademark of Burlington Industries) possesses a relatively high degree of elongation that makes it compatible with the higher elongation resin and reduces the potential for checking, crazing, and cracking in temperature cycling applications.

Nexus veiling exhibits excellent resistance to alcohols, bleaching agents, water, hydrocarbons, and aqueous solutions of most weak acids at boiling. It is not resistant to strong acids, such as 93% sulfuric acid.

Aramid Fibers

The most popular aramid fiber is Kevlar (trademark of E. I. duPont de Nemours). It is a high strength fiber. As such it is used in the manufacture of bullet-proof vests, canoe and boat manufacturing, and other areas where its high strength is required in the laminate. It is available both as a surfacing mat and in cloth form. Kevlar cloth is used occasionally as reinforcement for high stress areas in the corrosion industry, such as the vertical corners of rectangular tanks.

Polypropylene Fibers

Polypropylene fibers do not develop the strength of glass, but they do have a relatively wide range of corrosion resistance and they are less expensive.

Sec Refs. 7 and 10.

TIN COATINGS

Most of the tin plate (tin coating on steel) produced is used for the manufacture of food containers (tin cans). The nontoxic nature of tin salts makes tin plate an ideal material for the handling of foods and beverages.

An inspection of the Galvanic series will indicate that tin is more noble than steel and consequently the steel corrodes at the base of the pores. On the outside of a tinned container this is what happens—the tin is cathodic to the steel. However, on the inside of the container there is a reversal of polarity due to the complexing of the stannous ion, resulting in a change in the potential of tin in the active direction.

This change in polarity is absolutely necessary, since most tin coatings are thin and therefore porous. In order to avoid perforation of the can, the tin must act as a sacrificial coating. Figure T.1 illustrates this reversal of activity between the outside and the inside of the can.

Sec Refs. 11 and 12.

TITANIUM

Titanium is the ninth most abundant element on earth and the fourth most abundant metal. It is more plentiful than chromium, nickel, or copper, which are commonly employed as

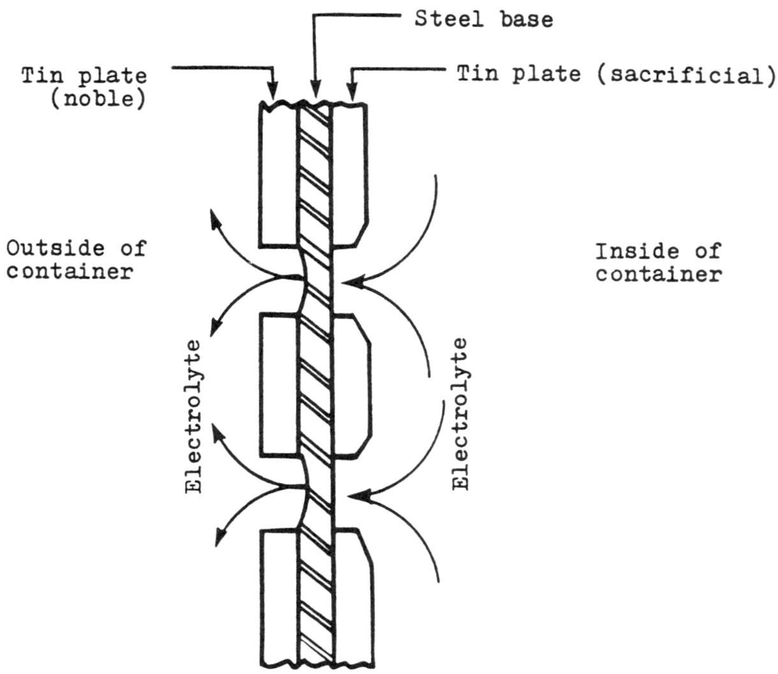

Figure T.1 Tin acting as both a noble and a sacrificial coating on a "tin" can.

alloys used to resist corrosion. In spite of this, little use was made of titanium until the commercialization of the Kroll process about 1950, which made titanium sponge available. Although it has the advantages of being highly corrosion resistant in oxidizing environments, of a low density (sp gr 4.5 approximately 60% that of steel), and of high tensile strength (60,000 psi), its widespread use has been limited somewhat by its cost. But as consumption has increased and new technologies have been developed to reduce the high cost, usage has increased and will probably continue to increase. At the present time it is competitive with nickel base alloys. Thinner sections, coupled with decreased maintenance requirements and longer life expectancy in many applications, permit titanium equipment installations to be cost effective, despite a higher initial cost.

Alloys

The primary initial applications of titanium were in the aerospace industry where mechanical properties were the primary consideration. During this period of time many titanium alloys were developed for these applications. In industrial applications, however, corrosion resistance is the most important property. Consequently this discussion of the titanium alloys will be limited to those used for their corrosion resistant properties.

ASTM grade 2, unalloyed titanium, is the single "alloy" most often used for corrosion resistance. If better formability or higher strength is required, ASTM grades 1, 3, and 4 are available containing more or less iron and oxygen.

ASTM grade 5 alloy is a general purpose alloy used in the aerospace industry for applications requiring higher strength or fatigue resistance. Its corrosion resistance is somewhat inferior to that of the unalloyed titanium.

Compared with unalloyed titanium grade 2, the titanium–palladium alloy grade 7 offers improved corrosion resistance. This alloy, as grade 11, is available with low oxygen and iron content for improved formability.

Ti-Code 12 is a low cost alternative for grade 7 in some applications.

Physical and Mechanical Properties

Titanium is a light metal with a density slightly over half that of iron or copper based alloys. The modulus of elasticity is also approximately half that of steel, while its specific heat and thermal conductivity are similar to those of stainless steel. Titanium has a low expansion coefficient and a relatively high electrical resistivity.

The mechanical and physical properties are shown in Table T.11.

Types of Corrosion

Titanium, like any other metal is subject to corrosion in certain environments. The corrosion resistance of titanium is the result of a stable, protective, strongly adherent oxide film. This film forms instantly when a fresh surface is exposed to air or moisture. Additions of alloying elements to titanium affect the corrosion resistance because these elements alter the composition of the oxide film.

The oxide film of titanium is very stable, though relatively thin, and is attacked by only a few substances, most notable of which is hydrofluoric acid. Because of its strong affinity for oxygen, titanium is capable of healing ruptures in this film almost instantly in any environment where a trace of moisture or oxygen is present.

Anhydrous conditions, in the absence of a source of oxygen, should be avoided because the protective film may not be regenerated if damaged.

The protective oxide film of most metals is subject to being swept away above a critical water velocity. Once this takes place, accelerated corrosion attack occurs. This is known as erosion corrosion. For some metals this can occur at velocities as low as 2 to 3 ft/sec. The critical velocity for titanium in seawater is in excess of 90 ft/sec. Numerous corrosion erosion tests have been conducted and all have shown that titanium has outstanding resistance to this form of corrosion.

General Corrosion

This form of corrosion is characterized by a uniform attack over the entire exposed surface of the metal. The severity of this kind of attack can be expressed by a corrosion rate. With titanium this type of corrosion is most frequently encountered in hot reducing acid

Table T.11 Mechanical and Physical Properties of Titanium (Grade 2)

Modulus of elasticity $\times 10^6$, psi	14.9
Tensile strength $\times 10^3$, psi	50
Yield strength 0.2% offset $\times 10^3$, psi	40
Elongation in 2 in., %	20
Density, lb/in.3	0.163
Specific gravity	4.48
Specific heat at 75°F, Btu/lb °F	0.125
Thermal conductivity at 75°F, Btu/ft^2/h/°F/in.	114
Coefficient of thermal expansion at 32–600°F, in./in. °F $\times 10^{-6}$	5.1

solutions. In environments where titanium would be subject to this type of corrosion, oxidizing agents and certain multivalent metal ions have the ability to passivate the titanium. Many process streams, particularly sulfuric and hydrochloric acid solutions, contain enough impurities in the form of ferric or cupric ions etc. to passivate titanium and give trouble free service.

Galvanic Corrosion

The coupling of titanium with dissimilar metals usually does not accelerate the corrosion of titanium. The exception is in reducing environments where titanium does not passivate. Under these conditions titanium has a potential similar to aluminum and will undergo accelerated corrosion when coupled to more noble metals.

For most environments titanium will be the cathodic member of any Galvanic couple. It may accelerate the corrosion of the other member of the couple, but in most cases the titanium will be unaffected. As a result of this, hydrogen will be evolved on the surface of the titanium proportional to the Galvanic current flow. This may result in the formation of surface hydride films that are generally stable and cause no problems. However, if the temperature exceeds 170°F (77°C), hydriding can cause embrittlement.

The surest way to avoid problems with Galvanic corrosion is to construct equipment of a single metal. If this is not practical, select two metals that are close in the Galvanic series. If contact of dissimilar metals with titanium is necessary, the critical parts should be constructed of titanium, since this is not usually attacked.

Hydrogen Embrittlement

The oxide film on titanium in most cases acts as an effective barrier to penetration by hydrogen. However, embrittlement can occur under conditions that allow hydrogen to enter titanium and exceed the concentration needed to form a hydride phase (about 100 to 150 ppm). Hydrogen absorption has been observed in alkaline solutions at temperatures above the boiling point. Acidic conditions that cause the oxide films to be unstable may also result in embrittlement under conditions in which hydrogen is generated on the titanium surface. In any event it appears that embrittlement occurs only if the temperature is sufficiently high above 170°F (75°C) to allow hydrogen to diffuse into the titanium. Otherwise, if surface hydride films do form, they are not detrimental.

Gaseous hydrogen has had no embrittlement effects on titanium. The presence of as little as 2% moisture effectively prevents the absorption of molecular hydrogen up to a temperature as high as 600°F (315°C). This may reduce the ability of the titanium to resist erosion, resulting in a higher corrosion rate.

Crevice Corrosion

Crevice corrosion of titanium is most often observed in hot chloride solutions. However, it has also been observed in bromide, iodide, and sulfate solutions.

Dissolved oxygen or other oxidizing species present in the solution are depleted in the restricted volume of solution in the crevice. These species are consumed faster than they can be replenished by diffusion from the bulk solution. As a result the potential of the metal in the crevice becomes more negative than the metal exposed to the bulk solution. This establishes an electrolytic cell with the metal in the crevice acting as the anode and the metal outside the crevice acting as the cathode. Metal dissolves at the anode under the influence of the resulting current. Titanium chlorides formed in the crevice are unstable and tend to hydrolyze, forming small amounts of hydrochloric acid. This reaction is very slow at first, but in the very restricted volume of the crevice it can reduce the pH of the solution to a value as low as 1. This reduces the potential still further until corrosion becomes quite severe.

Alloying with elements such as nickel, molybdenum, or palladium improves the crevice corrosion resistance of titanium. Consequently Ti-Code 12 and the titanium–palladium alloys are much more resistant to crevice corrosion than unalloyed titanium.

Stress Corrosion Cracking (SSC)

Unalloyed titanium with an oxygen content of less than 0.2% (ASTM grades 1 and 2) is susceptible to cracking only in absolute methanol and higher alcohols, certain liquid metals such as cadmium and possibly mercury, red fuming nitric acid, and nitrogen tetraoxide. The presence of halides in the alcohols accelerates cracking tendencies. The presence of water (> 2%) tends to inhibit stress cracking in alcohols and red fuming nitric acid. Titanium is not recommended for use in these environments under anhydrous conditions.

Corrosion Resistance

In general titanium offers excellent resistance in oxidizing environments and poor resistance in reducing environments.

It has excellent resistance to moist chlorine gas, chlorinated brines, and hypochlorites. Some corrosion rates for titanium in hypochlorite solutions are given in Table T.12. Titanium is not resistant to dry chlorine gas. It is attacked rapidly and can ignite and burn if the moisture content is sufficiently low. Approximately 1% water is required under static conditions at room temperature. Somewhat less is required if the chlorine is flowing. Approximately 1.5% water is required at 392°F (200°C).

Titanium is immune to all forms of corrosive attack in seawater and chloride salt solutions at ambient temperatures. It is also very resistant to attack in most chloride solutions at elevated temperatures.

Titanium offers excellent resistance to oxidizing acids such as nitric and chromic acids. It is not recommended for use in red fuming nitric acid, particularly if the water content is below 1.5% and the nitrogen dioxide content is above 2.5%. Pyrophoric reactions have occurred in this environment.

Titanium will be attacked by reducing acids such as hydrochloric, sulfuric, and phosphoric acids.

It is also quite resistant to organic acids that are oxidizing. Only a few organic acids

Table T.12 Corrosion of Titanium in Hypochlorite Solutions

Environment	Temperature (°F)	Test duration (days)	Corrosion rate (mpy)	Pitting
17% hypochlorous acid, with free chlorine and chlorine monoxide	50	203	< 0.1	—
16% sodium hypochlorite	70	170	< 0.1	None
18–20% calcium hypochlorite	70–75	204	Nil	None
1.5–4% sodium hypochlorite, 12–15% sodium chloride, 1% sodium hydroxide	150–200	72	0.1	None

are known to attack titanium; these are hot nonaerated formic acid, hot oxalic acid, concentrated trichloracetic acid, and solutions of sulfamic acid.

Titanium is resistant to acetic acid, teraphthalic acid, and adipic acids. It also exhibits good resistance to citric, tartaric, carbolic, stearic, lactic, and tannic acids. Good corrosion resistance is also shown to organic compounds. In anhydrous environments when the temperature is high enough to cause dissociation of the organic compound hydrogen embrittlement of the titanium is a consideration.

The compatibility of titanium with selected corrodents is given in Table T.13.

See Refs. 4 and 13.

Table T.13 Compatibility of Titanium, Zirconium, and Tantalum with Selected Corrodents[a]

	Maximum temperature (°F/°C)		
Chemical	Titanium	Zirconium	Tantalum
Acetaldehyde	300/104	250/121	90/32
Acetamide			
Acetic acid 10%	260/127	220/104	302/150
Acetic acid 50%	260/127	230/110	302/150
Acetic acid 80%	260/127	230/110	302/150
Acetic acid, glacial	260/127	230/110	302/150
Acetic anhydride	280/138	250/121	302/150
Acetone	190/88	190/88	302/150
Acetyl chloride		80/27	80/27
Acrylic acid			
Acrylonitrile	210/93	210/93	210/93
Adipic acid	450/232		210/93
Allyl alcohol	200/93	200/93	300/149
Allyl chloride		200/93	
Alum	200/93	210/99	90/32
Aluminum acetate			
Aluminum chloride, aqueous	10%	40%	302/150
	310/154	200/93	
Aluminum chloride, dry		37%	
	200/93	210/99	302/150
Aluminum fluoride	80/27	x	x
Aluminum hydroxide	190/88	200/93	100/38
Aluminum nitrate	200/93		80/27
Aluminum oxychloride			
Aluminum sulfate	210/99	210/99	302/150
Ammonia gas		100/38	
Ammonium bifluoride			
Ammonium carbonate	200/93		200/93
Ammonium chloride 10%	210/99	210/99	302/150
Ammonium chloride 50%	190/88	220/104	302/150
Ammonium chloride, sat.	203/93		302/150
Ammonium fluoride 10%	90/32	x	x
Ammonium fluoride 25%	80/27	x	x
Ammonium hydroxide 25%	80/27	210/99	302/150

Table T.13 Continued

	Maximum temperature (°F/°C)		
Chemical	Titanium	Zirconium	Tantalum
Ammonium hydroxide, sat.	210/99	210/99	302/150
Ammonium nitrate	210/99	210/99	210/99
Ammonium persulfate	80/27	220/104	90/32
Ammonium phosphate 10%	210/99	210/99	302/150
Ammonium sulfate 10–40%	210/99	210/99	302/150
Ammonium sulfide			90/32
Ammonium sulfite			210/99
Amyl acetate	210/99	210/99	302/150
Amyl alcohol	200/93	200/93	320/160
Amyl chloride		210/99	302/150
Aniline	210/99	210/99	210/99
Antimony trichloride	110/43		210/99
Aqua regia 3:1	80/27	x	302/150
Barium carbonate	80/27	210/99	90/32
Barium chloride 25%	210/99	210/99	210/99
Barium hydroxide	210/99	200/93	302/150
Barium sulfate	210/99	210/99	210/99
Barium sulfide	90/32	90/32	90/32
Benzaldehyde	100/38	210/99	210/99
Benzene	230/110	230/110	230/110
Benzene sulfonic acid 10%		210/99	210/99
Benzoic acid	400/204	400/204	210/99
Benzyl alcohol	210/99	210/99	210/99
Benzyl chloride			230/110
Borax	190/88		x
Boric acid	210/99	210/99	300/149
Bromine gas, dry	x	x	302/150
Bromine gas, moist	190/88	60/16	302/150
Bromine liquid	x	60/16	570/299
Butadiene			80/27
Butyl acetate	210/99	210/99	80/27
Butyl alcohol	200/93	200/93	80/27
n-Butylamine	210/99		
Butyl phthalate	210/99	210/99	210/99
Butyric acid	210/99	210/99	302/150
Calcium bisulfide			
Calcium bisulfite	210/99	90/32	80/27
Calcium carbonate	230/110	230/110	230/110
Calcium chlorate	140/60		210/99
Calcium chloride	310/154	210/99	302/150
Calcium hydroxide 10%	210/99	210/99	302/150
Calcium hydroxide, sat.	230/110	210/99	302/150
Calcium hypochlorite	200/93	200/93	302/150
Calcium nitrate	210/99		80/27
Calcium oxide			
Calcium sulfate	210/99	210/99	210/99
Caprylic acid	210/99	210/99	300/149

Table T.13 Continued

Chemical	Maximum temperature (°F/°C)		
	Titanium	Zirconium	Tantalum
Carbon bisulfide	210/99		210/99
Carbon dioxide, dry	90/32	410/210	310/154
Carbon dioxide, wet	80/27		300/149
Carbon disulfide	210/99		210/99
Carbon monoxide	300/149		
Carbon tetrachloride			302/150
Carbonic acid	210/99	210/99	300/149
Cellosolve	210/99	210/99	210/99
Chloracetic acid, 50% water	210/99	210/99	210/99
Chloracetic acid	210/99	210/99	302/150
Chlorine gas, dry	x	90/32	460/238
Chlorine gas, wet	390/199	x	570/299
Chlorine, liquid		x	300/149
Chlorobenzene	200/93	200/93	300/149
Chloroform	210/99	210/99	210/99
Chlorosulfonic acid	210/99		210/99
Chromic acid 10%	210/99	210/99	302/150
Chromic acid 50%	210/99	210/99	302/150
Chromyl chloride	60/16		210/99
Citric acid 15%	210/99	210/99	302/150
Citric acid, concentrated	180/82	180/82	302/150
Copper acetate		200/93	300/149
Copper carbonate	80/27		300/149
Copper chloride	200/93	x	300/149
Copper cyanide	90/32	x	300/149
Copper sulfate	210/99	210/99	300/149
Cresol	210/99		
Cupric chloride 5%	210/99	x	300/149
Cupric chloride 50%	210/99	190/88	90/32
Cyclohexane			
Cyclohexanol			
Dichloroacetic acid	280/138	350/177	260/127
Dichloroethane			
Ethylene glycol	210/99	210/99	90/32
Ferric chloride	300/149	x	302/150
Ferric chloride 50% in water	210/99	x	302/150
Ferric nitrate 10–50%	90/32		210/99
Ferrous chloride	210/99	210/99	210/99
Ferrous nitrate			
Fluorine gas, dry	x	x	x
Fluorine gas, moist	x	x	x
Hydrobromic acid, dilute	90/32	80/27	302/150
Hydrobromic acid 20%	200/93	x	302/150
Hydrobromic acid 50%	200/93	x	302/150
Hydrochloric acid 20%	x	300/149	302/150
Hydrochloric acid 38%	x	140/60	302/150
Hydrocyanic acid 10%			

Table T.13 Continued

Chemical	Maximum temperature (°F/°C)		
	Titanium	Zirconium	Tantalum
Hydrofluoric acid 30%	x	x	x
Hydrofluoric acid 70%	x	x	x
Hydrofluoric acid 100%	x	x	x
Hypochlorous acid	100/38		302/150
Iodine solution 10%	90/32		
Ketones, general	90/32		
Lactic acid 25%	210/99	300/149	302/150
Lactic acid, concentrated	300/149	300/149	300/149
Magnesium chloride	300/149		302/150
Malic acid	210/99	210/99	210/99
Manganese chloride 5–20%	210/99	210/99	210/99
Methyl chloride	210/99		210/99
Methyl ethyl ketone	210/99	210/99	210/99
Methyl isobutyl ketone	200/93	200/93	210/99
Muriatic acid	x		302/150
Nitric acid 5%	360/182	500/260	302/150
Nitric acid 20%	400/204	500/260	302/150
Nitric acid 70%	390/199	500/260	302/150
Nitric acid, anhydrous	210/99	90/32	302/150
Nitrous acid, concentrated			300/149
Oleum			x
Perchloric acid 10%	x		302/150
Perchloric acid 70%	x	210/99	302/150
Phenol	90/32	210/99	302/150
Phosphoric acid 50–80%	x	180/82	302/150
Picric acid	90/32		200/93
Potassium bromide 30%	200/93	200/93	90/32
Salicylic acid	90/32		210/99
Silver bromide 10%			90/32
Sodium carbonate	210/99	210/99	210/99
Sodium chloride	210/99	250/121	302/150
Sodium hydroxide 10%	210/99	210/99	x
Sodium hydroxide 50%	200/93	200/93	x
Sodium hydroxide, concentrated	200/93	210/99	x
Sodium hypochlorite 20%	200/93	100/38	302/150
Sodium hypochlorite, concentrated			302/150
Sodium sulfide to 10%	210/99	x	210/99
Stannic chloride 20%	210/99	210/99	300/149
Stannous chloride	90/32		210/99
Sulfuric acid 10%	x	300/149	302/150
Sulfuric acid 50%	x	300/149	302/150
Sulfuric acid 70%	x	210/99	302/150
Sulfuric acid 90%			302/150
Sulfuric acid 98%	x	x	302/150
Sulfuric acid 100%	x	x	300/149
Sulfuric acid, fuming	x		x
Sulfurous acid	170/77	370/188	300/149

Table T.13 Continued

Chemical	Maximum temperature (°F/°C)		
	Titanium	Zirconium	Tantalum
Thionyl chloride			300/149
Toluene	210/99	80/27	300/149
Trichloroacetic acid	x	x	300/149
White liquor		250/121	
Zinc chloride			210/99

[a]The chemicals listed are in the pure state or in a saturated solution unless otherwise indicated. Compatibility is shown to the maximum allowable temperature for which data are available. Incompatibility is shown by an x. A blank space indicates that data are unavailable. When compatible, corrosion rate is < 20 mpy.
Source: PA Schweitzer. *Corrosion Resistance Tables,* 4th ed. Vols. 1–3. New York: Marcel Dekker, 1995.

TRANSGRANULAR CORROSION

Transgranular corrosion is a form of localized corrosion in the form of subsurface attack where a narrow path is corroded at random across the grain structure of a metal, disregarding the grain boundaries.

TRIAX

Triax is the trademark for Bayer's polyamide/acrylontrile–butadiene–stryene thermoplastic alloy. It has high impact strength, excellent abrasion characteristics, good chemical resistance, and good fatigue performance. The range of physical and mechanical properties that can be achieved through formulation are shown in Table T.14.

Table T.14 Range of Physical and Mechanical Properties of Triax Based on Formulation

Property	Value	Units
Specific gravity	1.06–1.07	
Water absorption (immersion at 73°F (25°C) 24 h	1.1–1.5	%
Tensile stress at yield	4.3–6.3	psi × 10^3
Tensile elongation at break	1.4–3.6	psi × 10^3
Tensile stress at break	6.3–5.8	psi × 10^3
Tensile modulus	245–295	psi × 10^3
Flexural modulus	1.7–3.00	psi × 10^2
Impact strength, notched		
Izod, 0.125 in thickness at 73°F (25°C)	15–20	ft lb/in.
Deflection temperature		
0.125 in. thickness, 264 psi	207–210	ft lb/in.
0.125 in. thickness, 66 psi	183–144	ft lb/in.

REFERENCES

1. A Perkins. Corrosion monitoring. In: PA Schweitzer, ed. Corrosion Engineering Handbook. New York: Marcel Dekker, 1996, pp. 623–652.
2. PA Schweitzer. Tantalum. In: PA Schweitzer, ed. Corrosion and Corrosion Protection Handbook. 2nd ed. New York: Marcel Dekker, 1989, pp. 213–229.
3. M Schussler and C Pokross. Corrosion Data Survey on Tantalum. 2nd ed. North Chicago: Fansteel, 1985.
4. PA Schweitzer. Corrosion Resistance Tables. 4th ed. New York: Marcel Dekker, 1995.
5. J Lambert. Tantalum. In: PA Schweitzer, ed. Corrosion Engineering Handbook. New York: Marcel Dekker, 1986, pp. 165–194.
6. PA Schweitzer. Corrosion Resistance of Elastomers. New York: Marcel Dekker, 1990.
7. JH Mallinson. Corrosion Resistant Plastic Composites in Chemical Plant Design. New York: Marcel Dekker, 1988.
8. GT Murray. Introduction to Engineering Materials. New York: Marcel Dekker, 1993.
9. PA Schweitzer. Corrosion Resistant Piping Systems. New York: Marcel Dekker, 1994.
10. EL Liening and JM Macki. Aqueous corrosion of advanced ceramics. In: PA Schweitzer, ed. Corrosion Engineering Handbook. New York: Marcel Dekker, 1996, pp. 419–458.
11. I Suzuki. Corrosion Resistant Coatings Technology. New York: Marcel Dekker, 1989.
12. H Leidheiser Jr. Coatings. In: F Mansfield, ed. Corrosion Mechanisms. New York: Marcel Dekker, 1987, pp. 165–209.
13. LC Covington. Titanium. In: PA Schweitzer, ed. Corrosion and Corrosion Protection Handbook. New York: Marcel Dekker, 1989, pp. 187–211.

U

ULTRASONIC MEASUREMENT

Ultrasonic measurement is one method used to determine the effects of corrosion. Special equipment is required for ultrasonic measurements, of which there are several types. The three types of equipment available are the A-scan, the B-scan, and the C-scan systems. The A-scan provides a simple depth measurement from the exterior surface of a pipe or vessel to the next interface that reflects sound waves. Generally this measures wall thickness, but the A-scan can be fooled occasionally by midwall pipe flaws.

B-scan instruments are more powerful, since they produce cross-sectional images similar to X-rays.

The C-scan instruments produce a three-dimensional view of a surface using complex and expensive equipment. C-scan systems can be very useful for large critical surfaces such as aircraft skins but are less likely to be used in process plants at this time because of cost, speed of coverage, and the very large quantity of data produced.

ULTRAVIOLET STABILIZER

An ultraviolet stabilizer is any material added to a plastic that helps the plastic resist degradation from exposure to sunlight and ultraviolet radiation.

UNDERFILM CORROSION

Underfilm corrosion is corrosion that occurs under paint coatings and other organic films at exposed edges or due to filiform corrosion.

UNIFIED NUMBERING SYSTEM

Identification of metallic alloys is covered by the unified numbering system (UNS), whereby an alloy identification number begins with a letter followed by a five digit number. For stainless steels most of the old AISI designations were retained as the first three digits of the UNS number, such that the old type 304 stainless steel is designated S-30400, with S being used for all stainless steels. Copper and copper alloys are designated by C, high nickel alloys by N, titanium alloys by R, with digits in the 50 thousand range and zirconium alloys by R with digits in the 60 thousand range.

Typical UNS designations are as in the table.

Alloy	UNS number	Reference
Coppers (minimum of 99.3% copper)	C-10200 thru C-14200	1, 2
High copper alloys (minimum of 96% copper if in wrought form and 94% copper in cast form)	C-17000 thru C-19400	1, 2
Brasses	C-27000 thru C-68700	1, 2
Tin bronzes (phosphor bronzes)		
wrought	C-51000 thru C-54400	1, 2
cast	C-90300 thru C-94700	1, 2
Aluminum bronzes		
wrought	C-60800 thru C-63200	1, 2
cast	C-95200 thru C-95800	1, 2
Copper-nickel		
wrought	C-70600 thru C-71900	1, 2
cast	C-96200 thru C-96400	1, 2
Stainless steels (ferritic)		
405	S-40500	3
409	S-40900	3
430	S-43000	3
446	S-44600	3
Stainless steels (martensitic)		
410	S-41000	3
416	S-41600	3
440	S-44000	3
Stainless steels (austenitic)		
201	S-20100	3
303	S-30300	3
304	S-30400	3
304L	S-30403	4
321	S-32100	3
347	S-34700	3
316L	S-31603	4
316	S-31600	3
317	S-31700	4
317L	S-31703	4
321	S-32100	4
Duplex stainless steels		
329	S-32900	4
2205	S-31803	4
255	S-32550	4
2507	S-32750	3
Precipation hardening stainless steels		
17-4PH	S-17400	4
17-7PH	S-17700	4
15-7Mo	S-15700	4
Speciality grades of stainless steels		
317 LM	S-31725	4
317 LN	S-31753	4
254 SMO	S-31254	4
High nickel alloys		
800	N-08800	4
20Cb3	N-08020	4

Alloy	UNS number	Reference
367 XN	N-08367	4
825	N-08825	4
904 L	N-08904	4
alloy 28	N-08028	4
20 Mod	N-08320	4
20Mo4	N-08024	4
20Mo6	N-08026	4
alloy G	N-06007	4
alloy G-3	N-06985	4
alloy G-30	N-06030	4
Titanium alloys		
Ti 50A	R-50400	5
Ti 6A14V	R-56400	5
TiPd	R-52400	5
Ti code 12	R-53400	5
Zirconium alloys		
Zr702	R-60702	6
Zr705	R-60705	6
Zr704	R-60704	6
Zr706	R-60706	6
Zircalloy 2	R-60802	6
Zircalloy 4	R-60804	6
Zr2.5Nb	R-60901	6

UNIFORM CORROSION

A metal resists corrosion by forming a passive film on the surface. This film is formed naturally when the metal is exposed to air for a period of time. It can also be formed more quickly by a chemical treatment. For example, nitric acid if applied to an austenitic stainless steel will form this protective film. Such a film is actually a form of corrosion, but once formed it prevents the further degradation of the metal, as long as the film remains intact. It does not provide an overall resistance to corrosion, since it may be subject to chemical attack. The immunity of the film to attack is a function of the film compostion, the temperature, and the aggressiveness of the chemical. Examples of such films are the patina formed on copper, the rusting of iron, the tarnishing of silver, the fogging of nickel, and the high temperature oxidation of metals.

When exposed to a corrosive medium, metals tend to enter into a chemical union with the elements of the corrosive medium, forming stable compounds similar to those found in nature. When metal loss occurs in this manner, the compound formed is referred to as the corrosion product and the metal surface is referred to as being corroded. An example of such an attack is that of halogens, particularly chlorides. They will react with and penetrate the film on stainless steel, resulting in general corrosion. Corrosion tables are developed to indicate the interaction between a chemical and a metal. This type of chemical attack is termed uniform corrosion.

Uniform corrosion of a metal in contact with a specific corrodent is measured in metal loss in mils per year (mpy) or inches per year (ipy). Losses are generally classified into three groups as follows:

1. < 5 mpy (0.005 ipy). Metals in this category are considered as having excellent resistance against chemical attack.
2. 5–50 mpy. Metals in this group are considered satisfactory, providing that a high corrosion rate can be tolerated.
3. > 50 mpy. Usually not a satisfactory choice. In most corrosion tables, rates are given < 5 mpy, < 20 mpy, < 50 mpy, or > 50 mpy. A metal with a rating of < 5 mpy is an excellent choice for use in that specific media. Metals with ratings of < 20 mpy, while not as resistent as those with a lower rating, can still be used. Usually any metal with a loss greater than 20 mpy is not recommended for the application.

Those corrosion rates will also assist in determining how much corrosion allowance should be included in the design, based on the expected lifetime of the equipment.

To determine the corrosion rate, a prepared specimen is exposed to the test environment for a period of time and then removed to determine how much metal has been lost. The exposure time, weight loss, surface area exposed, and density of the metal are used to calculate the corrosion rate of the metal using the formula

$$\text{mpy} = \frac{WL \times 22{,}273}{D \times A \times T}$$

where

WL = weight loss, g

D = density, g/cm^3

A = area, in^2

T = time, days.

To convert from mpy to ipy divide the mpy value by 1000. On occasion the uniform rate of attack will be reported as milligrams per square decimeter per day (mdd). To convert from ipy to mdd or vice-versa requires knowledge of the metal density. Conversion factors are given in Table U.1.

URETHANE (AU) RUBBERS

The urethane rubbers are produced from a number of polyurethane polymers. The properties exhibited are dependent upon the specific polymer and the compounding. Urethane (AU) rubber is a unique material that combines many of the advantages of rigid plastics, metals, and ceramics yet still has the extensibility and elasticity of rubber. It can be formulated to provide a variety of products with a wide range of physical properties.

Composition with a Shore A hardness of 95 (harder than a typewriter platen) are elastic enough to withstand stretching to more than four times their own lengths.

At room temperature a number of raw polyurethane polymers are liquid, simplifying the production of many large and intricately shaped molded products. When cured, these elastomeric parts are hard enough to be machined on standard metalworking equipment. Cured urethane does not require fillers or reinforcing agents.

Table U.1 Conversion Factors from ipy to mdd[a]

Metal	Density (g/cc)	0.00144 / Density × 10–3	696 × Density
Aluminum	2.72	0.529	1890
Brass (red)	8.75	0.164	6100
Brass (yellow)	8.47	0.170	5880
Cadmium	8.65	0.167	6020
Columbium	8.4	0.171	5850
Copper	8.92	0.161	6210
Copper–nickel (70–30)	8.95	0.161	6210
Iron	7.87	0.183	5480
Duriron	7.0	0.205	4870
Lead (chemical)	11.35	0.127	7900
Magnesium	1.74	0.826	1210
Nickel	8.89	0.162	6180
Monel	8.84	0.163	6140
Silver	10.50	0.137	7300
Tantalum	16.6	0.0868	11550
Tin	7.29	0.198	5070
Titanium	4.54	0.317	3160
Zinc	7.14	0.202	4970
Zirconium	6.45	0.223	4490

[a]Multiply ipy by (696 × density) to obtain mdd. Multiply mdd by (0.00144/density) to obtain ipy.

Physical and Mechanical Properties

The urethane rubbers are known for their toughness and durability. Because of the high resistance to abrasion, urethane rubbers are used where severe wear is a problem. Urethane rubber, in actual service, has outworn ordinary rubbers and plastics by a factor of as high as 8 to 1. One such application is that of wearpads used to prevent the marring of multiton rolls of sheet metal.

Most applications use material with a hardness of from 80 Shore A to 75 Shore D. The D scale is used to measure hardnesses greater than 95 Shore A. Most elastomers have hardnesses between 30 and 80 on the A scale, while structural plastics begin at 55 on the D scale. Therefore it can be seen that the urethane rubbers bridge this gap.

Depending upon the formulation, resilience values as low as 15% or as high as 80% can be achieved. Urethane rubbers operate through the range of 50–212°F (10–100°C). This stability is valuable in certain shock mounting applications.

Specific polymers and/or formulations of these rubbers can produce appreciably better impact resistance than structural plastics. Standard compounds, including the hardest types, exhibit good low temperature impact resistance and low brittle points.

Urethane has a low unlubricated coefficient of friction that decreases sharply as hardness increases. This property, combined with its abrasion resistance and load-carrying ability, is an important reason why urethane elastomer is used in bearings and bushings. If necessary, special compounding can lower the coefficient of friction even further.

Urethane rubbers are readily machined using standard metalworking equipment. Operations such as drilling, routing, sawing, and turning are typical.

Some urethane polymers are naturally translucent. This offers several advantages in certain applications, such as potting or encapsulation. The flexibility and strength of these elastomers help protect electrical assemblies or sheathed glass elements from damage, while their translucence permits easy inspection. If desired, urethane rubbers can be pigmented, which is quite commonplace. Products made of the urethane rubbers have little or no resistance to burning. Some slight improvement can be made in this area by special compounding, but the material will still ignite in an actual fire situation. If fire safety is a factor in the selection of an elastomer, this property must be taken into account and the potential hazards must be evaluated.

Standard formulations do not support fungus growth and are generally resistant to fungus attack. However, certain of the special formulations are susceptible to fungus attack. To compensate for this, commercial fungicides must be added to these formulations.

Urethane products are stable under high vacuum conditions. Outgassing tests produce very little weight loss. This property, coupled with their excellent low temperature performance, makes the urethane rubbers suitable for aerospace applications.

Table U.2 lists the physical and mechanical properties of the urethane rubbers.

The urethane rubbers possess many desirable mechanical properties that are superior to those of other elastomeric materials. Their load-bearing capacity is greater than that of other rubbers of comparable hardness. Because of their strength and toughness they can be used in thin sections. They also resist cracking under repeated flexing.

These properties are retained over a wide temperature range. At very low temperatures the material remains flexible and has good resistance to thermal shock. The standard compositions can be used at temperature as low as –80°F (–62°C) without becoming brittle,

Table U.2 Physical and Mechanical Properties of Urethane Rubbers[a]

Specific gravity	1.02–1.20
Brittle point	–85 to –100°F (–65 to –73°C)
Dielectric strength, V/mil	450–500
Permeability to air	Good
Dielectric constant	
at 0.1 kHz	9.4
at 100 kHz	7.8
Tensile strength, psi	3000–8000
Elongation, % at break	270–400
Hardness, Shore A	30–80
Abrasion resistance	Excellent
Compression set, %	30–45
Machining qualities	Readily machined
Resistance to sunlight	Good
Effect of aging	Little effect
Resistance to heat	Good
Maximum temperature, continuous use	250°F (121°C)

[a]These are representative values since they may be altered by compounding.

although stiffness gradually increases as the temperature goes below 0°F (–18°C). Special compositions can be formulated that will permit some flexibility to be retained as low as –125°F (–87°C).

On the high temperature side, formulations can be provided that will permit continuous operation at temperatures of 200–250°F (93–121°C). Even at these temperatures there is only a minor decrease in physical and mechanical properties. However, caution is advised in arbitrarily applying urethane rubbers at these elevated temperatures. Each potential application should be evaluated beforehand by consultation with the supplier and by testing.

Under static, room-temperature conditions, time has little effect upon the urethane rubbers. Shelf aging is not a problem, based on samples stored for at least 15 years and then put into service. Neither is prolonged use in normal service.

There are other factors that can have an adverse effect on aging. Such factors as elevated temperature, dynamic forces, contact with chemical agents, or any combination of these factors will lead to a loss in physical and mechanical properties.

Typical mechanical properties are given in Table U.2.

Urethane rubbers exhibit excellent recovery from deformation. Parts made of urethane rubbers may build up heat when subjected to high frequency deformation, but their high strength and load-bearing capacity may permit the use of sections thin enough to dissipate the heat as fast as it is generated.

Resistance to Sun, Weather, and Ozone

The urethane rubbers exhibit excellent resistance to ozone attack and have good resistance to weathering. However, extended exposure to ultraviolet light will reduce their physical properties and will cause the rubber to darken. This can be prevented by the use of pigments or ultraviolet screening agents.

Chemical Resistance

The urethane rubbers are resistant to most mineral and vegetable oils, greases and fuels, and to aliphatic, aromatic, and chlorinated hydrocarbons. This makes these materials particularly suited for service in lubricating oils and automotive fuels.

Tests conducted at the Hanford Laboratories operation of General Electric Co. indicated that adiprene urethane rubber, as manufactured by DuPont, offered the greatest resistance to the effects of gamma-ray irradiation of the many elastomers and plastics tested. Satisfactory service was given even when the material was exposed to the relatively large gamma-ray dosage of 1×10^9 roentgens. When irradiated with gamma rays, these rubbers were more resistant to stress cracking than other elastomers and retained a large proportion of their original flexibility and toughness.

Aromatic hydrocarbons, polar solvents, esters, ethers, and ketones will attack urethane. Alcohols will soften and swell the urethane rubbers.

Urethane rubbers have limited service in weak acid solutions and cannot be used in concentrated acids. Neither are they resistant to steam or caustic, but they are resistant to the swelling and deteriorating effects of being immersed in water.

Applications

The versality of urethane has led to a wide variety of applications. Products made from urethane rubber are available in three basic forms: solid, cellular, and films and coatings.

Included under the solid category are those products that are cast or molded, comprising such items as large rolls, impellers, abrasion resistant parts for textile machines, O-rings, electrical encapsulations, gears, tooling pads, and small intricate parts.

Cellular products are such items as shock mountings, impact mountings, shoe soles, contact wheels for belt grinders, gaskets, and other similar items.

It is possible to apply uniform coatings or films of urethane rubber to a variety of substrate materials including metal, glass, wood, fabrics, and paper. Examples of products to which a urethane film or coating is often applied are tarpaulins, bowling pins, pipe linings, tank linings, and exterior coatings for protection against atmospheric corrosion. These films also provide abrasion resistance.

Filtration units, clarifiers, holding tanks, and treatment of municipal, industrial, and thermal generating station wastewater. In many cases, particularly in anaerobic, industrial, and thermal generating systems, urethane linings are used to protect the concrete from severe chemical attack and prevent seepage into the concrete of chemicals that can attack the reinforcing steel. These linings provide protection from abrasion and erosion and act as a waterproofing system to combat leakage of the equipment resulting from concrete movement and shrinkage.

The use of urethane rubbers in manufactured products has been established as a result of their high tensile and tear strengths, resiliency, impact resistance, and load-bearing capacity. Other products made from urethane rubbers include bearings, gear couplings, mallets and hammers, solid tires, conveyor belts, and many other miscellaneous items.

REFERENCES

1. PA Schweitzer. Corrosion of copper and copper alloys. In: PA Schweitzer, ed. Corrosion Engineering Handbook. New York: Marcel Dekker, 1996, pp. 89–97.
2. JM Cieslewicz. Copper and copper alloys. In: PA Schweitzer, ed. Corrosion and Corrosion Protection Handbook. 2nd ed. New York: Marcel Dekker, 1989, pp. 125–132.
3. PK Whitcraft. Corrosion of stainless steels. In: PA Schweitzer, ed. Corrosion Engineering Handbook. New York: Marcel Dekker, 1996, pp. 53–77.
4. CP Dillon. Corrosion of Stainless Steels. New York: Marcel Dekker, 1995.
5. PA Schweitzer. Corrosion of titanium. In: PA Schweitzer, ed. Corrosion Engineering Handbook. New York: Marcel Dekker, 1996, pp. 157–163.
6. TL Yau. Corrosion of zirconium. In: PA Schweitzer, ed. Corrosion Engineering Handbook. New York: Marcel Dekker, 1996, pp. 195–252.
7. PA Schweitzer. Resistance of Elastomers. New York: Marcel Dekker, 1990.
8. PA Schweitzer. Corrosion Resistance Tables. 4th ed. Vols. 1–3. New York: Marcel Dekker, 1995.

V

VAPOR

Vapor is the gaseous state of matter that normally exists in a liquid or solid state. It also applies to gas close to the liquid state, e.g., water in the atmosphere. The terms gas and vapor are often used interchangeably.

VAPOR BARRIER

A vapor barrier is a coating or layer that prevents the passage of vapor or moisture into a material or structure.

VAPOR CORROSION

Contaminants in the atmosphere such as sulfur dioxide, nitrogen-containing compounds, hydrogen chloride, phenol, hydrogen sulfide, nitric acid, ammonia, and other similar materials act as stimulants to atmospheric corrosion. This acceleration to corrosion is known as vapor corrosion.

VAPOR PHASE CORROSION INHIBITORS

Vapor phase corrosion inhibitors are solid or liquid volatile organic compounds with corrosion inhibitive behaviors. These compounds have a significantly high vapor pressure at room temperature. They are used within sealed enclosures to protect metallic articles. The most common organic compounds used are dicyclohexylamine nitrite and cyclohexylamine carbonate.

VERDIGRIS

Verdigris is the green patina of basic copper salts formed on copper due to atmospheric corrosion. The basic copper salts are primarily basic carbonate, basic sulfate, and in some cases the basic chloride.

VINYL ESTER RESINS

Also see "Polymers" and "Thermoset Polymers." There exists a wide variety of vinyl ester resins. As a result there can be a difference in the compatibility of formulations between manufacturers. When checking compatibility tables it must be kept in mind that all formulations may not act as shown. An indication that vinyl ester is compatible generally means that at least one formulation is compatible. The resin manufacturer must be checked to verify the resistance.

In general vinyl esters can be used to handle most hot, highly chlorinated, and acidic mixtures at elevated temperatures. They also provide excellent resistance to strong mineral acids and bleaching solutions. The vinyl esters excel in alkaline and bleach environments and are used extensively in the very corrosive conditions found in the pulp and paper industry. Refer to Table V.1 for the compatibility of the vinyl esters with a wider range of selected corrodents. Reference 1 provides a wide range of compatibilities.

See also Refs. 1–4.

VINYLIDENE FLUORIDE ELASTOMERS (HFP, PVDF)

Polyvinylidene fluoride (PVDF) is a homopolymer of 1, 1-difluorethene with alternating CH_2 and CF_2 groups along the polymer chain. These groups impart a unique polarity that influences its solubility and electrical properties. The polymer has the characteristic stability of fluoropolymers when exposed to aggressive thermal, chemical, and ultraviolet conditions.

In general PVDF is one of the easiest fluoropolymers to process, and it can be easily recycled without affecting its physical and chemical properties. As with other elastomeric materials, compounding can be used to improve certain specific properties. Cross-linking of the polymer chain and control of the molecular weight are also done to improve particular properties.

PVDF possesses mechanical strength and toughness, high abrasion resistance, high thermal stability, high dielectric strength, high purity, resistance to most chemicals and solvents, resistance to ultraviolet and nuclear radiation, resistance to weathering, and resistance to fungi. It can be used in applications intended for repeated contact with food per Title 21, Code of Federal Regulations, Chapter 1, Part 177.2520. PVDF is also permitted for use in processing or storage areas in contact with meat or poultry food products prepared under federal inspection according to the U.S. Department of Agriculture (USDA). Use is also permitted under "3-A Sanitary Standards for Multiple-Use Plastic Materials Used as Product Contact Surfaces for Dairy Equipment Serial No. 2000." This material has the ASTM designation of MFP.

Physical and Mechanical Properties

PVDF elastomers have high tensile and impact strengths. The ambient temperature tensile strength at yield of 4000–7000 psi (28–48 MPa) and the unnotched impact strength, 15–80 ft-lb/in. (800–4270 KJ/m), indicate that all grades of this polymer are strong and tough. These properties are retained over a wide temperature range.

Excellent resistance to creep and fatigue are also exhibited by PVDF, while thin sections such as films, filament, and tubing are flexible and transparent. PVDF wire

Table V.1 Compatibility of Vinyl Ester with Selected Corrodents[a]

Chemical	Maximum temp. °F	Maximum temp. °C	Chemical	Maximum temp. °F	Maximum temp. °C
Acetaldehyde	x	x	Barium hydroxide	150	66
Acetamide			Barium sulfate	200	93
Acetic acid 10%	200	93	Barium sulfide	180	82
Acetic acid 50%	180	82	Benzaldehyde	x	x
Acetic acid 80%	150	66	Benzene	x	x
Acetic acid, glacial	150	66	Benzene sulfonic acid 10%	200	93
Acetic anhydride	100	38	Benzoic acid	180	82
Acetone	x	x	Benzyl alcohol	100	38
Acetyl chloride	x	x	Benzyl chloride	90	32
Acrylic acid	100	38	Borax	210	99
Acrylonitrile	x	x	Boric acid	200	93
Adipic acid	180	82	Bromine gas, dry	100	38
Allyl alcohol	90	32	Bromine gas, moist	100	38
Allyl chloride	90	32	Bromine liquid	x	x
Alum	240	116	Butadiene		
Aluminum acetate	210	99	Butyl acetate	80	27
Aluminum chloride, aqueous	260	127	Butyl alcohol	120	49
Aluminum chloride, dry	140	60	*n*-Butylamine	x	x
Aluminum fluoride	100	38	Butyl phthalate	200	93
Aluminum hydroxide	200	93	Butyric acid	130	54
Aluminum nitrate	200	93	Calcium bisulfide		
Aluminum oxychloride			Calcium bisulfite	180	82
Aluminum sulfate	250	121	Calcium carbonate	180	82
Ammonia gas	100	38	Calcium chlorate	260	127
Ammonium bifluoride	150	66	Calcium chloride	180	82
Ammonium carbonate	150	66	Calcium hydroxide 10%	180	82
Ammonium chloride 10%	200	93	Calcium hydroxide, sat.	180	82
Ammonium chloride 50%	200	93	Calcium hypochlorite	180	82
Ammonium chloride, sat.	200	93	Calcium nitrate	210	99
Ammonium fluoride 10%	140	60	Calcium oxide	160	71
Ammonium fluoride 25%	140	60	Calcium sulfate	250	116
Ammonium hydroxide 25%	100	38	Caprylic acid	220	104
Ammonium hydroxide, sat.	130	54	Carbon bisulfide	x	x
Ammonium nitrate	250	121	Carbon dioxide, dry	200	93
Ammonium persulfate	180	82	Carbon dioxide, wet	220	104
Ammonium phosphate	200	93	Carbon disulfide	x	x
Ammonium sulfate 10–40%	220	104	Carbon monoxide	350	177
Ammonium sulfide	120	49	Carbon tetrachloride	180	82
Ammonium sulfite	220	104	Carbonic acid	120	49
Amyl acetate	110	38	Cellosolve	140	60
Amyl alcohol	210	99	Chloracetic acid, 50% water	150	66
Amyl chloride	120	49	Chloracetic acid	200	93
Aniline	x	x	Chlorine gas, dry	250	121
Antimony trichloride	160	71	Chlorine gas, wet	250	121
Aqua regia 3:1	x	x	Chlorine, liquid	x	x
Barium carbonate	260	127	Chlorobenzene	110	43
Barium chloride	200	93	Chloroform	x	x

Table V.1 Continued

Chemical	Maximum temp. °F	Maximum temp. °C	Chemical	Maximum temp. °F	Maximum temp. °C
Chlorosulfonic acid	x	x	Manganese chloride	210	99
Chromic acid 10%	150	66	Methyl chloride		
Chromic acid 50%	x	x	Methyl ethyl ketone	x	x
Chromyl chloride	210	99	Methyl isobutyl ketone	x	x
Citric acid 15%	210	99	Muriatic acid	180	82
Citric acid, concentrated	210	99	Nitric acid 5%	180	82
Copper acetate	210	99	Nitric acid 20%	150	66
Copper carbonate			Nitric acid 70%	x	x
Copper chloride	220	104	Nitric acid, anhydrous	x	x
Copper cyanide	210	99	Nitrous acid 10%	150	66
Copper sulfate	240	116	Oleum	x	x
Cresol	x	x	Perchloric acid 10%	150	66
Cupric chloride 5%	260	127	Perchloric acid 70%	x	x
Cupric chloride 50%	220	104	Phenol	x	x
Cyclohexane	150	66	Phosphoric acid 50–80%	210	99
Cyclohexanol	150	66	Picric acid	200	93
Dichloroacetic acid	100	38	Potassium bromide 30%	160	71
Dichloroethane (ethylene dichloride)	110	43	Salicylic acid	150	66
Ethylene glycol	210	99	Silver bromide 10%		
Ferric chloride	210	99	Sodium carbonate	180	82
Ferric chloride 50% in water	210	99	Sodium chloride	180	82
Ferric nitrate 10–50%	200	93	Sodium hydroxide 10%	170	77
Ferrous chloride	200	93	Sodium hydroxide 50%	220	104
Ferrous nitrate	200	93	Sodium hydroxide, concentrated		
Fluorine gas, dry	x	x	Sodium hypochlorite 20%	180	82
Fluorine gas, moist	x	x	Sodium hypochlorite, concentrated	100	38
Hydrobromic acid, dilute	180	82	Sodium sulfide to 50%	220	104
Hydrobromic acid 20%	180	82	Stannic chloride	210	99
Hydrobromic acid 50%	200	93	Stannous chloride	200	93
Hydrochloric acid 20%	220	104	Sulfuric acid 10%	200	93
Hydrochloric acid 38%	180	82	Sulfuric acid 50%	210	99
Hydrocyanic acid 10%	160	71	Sulfuric acid 70%	180	82
Hydrofluoric acid 30%	x	x	Sulfuric acid 90%	x	x
Hydrofluoric acid 70%	x	x	Sulfuric acid 98%	x	x
Hydrofluoric acid 100%	x	x	Sulfuric acid 100%	x	x
Hypochlorous acid	150	66	Sulfuric acid, fuming	x	x
Iodine solution 10%	150	66	Sulfurous acid 10%	120	49
Ketones, general	x	x	Thionyl chloride	x	x
Lactic acid 25%	210	99	Toluene	120	49
Lactic acid, concentrated	200	93	Trichloroacetic acid 50%	210	99
Magnesium chloride	260	127	White liquor	180	82
Malic acid 10%	140	60	Zinc chloride	180	82

[a]The chemicals listed are in the pure state or in a saturated solution unless otherwise indicated. Compatibility is shown to the maximum allowable temperature for which data are available. Incompatibility is shown by an x. A blank space indicates that data are unavailable.

Source: PA Schweitzer. *Corrosion Resistance Tables,* Vols. 1 and 2. New York: Marcel Dekker, 1991.

insulation has excellent resistance to cut-through. Where load bearing is important, these polymers are rigid and resistant to creep under mechanical load. Resistance to deformation under load is extremely good over the temperature range of –112 to 302°F (–80 to 150°C).

PVDF finds application as thin-wall primary insulation and as a jacket for control wiring as a result of its high dielectric strength and good mechanical properties. Because of its high dissipation factor, PVDF has limited use at high frequencies. However, this property becomes an advantage in components utilizing dielectric heating techniques.

In order for PVDF to burn, it is necessary to have a 44% oxygen environment. The Underwriters Laboratories give PVDF a vertical burn rating of 94-V-O. PVDF is fire resistant and self-extinguishing.

PVDF will not support the growth of fungi. However, if additives that will support the growth of fungi are used in compounding, then fungicides should also be added to overcome this problem.

Because of its extremely low weight loss when exposed to high vacuum, PVDF can be used in high vacuum applications. At 212°F (100°C) and a pressure of 5×10^{-6} torr, the measured rate of weight loss is 13×10^{-11} g/cm^2-s.

PVDF can also be pigmented.

The physical and mechanical properties of PVDF are given in Table V.2.

PVDF is suitable for applications where load-bearing characteristics are important, particularly at elevated temperatures. Long-term deflections at 140°F (60°C) with a load of 172 psi vary from 0.8 to 119%, while at 194°F (90°C) with a load of 530 psi the deflection ranges from 1.2 to 3.6%.

PVDF has a wide useful temperature range with a brittle point at –80°F (–62°C) and a maximum continuous operating temperature of 302°F (150°C). Over this entire range its properties remain virtually unaffected.

Special formulations are available for use in plenum cable applications. These formulations produce products having the impact strengths of 12–18 ft-lb/in. and elongations of 400–650%. They also have improved stress crack resistance.

Resistance to Sun, Weather, and Ozone

PVDF is highly resistant to the effects of sun, weather, and ozone. Its mechanical properties are retained while the percent elongation to break decreases to a lower level and then remains constant.

Chemical Resistance

In general PVDF is completely resistant to chlorinated solvents, aliphatic solvents, weak bases and salts, strong acids, halogens, strong oxidants, and aromatic solvents. Strong bases will attack the material.

The broader molecular weight of PVDF gives it a greater resistance to stress cracking than many other materials, but it is subject to stress cracking in the presence of sodium hydroxide.

PVDF also exhibits excellent resistance to nuclear radiation. The original tensile strength is essentially unchanged after exposure to 1000 Mrads of gamma irradiation from a cobalt-60 source at 122°F (50°C) and in high vacuum (10^{-6} torr). Because of cross-

Table V.2 Physical and Mechanical Properties of Vinylidene Fluoride (PVDF) Elastomer[a]

Specific gravity	1.76–1.78
Refractive index	1.42
Specific heat, Btu/lb-°F	0.30–0.34
Brittle point	–80°F (–62°C)
Coefficient of linear expansion	
per °F	7.8×10^{-5}
per °C	1.4×10^{-4}
Thermal conductivity	
BTU-in./hr-ft^2 °F	0.70–0.87
cal/cm-s-°C	3×10^{-4}
Dielectric strength, kV/mm	63–67
Dielectric constant at 77°F (23°C)	
at 100 Hz, ohm-cm	9.9
at 1 kHz, ohm-cm	9.3
at 100 kHz, ohm-cm	8.5
Tensile strength, psi	4000–7000
Elongation, % at break	25–650
Hardness, Shore A	77–83
Abrasion resistance, Armstrong (ASTM D 1242) 30-lb load volume loss, cm^3	0.3
Maximum temperature, continuous use	302°F (150°C)
Impact resistance, Izod notched, ft-lb/in.	3–18
Compression set	Good
Machining qualities	Excellent
Resistance to sunlight	Excellent
Effect of aging	Nil
Resistance to heat	Good

[a]These are representative values since they may be altered by compounding.

linking, the impact strength and elongation are slightly reduced. This resistance makes PVDF useful in plutonium reclamation operations.

Applications

PVDF finds many applications where its corrosion resistance, wide allowable operating temperature range, mechanical strength and toughness, high abrasion resistance, high dielectric strength, and resistance to weathering, ultraviolet light, radiation, and fungi are useful.

In the electrical and electronics fields PVDF is used for multiwire jacketing, plenum cables, heat-shrinkable tubing, anode lead wire, computer wiring, and cable and cable ties. Because of its acceptance in the handling of foods and pharmaceuticals, transfer hoses are lined with PVDF. Its corrosion resistance is also a factor in these applications.

In fluid-handling systems PVDF finds applications as gasketing material, valve diaphragms, and membranes for microporous filters and ultrafiltration.

As a result of its resistance to fungi and its exceptional corrosion resistance, it is also used as the insulation material for underground anode bed installations.

See Refs. 5 and 1.

VITON

See "Fluoroelastomers."

VITREOUS ENAMEL

See "Glass Coatings."

VITREOUS ENAMEL COATINGS

Vitreous enamels, glass linings, or porcelain enamels are all essentially glass coatings that have been fused on metals. Powdered glass is applied to a pickled or otherwise prepared metal surface and heated in a furnace at a temperature that softens the glass and permits it to bond to the metal. Several thin coats are applied to provide the required final thickness. These coatings are normally applied to steel, but some coatings can be applied to brass, aluminum, and copper.

There are many glass formulations, but those with very high silica ($> 96\%$ S_1O_2), aluminosilicate, and borosilicate compositions have the highest corrosion resistance to a wide range of corrosive environments. Glass is assumed to be inert to most liquids, but in reality it slowly dissolves.

The greatest danger of failure of a glass coating comes from mechanical damage or from cracking as a result of thermal shock. Thus care must be taken in handling glassed equipment so as not to damage the lining, and sudden temperature changes in the operation must be avoided, particularly cold shock, which poses a greater danger of failure than hot shock.

Cold shock is the sudden introduction of a cold material onto a hot glassed surface; hot shock is the reverse. Manufacturers of this type of equipment will specify the maximum allowable thermal shock. These precautions must be followed.

VITRIFIED CLAY PIPE

Vitrified clay pipe is virtually impervious to every chemical except hydrofluoric acid. It was and still is used for the handling of sewage. With the advent of the problems of cleaning up of toxic dumps and landfills, new applications have arisen. The primary reasons for using clay pipe in these new applications are that it

1. Is chemically inert, unaffected by sewer gases and acids
2. Is rigid and will not flatten or sag
3. Is rustproof
4. Is unaffected by harsh household cleaning compounds and solvents
5. Withstands the extra stresses of heavy backfill loads
6. Will not soften or swell under any conditions
7. Is durable, will not roughen, erode, or wear out
8. Is unaffected by gases and acids generated by ground garbage
9. Is made impervious through vitrification

Vitrified clay pipe is presently being used to conduct the leachate from Love Canal to holding tanks for processing.

Though salt-glazed pipe may be used for sanitary sewers, unglazed pipe should be employed when handling industrial wastes. Clay pipe is normally applied for gravity flow lines. However, they should be designed for heads of 5–10 feet to guard against the possibility of blockage or of the sudden or abrupt introduction of material into the line. Table V.3 provides the compatibility of vitrified clay with selected corrodents.

Table V.3 Compatibility of Vitrified Clay with Selected Corrodents[a]

Chemical	Maximum temperature (°F/°C)
Acetic acid 5%	150/66
Acetone	73/23
Aluminum chloride	x
Aluminum sulfate 5%	150/66
Ammonium chloride 5%	150/66
Ammonium chloride 10%	x
Ammonium chloride 25%	x
Ammonium hydroxide 5%	73/23
Ammonium hydroxide 10%	73/23
Aniline	73/23
Benzene	73/23
Borax 3%	150/66
Carbon tetrachloride	73/23
Chromic acid 40%	150/66
Citric acid 10%	150/66
Copper sulfate 3%	150/66
Ferric chloride 1%	150/66
Hydrochloric acid 10%	120/49
Hydrofluoric acid 30%	x
Hydrofluoric acid 70%	x
Hydrofluoric acid 100%	x
Nitric acid 1%	150/66
Nitric acid 10%	150/66
Nitric acid 20%	150/66
Sodium carbonate 20%	150/66
Sodium chloride 30%	150/66
Sodium hydroxide 10%	150/66
Sulfuric acid 20%	150/66
Sulfuric acid 30%	150/66
Toluene	120/49

[a]The chemicals listed are in the pure state or in a saturated solution unless otherwise indicated. Compatibility is shown to the maximum allowable temperature for which data are available. Incompatibility is shown by an x.

REFERENCES

1. PA Schweitzer. Corrosion Resistance Tables. 4th ed. Vols. 1–3. New York: Marcel Dekker, 1995.
2. GT Murray. Introduction to Engineering Materials. New York: Marcel Dekker, 1993.
3. JH Mallinson. Corrosion-Resistant Plastic Composites in Chemical Plant Design. New York: Marcel Dekker, 1988.
4. PA Schweitzer. Corrosion Resistant Piping Systems. New York: Marcel Dekker, 1994.
5. PA Schweitzer. Corrosion Resistance of Elastomers. New York: Marcel Dekker, 1990.

W

WATERLINE ATTACK

When a metal is partially submerged in an aqueous system with air above the metal it undergoes simultaneous differential aeration cell and crevice corrosion at the waterline junction. This is referred to as waterline attack.

WEATHERING

Weathering is the process of decomposition and disintegration from the chemical action of sunlight, frost, water, and heat resulting from exposure to the atmosphere.

It is also a method of removing mill scale from heavy structural steel members by exposing them outdoors for a period of several months to allow mill scale to crack off under the stress of expansion and contraction.

WEATHERING STEELS

When small amounts of copper, chromium, nickel, phosphorus, silicon, manganese, or various combinations thereof are added to conventional carbon steel, a low alloy carbon steel results that has an improved corrosion resistance. These steels are known as weathering steels. The corrosion resistance of these steels is dependent upon the climatic conditions, the pollution levels, the degree of sheltering from the atmosphere, and the specific composition of the steel.

Upon exposure to most atmospheres, the corrosion rate becomes stabilized within 3 to 5 years. A dark brown to violet patina, or protective film, develops over this period. This patina is a rust formation that is tightly adhered to the surface and cannot be wiped off. In rural areas with little or no pollution, a longer period may be required to form this protective film. In areas that are highly polluted with SO_2, the weathering steels exhibit a much higher corrosion rate, and loose rust particles are formed. Under these conditions, the film formed offers little to no protection.

These steels will not produce this protective film in marine environments where chlorides are present. Corrosion rates will be as high as for conventional carbon steels.

The formation of the patina is dependent upon a series of wet and dry periods, because periodic flushing followed by a period of drying are necessary. In areas where the steel is sheltered from the rain, the dark patina is not formed. In its place, a layer of rust in lighter colors forms, which has the same effect. With continued exposure to wetness, such as in water or soil, the corrosion rate for the weathering steels is equal to that of plain carbon steel.

Because the patina formed has a pleasant esthetic appearance, the weathering steels can be used without the application of any protective coating of antirust paint, zinc, or aluminum. This is particularly true in urban or rural areas.

When employing these weathering steels, design consideration should be given to the elimination of possible areas where water, dirt, and corrosion products can accumulate. If pockets are present, the time of wetness will be increased, which will lead to the development of corrosive conditions. The design should make maximum use of exposure to the weather, Sheltering from rain should be avoided.

The designer should also be aware that during the period over which this protective film is forming, rusting will proceed at a relatively high rate, during which time rusty water is produced. This rusty water may stain masonry, pavements, and the like. In view of this, precautions should be taken to prevent detrimental staining effects, such as coloring the masonry brown, so that any staining will not be obvious.

The weathering steels are used primarily for buildings, bridges, structures, and guard rails.

WELD RUSTING

Locally chromium depleted iron–nickel alloys are subject to slow rusting by mildly acidic liquids, such as the weld metal rusting of stainless steel by dew (containing dissolved carbon dioxide) condensing on the outside of a pipe. Even though this is primarily an aesthetic problem, in some applications it would be unacceptable. If chlorides are present, such corrosion could become aggresive because of the formation of ferric chloride.

WHITE IRON

When cooled, the carbon in white iron forms a hard, abrasion resistant, iron–chromium carbide, instead of forming free graphite. White irons are used primarily for abrasive applications. They are very brittle. There is very little difference in corrosion resistance between the white irons and gray irons. The high chromium content provides only slightly better corrosion resistance.

See Refs. 1 and 2.

WHITE RUST

White rust is the name applied to zinc corrosion products that appear as a white to dirty gray deposit found on galvanized steel surfaces below the water line, when exposed to recirculated water. The deposit is zinc carbonate, which does not form a protective film on the galvanized surface. Because of this the white rust corrosion will continue until all of the protective zinc is entirely removed from the base metal.

White rust occurs when galvanized steel or zinc is exposed to water having a pH value above 8.2. The alkalinity of the water and the presence of any accelerating agents such as phosphates or phosphonates will govern the rate of corrosion. Higher alkalinity and the presence of accelerating agents will increase the corrosion rate.

Prevention of white rust is possible by maintaining pH control to hold the alkalinity level below the critical 8.2 and by the use of inhibitors.

WOOD

Wood is a naturally formed organic material composed of cells arranged in a parallel manner. The chemical composition of the woody cell walls is approximately 40 to 50% cellulose, 15 to 30% lignin, less than 1% mineral, 25 to 35% hemicellulose, and the remainder extractable matter of various types. Softwoods and hardwoods both contain approximately the same cellulose content.

Timber is classified as hardwood and softwood. Hardwood comes from the broad-leaved trees such as oak, maple, and ash. Softwood is the product of coniferous trees such as pines, birch, spruce and hemlock. The terms hardwood and softwood have no relation to the actual hardness of the wood.

Sapwood is the living wood on the outside of the stem. Heartwood is the inner core of physiologically inactive wood in the tree. Heartwood is usually darker in color than sapwood.

Wood is relatively inert chemically but is readily dehydrated by concentrated solutions and consequently shrinks badly when subjected to the actions of such solutions. It is slowly hydrolyzed by acids and alkalies, particuarly when hot. In tank construction, if sufficient shrinkage takes place to permit crystals to form between the staves it becomes very difficult to make the tank tight again. Wood can be impregnated to resist acids or alkalies and the effects of high temperatures.

Wood deteriorates from two principal causes, chemical and biological attack. The chemical resistance of the wood is dependent upon the ability of the wood's cell walls to resist chemical action, and the extent that the chemical penetrates into the wood.

Wood is most resistant to chemical attack in the pH range of 2 through 9, and in certain conditions can be used up to a pH of 11. Wood is resistant to weak acids, but concentrated mineral acids tend to hydrolyze the cellulose and hemicellulose constituents. The lignin in the wood that binds the fibers together is attacked by oxidizing agents such as ozone. Strong oxidizing agents such as nitric acid, chromates, potassium permanganate, and chlorinated water can also oxidize the cellulose. Aqueous solutions of sodium hydroxide, nitric acid, sulfuric acid, and hydrochloric acid can cause the most damage, since swelling and degradation are simultaneous.

Wood can be used to handle dilute hydrochloric acid (< 5%) and sulfuric acids at ambient temperature, phosphoric acid up to 30% at ambient temperature, organic acids, aldehydes, alcohols, and acid salts.

Various types of wood are used depending upon the service. Cypress is used for general chemical service; redwood and fir are used for sulfite liquors from the pulp and paper industries; pine is used in acid mine waters, dilute mineral acids, and mildly alkaline solutions; maple because of its hardness is used for abrasive slurries; oak is used to store and age whiskey and wine; and redwood, red cedar, douglas fir, and various pines are generally used for cooling towers and many chemical exposures.

Biological deterioration of wood is caused by aquatic organisms, insects, and fungi. This deterioration can be delayed or prevented by pressure treating the wood with creosote or some other preservative.

WORM TRACK CORROSION

This is another name for "Filiform Corrosion."

WROUGHT IRON

Wrought iron is a highly refined form of iron containing less than 0.03% carbon and with 1 to 3% slag which is evenly distributed throughout the material in threads and fibers so that the product has a fibrous structure quite dissimilar to that of crystalline cast iron. It is a mechanical mixture of slag and low carbon steel. Wrought iron generally rusts less readily than other forms of metallic iron.

REFERENCES

1. GW George and PG Breig. Cast alloys. In: PA Schweitzer, ed. Corrosion and Corrosion Protection Handbook. New York: Marcel Dekker, 1989, pp. 289–290.
2. JL Gossett. Corrosion resistance of cast alloys. In: PA Schweitzer, ed. Corrosion Engineering Handbook. New York: Marcel Dekker, 1996, p. 259.

X

XENOY

Xenoy is G.E. Plastics' trademark for their polycarbonate/polybutylene terephthalate thermoplastic alloy. It is available in several blends with different impact strengths, temperature ratings, mechanical properties, and chemical resistances. See Table X.1 for the range of physical and mechanical properties depending upon the formulation.

Table X.1 Range of Physical and Mechanical Properties of Xenoy Dependent upon Formulation

Property	Range	Units
Specific gravity	1.2–1.51	
Water absorption, 24 h at 73°F (23°C)	0.08–0.11	%
Tensile strength, break type 1 0.125 in. (3.2 mm)	6.3–13.3	psi × 10^3
Tensile elongation, break type 1 0.125 in. (3.2 mm)	1.3–4	%
Flexural strength, break 0.125 in. (3.2 mm)	9.3–20	psi × 10^3
Flexural modulus 0.250 in. (6.4 mm)	2.5–8.5	psi × 10^5
Compressive strength	6.7–11.4	psi × 10^3
Sheer strength	5.3–6.9	psi × 10^3
Hardness, Rockwell R	108–116	
Izod impact unnotched 0.125 in. (3.2 mm) at 73°F (23°C)	12–14	ft lb/in.
Izod impact notched 0.125 in. (3.2 mm) at 73°F (23°C)	3.2–16	ft lb/in.

Y

YELLOW BRASS

Yellow brass is a copper–zinc alloy containing approximately 30% zinc. This alloy is subject to dealloying (dezincification), a type of corrosion in which the brass dissolves as an alloy and the copper constituent redeposits from solution onto the surface of the brass as a metal in a porous form. This can be prevented by adding 1% tin; the resulting alloy is known as admiralty metal. Yellow brass is also susceptible to stress corrosion cracking.

For corrosion resistance details, refer to "Copper–Zinc Alloys" in this book.

See also Refs. 1 and 2.

REFERENCES

1. PA Schweitzer. Corrosion Resistance Tables. 4th ed. New York: Marcel Dekker, 1995.
2. JM Cieslewicz and PA Schweitzer. Copper and copper alloys. In: PA Schweitzer, ed. Corrosion and Corrosion Protection Handbook. New York: Marcel Dekker, 1989, pp. 125–152.

Z

ZINC

Zinc coatings represent the single biggest use of zinc. There are a variety of zinc alloy coatings that are available. The specific alloy to be used and the thickness of coating will be dependent upon the coating process to be used. There are six methods of coating iron and steel in general use.

Hot Dipping

Hot dip galvanizing is the most important zinc coating process. All mild steels and cast irons can be coated. The thickness of the coating will depend upon the alloying elements. In this process the cleaned steel is immersed in molten zinc or zinc alloy, and reaction takes place to form a metallurgically bonded coating.

Zinc Electroplating

Electroplated zinc is very ductile and has a bright smooth finish. This process is used to provide a decorative effect and to protect delicate objects where rough or uneven surfaces cannot be tolerated. It is also used for continuous plating of strip and wire where severe deformation may be required. In this process zinc salt solutions are used in the electrolytic deposition of a layer of zinc on a cleaned steel surface.

Mechanical Coating

Properly prepared small steel parts are tumbled in zinc dust at room temperature. Deposition from vapor is also used to give a mechanically or physically adherent coating on steel sheet. All types of steel can be coated. Because the tumbling process reduces the coating thickness on the edges, the process is not recommended for parts heavier than about 250 g.

Sheradizing

In this process small steel parts are tumbled in zinc dust just below the melting point of zinc. This produces a metallurgically bonded coating. The uniformity of the coating over complicated shapes together with its abrasion resistance makes this process useful for nuts and bolts which, with proper allowance for the coating thickness, can be sheradized after manufacture and the threads will not have to be retapped.

Thermal Spraying

Gas or electric heating is used to melt zinc wire or powder, which is then sprayed on grit blasted surfaces.

Coatings Incorporating Zinc Dust or Flake

Zinc paints consist of very fine dust and/or flake suspended in an organic or inorganic medium. Zinc dust paints may be used alone for protection or as a primer followed by a conventional topcoat. Zinc rich paints are closest in behavior to other zinc coatings. They contain sufficient zinc powder to give a film capable of conducting electricity.

Porter's book (Ref. 1) provides extensive tables giving corrosion rates of zinc in the presence of organic and inorganic materials.

See also Ref. 2.

ZINCATING

This is an immersion coating of aluminum base materials with zinc to facilitate electroplating of other metals on the aluminum article. Zincating is a chemical replacement where aluminum ions replace zinc ions in an aqueous solution of zinc salts, thus depositing a thin adherent film of metallic zinc on the aluminum surface. Adhesion of the zinc depends on metallurgical bonding.

ZINC EMBRITTLEMENT

This is a form of liquid metal embrittlement of austenitic stainless steels. It most commonly occurs in fire exposure or welding of these steels while in contact with galvanized steel parts.

ZIRCALOYS

These are zirconium alloys, Zr2.5Nb and Zr-1Nb, which are hafnium free and are classified as nuclear grade.

ZIRCONIUM

Zirconium is more abundant than such common metals as nickel, chromium, and cobalt. It is ranked 19th in abundance of the chemical elements found in the earth's crust. The primary ore, in the form of beach sand, from which zirconium is recovered, is zircon ($ZrSiO_4$). Hafnium occurs naturally in ores with zirconium. It has chemical and metallurgical properties similar to those of zirconium. Grade 702 (ASTM designation R-60702) is an unalloyed zirconium that contains approximately 4.5% hafnium. Since this grade has the best overall corrosion resistance, it is the most widely used.

Zirconium exhibits excellent resistance to many chemical solutions, even at elevated temperatures and pressures. When exposed to an oxygen-containing environment, an adherent protective oxide film forms on its surface. This film is formed spontaneously in air or water at ambient temperatures and below, is self-healing, and protects the base metal from chemical and mechanical attack at temperatures up to 572°F (300°C).

As a result, zirconium resists attack by mineral acids, alkaline solutions, most organic and salt solutions, and molten alkalies. It has excellent oxidation resistance up to 752°F (400°C) in air, steam, carbon dioxide, sulfur dioxide, nitrogen, and oxygen. It will not be attacked by oxidizing media unless halides are present.

Hydrofluoric acid, even in amounts as low as 1 ppm, in an acidic solution will attack zirconium. When fluoride impurities are present in a solution having a pH < 3, they tend to form hydrofluoric acid, which will attack zirconium. Corrosion can be prevented by adding an inhibitor that will form a strong fluoride complex. Zirconium is immune to attack by the other halogen acids (e.g., hydrochloric, hydrobromic, and hydroiodic).

Of the common acids, hydrochloric is the most difficult for metallic materials to handle. Even small amounts present in a medium are capable of causing pitting and stress corrosion cracking of common metals and alloys. Zirconium has the important corrosion resistant property of exhibiting excellent resistance to hydrochloric acid at all concentrations and temperatures, even above boiling.

The corrosion resistance of metals to sulfuric acid is greatly dependent upon the concentration and temperature of the acid. Dilute solutions of sulfuric acid have a reducing character, while concentrated solutions are oxidizing in character. Consequently the corrosion resistance of many materials is restricted to certain specific conditions, such as low concentrations, high concentrations, or low temperatures.

Zirconium resists attack by sulfuric acid at all concentrations up to 70% and at temperatures to boiling and above. As the concentration increases above 70% the corrosion resistance of zirconium decreases.

Zirconium is resistant to attack by pure phosphoric acid at concentrations up to 55% with temperatures exceeding the boiling point. However, fluoride compounds are usually present in phosphoric acid, and if they are present in more than a trace, the zirconium will be attacked.

Most alkalies will not attack zirconium, either fused or in solution. It is resistant to potassium hydroxide solutions, and sodium hydroxide solutions, even under anhydrous conditions. It is also resistant to calcium hydroxide and ammonium hydroxide solutions at concentrations to 28% and temperatures to boiling.

Zirconium is also resistant to attack by such salt solutions as halogen, carbonate, nitrate, and sulfate. Corrosion rates are usually very low at all temperatures up to the boiling point. Refer to Table T.13 under Titanium for a compatibility chart of zirconium with selected corrodents.

More detailed corrosion resistance of zirconium in contact with a wide variety of corrodents can be found in Refs. 3 and 4.

ZYMAXX

Zymaxx is the registered trademark for DuPont's carbon fiber reinforced Teflon. This composite material has outstanding mechanical and corrosion resistant properties. It has an operating temperature range of minus 350°F to 550°F. At 550°F Zymaxx has four times the flexural strength of filled PTFE at 400°F and with less deflection. Its compressive creep is less than 1% after 100 hours at 500°F and 6000 psi as compared to other polymeric materials that soften and "cold flow." In addition, it offers low friction, low wear resistance, a coefficient of thermal expansion less than that of steel, and virtually no water absorption. Listed in the table is a comparison of the major properties of Zymaxx, filled PTFE, and reinforced PEEK.

Property	Zymaxx	Filled PTFE	Reinforced PEEK
Chemical resistance	E	E	G
Intrinsic mech. properties	G+	P	G+
Impact resistance	E	G	P
Wear performance	G	F	F
Low coeff. of thermal exp.	E	P	G+
Low water absorption	E	E	G+

E = excellent; G = good; F = fair; P = poor.

REFERENCES

1. FC Porter. Corrosion Resistance of Zinc and Zinc Alloys. New York: Marcel Dekker, 1994.
2. I Suzuki. Corrosion Resistant Coatings Technology. New York: Marcel Dekker, 1989.
3. PA Schweitzer. Corrosion Resistance Tables. 4th ed. Vols. 1–3. New York: Marcel Dekker, 1995.
4. TL Yau. Zirconium. In: PA Schweitzer, ed. Corrosion Engineering Handbook. New York: Marcel Dekker, 1996, pp. 195–252, 231–281.

Index